Lentil

Lentil

An Ancient Crop for Modern Times

Edited by

Shyam S. Yadav
Indian Agricultural Research Institute, New Delhi, India

David L. McNeil
University of Tasmania, Tasmania, Australia

and

Philip C. Stevenson
Natural Resource Institute, Kent, and Royal Botanic Gardens, Kew, U.K.

Springer

A C.I.P. Catalogue record for this book is available from the Library of Congress.

ISBN 978-1-4020-6312-1 (HB)
ISBN 978-1-4020-6313-8 (e-book)

Published by Springer,
P.O. Box 17, 3300 AA Dordrecht, The Netherlands.

www.springer.com

Printed on acid-free paper

CONTENTS

FOREWORD

On behalf of the United States Department of Agriculture, I am pleased to introduce the book *Lentil: An Ancient Crop for Modern Times*. The articles and essays in this volume, submitted by nearly 100 researchers, educators, and other experts, contain comprehensive information on a variety of topics of significance for lentil growers, researchers, and consumers worldwide.

Cultivated lentils (*Lens culinars*), an annual legume crop, have been grown as an important food source for over 8,000 years. They come in two main varieties: macrosperma (with large seeds and little pigmentation), and microsperma (with small seeds and some pigmentation). Depending on their variety and breed, however, lentil seeds can range in color from red-orange, to yellow, green, brown, or black. They are cultivated and consumed throughout the world, with Canada, Turkey and India being the top producers.

Although the production of lentils and other pulse legume crops lags far behind cereal production in most nations, including the United States, production remains highly important because of its benefits for producers and consumers alike. Lentil seeds provide high levels of protein and, when consumed in combination with cereals, they provide adequate amounts of essential amino acids for the human diet. Their relatively short cooking time provides an additional advantage. Lentil production is equally beneficial for producers, as lentils have a high tolerance for extreme environmental conditions such as drought and hot temperatures, and can be grown in semiarid regions without irrigation. Moreover, the crop can be grown in rotation with cereal crops to reduce soil erosion, improve disease and weed control, and reduce demand for nitrogen fertilizers. Beyond their longstanding food and agronomic attributes, there is increasing interest in using lentils as a biomass energy crop and for other industrial non-food uses.

USDA scientists are among many participants in a global effort to enhance lentil quality and promote its growth as a crop. The Agricultural Research Service partners with university researchers and other cooperators worldwide to improve crop disease resistance, develop effective insect and weed control practices, and identify other effective crop production strategies for lentils. This effort also includes international-scale lentil breeding and improvement programs to collect, introduce, maintain, and exchange germplasm. These programs are essential for improving lentils' genetic diversity, which, in turn, is important for increasing the yield and crop quality of the legume as well as for limiting the impact that diseases, weeds, and insect pests have on lentil crops. Although lentils tend to suffer less from disease than do other legume crops, they are still impacted by root rots and wilts, rusts, blights, and viruses. Lentils are also susceptible to damage from weeds, which can reduce yields by up to 75 percent, as well as from aphids, beetles, maggots, wireworms, grasshoppers, and other insects. Ongoing efforts to develop improved technologies to manage these threats as well as to enhance the end-use attributes of lentils for food, nutrition, and industrial uses will be critical to ensuring the long-term sustainability of these crops.

Lentil, edited by Shyam S. Yadav, David McNeil, and Philip C. Stevenson, provides a valuable overview of the history and background behind lentils as well as a detailed analysis of the research that has been conducted on lentil breeding and production strategies. As such, it will be useful to breeders, producers, researchers, educators, nutritionists, and anyone interested in obtaining an insight into the world of lentils.

Edward B. Knipling
Administrator, Agricultural Research Service
United States Department of Agriculture

ABOUT THE EDITORS

Dr. Shyam Singh Yadav

Shyam Singh Yadav is a principal legume breeder at Division of Genetics, Indian Agricultural Research Institute, New Delhi, India. He received his B Sc degree in agricultural sciences from University of Agra, UP, his M Sc degree in plant breeding from University of Meerut, UP, before completing his PhD in Genetics & Plant Breeding from the Indian Agricultural Research Institute, New Delhi, India in 1986. During his PhD research program he worked on the genetic & physiological basis of plant architecture of chickpea (*Cicer arietinum* L.) He started his scientific journey as assistant wheat breeder in January 1969 at the Division of Genetics, Indian Agricultural Research Institute (IARI) New Delhi, India. Since then he has worked on various breeding positions in India and abroad till April 2007. As Principal Legume Breeder he used different breeding options/approaches for genetic enhancement, varietal development, germplasm enrichment and participatory breeding etc. More than 20 high yielding, widely adapted multiple resistant chickpea varieties were released for commercial cultivation in different eco-geographical under his leadership. He has also developed excellent germplasm lines and distributed to many international legume breeders around the world. These lines are being utilized in many countries by legume breeders. Along with this, he worked on the development of integrated crop production and management technologies and its dissemination and popularization in farmers fields. Professional collaborations were also developed under his leadership with international organizations like ACIAR, Australia, USDA, ICRISAT, ICARDA etc. during this period. The post graduate school at IARI, New Delhi provides Graduate and Doctoral research programs to national and international students. During the last 30 years, Yadav has been a faculty member at post graduate school of IARI and taught various professional courses in Genetics and Plant Breeding at Graduate and Doctoral level. He has published more than 125 research papers in national and international journals and written 10 book chapters for international books. In March 2007, a book on chickpea breeding and management was published by CABI, UK for which Yadav was senior editor.

Prof. David Leslie McNeil

David McNeil started his career in agricultural science in 1971 as a trainee crop physiologist with the New South Wales (NSW) Department of Agriculture followed by PhD on lupin physiology at the University of Western Australia. Since then he has swung between Departments of Agriculture and Universities with a strong involvement in pulse and grain legume crop research, development and extension. A key area of effort has been to expand scientific understanding of new crops and develop new productive, viable and sustainable industries around these new crops with pulses as a major area of effort. He has published well over 100 scientific papers as well as a similar number of extension publications. Professor McNeil's research publications have covered a wide range; from molecular mapping, GM, mutation and traditional breeding through crop/plant agronomy and physiology to market testing and consumer evaluation of new crops. David McNeil's work with legumes has included a period with NifTAL in Hawaii promoting N fixation world wide, including consulting for the UNEP program on N fixation. He has also developed super nodulation in soybeans as well as led major Australian breeding programs for a range of temperate pulses including lentils. Professor McNeil's career has included acting as researcher, lecturer, program manager and extension expert based at locations in the USA, New Zealand and Australia, usually with goals spanning both developing and developed country agricultural systems. Professor McNeil has worked at the Boyce Thompson Institute at Cornell, the University of Hawaii, the Australian National University, the WA Department of Agriculture, Lincoln University in New Zealand, the Victorian Department of Primary Industries, with the University of Melbourne in the Joint Centre for Crop Innovation. Presently he occupies the Chair of Agricultural Sciences in the School of Agricultural Science at the University of Tasmania. This school incorporates the Tasmanian Institute of Agricultural Sciences and as such is the predominant source of research in Tasmania for horticulture, dairy, vegetable, cropping and food safety. Professor McNeil's commitment to linking research and industry development continues with a strong interest in retaining pulses as a major component of a total cropping system particularly in high rainfall zone cropping. His interest in other areas of plant physiology, breeding and agronomic research continue including attempts to use biotechnological approaches in lentil, pea and chickpea breeding. Thus from his present position Professor McNeil continues his interest in combining detailed science with industry development of pulse, horticultural and other new crops in the Tasmanian, Australian and global environment.

Dr. Philip. C. Stevenson

Phil Stevenson is a Reader in Plant Chemistry at the Natural Resources Institute (NRI), University of Greenwich, UK, and holds a joint position between NRI and the Jodrell Laboratory at the Royal Botanic Gardens, Kew, UK. He received his B.Sc degree in Applied Biology from Brunel University of West London, before completing his PhD at University of London in 1992. During his PhD he worked on the chemical basis of resistance in wild species of groundnut (*Arachis* spp.) to the tobacco armyworm (*Spodoptera litura*) and so began a transformation from plant biologist to plant chemist. The compounds identified during this work were demonstrated to inhibit development of *S. litura* larvae and have subsequently been used as markers for breeding resistant groundnut cultivars in India. His interest in natural resistance in crop plants continued into his postdoctoral work at NRI when he studied Fusarium wilt and Botrytis grey mould of chickpea (*Cicer arietinum*) and extended this to other non-cultivated (wild) species of *Cicer*. He has also studied the chemistry of resistance in wild and cultivated pigeon pea and chickpeas to the pod borer (*Helicoverpa armigera*) and in *Cedrela odorata* to leaf weevils (*Exophthalmus* spp.). Much of this work has been in collaboration with the ICRISAT, India. He is working presently on the biological activity of pesticidal plants against storage insect pests developing ways to optimise their use, collection and even cultivation. He is also exploring resistance on Sweet potato (*Ipomoea batatas*) to sweetpotato weevils (*Cylas* spp). Phil has now published over 50 peer reviewed papers and book chapters on this work and other aspects of natural product chemistry and the role of plant compounds in biological systems, agriculture and medicine. He is a member of the Editorial Board of *Crop Protection*; the official journal of the International Association of Plant Protection Services.

LIST OF CONTRIBUTORS

Maqbool Ahmad, South Australian Research and Development Institute (SARDI), GPO Box 397, Adelaide, SA 5001, Australia. E-mail: ahmad.maqbool@saugov.sa.gov.au

M. Andrews, School of Sciences, University of Sunderland, Sunderland SR1 3SD, UK. E-mail: mitchell.andrews@sunderland.ac.uk

P. N. Bahl, A-9, Nirman Vihar, New Delhi 110092, India. E-mail: pnbahl@hotmail.com

G. Bejiga, Green Focus Ethiopia, PO Box 802, Code No. 1110, Addis Ababa, Ethiopia. E-mail: geletub@hotmail.com

Zewdie Bishaw, International Center for Agricultural Research in the Dry Areas (ICARDA) PO Box 5466, Aleppo, Syria. E-mail: z.bishaw@cgiar.org

Jason Brand, Department of Primary Industries, Victoria, PB 260, Horsham, Victoria, 3401, Australia. E-mail: jason.brand@dpi.vic.gov.au

Chengci. Chen, Department of Agronomy, Montana State University, Bozeman, Montana, USA. E-mail: cchen@montana.edu

W. Chen, USDA-ARS, Grain Legume Genetics and Physiology Research Unit, Washington State University, Pullman, WA 99164–6402, USA. E-mail: w-chen@wsu.edu

C. J. Coyne, Plant Introduction Unit, 303 Johnson Hall, Washington State University, Pullman, WA 99164, USA. E-mail: coynec@wsu.edu

Philip A Davies, South Australian Research and Development Institute (SARDI), GPO Box 397, Adelaide, SA 5001, Australia. E-mail: davies.phil@saugov.sa.gov.au

M. K. Dhillon, Genetics Resources Divisions, International Crops Research Institute for the Semi-Arid Tropics (ICRISAT), Hyderabad 502324, Andhra Pradesh India. E-mail: m.dhillon@cgiar.org

M. El Bouhssini, International Center for Agricultural Research in the Dry Areas (ICARDA), Aleppo, Syria. E-mail: m.bohssini@cgiar.org

Rebecca Ford, BioMarka, Faculty of Land and Food Resources, The University of Melbourne, Victoria, Australia. 3010 E-mail: rebecca.ford@landfood.unimelb.edu.au, rebeccaf@unimelb.edu.au

Juana Frías, Instituto de Fermentaciones Industriales, Consejo Superior de Investigaciones Científicas, Juan de la Cierva 3, Madrid 28006, Spain. E-mail: frias@ifi.csic.es

Bonnie Furman, Genetic Resource Unit, International Centre for Arid Research in the Dry Areas, PO Box 5466, Aleppo, Syria (ICARDA). E-mail: b.furman@cgiar.org

Stephanie Gailing, The Foundation, PO Box 25801, Seattle, Washington 98165, USA. E-mail: Stephanie.gailing@gmail.com

Yantai Gan, Agriculture and Agri-Food Canada, Box 1030, Swift Current, SK, S9H 3X2, Canada. E-mail: gan@agr.gc.ca

P. K. Ghosh, Department of Biosystems Engineering, University of Manitoba, Winnipeg, Manitoba, Canada, R3T 5V6. E-mail: umghoshp@cc.umanitoba.ca

G. D. Hill, Agriculture Group, Agriculture and Life Science Division, PO Box 84, Lincoln University, Canterbury, New Zealand. E-mail: Hill1@lincoln.ac.nz

Kristy Hobson, Grains Innovation Park, The Department of Primary Industries, Private Bag 260, Horsham, Victoria 3401, Australia. E-mail: kristy.hobson@dpi.vic.gov.au

Prahbpreet Inder, BioMarka, Faculty of Land and Food Resources, The University of Melbourne, Victoria, Australia 3010. E-mail: p.inder@pgrad.unimelb.edu.au

D. S. Jayas, Department of Biosystems Engineering, University of Manitoba, Winnipeg, Manitoba, Canada, R3T 5V6. E-mail: Digvir_Jayas@ umanitoba.ca

A. N. Jha, Division of Entomology, Indian Agricultural Research Institute, India, New Delhi 110012, India. E mail: chitrasriv@gmail.com

Kurt Lindbeck, Grains Innovation Park, The Department of Primary Industries, Private Bag 260, Horsham, Victoria 3410, Australia. E-mail: Kurt.Lindbeck@dpi.vic.gov.au

Monika M. Lülsdorf, Crop Development Centre, University of Saskatchewan, 51 Campus Drive, Saskatoon, SASK S7N 5A8, Canada. E-mail: monika.lulsdorf@usask.ca

M. Manohar, Pulse Laboratory, Division of Genetics, Indian Agricultural Research Institute, New Delhi 110012. E-mail: manohar.iitr@gmail.com

George Mateljan, The Foundation, PO Box 25801, Seattle, Washington 98165, USA. E-mail: gmf@mauigateway.com

Michael Materne Grains Innovation Park, The Department of Primary Industries, Private Bag 260, Horsham, Victoria 3401, Australia. E-mail: Michael.materne@dpi.vic.gov.au

N. Maxted, School of Biological Sciences, University of Birmingham, Edgbaston, Birmingham, B15 2TT, UK. E-mail: n.maxted@bham.ac.uk

B. A. McKenzie, Agriculture Group, Agriculture and Life Science Division, PO Box 84, Lincoln University, Canterbury, New Zealand. E-mail: mckenzie@lincoln.ac.nz

Larn McMurray, South Australian Research and Development Institute, PO Box 822, Clare, South Australia, 5453, Australia. E-mail: cmurray.larn@saugov.sa.gov.au

David L. McNeil, School of Agricultural Science, University of Tasmania, Private Bag 54, Hobart, Tasmania 7001, Australia. E-mail: david.mcneil @utas.edu.au

S. K. Mishra, Division of Germplasm Evaluation, National Bureau of Plant Genetic Resources, Pusa Campus, New Delhi 110012, India. E-mail: skmishra_gene@rediffmail.com

Barkat Mustafa, BioMarka, Faculty of Land and Food Resources, The University of Melbourne, Victoria, 3010 Australia. E-mail: b.Mustafa@pgrad. unimelb.edu.au

Abdul A. Niane, International Center for Agricultural Research in the Dry Areas (ICARDA), P.O. Box 5466, Aleppo, Syria. E-mail: a.niane@cgiar.org

Jesús M. Porres, Departamento de Fisiología, Instituto de Nutrición, Universidad de Granada. Campus Universitario de Cartuja s/n, Granada 18071, Spain. E-mail: imporres@ugr.es

Hari Ram, Department of Plant Breeding, Genetics and Biotechnology, Punjab Agricultural University, Ludhiana 141004, India. E-mail: hr_saharan@yahoo.com

Bob Redden, Australian Temperate Field Crops Collection, Department of Primary Industries, Victorian Institute for Dryland Agriculture, Horsham, Victoria 3401, Australia. E-mail: bob.redden@dpi.vic.gov.au

A. Amarender Reddy, Centre for Poverty and Rural Development Administrative Staff College of India, Hyderabad. E-mail: aareddy@asci.org.in

A. H. Rizvi, Pulse Laboratory, Division of Genetics, Indian Agricultural Research Institute, New Delhi 110012, India. E-mail: hasan_ra@rediffmail.com

J. S. Sandhu, Department of Plant Breeding, Genetics and Biotechnology, Punjab Agricultural University, Ludhiana 141004, India. E-mail: js_sandhuin@yahoo.com

H. S. Sekhon, Department of Plant Breeding, Genetics and Biotechnology, Punjab Agricultural University, Ludhiana 141004 India. E-mail: sekhonhsd@yahoo.com

Rubeena Shaikh, Department of Crop and Soil Sciences, Washington State University, 291 Johnson Hall, PO Box 646420, Pullman, WA 99164, USA. E-mail: rubeena25@yahoo.co.in

B. Sharma, Division of Genetics, Indian Agricultural Research Institute, New Delhi 110012, India. E-mail: drbrsh@yahoo.co.in

H. C. Sharma, Genetics Resources Divisions, International Crops Research Institute for the Semi-Arid Tropics (ICRISAT), Hyderabad 502324, Andhra Pradesh, India. E-mail: h.sharma@cgiar.org

S. K. Sharma, National Bureau of Plant Genetic Resources, Pusa Campus, New Delhi 110012, India. E-mail: skspbg@yahoo.co.in

B. G. Shivakumar, Division of Agronomy, Indian Agricultural Research Institute, Pusa, New Delhi 110012, India. E-mail: bgskumar@yahoo.com

Renuka Shrestha, Division of Agronomy, Nepal Agricultural council, GPO box 404, Kathmandu, Nepal. E-mail: renuka_shrestha@hotmail.com

Guriqbal Singh, Department of Plant Breeding, Genetics and Biotechnology, Punjab Agricultural University, Ludhiana 141004 India. E-mail: singhguriqbal@rediffmail.com

Sarvjeet Singh, Department of Plant Breeding, Genetics and Biotechnology, Punjab Agricultural University, Ludhiiana, Punjab, India. E-mail: sarvjeetm@rediffmail.com

I. S. Solanki Department of Plant Breeding, Chaudhary Charan Singh Haryana Agricultural University Hisar 125004, Haryana, India. E-mail: solanki255@rediffmail.com

Chitra Srivastava, Division of Entomology, Indian Agricultural Research Institute, New Delhi 110012, India. E mail :chitrasriv@gmail.com

Philip C. Stevenson, Natural Resources Institute, University of Greenwich, Chatham, ME4 4TB, UK and Royal Botanic Gardens, Kew, Surrey, TW9 3AB, UK. E-mail: p.c.stevenson@gre.ac.uk

Paul Taylor, FLFR Associate Dean (International) Center for Plant Health/BioMarka School of Agriculture and Food Systems, Faculty of Land and Food Resources, The University of Melbourne, Victoria 3010 Australia. Email: paulwjt@unimelb.edu.au

C. Toker, Department of Field Crops, Faculty of Agriculture, Akdeniz M080, The University, TR-07059 Antalya, Turkey. E-mail: toker@akdeniz.edu.tr

Gloria Urbano, Departamento de Fisiología, Instituto de Nutrición, Universidad de Granada. Campus Universitario de Cartuja s/n, Granada 18071, Spain. E-mail: gurbano@ugr.es

A. K. Verma, Pulse Laboratory, Division of Genetics, Indian Agricultural Research Institute, New Delhi 110012. E-mail: shyamsinghyadav@yahoo.com

Concepción Vidal-Valverde, Instituto de Fermentaciones Industriales, Consejo Superior de Investigaciones Científicas, Juan de la Cierva 3, Madrid 28006, Spain. E-mail: ificv12@ifi.csic.es

M. Yadav, Weatherford College, Weatherford, TX 76086, USA. E-mail: manav.yadav@gmail.com

Shyam S. Yadav, Pulse Laboratory, Division of Genetics, Indian Agricultural Research Institute, New Delhi 110012. E-mail: shyamsinghyadav@yahoo.com

N. T. Yaduraju, National Agricultural Innovation Programme, KAB II, Pusa, New Delhi 110 012, India. E-mail: yaduraju@yahoo.co.uk

PREFACE

Lentils are believed to have originated and been consumed since pre-historic times. They are one of the first crops to have been cultivated. Lentil seeds dating back 8,000 years have been found at archeological sites in the Middle East. Lentils were mentioned in the Bible, both as the item that Jacob traded to Esau for his birthright and as a part of bread that was made during the Babylonian captivity of the Jewish people. In the modern and technologically advanced world, they are under cultivation in more than 35 countries of 5 different continents. It is this ancient origin and modern diversity that makes lentil an "ancient crop for modern times".

Lentils are an important cool season food legume. Traditionally they are a low input crop grown extensively for subsistence or local consumption in rainfed agro-ecosystems. However, in the last 20 years they have also increasingly been grown in extensive, export oriented, production systems in north America. The increasing world interest in legumes has stimulated the need to document what is known about them in order to develop efficient agronomic production systems and breed widely adapted multiple resistant cultivars for wider ecologies. This book provides a comprehensive review of current constraints, achievements and future prospects for lentil crop improvement, production, protection and management technologies. The chapters, each written by specialists help teachers, scientists, students, extension workers, farmers, consumers, traders and administrators in increasing their understanding of the lentil crop.

This book on lentil comprises 23 chapters. Chapters 1–6 present the history, origin, biodiversity, ecology, consumption pattern, nutritional value along with geographic distribution and world trade. Chapters 7–11 explain the role of lentils in the cropping system, rhizobium management and nitrogen fixation, nutrient, weed, irrigation management and profitability in cultivation. Chapters 12–17 explain genetics, cytogenetics, mutation breeding, wild relatives, breeding methods, varietal adaptation and new plant type, genomics and molecular approaches and quality seed production. Chapters 20–23 highlight biotic and abiotic stresses, insect-pest management, post harvest management and lentil growers around the world.

In the modern world knowledge tends to be interdisciplinary and global and lentil systems are no exception. Therefore, most chapters have involved collaboration of 3 or more diverse international authors. Thus the book represents a truly global perspective consistent with the nature of world production, trade and consumption. The book also stress the interactions that have arisen globally for lentil technologies for international marketing, breeding, production and protection approaches, domestic consumption and economic issues arising internationally. This book offers the latest reviews of lentil technologies and publications as well as presenting new findings direct from leading researchers for use by researchers, professionals, technologists, economists, students, traders, consumers and growers. We are certain you will find it both a timely, interesting and valuable addition to the literature on an extraordinary crop.

Shyam S. Yadav
David Mc Neil
Philip C. Stevenson

ACKNOWLEDGEMENTS

The editorial team would like to express our sincere thanks to all the contributors for their valuable professional contributions, patience, efficiency, dedication and devotion. The editing of multi-author texts is not always easy. In this case, it was painless, encouraging and enjoyable. All the authors and co-authors responded speedily, effectively and efficiently to the collective pressure exerted by the editors, with the consequences that the manuscripts were delivered on time and without any difficulty. This made the job of the editors easier and the job of collecting the scripts and preparing the final text for the publisher relatively straight forward.

We express our deep gratitude to several people who have rendered invaluable assistance in making this publication possible. Shyam S. Yadav, Senior Editor expresses his gratitude to Dr. S. A. Patil, Director, Indian Agricultural Research Institute, New Delhi, India for providing excellent technical support in completion of this book. Lastly, we thank Springer for publishing this book.

The Editors would like to express their sincere thanks to Manav Yadav, who agreed to work as Project Manager for this book. He has been working for this project right from the beginning when it was just a proposal until the final stages of book publication. He worked and managed the communications with Editors, Springer, Authors, and Technical Editor Queries. His able leadership and sincerity helped everyone involved to work as a team and finish this daunting task in a timely manner.

The Editors would like to thank Pulse Canada, who agreed to provide the front cover photograph of Lentils. More information about Pulse Canada can be accessed by visiting *www.pulsecanada.com* or by calling (204) 925–4455. The Editors are thankful to Pulse Canada for their assistance and help provided in the completion of this book.

GENERAL NOTE

References to any chemical control products, uses and operations in this book should not be considered an endorsement of their use in areas for which they have not been approved. Their incorporation here is to provide information on research that has been carried out and not to propose their use where not registered.

CHAPTER 1

HISTORY AND ORIGIN

J. S. SANDHU AND SARVJEET SINGH

Department of Plant Breeding, Genetics and Biotechnology, Punjab Agricultural University, Ludhiiana, Punjab, India

E-mail: js_sandhuin@yahoo.com

Abstract: Lentil (*Lens culinaris* Medikus) is the oldest pulse crop with remains found alongside human habitation up to 13,000 years BC. Its domestication is equally old and was probably one of the earliest crops domesticated in the Old World. It is mainly grown in India, Bangladesh, Pakistan, Egypt, Greece, Italy, countries in the Mediterranean region and North America. It is also being cultivated in the Atlantic coast of Spain and Morocco. The crop has a high significance in cereal-based systems because of its nitrogen fixing ability, its high protein seeds for human diet and its straw for animal feed. It is widely used in a range of dishes and reputed to have many uses in traditional medicine. There are a range of wild lentils but *L. orientalis* is believed to be the progenitor of the cultivated lentil

1. INTRODUCTION

Lentil, the plant varies from 6 to 18 inches in height, and has many long ascending branches. The leaves are alternate, with six pairs of oblong-linear, obtuse, mucronate leaflets. The flowers, two to four in number, are of a pale blue colour, and are borne in the axils of the leaves, on a slender footstalk nearly equaling the leaves in length. The period are about 1.5 cm long, broadly oblong, slightly inflated, and contain two seeds, which are of the shape of a doubly convex lens, and about 0.5 cm in diameter.

There are several cultivated varieties of the plant, differing in size, hairiness and colour of the leaves, flowers and seeds (Figure 1). The last may be more or less compressed in shape, and in colour may vary from yellow or grey to dark brown; they are also sometimes mottled or speckled. In English commerce two kinds of lentils are principally met with, French and Egyptian. The former are usually sold entire, and are of an ash-grey colour externally and of a yellow tint within; the latter are usually sold like split peas, without the seed coat, and

S.S. Yadav et al. (eds.), Lentil: An Ancient Crop for Modern Times, 1–9.

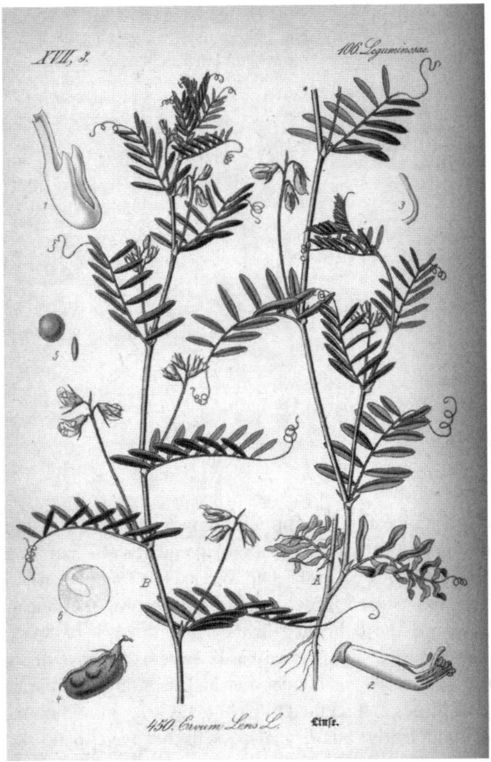

Figure 1. Illustration of the lentil plant, 1885

consist of the reddish-yellow cotyledons, which are smaller and rounder than those of the French lentil; the seed coat when present is of a dark brown colour. (www.1911encyclopedia.org,2006)

Popular in parts of Europe and a staple throughout much of the Middle East and India, this tiny, lens-shaped pulse has long been used as a meat substitute. There are three main varieties of lentils, the French or European lentil, sold with the seed coat on, has a grayish-brown exterior and a creamy yellow interior. The reddish orange Egyptian or red lentil is smaller, rounder and sans seed coat and yellow lentil. None of these varieties are used fresh but are dried as soon as they're ripe. The regular brown lentils are commonly found in supermarkets whereas the red and yellow lentils, though available in some supermarkets, must usually be purchased in Middle Eastern or East Indian markets. Lentils should be stored airtight at room temperature and will keep up to a year. They can be used as a side dish (puréed, whole and combined with vegetables), in salads, soups and stews. One of the most notable showcases for the lentil is the spicy Indian dhal. Lentils have a fair amount of calcium and vitamins A and B, and are a good source of iron and phosphorus.

2. COMMON NAMES

Lentil is known by various names in different parts of the world. The most common names are lentil (English), adas (Arabic), mercimek (Turkey), messer (Ethopia), masser or massur (India), heramame (Japanese). Other names mentioned in literature are mangu or margu (Persian), masura, renuka, mangalaya (Sanskrit).

3. USES

Lentil seeds are consumed as whole grains and as dehulled *dhal*. There are two types of lentil, the large seeded (*macrosperma*) and the small to medium sized seeded called (*microsperma*). The colour of seeds also varies with lines being brown, red, green or white. The lentil seeds are relatively higher in protein content (25%), carbohydrates and calories than other legumes (Muehlbauer *et al.* 1985). Its seeds are also a good source of essential minerals like calcium, phosphorous, iron and vitamin B. Lentil seeds are used for various cuisines worldwide and most commonly used as main dishes, side dishes, as sprouted grain in salads with *rotis* and rice. Seeds can be fried and seasoned for other uses. It is used as a staple of the diet in many Middle Eastern countries and India. Lentil flour can be used to prepare dishes such as soups, stews and purees. The flour can be mixed with cereals to make breads and cakes and as a food for infants (Williams and Singh 1988). Besides, being highly nutritious lentil seeds also contain anti-nutritional factors such as protease inhibitors, lectins or phytohaemoglutins and oligosaccharides that cause flatulence. These anti-nutritional factors can be minimized by heating, water soaking and germination (Jumbunathan *et al.* 1994). Williams *et al.* (1994) reported that lentils have the least while fababean generally have the highest concentrations of these anti-nutritional factors. Tannins are another set of anti-nutritional compounds found in the seed coat which is removed during dehulling while processing (Williams *et al.* 1994). Plant residues like dried leaves, stalks, husk, podwall etc. left after threshing are a good source of cattle feed. These residues contain about 10.2% moisture, 1.8% fat, 4.4% protein, 50% carbohydrate, 21.4% fibre and 12.2% ash (Muehlbauer *et al.* 1985). Lentil seeds are also used by industry as a source of commercial starch for textiles and printing (Kay 1979).

4. MEDICINAL VALUES

Lentil soups have a place in traditional medicines. They are claimed to improve digestion and are prescribed by traditional physicians during convalescence and also claimed to be as a blood purifier. An old traditional medicinal practice was the application of lentil paste to the skin to erase old skin disorders. It also said to alleviate peptic or duodenal ulceration and other intestinal afflictions that are seen in rice eating people. “In India, lentils are poulticed onto ulcers that follow smallpox and slow healing sores” (Duke 1981). In the sixth century, chickpeas were believed to be an aphrodisiac while lentils were considerd to have the opposite effect. Probably, this was the reason that lentil was a part of the diet of monasteries during seasons when there was a low availability of meat (van der Maesen 1972).

5. HISTORY

The history of lentil is as old as Agriculture (Helbaek 1963). The carbonized remains of lentil dated to 11,000 BC from Greece's Franchthi cave are the oldest known remains. Small seeded (2–3 mm) types were found at Tell Mureybit in Syria and have been dated to 8500–7500 BC (Van Zeist 1971, Zohary 1972, Hansen and Renfrew 1978). Lentil remains have been found in Neolithic, aceramic farming villages which were occupied in the 7th millennium BC in the near east arc (Helbaek 1959). The type of agriculture surround these lentils cannot be determined as during this period small seeded cultivated lentil could not be differentiated from wild lentil seeds. In an archaeological site in northern Israel the presence of a large storage of lentils clearly established that by 6800 BC lentil was a part of farming. Carbonized lentil seeds have been recovered from widely dispersed places such as Tell Ramand in Syria (6250–5950 BC), aceramic Beidha in Jordan, ceramic Hacilar in Turkey (5800–5000 BC) and Tepe Sabz in Iran (5500–5000 BC) (Van Zeist and Bottema 1971, Helbaek 1970). In Greece, lentils dating back to 6000–5000 BC have been found in Neolithic settlements such as Argissa-Magula Tessaly (Hopf 1962) and Nea Mikomedeia, Macedonia (Renfrew 1969, Van Zeist and Bottema 1971) and in the same period lentil remains were also seen in Egypt (Matmur, El Omari late 4th millennium, Helbaek 1963). The archaeobotanical remains of lentil have been found in the excavations of the Harappan civilization covering the period of 3300–1300 BC.

6. ORIGIN

Lens culinaris is indigenous to the near East and Central Asia. The putative progenitor of cultivated species *Lens culinaris* subsp. *orientalis* (Bioss.) Ponert is found in Turkey, Syria, Lebanon, Israel, Jordan, Iraq, Iran, Afghanistan, Greece, Uzbekistan (Ladizinsky 1979a, Cubero 1981, Zohary 1973). Most of the West Asian lentils have a flattened lens-like appearance while South Asian lentils have a convex shape on both sides. The electrophoretic studies of seed protein profiles in 22 lines comprising of 11 of *L. culinaris,* six of *L. orientalis,* four of *L. ervoides* and one line of *L. nigricans* belonging to different regions of distributions showed that *L. culinaris, L. orientalis* and *L. nigricans* were related to each other while *L. ervoides* was different (Ladizinsky 1979b). Further, Renfrew (1973) suggested that *L. nigricans* is the progenitor of the cultivated species *L. culinaris* based on his belief of the domestication of lentil in southern Europe. Other groups of workers more recently (Ladizinsky 1979a, Barulina 1930, Zohary 1972 and Williams *et al.* 1974) have claimed that *L. orientalis* is the progenitor of the cultivated species based on the fact that the wild species were found in the fields of the farmers where lentil crops were cultivated in the Middle East. Secondly, plant characteristics and pollen grain morphology were found to be quite close. Further, Ladizinsky (1979c) attempted the cross between *L. culinaris* and *L. orientalis* and studied the behaviour of F_1 and F_2 populations. He also concluded that pod indehiscence is governed by single recessive gene. Together these data gave a reason to believe that domestication of lentil might have started with selection of the variants from wild populations which were non-pod dehiscence (Ladizinsky 1979b). Non-pod dehiscence variants made the harvesting easy due to retention of pods till harvest. *L. orientalis* might have originated first from perennial species and at the secondary level become the progenitor of cultivated species (Singh 2001). According to Ladizinsky (1979a), *L. orientalis* is the progenitor of cultivated species based on the cytoplasmic studies. In the cytogenetic analysis of interspecific hybrids, three chromosome interchanges were found between the cultivated *L. culinaris* and *L. nigricans* while only one was found between the cultivated species and *L. orientalis*, which accentuates the concept that *L. orientalis* is the most probable progenitor of the domestic lentil. *Lens orientalis* is the presumed progenitor of *Lens culinaris* and the two species are crossable and produce fully fertile progeny (Muehlbauer *et al* 2006).

7. DOMESTICATION

Domestication is likely to have started with selection of plants from wild species that retain their seeds in pods before harvesting and continuous selection for large seed size. Lentil is a self-pollinated crop species and this would have helped greatly in maintaining line identity in the process of domestication. Archaeological studies, presented above, have confirmed the presence of lentil in the Turkey-Syria-Iraq region as far back as 8500–600 BC. This region probably played an important role in lentil domestication and starting the further spread to the Nile, Greece, Central Europe and eastwards to South Asia (Nene 2006). At the same time, lentil also spread to Ethopia, Afghanistan, India, Pakistan, China and later to the New World including the Latin America (Cubero 1981, Duke 1981, Ladizinsky 1979a). Bahl et al. (1993) suggested that probably lentil's domestication was the oldest among grain legumes. Lentil was thus an important part right from the start of the agricultural revolution in the Old World alongside the domestication of wheat, barley, pea, flax, einkorn and emmer wheats (Zohary 1976). The crop was part of the assemblage of Near Eastern grain crops introduced to Ethopia by the invaders of the Hamites. From the Bronze age onward, lentils were grown wherever wheat and barley were cultivated throughout the expanding realm of Mediterranean-type agriculture (Erskine 1989). This indicates a specific demand for lentil in the social system as the yield of lentil was quite low in comparison to wheat and barley. Thus along with other grain crops, lentil cultivation as part of a farming system was probably initiated in late 5th or early 4th millennia BC. The Harappan civilization (3300–1300 BC) remains are indicative of domestication of lentil starting prior to 2500 BC in India. De Candolle (1882) dated the start of lentil cultivation on linguistic grounds. "It may be supposed that the lentil was not in this country (India) before the invasion of the Sanskrit speaking race." This probably occurred before 2000 BC and is consistent with the other evidence presented above.

8. BOTANICAL DESCRIPTION

Lentil is annual bushy herb with erect, semi-erect or spreading and compact growth habit. It has many branches with soft hairs. Its stems are slender, angular, light green in colour about 15–75 cm in height (Duke 1981, Muehlbauer *et al.* 1985). The rachis is 4–5 cm in length bearing 10–15 leaflets in 5–8 pairs. Generally, the upper leaves are converted into tendrils or bristle, whereas the lower leaves are mucronate (Muehlbauer *et al.* 1985). The leaves are alternate, compound, pinnate and yellowish green, light yellow green, dull green, dark green or dark bluish green in colour. The stipules are small or absent. The axillary racemes generally bear 1–4 flowers on short peduncles having 2.5–5.0 cm length. The flowers are small, white, pink, purple, pale purple, pale blue in colour (Muehlbauer *et al.* 1985). The flowering proceeds acropetally. The lowermost buds open first and flowering proceeds upward and it takes about two weeks to complete opening of all the flowers on the single branch (Nezamudhin 1970). The opening of flower occurs between 8.00 to 10.00 hrs and continues till noon and each flower remains open for about 16–24 hrs. At the end of the second day and on the third day all the opened flowers close completely and the colour of the corolla begins to fade. The setting of pods occurs after 3–4 days. The flowers have small ovaries with one or two ovules. The style is covered with a hairy inner surface. The pods are oblong, flattened or compressed, smooth with 1–2 cm in length. Pods have a curved beak, persistent calyx and contain 1–3 seeds. The seeds are biconvex, round, small, lens-shaped and weigh between 2–8 g per 100 seeds. The colour of the testa varies from tan to brown black, purple and black. The mottling and speckling of seed is a common feature in some germplasm lines. The cotyledons are red, orange, yellow or green, bleaching to yellow (Kay 1979, Duke 1981, Muehlbauer *et al.* 1985). The germination of seed is hypogeal.

9. TAXONOMY

The genus *Lens* comprises six species (Ferguson 1998; Ferguson *et al.*, 2000) as *Lens orientalis* is generally now classified as *Lens culinaris* subsp *orientalis.* Only one species, *L. culinaris* Medikus is cultivated. Among the wild species namely *L. montbretii* (Fisch and Mey.) Davis and Plit., *L. ervoides* (Brign.) Grande; *L. nigricans* (Bieb.) Godr., and *L. orientalis* (Boiss.) M. Popov., the latter two species possess morphological similarities to the cultivated lentil (Ladizinsky 1979b). However, Ladizinsky and Sakar (1982) suggested that *L. montbretii* should be placed in the genus *Vicia* and named as *Vicia montbretii* (Fisch & Mey.) based on morphological and cytological data and breeding experiments. The cultivated lentil originated from *L. orientalis* (Barulina 1930) and chromosome number of cultivated lentil, its progenitor species, *L. orientalis* and

L. nigricans are same, i.e. $2n = 14$. The cultivated species, *L. culinaris* has been divided into two sub-species (Barulina 1930) namely *macrosperma* (seed diameter, 6–9 mm) and *microsperma* (seed diameter, 2–6 mm). The *macrosperma* type have yellow cotyledons and very light or no pigmentation in their flowers and other plant parts, whereas the *microsperma* types have red, orange or yellow cotyledons and pigmented flowers and other plant parts. Williams *et al.* (1974) classified *L. culinaris* in the order Rosales, sub-order Rosineae, family Leguminosae and sub-family Papilionaceae. The genus *Lens* occupies an intermediate position between *Vicia* and *Lathyrus*, the two other members of papilionaceae. However, it is more closely related to the genus *Vicia.* Of the annual *Lens* species, *L. nigricans* can be separated from *L. culinaris* and *L. orientalis* from the stipule shape (Barulina 1930, Ball 1968, Davis and Plitmann 1970, Williams *et al.* 1974). The stipules of *L. culinaris* and *L. orientalis* (*Lens culinaris* subsp *orientalis*) are oblong or elliptic, lanceolate, entire, whereas stipules of *L. nigricans* are semi-hastate, entire or dentate. The classical taxonomy is based on morphological characteristics, it does not necessarily represent biological entities. Hybridization within the genus indicates the classification of species based on morphology is not valid and accessions based on morphology classified in the same species sometimes turn out to be cross – incompatible. Due to this reason, *L. odemensis*, formerly a member of *L. nigricans*, has been described as a new species (Ladizinsky 1986). The stipules of *L. nigricans* are semi-hastate with up right position, whereas in *L. odemensis* the stipules are less dentate, semi-hastate with horizontal position. Both these species are cross incompatible (Ladizinsky 1993). The detailed characteristic features of various *Lens* species are as follows: The wild species *L. nigricans* is a slender, densely pilose having semi-hastate stipules, conspicuously aristate peduncles and mauve flowers. The pods are glabrous and small usually with two tiny lenticular seeds. It is morphologically more closely related to cultivated species *L. culinaris.* However, it can be differentiated from cultivated lentil by many characters like toothed semi-hastate stipules and the strong aristate peduncles. The wild species *L. ervoides* is very slender (10–30 cm tall), with semi-hastate stipules and long filiform penduncles. It has very small puberulent pods with lenticular seeds. Like *L. nigricans,* it is also morphologically closely related to cultivated lentil. However, it can be separated from cultivated lentil by traits like structure of stipules and peduncle, size of pod and seed and flower shape.

The wild species *L. orientalis* (*Lens culinaris* subsp *orientalis*) is slender, pilose (10–30 cm tall) and has a very strong resemblance to *L. culinaris* with respect to vegetative growth and structure of the flower and pod. The stipules are entire, obliquely lanceolate and unappendaged. The calyx is 4–6 mm long with teeth much longer than tube. The pods are glabrous with small lenticular seeds. Overall, *L. orientalis* (*Lens culinaris* subsp *orientalis*) looks like a miniature version of *L. culinaris* and morphological boundaries between both the species are occasionally intermixed. Some intergradation between the two species has been reported by Davis and Plitmann (1970) and some intermediates were also found in several localities in the Judean hills and in Galilee, Israel (Zohary 1972). The wild species

L. odemensis is decumbent-ascending with single or branched column. It has small rachis (8–20 mm) with 6–8 leaflets per leaf and ending in a tendril. The stipules are semi-hastate, slightly dentate at the base and horizontal to the stem. The calyx is 4–6 mm with long teeth. Pods are glabrous, rhomboid with 1–2 motted gray-brown small seeds.

Lentil is one of the oldest grain legumes. It is a short statured, annual, self-pollinated, high valued crop species. It is native to the Near East and Central Asia and rapidly spread to other parts of world. *L. orientalis* is the most probably a secondary level progenitor of the cultivated species *L. culinaris* Medikus. *L. orientalis* might had originated from the wild perennials through natural selection. The wild lentil species are potential resources yet to be tapped.

REFERENCES

Bahl P N, S Lal and B M Sharma (1993) An overview of the production and problems in southeast Asia. P.1–10.In: W Erksine and M C Saxena (eds.) Lentil in South Asia. *Proceedings of the seminar on lentils in South Asia.* ICARDA, Aleppo, Syria.

Ball P W (1968) Lens. *In*: Flora Europaea (ed; T G Tutin). Cambridge. pp. 136.

Barulina E I (1930) The lentils of the USSR and other countries. *Tr Po Prikl Bot Genet Sel [Bull Appl Genet & Select*] Suppl **40** :1–319 [Russian]Cambridge.

Cubero J I (1981) Origin, taxonomy and domestication. p. 15–38, Lentils. In: C. Webb and G. Hawtin (eds), CAB, London, UK.

Davis P E and U Plitmann (1970) Lens MILLER, In: (Ed. Davis, P.E.) Flora of Turkey. **3**: 325–328. Edinburgh Univ. Press, Edinburgh.

De Candolle A P (1982) Origins of cultivated species. Reprint 1967. Hafner, London.

Duke J A (1981) Handbook of legumes of world economic importance. Plenum Press, NewYork. P. 52–57.

Erskine W, Y Adham and L Holly (1989) Geographic distribution of variation in quantitative characters in a world lentil collection. *Euphytica* **43**:97–103.

Ferguson, M. (1998) PhD Thesis: *Studies of genetic variation within the genus lens*. School of Biological Sciences, University of Birmingham, Birmingham, UK.

Ferguson M E, Maxted N, van Slageren M and Robertson L D (2000) A re-assessment of the taxonomy of *Lens* Mill. (Leguminosae, Papilionoideae, Vicieae*). Bot J Linnean Soc* **133**: 41–59.

Hansen J and J M Renfrew (1978) Paleolithic-Neolithic seed remains at Franchthi cave, Greece. *Nature* **71**: 349–352.

Helbeck H (1959) Domestication of food plants in old World. *Science* **130**: 365–372.

Helbeck H (1963) Late Cypriote vegetable diet in Apliki. *Act. nstit. Athen. Reg. Sueciae. Ser.* **4**: VIII: 171–186.

Helbeck H (1970) The plant industry in Hacilar. *In:* Mellart, J. Excavations at Hacilar Vol. I. Occasional Publications No.9 of the British Institute of Archaelogy, Ankara. Edinburgh Univ.Press, pp. 189–244.

Hopf M (1962) Bericht Uber die Untersuchung von Samen und Holzkohlenresten von der Argissa-Magula aus den prakermischen bis mittelbronzezeitlichen Schichten. In: V. Milojcic, J Boessneck und M Hopf : Die Deutschen Ausgrabungen auf der Argissa-Magula in Thessalien.I. Rudolf Habelt Verlag Bonn. PP. 101–119.

Jumbunathan R, H L Blain, K S Dhindsa, L A Hussein, K Kogure, L Li-Juan and M M Youseef (1994) Diversifying use of cool season food legumes through processing. pp. 98–112. In: F J Muehlbauer and W J Kaiser (eds.) Expanding the Production and Use of Cool Season Food Legumes. Kluwer Academic Publishers. Dordrecht, The Neetherlands.

Kay D (1979). Food legumes. Tropical Development and Research Institute (TPI). TPI Crop and Product Digest No. 3. p. 48–71.UK.

Ladizinsky G (1979a) The Origin of lentil and its wild genepool. *Euphytica*. **28**: 179–187.
Ladizinsky G (1979b) Species relationships in the genus *Lens* as indicated by seed protein electrophresis. *Bot Gaz* **140**:449–451
Ladizinsky G (1979c) The genetics of several morphological traits in the lentils. *J Hered* **70**:135.
Ladizinsky G (1986) A new *Lens* species from the Middle East, *Notes R Bot Gard, Edinburgh* **643**: 489.
Ladizinsky G (1993) Wild Lentils. *Critical Reviews in Plant Sciences*. **12**:169–184.
Ladizinsky G and D Sakar (1982) Morphological and cytogenetical characterization of *Vicia montbretii* Fisch & Mey. Synonym: *Lens montbretii* (Fisch & Mey.) Davis and Plitman, Bot J Lin Soc.**85**: 209.
Muehlbauer F J, J I Cubero and R J Summerfield (1985) Lentil (*Lens culinaris Medik.*).p. 266–311. In: R J Summerfield and E I I Roberts (eds) Grain Legume Crops. Collins, 8 Grafton Street, London, UK.
Muehlbauer F J, S Cho, A Sarker, K E McPhee, C J Coyne, P N Rajesh and R Ford (2006). Application of biotechnology in breeding lentil for resistance to biotic and abiotic stress. *Euphytica* **147** (1–2): 149–165.
Nene Y L (2006) Indian Pulses through the Millennia. *Asian Agri. History*.**10**: 179–202.
Nezamudhin S (1970) Miscellanceous, Masour. In: Pulse Crops of India (ed. P Kachroo). Indian Council of Agricl Res, Krishi Bhawan, New Delhi, India. pp. 306–313.
Renfrew J M (1969) The archaeological evidence for the domestication of plants: methods and problem. In: Ucko P J and G W Dimbleby (eds), The Domestication and Exploitation of plants and animals. Aidine, Chicago.
Renfrew J M (1973) Palaeoethnobotany. Columbia Univ. Press, New York.
Singh D P (2001) Genetics and Breeding of Pulse Crops. Kalyani Publishers, New Delhi.
van der Maesen L J G (1972) *Cicer* L., a monograph of the Genus, with special reference to the chickpea (*Cicer arietinum L.*), its ecology and cultivation, commun. Agric University, Wageningen.
Van Zeist W and S Bottema (1971) Plant husbandry in early neolithic Nea Nikomedeia, Greece.*Acta Bot Neerl*. **20**: 521–538.
Williams J T, A M C Sanchez and M T Jackson (1974) Studies on lentils and their variation .I. The taxonomy of the species. *SABRAO Journal* **6**:133–145.
Williams P C and U Singh (1988) Quality screening and evaluation in pulse breeding.p. 445–457. In: R J Summerfield (ed.), World Crops : Cool Season Food Legumes. Kluwer Academic Publishers, Dordrecht. The Netherlands.
Williams P C, R S Bhatty, S S Deshpande, L A Hussein and G P Savage (1994) Improving nutritional quality of cool season food legumes.p. 113–129. In: F J Muehlbauer and W J Kaiser (eds.), Expanding the Production and Use of Cool Season Food Legumes. Kluwer Academic Publishers, Dordrecht. The Netherlands.
Zohary D (1972) The wild progenitor and place of origin of the cultivated lentil *Lens culinaris*. *Econ Bot* **26**: 326–332.
Zohary D (1976) Lentil. Pages 163–164 *in* Evolution of crop Plants (N.W.Simmonds, ed.). Longman, London, UK.

CHAPTER 2

LENS BIODIVERSITY

BOB REDDEN[1], NIGEL MAXTED[2], BONNIE FURMAN[3], AND CLARICE COYNE[4]

[1] *Curator, Australian Temperate Field Crops Collection, Department of Primary Industries Victoria, Private Bag 260, Horsham, Victoria 3401, Australia*
[2] *Senior Lecturer, School of Biosciences, University of Birmingham, Birmingham, UK*
[3] *Curator, Genetic resource Unit, International Centre for Arid Research in the Dry Areas, PO Box 5466, Aleppo, Syria*
[4] *Curator, Plant Introduction Unit, 303 Johnson Hall, Washington State University, Pullman, WA 99164, USA*
E-mail: bob.redden@dpi.vic.gov.au

Abstract: The *Lens* genus includes the cultivated *L. culinaris,* and wild subspecies *orientalis* - the progenitor, *tomentosus,* and *odemensis,* are in the primary genepool, while *L. ervoides, L. nigricans* and *L. lamottei* are in the secondary – tertiary gene pool. The Middle East is the primary centre of diversity for the primary genepool, with distribution of *L. orientalis* extending to central Asia, and of *L. ervoides* extending along the Mediterranean to Spain. The largest *Lens* collection is held at ICARDA. *In situ* reserves of *Lens* diversity are in Turkey and Syria. Documentation and storage of *Lens* germplasm is discussed. An evaluation database covering a number of genebanks has been developed for lentil germplasm. Core collections are discussed in the context of the generation Challenge program. Application of DNA characterisation is outlined, along with the potential for allele mining for variation in key traits, the study of relationships within *Lens* and the use of mapping populations. Reference is made to the International treaty for Plant Genetic Resources for Food and Agriculture

1. SOURCES OF GENETIC DIVERSITY

The genus *Lens* is a member of the legume tribe Vicieae which includes the major legume crops of the classical Mediterranean civilizations, faba bean, pea along with lentil. The precise generic boundaries between *Lens* and the related Vicieae genera (*Vicia* L., *Lathyrus* L., *Pisum* L. and *Vavilovia* A. Fedorov) have been much debated, but *Lens* appears most closely related to *Vicia* (Kupicha 1981). *Lens* is a small Mediterranean genus which contains the cultivated lentil (*L. culinaris* Medikus subsp. *culinaris*) and six related taxa (Ferguson 2000), as shown:

S.S. Yadav et al. (eds.), Lentil: An Ancient Crop for Modern Times, 11–22.

L. culinaris Medikus
subsp. *culinaris*
subsp. *orientalis* (Boiss.) Ponert
subsp. *tomentosus* (Ladizinsky) Ferguson, Maxted, van Slageren & Robertson
subsp. *odemensis* (Ladizinsky) Ferguson, Maxted, van Slageren & Robertson
L. ervoides (Brign.) Grande
L. nigricans (M.Bieb.) Godron
L. lamottei Czefr.

The cultivated taxon, *Lens culinaris* Medikus subsp. *culinaris*, includes two varietal types: small-seeded *microsperma* and large-seeded *macrosperma* (Barulina 1930).

Lens culinaris subsp. *orientalis,* in the primary genepool (Ladizinsky and Alder 1976), is fully cross compatible with cultivated lentil (Muehlbauer and Slinkard 1981, Robertson and Erskine 1997). *L. nigricans,* in the secondary genepool, can also cross with the cultigen, yet seed set in hybrids is considerably lower than a *L. culinaris* subsp. *culinaris* x *L. culinaris* subsp. *orientalis* cross (Muehlbauer and Slinkard 1981). *L. culinaris* subsp. *orientalis* accessions have been found to contain resistance to drought (Hamdi and Erskine 1996), cold (Hamdi *et al.* 1996, Robertson et al. 1996), wilt (Bayaa et al. 1995), and Ascochyta blight (Bayaa et al. 1994, Robertson et al. 1996).

The Middle East is the primary centre of diversity for both the domestic *L. culinaris* and its wild progenitor *L. culinaris subsp. orientalis* (Cubero 1981, Zohary 1972), but taxa are found from Spain to Tajikistan. Although Ferguson et al. (1998) showed that for *L. culinaris* subsp. *orientalis* two centres exist, (a) south-eastern Turkey and north-western Syria, and (b) southern Syria and northern Jordan. The southern centre overlaps with the centre of diversity for *L. culinaris* subsp.*odemensis*. A region of diversity exists for *L. ervoides* along the eastern Mediterranean coast and for *L. nigricans* in south-western Turkey. Ferguson et al. 1998 also showed that *L. culinaris* subsp. *orientalis* accessions from Iran, central Asia and northern Turkey are all very similar to each other, and correspond to the common cytotype identified by Ladizinsky et al. (1984). *L. ervoides* populations from the coastal region of the former Yugoslavia have a narrow genetic base, as do *L. nigricans* accessions from the same region as well as those from France and Spain. Maps illustrating the distribution of variation can be found in Ferguson and Erskine (2001).

Wild populations of *Lens* are generally poor competitors and highly palatable to grazing animals (Ferguson and Erskine 2001). They are found predominantly in primary, ungrazed habitats where they are not subject to competition by aggressive coloniser plants. They usually form small disjunct populations. The density of plants may vary dramatically between years apparently because of climatic conditions. *L. culinaris* subsp. *orientalis*, subsp. *tomentosus*, subsp. *odemensis* and *L. nigricans* are generally found in open or partially shaded habitats, on shallow stony soils originating from calcareous, metamorphic or basalt rocks. *L. nigricans* and *L. lamottei* have also been found in habitats showing signs of earlier human disturbance such as abandoned terraces or plantations and ruins. Both *L. nigricans* and *L. ervoides* have been found from sea level to over 1000 m.*L. culinaris* subsp. *orientalis* and subsp. *odemensis* have been found at even higher altitudes from 500 m up to 1700 m and from 700 m up to 1400 m respectively. Populations of *L. lamottei* have been located between 100 and 800 m. Although all wild species (except *L. ervoides*) have similar habitat preferences, they are rarely found together. *L. ervoides* usually inhabits shady places and is frequently associated with pine forests often growing on calcareous bedrock (Ferguson and Erskine 2001).

2. *EX SITU* COLLECTIONS

The International Center for Agricultural Research in Dry Areas (ICARDA) has a global mandate for research on lentil improvement. As such, ICARDA houses the world collection of *Lens*, totalling 10,578 accessions (Figure 1). The ICARDA lentil genetic resources collection includes 8858 accessions of landraces and cultivars from 69 different countries representing 4 major geographic regions (Figure 2), 1137 ICARDA breeding lines, and 583 accessions of 6 wild *Lens* taxa representing 24 countries (Table 1). The majority of the collection (48%) consists of accessions

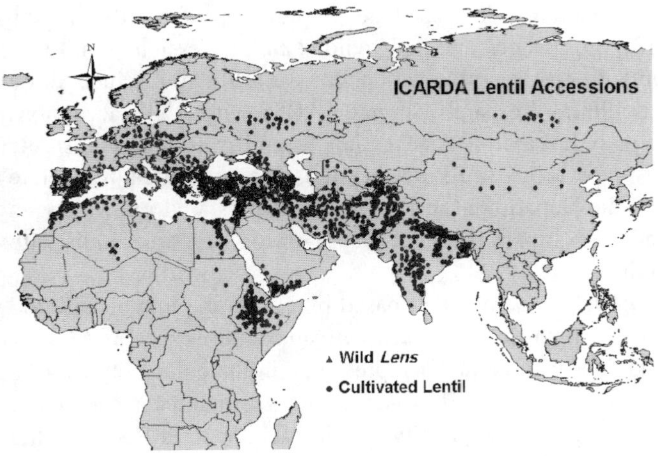

Figure 1. Distribution of ICARDA *Lens* accessions

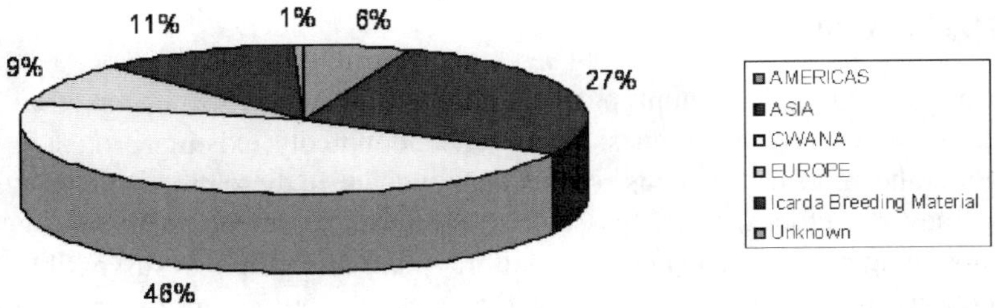

Figure 2. Geographic regions represented in ICARDA lentil collection

Table 1. Wild *Lens* accessions maintained at ICARDA

Taxonomic name	Number of countries represented	Number of accessions
Lens culinaris subsp. odemensis	4	65
Lens culinaris subsp. Orientalis	14	268
Lens culinaris subsp. tomentosus	2	11
Lens ervoides	15	166
Lens lamottei	3	10
Lens nigricans	8	63
Total	**46**	**583**

from Central and West Asia and North Africa, the centre of origin and primary diversity (Zohary and Hopf 1988; Ferguson and Erskine 2001), while South Asia represents an additional 25%.

Lentil accessions in the ICARDA collection were obtained from 113 ICARDA collection missions (46%), 56 donor institutions (44%) and ICARDA's breeding program (11%). Collection missions have yielded 1734 cultivated and 374 wild accessions. Recent missions (13 missions since 1999) to Central Asia and the Caucasus has substantially decreased geographic gaps from this region yielding a total of 122 cultivated and 27 wild *Lens* accession (Dr. Ken Street pers. Com. 2006.). Other important collections worldwide include those at Pullman USDA Agricultural Research Service (ARS) with 2797 accessions (http://www.ars.grin.gov/), N.I. Vavilov All-Russian Scientific Research Institute of Plant Industry (http://vir.nw.ru/) with 2396 accessions, the Australian Temperate Field Crops Collection in the Department of Primary Industries Victoria, Australia with 5250 accessions and the National Board of Plant Genetic Resource, India with 2212 accessions (Dwivedi et al. 2006).

Ferguson and Erskine (2001) conclude that based purely on the number of wild *Lens* accessions available, germplasm from North African countries such as Algeria, Libya, Sudan and Tunisia appear to be under-represented in the collection, as does germplasm from the new central and west Asian republics of the former Soviet Union. However the overall collection priority for the wild species based on the distribution of genetic variation (Ferguson et al. 1998a) remains south-west Turkey, particularly the provinces of Burdur, Isparta and Afyon.

3. *IN SITU* CONSERVATION

There has been no systematic attempt to conserve *Lens* diversity *in situ* using either genetic reserves or on-farm techniques. However, undoubtedly existing protected areas throughout the range of the genus contain *Lens* species. In these protected areas conservation may be referred to as 'passive' (*Lens* species presence and genetic diversity is not being actively managed or monitored) and therefore it is susceptible to further unobserved genetic erosion. More active conservation of Lens diversity is found in the managed genetic reserves of the Eastern Mediterranean (e.g. Ammiad in Eastern Galilee, Israel; Kaz Dag, Aegean Region, Amanos, Mersin in Turkey and Ceylanpinar of South-east Turkey), all of which contain *Lens* diversity. The latter reserve is particularly important as Ferguson et al. (1998) showed that this area is one of the two centres of genetic diversity for *L. culinaris* subsp. *orientalis*. The second centre identified by Ferguson et al. (1998) is in southern Syria and Maxted (1995) proposed the establishment of a genetic reserve for Vicieae species in the region of Mimas, Djebel Druze in Southern Syria. Recently this area has been designated as a genetic reserve by the Syrian Scientific Agricultural Research Commission, as part of their Global Environment Facility funded "Conservation and Sustainable Use of Dryland Agrobiodiversity" project (Maxted et al. 2003). So although some progress has been made recently in *in situ* conservation in the areas of highest *Lens* genetic diversity, there remains an urgent need to systematically establish both reserves for the wild species of *Lens* and on farm projects to conserve the ancient land races of cultivated *Lens* species.

4. CONSERVATION AND STORAGE

The recommended storage facilities for genebanks (FAO/IPGRI 1994), would contain a Base collection of over 100 seed (FAO/IPGRI recommend over 1,000 seed) in −18°C–20°C, and an Active collection of up to 1,000 seed at 2°C – 5°C, respectively, for long term conservation and for medium term storage and distribution. Drying of seed at 10 – 25°C and 10–15% relative humidity (RH) is recommended before storage. The Base collection should aim to maintain the genetic integrity of the seed samples of accessions as originally received, and provide very long term conservation of genetic diversity. It also can replenish seed stocks in the Active collection in every 4th regeneration cycle, a precaution against genetic drift within samples over repeat cycles of regeneration (FAO/IPGRI 1994). A safety duplicate collection for long term storage in another location is recommended, also referred to as a Back-up collection.

Viability will decline in all stored seed, as a function of initial viability, seed moisture and storage temperature (Roberts and Ellis 1984). Seed harvest, drying, processing and pre-storage temperature and humidity conditions may all affect the initial viability, which is usually close to 100% for freshly harvested seed from well grown plants. An investigation of lentil seed longevity was initiated at ATFCC in 2003, with testing seed of 2 varieties at 5 temperatures (40°C, 22°C, 15°C, 2°C and −18°C) and 3 seed moisture levels (6–7%, 10–11%, 12–13%). An additional 8 genotypes are being tested at 22°C. This study will estimate the seed moisture and temperature storage constants for *Lens*, and examine whether these conform to a logarithmic formula as per the Roberts and Ellis (1984) prediction. Results are available for the 40°C treatments, in which the lentil variety Cumra with near 100% initial viability, at high (12.3%), medium (9.5%) and low (7.6%) seed moisture levels, took 200, 685 and 1060 days respectively to fall 50% seed viability.

These studies are important to enable better planning of intervals between regeneration cycles, and more strategic monitoring of viability with germination tests.

Seed in the Active 2°C collection may retain high viability for at least 30 years, given high initial viability, reduction of seed moisture and immediate storage post-harvest. Seeds should be stored in moisture proof containers. Vacuum and heat sealing of plastic lined foil envelopes is one of the common genebanks procedures. Opening and resealing of these packets for seed removal and distribution is best carried out in a drying room, e.g. 15°C and 15% RH as above. Alternately, 1–2 additional small seed samples per accession may be prepared in anticipation of requests for germplasm. A decision guide for seed regeneration, and precautions to maintain genetic integrity, are outlined by Sackville Hamilton and Chorlton (1997).

5. DOCUMENTATION

There are 3 main activities, described in the IPGRI handbook for germplasm collections (Reed 2004);

1. germplasm and site characterisation (passport) for an accession collected as a traditional landrace *in-situ*, for placement in an *ex-situ* genebank elsewhere,
2. genebank inventory with accession identifiers, taxonomic descriptors, country and site passport data, dates of accession, collector/donor details, synonyms for accessions acquired from other genebanks,
3. accession characterisation data. In the case of lentil standard trait descriptors are provided in IBPGR 1985. These describe discrete classifications for morphological traits such as colours of seed, stems and flowers; scales for disease susceptibility from 1 = none to 9 = lethal; and standards for quantitative traits such as flowering time, 100 seed weight, and seed yield, and others.

Full records are very important to check through synonyms for duplication of accessions, for precise collection site data with referencing by latitude, longitude and altitude whenever possible, and for quality assurance on identity. This can be an issue in genebanks with multiple sources of accessions, and large numbers of accessions and species.

Equally important are characterisation and evaluation data. Better targeted and more efficient utilisation of germplasm by plant breeders/researchers can be achieved if the trait characteristics of accessions are known. This can include the agronomic, disease reaction, yield and quality data of accessions in a particular study, and over the different studies for each accession. Evaluation data for lentil germplasm can now be queried for multiple traits from combined databases, such as constructed by ATFCC, using ICIS (International Crop Information Systems) platforms for digital search and retrieval across relational databases over countries and years. This database, the International Lentil Information System (ILIS) combines databases of ICARDA, USDA and ATFCC, enabling a more comprehensive search of the combined genetic resources over large collections for multiple trait expressions; http://149.144.200.51:8080/QMWebRoot

As far as possible, the internationally standard IPGRI descriptors are used for trait records in ILIS, as provided in IPGRI (1985). Additionally ILIS has options for discretisation, with quantitative traits converted to a 1–9 scale based on arithmetic division of the range of values (outlying values quality checked and filtered out), or on statistical normalisation of data from each location/year. These conversions eliminate the environmental component from observed values, as well as the genotype X environment interaction, to provide genetic comparisons of landraces for trait expressions across locations, years, and genebanks. Breeders can request selected accessions on-line from genebanks, for validation in the respective target environments for farmer clients, and choice of parents for the breeding program.

There is increasing interest by breeders in exploiting the wider diversity usually available in the primary, secondary and tertiary gene pools of wild relatives. A number of genebanks provide well documented and up to date taxonomic guides for the *Lens* genus, eg. The GRIN database of USDA ARS; http://www.ars-grin.gov/cgi-bin/npgs/html/genform.pl

6. MOLECULAR DIVERSITY, INTRA-SPECIFIC AND INTER-SPECIFIC VARIATION

Two important new developments in molecular genetics have applicability to genebanks. One is the use of molecular markers for DNA 'fingerprint' characterisation of individual accessions. Spooner et al. (2005) reviewed choice of markers for particular investigations. These tools enable the identity of accessions to be confirmed between collections, detection of duplicates within collections, and analyses of genetic relatedness amongst accessions to define the structure of genetic diversity within a species/genus. This knowledge assists breeders seeking allelic diversity for particular traits. A second application is the capacity of allele mining for various traits using association genetics, both within species/wild relative collections, and across species with comparative genomics (Spooner et al. 2005). These approaches are scheduled for phases 2 and 3 of GCP with the lentil composite collection.

Molecular analysis corrected the taxonomy of *Lens* and resolved the relationships between the species and subspecies of cultivated and wild lentil (Ferguson 2000). This study confirmed the relationship results using morphology, isozymes and RAPDs on 404 accessions (100 *L. culinaris* subsp. *culinaris*, 128 *L. culinaris* subsp *orientalis*, 32 *L. odemensis*, 30 *L. nigricans*, 118 *L. evoides*, 32 *L. nigricans*, 7 *L. lamottei*, 8 *L. tomentosus)*. Numerous previous studies also used molecular markers and proposed a mixture of different phylogenies for *Lens* species. These earlier studies used isozymes (Hoffman et al 1986; Ferguson and Robertson 1996); RFLPs (Harvey and Muehlbauer 1989); cpDNA (Muench et al. 1991; van Oss et al. 1997) and RAPDs (Abo-Alwafa et al. 2005; Sharma et al. 1995; Ford 1997 and 1999). However, far fewer accessions were used than the Ferguson (2000) study used and therefore the Ferguson (2000) classification, shown above, is now widely accepted (Sarker and Erskine 2006).

The *ex situ* collections of lentil and its wild relatives are serving as mapping populations using association mapping statistics as first demonstrated in maize (Thornsbury et al. 2001). First, the genetic structure of the germplasm mapping population needs to be determined (Pritchard et al. 2000), and then an understanding of the amount of linkage disequlibrium (LD) needs to be determined as each crop species will vary. Large scale genotyping of the lentil germplasm collection held at ICARDA is complete (Furman 2006) and is underway on the lentil core collection held by the USDA Agriculture Research Service (ARS) in Pullman, WA USA using 39 mapped SSRs (Coyne personal communication). The ARS studies are for fine mapping of disease resistance QTL identification (Muehlbauer personal communication). The ARS plan is to sequence candidate genes in a core subset (32 pure lines) of the single plant core accessions and arrive at an estimate of the linkage disequilibrium in lentil.

7. CORE COLLECTIONS

The International Center for Agricultural Research in Dry Areas (ICARDA) is participating in the lentil project for the Generation Challenge Program (GCP). This project will identify a 'composite collection' of germplasm for individual crops, representing the range of diversity of each crop species and its wild relatives, and characterize each composite set using anonymous molecular markers, mainly SSRs. ICARDA was responsible for creating the composite (core) collection for lentil as part of the CGP, that aims to explore the genetic diversity of the global germplasm collections held by international research centers (http://www.generationcp.org). A global composite collection of 1000 lentil accessions was established representing the overall genetic diversity and the agro-climatological range of lentil. Entries included landraces, wild relatives, elite germplasm and cultivars representing the overall ICARDA collection in both distribution and type.

The methodology for establishing the composite collection combined classical hierarchical cluster analyses using agronomic traits and two-step cluster analyses using agro-climatological data linked to the geographical coordinates of the accessions' collection sites (Furman 2006). The hierarchical cluster analysis ensured that the level of variation found in the larger collection was maintained in the composite collection. In addition, scientists at ICARDA suggested 64 accessions of landraces, released cultivars, and breeding materials for their resistances to a number of stresses affecting lentil production to be included in the composite collection.

The composite core lentil collection for the GCP is both being evaluated for morphologic and agronomic traits, and analysed for molecular diversity. This collection has 522 landraces with latitude and longitude passport data. These are matched to climatic and soil maps in selecting for adaptation to crop moisture and abiotic stresses. This will enable trait association genetic mapping based on the combined phenotypic/molecular analyses of the consensus collection. The strategy is to identify sources of desired traits for development of drought tolerant varieties. The USDA ARS lentil collection totals 2839 accessions with 123 wild taxa. A core of 234 accessions was selected based on country of origin (Simon and Hannan 1995). Recently, the core was extended (384 accessions) to add mapping population parents, cultivars and wild accessions, and a subset of pure lines was created.

This pure line subset will be distributed to scientists interested in LD mapping in lentil.

8. UTILISATION OF LENTIL GERMPLASM

With provision of an evaluation/passport database covering the major world collections, targeted multiple trait searching of most of the available lentil germplasm can now be utilised more efficiently and effectively, with strategies such as allele pyramiding in key traits and sourcing of novel genes from wild relatives (see chapters 13, 16 and 17). Generally, breeders narrow the genetic diversity in their breeding populations in the process of selecting the required trait combinations for outputs of improved varieties (Maxted et al. 2000). Genebanks try to conserve the original landrace diversity for both current and for unforseen future needs in breeding, enabling survey and enhancement of germplasm diversity for key traits, and identification of novel genes from wild relatives.

The standard practice for genebanks is to document passport data on the origin including latitude/longitude/altitude (GPS data), synonyms, and collection data with details of source location and associated agriculture (IPGR 1985). This enables traceability of the accession with details of collection procedure and evolutionary status. However, without knowledge of the growth characteristics and reactions to abiotic and biotic stresses of accessions there is a risk that genebanks may become museums, unless conservation is linked to utilisation (Maxted et al. 2000). The example given is the collection of 30,000 pulse accessions at one national genebank with 'only 2–3% used in crosses, because of the lack of characterisation and evaluation data'. A breeder can only utilise a small number of accessions in practice, and is in need of all available characterisation and evaluation data. Given (1994) indicated that in general 80% of germplasm lacks characterisation data and 95% lack evaluation data. While FAO (1998) reports that many country's collections require further characterisation and the characterisation that has occurred is restricted to a few species, most commonly the major food or commodity crops.

ICARDA evaluates lentil germplasm for morpho-agronomic traits as per IBPGR (1985) descriptors, and has published a Lentil Germplasm Catalogue (Erskine and Witcombe 1984). This provides a listing of 20 trait evaluations for 4,550 lentil accessions. There has been a large increase in the size of the lentil collections in various genebanks, and also in the associated evaluation data which is often recorded with the regeneration of germplasm in various nurseries over many years. The combining of this information from many genebank sources is now possible with suitable electronic software for relational databases (Balachandra et al. 2006), as exemplified by the ILIS web site in section 2.5.

The ILIS database allows on-line users to run search-query interrogations for germplasm with multiple trait expressions, currently from the combined ICARDA, USDA and ATFCC collections. This is a very powerful search engine, to assist genebank clients, to add value to the collections, and to facilitate utilisation for crop improvement. In addition, GPS passport data linked to climatic and stress maps may indicate adaptation characteristics of an accession. It is important to justify the expense of genebank operations, in both the short term with current crop improvement outcomes, and in the long term where a genetic insurance strategy may be very prescient to cope with a looming climate change.

Close linkage of lentil collections with breeders, pathologists and students, for evaluating key traits is becoming a significant genebank activity. ATFCC staff actively work with lentil breeders in germplasm enhancement for evaluation of tolerances to salinity, frost, herbicides, and disease (Redden et al. 2006). The ATFCC provided the Australian lentil breeding program with accessions ILL 2024 and ILL 213A identified for tolerance of high levels of boron. ICARDA has identified sources of resistance in the domestic genepool to rust, vascular wilt and to aschochyta, and in the wild progenitor *L.culinaris* subsp. *orientalis,* has found additional sources of resistance to vascular wilt and aschochyta and greater cold tolerance (Ferguson 2000).

9. INTERNATIONAL TREATY FOR PLANT GENETIC RESOURCES FOR FOOD AND AGRICULTURE (PGRFA)

Over 80 countries have now formally endorsed PGRFA, which is rapidly becoming the relevant international treaty for genetic resources of major food crops including lentil; http://www.fao.org/ag/cgrfa/itpgr.htm

The objectives of PGRFA are;

- Conservation and sustainable use of plant genetic resources – especially landraces and wild relatives as prime sources of diversity,
- Facilitated multi-lateral access and sharing of germplasm among genebanks,
- Associated provision of passport and evaluation data for better informed utilisation of germplasm,
- Fair and equitable sharing of benefits derived from use of germplasm –if there is restricted Intellectual Property (IP) access to derived commercial varieties,
- Harmony with the CBD, and benefit sharing principles for non-PGRFA crops.

The creation of ILIS in consistent with PGRFA, with collated evaluation data both adding value to collections and enabling efficiencies in germplasm utilisation with targeted search and request of germplasm for enhancement/breeding.

There are CGIAR guidelines for the application of a Standard Material Transfer Agreement under the PGRFA treaty, relevant to the world's largest collection of lentil landraces and wild relatives held by ICARDA.

It is hoped that this under-pinning framework for genetic resource diversity will increase the attraction of using available lentil germplasm to solve the need for novel genes and new QTL expressions, in lieu of transgenic solutions.

REFERENCES

Abo-Alwafa A, Murai K, & Shimada T (1995) Intra-specific and inter-specific variations in *Lens,* revealed by RAPD markers. Theor Appl Genet 90:335–340

Bayaa B, Erskine W, and Hamdi A (1994) Response of wild lentil to *Ascochyta fabae* fsp *lentis* from Syria. *Genetic Resources and Crop Evolution* 41: 61–65.

Bayaa,B, Erskine W, and Hamdi A (1995) Evaluation of a wild lentil collection for resistance to vascular wilt. *Genetic Resources and Crop Evolution* 42: 231–235.

Cubero JI, (1981) Origin, Taxonomy and Domestication. In:Webb C, Hawtin G (Eds), Lentils, pp 15–38. CAB, Slough, UK.

Balachandra R, Redden B, and Enneking D (2006) Development of an integrated database system to facilitate the storage and retrieval of germplasm data, as well as sourcing for germplasm. 13th Australasian Plant Breeding Conference, Christchurch, New Zealand: p 182.

Erskine W, and Witcombe JR (1984) Lentil Germplasm Catalogue. ICARDA:p 363.

FAO (1998).The State of the World's Plant Genetic Resources for Food and Agriculture. FAO, Rome.

FAO/IPGRI (1994) Genebank Standards. Food and Agriculture Organisation of the United Nations, Rome, International Plant Genetic Resources Institute, Rome, p 13.

Furman BJ, (2006) Methodology to establish a composite collection: Case study in lentil. *Plant Genetic Resources: Conservation and Utilization*, 4(1): 2–12.

Ferguson M (2000) Lens spp: Conserved resources, priorities for collection and future prospects. *In*: R. Knight (Ed.), Proceedings of the Third International Food Legumes Research Conference: Linking Research and Marketing Opportunities for Pulses in the 21st Century (Current Plant Science and Biotechnology in Agriculture; V. 34), Adelaide, Australia. Sept. 22–26, 1997. Kluwer Academic Publishers, Dordrecht, The Netherlands. pp 613–620.

Ferguson ME, Erskine W (2001) Lentiles (*Lens* L.). *In:* Maxted N and Bennett SJ (Eds), Plant Genetic Resources of Legumes in the Mediterranean. Kluwer Academic Publishers, Dordrecht, The Netherlands. pp 132–157.

Ferguson M.E, Ford-Lloyd B.V, Robertson L.D, Maxted N, Newbury HJ (1998) Mapping of geographical distribution of genetic variation in the genus *Lens* for enhanced conservation of plant genetic diversity. *Molecular Ecology*, 7: 1743–1755.

Ferguson ME, Maxted N, Van Slageren M, Robertson LD (2000) A re-assessment of the taxonomy of *Lens* Mill. (Leguminosae, Papilionoideae, Vicieae) Botanical Journal of the Linnean Society 133: 41–59.

Ferguson ME, Robertson LD (1996) Genetic diversity and taxonomic relationships within the genus *Lens* as revealed by allozyme polymorphism. Euphytica 91:163–172.

Ford R (1997) Diversity analysis and species identification in *Lens* using PCR generated markers. Euphytica 96:247–255.

Ford R, Pang ECK, Taylor PWJ. (1999) Diversity analysis and species identification in *Lens* using PCR generated markers. Euphytica 96:247–255.

Hamdi A, Erskine W (1996) Reaction of wild species of the genus *Lens* to drought. *Euphytica* 91: 173–179.

Hamdi A, Küsmenoplu I, Erskine W. (1996) Sources of winter hardiness in wild lentil. *Genetic Resources and Crop Evolution* 43: 63–67.

Harvey MJ, Muehlbauer FJ (1989) Variation for restriction fragment length and phylogenies in lentil. Theor Appl Genet 77:839–843.

Hoffman DL, Soltis DE, Muehlbauer FJ, Ladizinsky G (1986) Isozyme polymorphisms in Lens (Leguminosae). Systematic Botany 11:392–402.

IBPGR (1985) Lentil descriptors. International Board for Plant Genetic Resources and International center for agricultural research in the Dry Areas, IPGR Secretariat, Rome, Italy, pp 15.

IPGRI (2006) Law and policy of relevance to the management of plant genetic resources. Newsletter No 48 (IPGRI Rome): p 10.

Kupicha FK (1981) Vicieae. In: Advances in Legume Systematics. Polhill RM and Raven PM (eds). pp 377–381. Royal Botanic Gardens, Kew.

Ladizinsky G, Alder A (1976) Genetic relationships among the annual species of *Cicer L.* Theoretical and Applied Genetics 48:197–203.

Maxted N (1995) An ecogeographic study of *Vicia* subgenus*Vicia*. Systematic and Ecogeographic Studies in Crop Genepools 8. IBPGR, Rome. p 184.

Maxted N, Erskine W, Singh DP, Robertson LD. Asthana AN (2000) Are our germplasm collections museum items?. *In*: R. Knight (Ed.), Proceedings of the Third International Food Legumes Research Conference: Linking Research and Marketing Opportunities for Pulses in the 21st Century (Current Plant Science and Biotechnology in Agriculture; V. 34), Adelaide, Australia. Sept. 22–26, 1997. Kluwer Academic Publishers, Dordrecht, The Netherlands. P. 589–602.

Maxted N, Guarino L, Shehadeh A (2003).*In Situ* Techniques for Efficient Genetic Conservation and Use: A Case Study for *Lathyrus*. In: XXVI International Horticultural Congress: Plant Genetic Resources, The Fabric of Horticulture's Future. Forsline PL, Fideghelli C, Knuepffer H, Meerow A, Nienhus J, Richards K, Stoner A, Thorn E, Tombolato AFC, Williams D (eds). Acta Horticulturae

Muehlbauer FJ, Slinkard AE (1981) Genetics and Breeding Methodology. In: Webb C and Hawtin G (eds.) *Lentils*. Slough, UK: Commonwealth Agricultural Bureaux, pp 69–90.

Muench DG, Slinkard AE, Scoles GJ (1991) Determination of genetic variation and taxonomy in lentil (*Lens* Miller) species by chloroplast DNA polymorphisms. Euphytica 56:213–218.

Pritchard JK, Stephens M, Donnelly P (2000) Association mapping in structured populations. Am J Hum Genet 67: 170–181.

Redden B, Enneking D, Balachandra R, Murray K, Smith L, Clancy T (2006) The Australian Temperate Field crops Collection of genetic resources for pulses and oilseeds. 13th Australasian Plant Breeding Conference, Christchurch, New Zealand:p 84.

Reed BM, Engelmann F, Dulloo ME, and Engels JMM (2004) *Technical guidelines for the management of field and in vitro germplasm collections*, IPGRI Technical Handbook No 7, p 106.

Roberts EH and Ellis RH (1984) The implications of the deterioration of orthodox seeds during storage for genetic resources conservation. In: Holden JHW, Williams JT (eds) *Crop Genetic Resources: Conservation & Evaluation*, Ch2, pp 18–37.

Robertson LD, Erskine W (1997) Lentil. In: Fuccillo D, Sears L and Stapleton P (eds) *Biodiversity in Trust*. Cambridge: Cambridge University Press, pp 128–138.

Robertson LD, Singh KB, Erskine W, Abd El Moneim AM (1996) Useful genetic diversity in gemplasm collections of food and forage legumes from West Asia and North Africa. *Genetic Resources and Crop Evolution* 43: 447–460.

Sackville Hamilton NR and Chorlton KK (1997) Regeneration of accessions in seed collections: A Decision guide, In: Engels J (ed) handbook for genebanks No 5, pp 1–72.

Sangam L. Dwivedi SL, Blair MW, Upadhyaya HD, Serraj R, Balaji J, Buhariwalla HK, Ortiz R, Crouch JH (2006) Using Genomics to Exploit Grain Legume Biodiversity in Crop Improvement. In: Janick J. (ed.) *Plant Breeding Reviews Volume 26*, John Wiley & Sons, Inc, UK.

Sarker A, Erskine W (2006) Recent progress in ancient lentil. J Agri Sci 144:19–29.

Sharma SK, Dawson IK, Waugh R (1995) Relationships among cultivated and wild lentils revealed by RAPD analysis. Theor Appl Genet 91:647–654.

Spooner D, van Trueren R, de Vicente MC (2005) Molecular markers for genebank management. IPGR technical Bulletin No. 10 (IPGRI Rome):pp 126.

Simon C, Hannan R (1995) Development and use of core subsets of cool-season food legume germplasm collections. HortScience 30:907.

Thornsbury JM, Goodman MM, Doebley J, Kresovich S, Nielsen D, Buckler ES (2001) Dwarf polymorphisms associate with variation in flowering time. Nat Genet 28: 286–289.

van Oss H, Aron Y, Ladizinsky G (1997) Chloroplast DNA variation and evolution in the genus *Lens* Mill. Theor Appl Genet 94:452–457.

Zohary D (1972) The wild progenitor and the place of origin of the cultivated lentil. Economic Botany 26: 326–332.

CHAPTER 3

ADAPTATION AND ECOLOGY

M. ANDREWS[1] AND B. A. McKENZIE[2]

[1] *School of Sciences, University of Sunderland, Sunderland SR1 3SD, UK*

[2] *Agriculture Group, Agriculture and Life Science Division, PO Box 84, Lincoln University, Canterbury, New Zealand*

E-mail: mitchell.andrews@sunderland.ac.uk

Abstract: Worldwide, the major abiotic restrictions on yield of lentil are drought (usually linked with high temperature) and low temperature. LENMOD, a lentil crop growth model, gives greater understanding of how different climatic factors, including water availability and temperature, interact to determine lentil crop growth and yield. This model has considerable potential in predicting where lentil may be grown successfully. Breeding programmes are underway with the objectives of increasing adaptation of lentil to stress environments. The strategy used to combat drought has been to match the crop's development with the period of soil moisture availability. Genotypes with early seedling establishment, early and rapid biomass development and early flowering and maturity have been selected in sites of extremely low rainfall. Also, seed has been sown earlier in the spring or in the autumn. Success in the production of cold tolerant cultivars has been achieved by field screening of lentils in areas prone to extreme cold. High yielding varieties have been released for use in sites which experience over winter temperatures of −12 to −30° C

1. INTRODUCTION

Lentil is a self-pollinating, quantitative long day or day neutral, diploid, annual grain legume suitable for cultivation in warm temperate, subtropical and high altitude tropical regions of the world (Muehlbauer et al. 1995). Currently, it is under cultivation on around 4 million hectares in more than 40 countries and in all continents except Antarctica. Lentil is usually grown alone but can be intercropped with a range of species including wheat, barley, rice, sugarcane, mustard, castor bean and linseed. Lentil seed can be surface broadcast or sown by drilling: it is suitable for direct drilling in areas of relatively high erosion potential (McPhee et al. 2004). 'Optimum' populations in relation to yield are around 100 plants m^{-2} but plants can initiate lateral branches to compensate for poor emergence or thin

S.S. Yadav et al. (eds.), Lentil: An Ancient Crop for Modern Times, 23–32.

stands (McKenzie 1987; Muehlbauer et al. 1995). Lentil can grow on a wide range of soil types and soil pH although there is evidence that, in comparison with a range of other grain legumes including pea, faba bean, chickpea and white lupin, it is more sensitive to waterlogging and soil pH < 6.5 (Tang and Thomson 1996). However, in comparison with these legumes, it would be considered drought and low temperature tolerant (McKenzie and Hill 1990; see below).

The largest collection of lentil genotypes ($> 10,000$) is held at ICARDA (International Centre for Agricultural Research in the Dry Areas, www.icarda.org), Syria (Sarker et al. 2002a). Substantial collections of greater than 2000 genotypes are also held in India, the USA and Russia, while several countries including Bulgaria, China and Spain have much smaller collections (Muehlbauer et al. 1995; Stoilova and Pereira 1999; De La Rosa et al. 2005). Several of these collections have been evaluated for variation in phenology, morphology and growth parameters. Positive correlations have been found between seed size, plant height, plant canopy width, number of productive branches/ pods/ seeds per plant, total biomass, seed yield and residual amounts (Erskine 1985, 1996, 1997; Stoilova and Pereira 1999; Tullu et al. 2001). Genotypic differences were also found in time to flowering and harvest, flowering response to photoperiod, sensitivity to low and high temperatures and disease resistance (Erskine 1985, 1997).

Worldwide, the major restrictions on yield of lentil are drought (usually linked with high temperature), low temperature and disease although other factors such as salt stress, nutrient deficiency and nutrient toxicity are important locally (Muehlbauer et al. 2006; Tivoli et al. 2006). Several breeding programmes are underway in the main lentil growing regions of the world with the objectives of increasing yield and yield reliability, adaptation to stress environments and disease resistance. In particular, ICARDA, which has a worldwide franchise for lentil, is involved with many countries in the development of lentil lines for different agroecological niches (Sarker et al. 2002a,b; ICARDA 2005). Here we focus on abiotic effects on lentil growth. Firstly, we take an integrated view of environmental effects on lentil crop growth via a discussion of LENMOD, a lentil crop model, then we consider recent work on selection of plants for drought/high temperature and low temperature tolerance.

2. MODELLING CROP GROWTH AND DEVELOPMENT UNDER DIFFERENT CLIMATIC CONDITIONS

Plant growth is greatly dependent on weather conditions, with physiological processes responding to changes in air and soil temperature, solar radiation, moisture availability and wind speed (Monteith 1981). The effects of individual climatic elements on crop growth during distinct phases of plant development can be quantified allowing the calibration of mechanistic numerical models of crop growth. Such models give greater understanding of how different climatic factors interact to determine crop yield. A lentil crop growth model (LENMOD) was developed and calibrated in experiments carried out on a silt loam soil at Lincoln, Canterbury,

New Zealand (43.38° S, 172.30° E, 11 m above sea level) using cv. Titore, a small seeded, red variety of lentil (McKenzie 1987; McKenzie and Hill 1989; McKenzie et al. 1994). The experiments utilized spring, summer, autumn and winter sowing dates and a range of irrigation treatments over different years. The model requires the input of daily values of maximum and minimum air temperature, solar radiation, precipitation, potential evapo-transpiration and daylength and assumes that soil fertility is non-limiting and the crop is free of weeds and disease throughout all stages of growth.

All developmental stages in LENMOD except emergence to flowering depend on accumulated thermal time (TT). Sowing to emergence, flowering to physiological maturity and physiological maturity to harvest date require 115° C, 546° C and 270° C days above the critical temperature respectively. Emergence to flowering is dependent upon accumulated photothermal time and requires 278° C days (photothermal) above the critical temperature. The critical temperature below which growth and development stop is 2° C up to flowering and 6° C after flowering. The equations for relative daily leaf growth (RDLG), leaf area index (LAI) and crop growth rate (CGR) are:

(1) $\text{RDLG} = -0.0174 + 0.00829 \times \text{daily TT}$

(2) $\text{LAI} = \text{previous LAI} \times \text{RDLG} + \text{previous LAI}$

(3) $\text{CGR} = 0.5 \times \text{IR} \times \text{fraction IR intercepted} \times \text{RUE} \times \text{drought factor}$

In Equation(2), LAI is not allowed to exceed seven. In Equation(3), IR equals incident radiation and the drought factor is a ‘switch’ that turns off growth when the limiting deficit is passed. Radiation use efficiency (RUE) is set at 1.7 g DM MJ^{-1} of intercepted radiation. The equation for the fraction of IR intercepted is:

(4) $\text{Fraction IR intercepted} = 1.0 - \exp(-k \times \text{LAI})$

where –k is the extinction coefficient which is usually set at 0.32. The limiting soil moisture deficit is calculated from the relationship between relative yield and maximum potential soil moisture deficit (Penman 1948). When all plant available soil water is depleted, the drought factor becomes zero and thus the CGR becomes zero. After achieving maximum LAI, LAI declines to zero as a parabolic function of TT, based on 650° C days as follows:

(5) $$\text{LAI} = \text{previous LAI} - \text{max LAI} \times [(\text{accumulated TT}/2.5) \times \text{daily TT} \times \text{leaf killer} \times (2/650^2)]$$

Leaf killer is dependent on soil moisture. If the soil moisture goes above the limiting deficit, leaf killer is five, but if it is below the limiting deficit, then leaf killer is one. Soil moisture deficit is based on Penman’s potential evapotranspiration. Total dry matter (TDM) is calculated from daily crop growth rate and the model assumes a stable harvest index of 40% for the calculation of seed yield.

Crop growth models have several practical applications including the prediction of where previously untested crops might be grown (Siddons et al. 1994). LENMOD was used to assess the potential of lentil as a grain legume crop in the UK (Andrews et al. 2001; Joyce et al. 2001). Firstly, the model was validated on one site (Durham, 54.77° N, 1.58° W, 40 m) over different studies. In the main study, predicted and actual time to flowering, and seed yield were determined for five spring sowing dates in 1999 (Table 1).

For the four sowing dates from 21 April to 12 May 1999, predicted flowering date was within 3 days of actual flowering date (Table 1). For the final sowing date (26 May), predicted flowering date was 3–7 days later than actual flowering date. For all sowing dates, predicted seed yields were within 9% of actual seed yields which ranged from 1.40 to 1.65 t ha^{-1}. These yields are substantially greater than average yields obtained in the major lentil producing countries (Muehlbauer et al. 1985). Interestingly, individual seed weight decreased steadily with sowing date from 29.1 mg for 21 April to 24.9 mg for 26 May but seed nitrogen content was similar (4.26–4.41%) for all sowing dates.

LENMOD was then used to predict maximum CGR, flowering date, maximum LAI, radiation intercepted, TDM produced, harvest date and seed yield for spring and autumn sowings of lentil over the period 1987–95 for eight sites selected from the UK Meteorological Office network of climate stations along a transect from NW Scotland to SE England (Andrews et al. 2001). This transect spans 7 degrees of latitude, corresponding to a difference in day length of approximately 1.5 h in mid-summer and is likely to capture the major spatial variability of mean temperature, rainfall and sunshine intensity throughout the year in the UK. Solar radiation and potential evapotranspiration are also likely to vary systematically over the length of the transect. Predicted data for three sites, Fort Augustus (57.13° N 4.68° W, 40 m), Eskdalemuir (55.32° N 3.20° W, 242 m) and East Malling (51.28° N0.45° E, 40 m) are presented to highlight how different climatic conditions would be expected to interact to determine crop growth and yield. In general, over the period 1987–1995, monthly mean daily solar radiation increased in the order Fort Augustus <

Table 1. Actual and predicted flowering date and seed yield and actual individual seed weight of lentil cv Titore for five sowing dates at Durham in 1999

Sowing date	Flowering date		Seed yield (t ha^{-1})		Seed wt (mg)
	Actual	Predicted	Actual	Predicted	Actual
21 April	24 June–28 June	26 June	1.65	1.55	29.1
28 April	29 June–1 July	30 June	1.58	1.47	28.7
5 May	2 July–5 July	4 July	1.55	1.46	26.6
12 May	6 July–10 July	9 July	1.40	1.47	25.4
26 May	11 July–15 July	18 July	1.62	1.48	24.9
SE (10 df)			0.109		0.55

Taken from Andrews et al. (2001)

Eskdalemuir < East Malling but monthly mean daily air temperature during the growing season (May to September) increased in the order Eskdalemuir (11.6° C) < Fort Augustus (12.6° C) < East Malling (15.4° C) (Joyce et al. 2001). Temperatures were lowest at Eskdalemuir because of its greater elevation. Monthly rainfall was generally greater for Fort Augustus and Eskdalemuir than for East Malling. In comparison with mean monthly rainfall, mean monthly potential evapotranspiration was generally less variable across the sites. During May and June for Eskdalemuir, from May to July for Fort Augustus and from April to August for East Malling, mean monthly potential evapotranspiration was substantially greater than mean monthly rainfall.

For the May sowing with 150 or 250 mm PAW, predicted mean values for maximum CGR, maximum LAI, radiation intercepted, TDM and seed yield increased with site in the order Fort Augustus < Eskdalemuir < East Malling (Table 2).

These effects were related to differences in average air temperature and radiation interception. An increase from 150 to 250 PAW with the May 1 sowing had only small effects on growth and yield at Fort Augustus and Eskdalemuir (Table 2). However, for East Malling where potential evapotranspiration was substantially greater than mean monthly rainfall for the longest period, this increase in PAW caused 54% increases in TDM and seed yield. A switch in sowing date from 1 May 250 PAW to 1 October gave small increases in crop growth and yield at Fort Augustus, substantial increases in crop growth and yield at East Malling but substantial decreases in crop growth and yield at Eskdalemuir. The positive effects of autumn sowing at East Malling were due, in part, to greater leaf area duration (LAI × time) and hence greater radiation interception. Also, autumn sowing reduced

Table 2. Predicted mean values for crop growth rate (CGR), flowering date, maximum leaf area index (LAI), radiation intercepted, total dry matter (TDM) produced and seed yield for lentil cv Titore with spring (1 May) and autumn (1 October) sowings at three sites in the UK over the period 1987–1995

	Fort Augustus			Eskdalemuir			East Malling		
	1 May[1]	1 May[2]	1 Oct[1]	1 May[1]	1 May[2]	1 Oct[1]	1 May[1]	1 May[2]	1 Oct[1]
Max CGR ($kg\,ha^{-1}\,d^{-1}$)	93	93	117	100	117	71	162	186	213
Flowering date	3 July	3 July	7 June	9 July	9 July	14 June	28 June	28 June	20 May
Maximum LAI	2.18	2.18	2.66	2.23	2.23	1.08	3.94	3.94	7.35
Radiation intercepted ($MJ\,m^{-2}$)	208	210	262	246	247	166	317	363	704
TDM ($t\,ha^{-1}$)	2.50	2.64	2.96	3.22	3.60	2.42	3.08	4.75	6.94
Seed yield ($t\,ha^{-1}$)	1.00	1.10	1.18	1.29	1.44	0.97	1.23	1.90	2.78

[1] 150 mm plant available water
[2] 250 mm plant available water
Taken from Andrews et al. (2001)

the period of time the crop was exposed to water stress due to its earlier maturation. The negative effects of autumn sowing at Eskdalemuir were due to low temperatures over-winter and in spring which restricted leaf development and hence reduced radiation interception. At all sites, flowering date was unaffected by an increase from 150 mm to 250 mm PAW with the May sowing but was earlier with the October sowing due to photothermal effects. For a 1 May sowing at 150 mm PAW, seed yield was similar at Eskdalemuir and East Malling but for the 1 October sowing, seed yield was three times greater at East Malling. In the case of East Malling, predicted yields for autumn sowing (2.78 t ha^{-1}) are exceptional but not unrealistic as yields of around 2.8 and 2.5 t ha^{-1} were obtained for autumn-sown lentil at Reading in S England over different years (Crook et al. 1999).

3. SELECTION FOR INCREASED DROUGHT TOLERANCE

Drought stress is the major restriction on lentil yield in lentil growing regions, worldwide (Muehlbauer et al. 1995, 2006; McKenzie and Hill 2004). For example, Erskine and El Ashkar (1993) reported that 80% of the variation in lentil seed yield in Mediterranean climates was accounted for by differences in seasonal rainfall. In the Mediterranean regions of West Asia and North Africa, lentil is usually grown in areas of 300–400 mm rain $year^{-1}$ (Erskine et al. 1994). In these areas, the major proportion of the rain falls in winter, and from March until crop maturity in May, the crop experiences water and high temperature stress to an extent that restricts yield.

Lentil is drought tolerant in comparison with other temperate grain legumes. For example, the limiting soil moisture deficit for lentil is similar to that for chickpea and substantially greater than that for pea or faba bean (McKenzie and Hill 2004). However, the amount of yield loss experienced for each mm of drought past the limiting soil moisture deficit was greater for lentil than chickpea or faba bean. Differences in limiting soil moisture deficits across the species are probably due to variation in their rooting depth and root proliferation (McKenzie and Hill 2004). Lentil roots are capable of extracting water from at least 90 cm depth (Sharma and Prasad 1984; McKenzie 1987). The variation in yield loss per mm of drought above the limiting soil moisture deficit was probably caused by factors which can influence water use, such as leaf size and orientation, stomatal number and orientation, and radiation use efficiency. Also, differences in osmotic adjustment could be important (Nielsen 2001). There are reports in the literature that several grain legume species including lentil have a critical period of sensitivity to water stress around time of flowering (Yusuf et al. 1979; McKenzie and Hill 2004). However, McKenzie and Hill (2004) report results from a series of experiments on lentil, chickpea, pea and faba bean which provide evidence that this is not the case. It was found that, throughout the growth of the crops, it was important to ensure that soil moisture was kept above critical deficit levels at which point crop growth stops.

Genetic variation exists in drought tolerance of lentil. For example, small seeded varieties in comparison with large seeded varieties were found to be better adapted to dry environments (Erskine 1996). Also, land races and wild accessions with high levels of drought tolerance but low biomass have been described (Muehlbauer et al. 1995). However, the key strategy used to combat drought has been to match the crop's development with the period of soil moisture availability. This has been achieved in two ways. Firstly, genotypes with early seedling establishment, early and rapid biomass development and early flowering and maturity have been selected (Kusmenoglu and Muehlbauer 1998; Sarker et al. 2002a,b; Muehlbauer et al. 2006). In the case of ICARDA, this has involved the selection of drought tolerant varieties in sites with an average annual rainfall of less than 300 mm (Sarker et al. 2002b). Secondly, seed has been sown earlier in the spring or in the autumn (Kusmenoglu and Muehlbauer 1998). Autumn sowing can result in increased leaf area duration and hence greater radiation interception and it can reduce the period the crop is exposed to water stress due to its earlier maturation (see discussion of LENMOD above). However, low temperatures over winter can restrict leaf development and hence reduce radiation interception (Table 2). Also, if temperatures go below −10° C, leaf damage can occur. In relation to this, selection for drought tolerance is often linked with selection for low temperature tolerance.

4. SELECTION FOR INCREASED LOW TEMPERATURE TOLERANCE

In comparison with other temperate grain legumes, lentil would be considered tolerant of low temperatures. Murray et al. (1988) ranked lentil similar to faba bean and greater than pea or chickpea in relation to winter hardiness. Also, lentil can be autumn sown in areas such as the lowlands of West Asia (< 850 m), which experience minimum temperatures of around −10° C (Sarker et al. 2002a). Agronomic practices, in particular establishment of high stand densities (~ 400 plants m^{-2}), can to some extent increase low temperature tolerance (Kusmenoglu and Aydin 1995; Crook et al. 1998). However, in the highlands of West Asia which experience overwinter temperatures of around −20° C, lentil is traditionally sown in spring as the risk of low temperature damage is high. Programmes are underway to select lentil varieties for autumn sowing in these environments (Sarker et al. 2002a, 2004).

As for drought stress, genetic variation exists in low temperature tolerance of lentil. For example, photothermally sensitive genotypes are more tolerant of low temperatures (Keatinge et al. 1996). Also, large seeded varieties in comparison with small seed varieties had greater cold tolerance (Erskine 1996). Low temperature tolerance of faba bean is also positively correlated with seed size which in part can be related to greater seed reserves (Andrews et al. 1986). Winter hardiness in lentil has been reported to have low to moderate heritability (Ali and Johnson 2000) and to be conferred by several genes (Kahraman et al. 2004a). Kahraman et al. (2004b) reported that 42% of the variation in winter survival could be explained by the cumulative effects of several quantitative trait loci (QTL). A project is now underway to develop a molecular map of the lentil genome and determine the location of the genes that confer winter hardiness

(Muehlbauer et al. 2004, 2006). An objective of the project is to identify suitable PCR based markers for use in a high throughput procedure for improved winter hardiness. However, it will be several years before this project could result in the release of field material and recent success in the production of cold tolerant cultivars has been achieved by more traditional approaches. In particular, ICARDA carries out field screening of lentils in areas prone to extreme cold and these trials have identified cold tolerant genotypes. As a result of this work, cold tolerant cultivars suitable for early spring sowing have been released in several countries including Afghanistan, Syria, Iran and Turkey; and varieties suitable for autumn sowing have been released in Pakistan, Turkey and Iran (Sarker et al. 2002a). In Turkey, high yielding varieties have been released for use in sites at altitudes of 600 to 1400 m and which experience temperatures of -12 to $-30°$ C in winter (Sarker et al. 2002b, 2004). A complication with autumn sowing is the greater probability of disease, in particular, ascochyta blight. This can be countered by combining high resistance to asochyta blight and other diseases with low temperature tolerance in lentil and/or improved agronomic management (Sarker et al. 2002a, 2004; Ye et al. 2002; Muehlbauer et al. 2006; Tivoli et al. 2006).

5. CONCLUSIONS

Plant growth is greatly dependent on weather conditions, with physiological processes responding to changes in air and soil temperature, solar radiation, moisture availability and wind speed. LENMOD, a lentil crop growth model developed and calibrated in NZ gives greater understanding of how different climatic factors interact to determine lentil crop growth and yield. This model has considerable potential in predicting where lentil may be grown successfully. Worldwide, the major abiotic restrictions on yield of lentil are drought (usually linked with high temperature) and low temperature. The strategy used to combat drought has been to match the crops development with the period of soil moisture availability. This has been achieved in two ways. Firstly, genotypes with early seedling establishment, early and rapid biomass development and early flowering and maturity have been selected in sites of extremely low rainfall. Secondly, seed has been sown earlier in the spring or in the autumn. Success in the production of cold tolerant cultivars has been achieved by field screening of lentils in areas prone to extreme cold. High yielding varieties have been released for use in sites which experience over winter temperatures of -12 to $-30°$ C.

REFERENCES

Ali A, Johnson DL (2000) Heritability estimates for winter hardiness in lentil under natural and controlled conditions. Plant Breeding 119: 283–285

Andrews M, McInroy S, Sprent JI, Taylor H (1986) Early growth and final yield of autumn sown *Vicia faba* L. cultivars given different forms of fertilizer N over winter. Plant and Soil 96: 421–427

Andrews M, McKenzie BA, Joyce A, Andrews ME (2001) The potential of lentil (Lens culinaris) as a grain legume crop in the UK: an assessment based on a crop growth model. Annals of Applied Biology 139: 293–300

Crook DG, Ellis RH, Summerfield RJ (1999) Winter-sown lentil and its impact on a subsequent cereal crop. Aspects of Applied Biology 56: 241–248

Crook DG, Summerfield RJ, Ellis RH, Smith NO (1998) Plant population density in autumn affects winter survival in lentil. In: Proceedings of the 3rd European Conference on Grain Legumes, AEP – l'Association Européenne de Recherche sur les Protéagineuse, Paris, France, pp 154–155

De La Rosa L, Martin I, Varela F, De La Cuadra C (2005) Genetic diversity in Spanish grain legume collections. Grain Legumes 42: 10–11

Erskine W (1985) Lentil genetic resources. In: Saxena MC, Verma S (eds) Fababeans, Kabuli chickpeas and lentils in the 1980s. ICARDA, Aleppo, Syria, pp 29–33

Erskine W (1996) Seed size effects on lentil (Lens culinaris) yield potential and adaptation to temperature and rainfall in West Asia. Journal of Agricultural Science, Cambridge 126: 335–341

Erskine W (1997) Lessons for breeders from land races of lentil. Euphytica 93: 107–112

Erskine W, El Ashkar F (1993) Rainfall and temperature effects on lentil (Lens culinaris) seed yield in the Mediterranean environment. Journal of Agricultural Science, Cambridge 121: 347–354

Erskine W, Tufail M, Russell A, Tyagi MC, Rahman MM, Saxena MC (1994) Current and future strategies in breeding lentil for resistance to biotic and abiotic stresses. Euphytica 73: 127–135

ICARDA (2005) Lentil research benefits Nepal's farmers. ICARDA Caravan 22: 7

Joyce AN, McKenzie BA, Griffin P and Andrews M (2001) Climatic control of spring sown lentil (*Lens culinaris*) in the UK. In: Proceedings of the 4th European Conference on Grain Legumes, AEP – l'Association Européenne de Recherche sur les Protéagineuse, Paris, France, pp 50–51

Kahraman A, Kusmenoglu I, Aydin N, Aydogan A, Erskine W, Muehlbauer FJ (2004a) Genetics of winter hardiness in 10 lentil recombinant inbred line populations. Crop Science 44: 5–12

Kahraman A, Kusmenoglu I, Aydin N, Aydogan A, Erskine W, Muehlbauer FJ (2004b) QTL mapping of winter hardiness genes in lentil. Crop Science 44: 13–22

Keatinge JDH, Qi A, Kusmenoglu I, Ellis RH, Summerfield RJ, Erskine W, Beniwal SPS (1996) Using genotypic variation in flowering responses to temperature and photoperiod to select lentil for the west Asian highlands. Agricultural and Forest Meteorology 78: 53–65

Kusmenoglu I, Aydin N (1995) The current status of lentil germplasm exploitation for adaptation to winter sowing in the Anatolian highlands. In: Keatinge JDH, Kusmenoglu I (eds) Autumn sowing of lentil in the Highlands of West Asia and North Africa. Central Research Institute for Field Crops (CRIFC), Ankara, Turkey, pp 63–71

Kusmenoglu I, Muehlbauer FJ (1998) Genetic variation for biomass and residue production in lentil (Lens culinaris Medik.). II. Factors determining seed and straw yield. Crop Science 38: 911–915

McKenzie BA (1987) The growth, development and water use of lentils (Lens culinaris Medik.). Ph.D. Thesis, Lincoln University, Canterbury, New Zealand

McKenzie BA, Hill GD (1989) Environmental control of lentil (Lens culinaris) crop development. Journal of Agricultural Science, Cambridge 113: 67–72

McKenzie BA, Hill GD (1990) Growth, yield and water use of lentils (*Lens culinaris*) in Canterbury, New Zealand. Journal of Agricultural Science, Cambridge 114: 309–320

McKenzie BA, Hill GD (2004) Water use in grain legumes. In: Proceedings of the 5th European Conference on Grain Legumes/ 2nd International Conference on Legume Genomics and Genetics, AEP – l'Association Européenne de Recherche sur les Protéagineuse, Paris, France, pp 61–62

McKenzie BA, Hill GD, Gallagher JN (1994) Computer simulation model of lentil growth and development. Lens 21: 31–35

McPhee K, Miller P, Chen C, Muehlbauer F (2004) Adaptation of winter legumes to direct seeding in northern climates. In: Proceeding of the the 5th European Conference on Grain Legumes/ 2nd International Conference on Legume Genomics and Genetics, AEP – l'Association Européenne de Recherche sur les Protéagineuse, Paris, France, pp 55–56

Monteith JL (1981) Coupling of plants to the atmosphere. In: Grace J, Ford ED, Jarvis PG (eds) Plants and their atmospheric environment. Blackwell, London, pp 1–29

Muehlbauer FJ, Cho S, Sarker A, McPhee KE, Coyne CJ, Rajesh PN, Ford R (2006) Application of biotechnology in breeding lentil for resistance to biotic and abiotic stress. Euphytica 147: 149–165

Muehlbauer FJ, Cubero JI, Summerfield RJ (1985) Lentil (Lens culinaris Medik.) In: Summerfield RJ, Roberts EH (eds) Grain legume crops. Collins, London, pp 266–311

Muehlbauer F, Kahraman A, Kusmenoglu I, Aydin N, Aydogan A, Erskine W (2004) A molecular marker map of the lentil genome and location of quantitative trait loci for tolerance to winter injury. In: Proceedings of the 5th European Conference on Grain Legumes/ 2nd International Conference on Legume Genomics and Genetics, AEP – l'Association Européenne de Recherche sur les Protéagineuse, Paris, France, pp 143–146

Muehlbauer FJ, Kaiser WJ, Clement SL, Summerfield RJ (1995) Production and breeding of lentil. Advances in Agronomy 54: 283–332

Murray GA, Eser D, Gusta LV, Eteve G (1988) Winterhardiness in pea, lentil, faba bean and chickpea. In: Summerfield RJ (ed) World crops: cool season food legumes. Kluwer Academic Publishers, Dordrecht, The Netherlands, pp 831–843

Nielsen DC (2001) Production functions for chickpea, field pea, and lentil in the Central Great Plains. Agronomy Journal 93: 563–569

Penman HL (1948) Natural evaporation from open water, bare soil and grass. Proceedings of the Royal Society 193: 120–125

Sarker A, Aydin N, Aydogan A, Sabaghpour SH, Ketata H, Kusmenoglu I, Erskine W (2002a) Winter lentils promise improved nutrition and income in West Asian Highlands. ICARDA CARAVAN 16: 14–16

Sarker A, Aydogan A, Sabaghpour SH, Kusmenoglu I, Sakr B, Erskine W, Muehlbauer FJ (2004) Lentil improvement for the benefit of highland farmers. www.cropscience.org.au

Sarker A, Neupane RK, Sakr B, El Ashkar F, Lutfir A, Erskine W (2002b) More grain from less rain: ICARDA'S strategy to improve lentil for resource-poor farmers in dry areas. ICARDA Caravan 17: www.icarda.org/publications/

Sharma SN, Prasad R (1984) Effect of soil moisture regimes on the yield and water use of lentil (*Lens culinaris* Medic.). Irrigation Science 5: 285–293

Siddons PA, Jones RJA, Hollis JM, Hallet SH, Milford GFJ, Scott T (1994) Land suitability for autumn-sown determinate lupins. Soil Survey Research Report No 1. Cranfield University, Silsoe, The Soil Survey and Land Research Centre, Cranfield University

Stoilova T, Pereira MG (1999) Morphological characterization of 120 lentil (*Lens culinaris* Medik.) accessions. Lens Newsletter 26: 7–9

Tang C, Thomson BD (1996) Effect of solution pH and bicarbonate on the growth and nodulation of grain legume species. Plant and Soil 186: 321–330

Tivoli B, Baranger A, Avila CM, Banniza S, Barbetti M, Chen W, Davidson J, Lindeck K, Kharrat M, Rubiales D, Sadiki M, Sillero JC, Sweetingham M, Muehlbauer FJ (2006) Screening techniques and sources of resistance to foliar diseases caused by major necrotrophic fungi in grain legumes. Euphytica 147: 223–253

Tullu A, Kusmenoglu I, McPhee KE, Muehlbauer FJ (2001) Characterisation of core collection of lentil germplasm for phenology, morphology, seed and straw yields. Genetic Resources and Crop Evolution 48: 143–152

www.icarda.com

Ye G, McNeil DL, Hill GD (2002) Breeding for resistance to lentil Ascochyta blight. Plant Breeding 121: 185–191

Yusuf M, Singh NP, Dastane NG (1979) Effect of frequency and timings of irrigation on grain yield and water use efficiency of lentil. Annals of Arid Zone 18: 127–134

CHAPTER 4

USES AND CONSUMPTION

SHYAM S. YADAV[1], PHILIP C. STEVENSON[2], A. H. RIZVI[1], M. MANOHAR[1], S. GAILING[3], AND G. MATELJAN[3]

[1] *Pulse Laboratory, Division of Genetics, Indian Agricultural Research Institute, New Delhi 110012, India*

[2] *Natural Resources Institute, University of Greenwich, Chatham, Kent ME4 4TB and Royal Botanic Gardens, Kew, Surrey, TW9 3AB, UK*

[3] *The George Mateljan Foundation, PO Box 25801, Seattle, Washington 98165, USA*

Email: shyamsinghyadav@yahoo.com

Abstract: Lentils are one the earliest known crops to be cultivated and archaeological evidence goes back at least 7000 years. They have been in constant use in different societies since then and their consumption has been widespread in developed and developing countries alike. It is consumed for its flavour, its versatility and its high nutritive value and other health benefits which are briefly discussed here and elsewhere in this book. In most of the Asian countries and particularly in the Indian subcontinent the major use for lentil is for making *dhal* for which the red lentils are preferred. The types of lentil soups prepared in different countries and regions throughout the world vary enormously depending on local tradition and palate from the spicy Indian *dhals* to the more aromatic north African lentil soups to the meat based European dishes and several recipes are included here. *Dhal* alone describes a whole group of dishes which vary enormously from the different regions of India and the other countries of the subcontinent. They are also used uncooked; either soaked crushed and moulded to make cakes or sprouted as an ingredient in salads in some parts of India and as such provide better nutrient value

1. INTRODUCTION

Lentil has a variety of different names in different countries and languages including *Masoor* (India), *Adas* (Arabic), *Mercimek* (Turkey), *Messer* (Ethiopia) and *Heramame* (Japanese) giving some indication of the breadth of their importance. Lentils are believed to have originated in central Asia, having been an important food since prehistoric times and are one of the first foods to have been cultivated (Sarker and Erskine, 2006). Lentil seeds dating back 7000 years have been found at archeological sites in the Middle East and from the Iberian peninsula (Zapata et al., 2004). Chemical analysis of the contents of a tomb discovered in central

S.S. Yadav et al. (eds.), Lentil: An Ancient Crop for Modern Times, 33–46.

Turkey at what was Gordion, the capital of the then-powerful Phrygian kingdom and reportedly dating back to 2500 BC indicates that wealthy people dined on barbecued lamb with a spicy stew (olive oil, honey and wine) and lentils. A king aged about 60 was buried in the tomb in a wooden coffin under a huge mound and is considered by some to be the tomb of King Midas (McGovern et al., 1999). Lentils are even mentioned several times in the Bible. For example, Esau traded his birthright with Jacob for a plate of food made from lentil (Genesis 25:34) and they were an ingredient in bread that was made during the Babylonian captivity of the Jewish people (Ezekiel 4:9). Lentils were part of the provisions brought to David when fleeing from Absalom (2 Samuel, 17:28).

For millennia, lentils have been eaten traditionally with barley and wheat, three foodstuffs that originated in the same regions and spread throughout Africa and Europe during similar migrations and explorations of cultural tribes. Before the 1st century AD, they were introduced into India, a country whose traditional cuisine still bestows high regard for the spiced lentil dish known as dhal. In many Catholic countries, lentils have long been used as a staple food during Lent although there appears to be no common origin of the two words. Currently, the leading commercial producers of lentils are India, Turkey, Canada, China and Syria.

The lentil (*Lens culinaris*: Leguminosae) is a bushy annual grown for its seeds that are characteristically lens-shaped; hence the name. It grows to about 40 cm tall and the seeds develop in the pods, usually with two seeds in each. With 25% protein it is the vegetable with the highest level of protein after soybeans (Bhattacharya et al., 2005), and because of this it is a very important part of the diet in many parts of the world, and especially South Asia which has a large vegetarian population (Singh, 1999).

The cooked seeds have a distinctive earthy flavor and importantly a short cooking time (especially for the small varieties with the husk removed such as the common red lentil). This can, however, be affected by harvesting procedure, storage time and even climatic conditions (Iliadis, 2001). Lentils are used throughout the Mediterranean regions and the Middle East as the basis of many meals but they are also used to make inexpensive and nutritious soups that are popular in Europe and North America as well. Indeed, in Europe they are often combined with some form of pork but in Asia they are most frequently combined with rice, which, conveniently, has a similar cooking time. Seeds are also reportedly used as a source of starch for textile and printing (Kay, 1979).

2. USE OF LENTIL AS A NUTRITIOUS AND HEALTHY FOOD

Lentil is grown for its seed and then eaten primarily as dhal. Dhal describes seeds (usually pulses) that are decorticated and split leaving just the lens shaped cotyledon which in lentil has higher protein, carbohydrate and calorie content than many other legumes. It is the most desired crop in many lentil producing regions particularly for its high protein content and fast cooking characteristic (Iqbal et al., 2006; Muehlbauer et al., 1985). Grain is fried and seasoned as well as being boiled. Lentil flour is also a basic ingredient in many soups, stews, purees and is even mixed with cereals to make bread and cakes and as a food for infants (Williams and Singh, 1988). While lentils are considered to be highly nutritious, they also contain anti-nutritional factors such as, trypsin inhibitors, hemagglutinins, and oligosaccharides to which the flatulence often associated with pulses is attributed (Jambunathan et al., 1994; Abdel-gawad, 1993). However, Williams et al. (1994) reported that lentils have relatively low levels of these anti-nutritional factors compared to other pulse legumes such as faba bean which is considered to have the highest concentrations. Tannins can impart a bitter flavour to the food as they are present in high concentrations in the seed coat of lentils. However, they can be removed by processing (Williams et al., 1994), most often and easily by removing the seed coat. These husks along with dried leaves, stems, fruit walls and bran (residues), can be fed to livestock. Lentil residues contain about 10.2% moisture, 1.8% fat, 4.4% protein, 50% carbohydrate, 21.4% fibre, and 12.2% ash (Muehlbauer et al., 1985) but nutrient levels can vary greatly according to the variety (Wang and Daun, 2006). The nutritional aspects of the seed are discussed in more detail elsewhere in this book. Lentils not only provide valuable food for humans but also for livestock. According to Muehlbauer et al. (1985), when production of forage crops falls below the level required in the market, lentil residue commands an equal or a better price than lentil seeds in some Middle Eastern countries. Green plants also make valuable green manure.

Lentils are quick and easy to prepare compared to other dried pulses and are also a very healthy food for a number of reasons and there are several websites dedicated to healthy eating that extol the virtues of this pulse such as www.whfoods.org. These health and cooking qualities are enhanced by the fact that the split peas easily absorb flavors from spices, other foods and seasonings and are usually available throughout the year. Lentils are sold whole or split into halves with the brown and green varieties retaining their shape well after cooking compared to other varieties.

Lentils, are also a very good source of cholesterol-lowering fibre. Lentils can help to lower cholesterol, are reportedly beneficial in managing blood-sugar disorders as mentioned above since their high levels of soluble fibre prevents blood sugar levels from rising rapidly after a meal (Araya et al., 2002; Brand-Millar et al., 2003) and could be a route to reducing cholesterol (Roberts et al. 1994). But lentils have still more to offer. They provide excellent amounts of six important minerals, two B-vitamins, and protein with virtually no fat. The calorie cost of all this nutrition? Just 230 calories for a whole cup of cooked lentils. This tiny food giant fills you up–not out.

Pulses, including lentils, are used increasingly in health-conscious diets to promote general well-being and reduce the risk of illness. As mentioned above lentils are particularly low in fat but high in protein (Iqbal et al., 2006), and are an excellent source of both soluble and insoluble fibre, complex carbohydrates, vitamins (especially B vitamins) and minerals (especially potassium, phosphorus, calcium, magnesium, copper, iron and zinc) yet lentils are inexpensive compared to other food sources with similar properties. Eating lentils may help lower blood cholesterol levels due to their high content of soluble fibre and vegetable protein (Anon, 2006; Rani and Kawatra, 1994).

Lentils can be beneficial in the management of type-2 diabetes since they have a low glycemic index (<55) suggesting that their impact on blood glucose levels is lower than that of many other carbohydrate containing foods (Anon, 2006; Araya et al., 2002). They have been shown to have low glycemic potential even when mixed with other grains (Hardacre et al., 2006). Lentils also reduce blood lipids that may help some serious complications of diabetes.

Flour made from lentils is gluten free and is a nutritious alternative to wheat based products for people with celiac disease a condition which damages the small intestine, interferes with absorption and makes sufferers intolerant of gluten. Lentil is very well suited to vegetarian diets as they are a good source of protein and iron, and complement the amino acid profile of cereals and nuts (Anon, 2006; Iqbal et al., 2006). The consumption of insoluble fibre can help maintain a healthy colon and so reduce the risks of colon cancer. Diets high in fibre also help with weight loss because they deliver more bulk and less carbohydrate. Lentils are also an excellent source of vitamin B9 (folic acid) which is an essential nutrient especially during pregnancy where it has been shown to reduce the risk of neural tube defects.

Lentils are good for the heart. The traditional Mediterranean diet encompasses dietary characteristics that include high levels of fruits, vegetables, legumes such as lentil and whole grains as well as fish, nuts, and low-fat dairy products. These have been shown to have protective health effects including being associated with a reduced risk of coronary heart disease (Giuliano and Esposito, 2005, Flight and Clifton, 2006). The contribution of lentils to the health of our heart also lies in the substantial content of folic acid and magnesium. As well as being an important vitamin in pregnancy folic acid helps lower levels of homocysteine, an amino acid that is an intermediate product in an important metabolic process called the methylation cycle (Araki et al., 2006) which converts homocysteine into the benign amino acids cysteine or methionine. High levels of homocysteine in the bloodstream damage artery walls and contribute to heart disease.

The magnesium in lentil acts as a calcium channel blocker, which helps to relax, blood vessels and improve blood flow and the transport of oxygen and nutrients throughout the body. Magnesium deficiency can also promote free radical injury to the heart immediately following a heart attack. Lentils are also a good source of iron although the available levels can be affected by cooking (Viadel et al., 2006) so eating them uncooked for instance after soaking or sprouted as described later in this chapter should be encouraged. Iron, of course, is an integral component of haemoglobin, the metalloprotein that transports oxygen from the lungs around the body. Lentils can therefore enhance energy by topping up iron. This is particularly important during menstruation, pregnancy or for those who are at risk of iron deficiency such as those with sickle cell and thallassaemia trait. And unlike red meat, perhaps the most obvious source of iron, lentils are low in fat and calories.

As mentioned throughout this chapter, legumes are rich in important nutrients and are thus excellent food. Pulses have been classified by different food based organizations into different categories on the basis of the percent of daily value of food that they contribute to the diet. According to www.whfoods.org lentils have excellent levels of molybdenum and folic acid, very good levels of tryptophan, fibre and manganese and good levels of Iron, Protein, Phosphorus, Copper, vitamin B1 (thiamin) and Potassium. Nutritional qualities of lentil will be covered in detail in the following chapter.

Lentils are purported to remedy constipation and other intestinal afflictions (Muehlbauer and Tullu, 1997). According to Duke (1981) lentils were used in India as poultices on the ulcers that follow smallpox and other slow-healing sores. In the 6th century, chickpeas were believed to be an aphrodisiac; while curiously enough, lentils were considered to have the opposite effect, and this was probably the reason why the lentil was included in the diet in monasteries on meatless days (Van der Maesen, 1972).

3. NUTRITIONAL QUALITY PARAMETERS

The amount of protein in lentils reportedly ranges from 22–30% (Wang and Daun, 2006), and 100 g of dried seeds contain 340–346 g calories, 12% moisture, 20.2 g protein, 0.6 g fat, 65.0 g total carbohydrate, about 4 g fibre, 2.1 g ash, 68 mg Ca, 325 mg P, 7.0 mg Fe, 29 mg Na, 780 mg K, 0.46 mg thiamine, 0.33 mg riboflavin, 1.3 mg niacin (Adsule et al., 1989; Muehlbauer et al., 1985). Among the cool season legume crops, lentil is the richest in the amino acids that are low in vegetarian diets (e.g., lysine, arginine, leucine, and sulphur containing amino acids) that are chosen by or imposed by poverty on different people throughout the world (Iqbal et al., 2006; Williams et al., 1994). The starch content ranges from 35–53% of which amylose comprises between 20 and 38.5% while dry matter constitutes 42% (Huisman and van der Poel, 1994; Hulse, 1994). In addition to thiamine and riboflavin, lentils are a good source of the other vitamins and 100 g lentil contains 1.7 mg nicotinic acid, 223 mg choline, 107 mg folic acid, 130 mg inositol, 1.6 mg pantothenic acid, 13.2 mg biotin, and 0.49 mg pyridoxine. With the exception of folic and pantothenic acids, vitamins increase markedly during sprouting thus this is a valuable way of consuming the seeds. Dry lentil husks contain 11.1% protein (1.3% digestible), 0.7% fat, 47.5% carbohydrate, 25.6% fibre, and 3.1% ash (Duke, 1981). The majority of the protein present in the cotyledons consists primarily of albumins and globulins, and lentil digestibility coefficients are relatively high and range from 78–93%. Oleic, palmitic and linoleic are the dominant fatty acids (Hulse, 1994) although the nutritional characters for the seeds are discussed in more detail elsewhere.

4. USE AND CONSUMPTION OF LENTILS: A PREPARATION GUIDE

Unlike some pulses, lentils do not need to be presoaked so can be prepared the day of serving since they are fairly small. Some dishes are enhanced by soaking though. Before washing lentils in clean cold water its worth spreading them out on a table top to check for small stones and other unwanted debris before cooking. There's nothing more off-putting than the crack of teeth against a small stone in food. To boil lentils, use three cups of liquid for each cup of lentils. Lentils placed in water that is already boiling will be more easily digested than if brought to a boil with the water. Green lentils usually take about 10 minutes more than red ones. But both are cooked fairly quickly compared to other pulses which make them especially favorable pulse in poor households since they require less fuel although this does not mean lentils are restricted to poor households.

4.1. Cooked Lentils

Lentils cook in about 20 to 25 minutes and can give a firm, but tender, seed to use in salads. Use about 1 cup of lentil in 3 cups of water. Bring water to boil and turn heat down to simmer until tender or until they are the required consistency. Seasoning with salt and pepper in the last few minutes of cooking helps with their flavour. Drain off water and reserve. (Reserved water can be used to serve over baked potatoes or rice.) To finish your salad, simply add chopped vegetables of your choice, lemon juice, a little olive oil, garlic, and seasonings for nutritious entree salad. Lentils can also be added to vegetable soups or even potato soups to encouraged thickening. This requires the lentils to be cooked for about 45 minutes so they will be completely broken down. Lentil soup with added lemon is a favorite of Middle Eastern households whereas in India aromatic spices are added. Cooking times can be reduced substantially by using a pressure cooker which precludes the need to soak many other pulses. Lentils are still the quickest to cook even by this method and usually take no more than 7 minutes. Lentils are most commonly consumed as dhal, the spicy and highly variable soup from South Asia that can be made with many different pulses including Lentil (Masoor – *Lens culinaris*), Moong bean (green gram – *Phaseolus aureus*), chickpea (channa/chole – *Cicer arietinum*), urad dhal (black gram – *Phaseolus mungo*) and Pigeonpea (Arhar – *Cajanus cajan*).

4.2. Uncooked Lentils

A less common way to consume Masoor lentils is as bean sprouts. A simple procedure to sprout lentils is to take between 6 & 8 table spoons of seeds (with husks) and soak them overnight in plenty of water ensuring the lentils are completely covered. The following day, the seeds can be drained and rinsed and then placed into a wide mouthed jar covered with muslin cloth that has been secured with a rubber band. The jar can be laid on its side and covered with a towel and set aside for the day, rinsing them occasionally (once or twice in 24 hours). In colder environments the seeds may take longer to sprout. The sprouts have a characteristic nutty flavour. White sprouts are more likely to be produced if kept in the dark but for green sprouts then sprout in light but not direct sunlight. The green and white give two slightly different textures and tastes and also provide two different colour options for a variety of aesthetic looks. Sprouted seeds store well in any unsealed container including plastic bags or plastic pots and will keep for 3–4 days if rinsed regularly. They are a fresh alternative in salads and also stir fries.

4.3. Tempering Dhal

Tempering or seasoning is very important for dhal both for appearance and flavour. For example, turmeric gives dhal the lovely golden color. There are countless variations of simple seasoning from ingredients that include salt and pepper, mustard seeds, red chilies or chili powder, asafetida (*hing*), onions, green chilies, ginger, garlic, tomatoes, Garam Masala and curry leaves (*Murraya koenigii*). The oil most commonly used for cooking lentils is clarified butter (ghee) or sunflower oil for vegans. The oil is important for frying the spices and seasonings which are added to the cooked dhal. Dhal also benefits from a generous handful of freshly chopped coriander.

5. UTILIZATION IN INDIA

Indians consume more lentils than any other country and the country produces more then 50 varieties of the crop. Despite the wide variety of different people who consume them the most common methods of preparation are more or less the same in these countries and include dhals and thinner lentil soups, but local ecologies and food habits differ further afield towards the middle east and Europe where they are often eaten as an accompaniment to or sauce for meats. Dhal is the staple food in every Indian home with rich and poor equally enjoying it. It is India's comfort food and usually accompanies every meal eaten along with either hot steaming rice or fresh *chapati* (leavened bread) straight off the griddle or *dhaba* (dry frying pan). It is also dried with spices to produce spice capsules to add to foods. The simplest may be boiled with turmeric and ginger and then seasoned with sautéed onion and tomatoes. Roasted or fried cumin seeds add an extra dimension to dhals and aids in their digestion.

5.1. Some Cooking Ideas for Dhal – Indian Style

Below are 3 different ideas for how to cook lentils as you might find them in an Indian home. They have been provided by P. Stevenson's aunt-Sudershan Kumari from Peterborough, (UK) – and from experience he guarantees their quality and excellent flavour. In each case the lentil seeds need to be sorted for small stones and other unwanted debris and washed thoroughly before use.

MIXED DHAL SOUP

The following dish is Punjabi and includes a mixture of 3 types of lentils (which can vary even from those presented below) and the ingredients are

urid/urad (black) dhal (*Phaseolus mungo*) – 1 cup
mung (green) dhal (*Phaseolus aureus*) – 1 cup;
masoor (pink) dhal (*Lens culinaris*) – $^1/_2$ cup
water – 7 $^1/_2$ cups

2 medium onions, chopped finely
2 garlic cloves, chopped
fresh ginger, 1 inch cube, chopped finely
oil, 2–3 tbsp
cumin seeds, 1 tsp
1–2 fresh green chillies
turmeric, 1 tsp
tomatoes (2–3 fresh chopped or $^{1}/_{2}$ tin)
fenugreek powder, $^{1}/_{2}$ tsp
garam masala, 1 teaspoon
salt, 1 $^{1}/_{2}$ tsp
fresh coriander, handful, chopped

Wash dhals and soak for at least 4 hrs in 7 $^{1}/_{2}$ cups of water. Then add 1 $^{1}/_{2}$ tsp salt and boil until soft (1/2 hr if using a pressure cooker, longer if not). Heat oil in pan on a medium heat and when hot add cumin seeds. When seeds pop, then add onions and garlic and fry until golden brown. Then add ginger and chilies and fry for a few more minutes. Add turmeric, fenugreek powder and tomatoes. Add this mixture to the dhal and finally add garam masala and fresh roughly cut coriander. Delicious!

LENTIL PAKORA

Masoor (Pink) lentils (*Lens culinaris*) , 1 cup
Moong (green) lentils (*Phaseolus aureus*), 1 cup
Chillies, 1 fresh green or $^{1}/_{2}$ tsp red powder
Garam masala, 1 tbsp
1 medium onion, chopped
salt, 1 tsp
besan (chickpea) flour (*Cicer arietinum*), 2 tbsp (optional)
1–2 medium potatoes, very thinly sliced into discs

Soak dhal for 2–3 hours, until soft and then grind into a soft batter (add water if needed). Add salt, chilies, garam masala, onions and potatoes. Other vegetables can also be added if desired (e.g. cauliflower, spinach). Mix all the ingredients into a thick lumpy paste. At this stage the *besan* (chickpea) flour can also be added if desired to help bind the paste. Divide the paste into small portions (tablespoon size) and deep fry over a medium heat until golden brown. Serve the pakoras with a mint or tamarind chutney.

KICHARI (KICHERI) – Punjabi style again. This dish is recommended in Ayurvedic practice to be pretty much a cure-all for digestive complaints.

Rice, 1.5 cups
Masoor dhal (*Lens culinaris*) (split, washed), 1 cup
1 small onion
cumin seeds 1 tsp
turmeric 1 tsp
garam masala 1 tsp
tomatoes, 2 chopped
ginger, 1 inch cube, chopped
oil, 1 tbsp

Boil rice and lentils together in 4 cups of water until soft (10 mins in a pressure cooker, longer in a normal pan). Fry onions and cumin seeds in oil (medium heat) until onions golden. Add ginger and fry for a couple more minutes, and then add tumeric and tomatoes and cook for 5 mins more. Finally add this mixture to the dhal/rice and finish by adding the garam masala. Strict yogis may prefer to leave out the onions from this one as these are considered to be firey in the ayurveda tradition. Onions can be replaced with asafetida. Similarly, in Gujurat, dishes tend to use fewer onions.

There are countless other recipes available on the world wide web including an excellent selection at http://www.icarda.org/Publications/Cook/12/12.html (accessed 16 March 2007). Coconut lentils (*Amati*) are a popular dish of Maharashtra and other Southern States of India and are quite different from the northern Indian types by virtue of the fact that they contain grated coconut, jaggery and tamarind pulp as major ingredients (4 oz/$^1/_2$ cup/45 g lentils; 2 tbsp. salt $^1/_2$ tsp. turmeric; 5 cups/2 pints/1 liter water; 2 tsp. crushed jaggery (or sugar); 1 tbsp. tamarind pulp; 2 garlic cloves, crushed; $^1/_2$ tsp. mustard seeds, 3–4 green chilies, chopped, a pinch asafetida, 3 tbsp. ghee or oil, 2$^1/_2$ tbsp. grated coconut (fresh, if possible), 1 tsp. coriander leaves (optional), 8 oz/225 g whole black lentils). In Himachal Pradesh, Raj Roopa or Black lentil dhal is popular which differs again by the addition of bay leaves, and the garnish which contains a pinch of mace, cinnamon and ground cloves. Tuver Dhal or Piquant Lentils are popular in Uttar Pradesh, India, and differ again from the base ingredients by the addition of tamarind. In Punjab, lentils are sometimes made into dhal with the addition of Mango juice, fenugreek and nigella seeds to give a local character to the dhal, known as Mahani.

Some of the more unusual preparations include lentil cakes made from raw but soaked and crushed lentils known as *Varhia*. These require 1 teacup red lentils (*Masoor* dhal), 2 tbsp. lentil flour, 1 tsp. caraway seeds, 1$^1/_2$ tsp. garam-masala, 1 tbsp. coriander seeds, 2 tsp. salt, $^1/_2$ tsp. turmeric, 1 tsp. chili powder (optional) small haricot bean-sized lump of asafetida (optional). After sorting and washing the lentils they need on this occasion to be soaked overnight. After draining them the following day they need to be crushed with a mortar and pestle. The crushed lentils are then mixed with the lentil flour (made by grinding lentils in a coffee mill) and the other ingredients in a mixing bowl, kneaded for a few minutes and then left for upto 4 hours in warm place. The mixture is kneaded again before making the mixture into small palm sized cake shapes which can be dried in the sun and stored.

Mongorhis are a similar food but are deep fried in hot oil after making into cakes. Alternatively the entire lentil component can be ground to a flour and water used with other ingredients to make into a batter and used to make individual lentil cake or used to coat other vegetables. See http://www.icarda.org/Publications/Cook/12/12.html for more details.

5.2. Uses of Lentil Elsewhere in the World

Lentil consumption in Pakistan is estimated at 120,000 tons a year but its production now varies from only 25,000–40,000 tons. Thus Pakistan needs to import up to

95,000 tons every year primarily from Canada, Australia, India and Turkey. The utilization of lentil in Pakistan, Nepal, Sri Lanka and Bhutan are similar to that of India for various preparations like dhal, soups, salads, mixed vegetables and concentrations for animal feeds etc. In morocco lentil soups are considered as integral a part of the daily diet as they are in South Asia. While outside of South Asia the soups and dhals tend to be more simple in their ingredients and perhaps for the unaccustomed palate, less challenging in flavour than some of the firey spicey and exciting recipes of the Indian sub-continent, the Moroccans have an equivalent dish in Spiced Bean and Lentil Soup. This can be made from 1 tbsp Olive Oil, 2 chopped Onions, 2 crushed Garlic Cloves, 1 tsp freshly grated Ginger, 1.5 L. Water, 200 g or 7oz Red Lentils, 1 × 400 g/14 oz tin Chickpeas, 1 × 400 g/14 oz tin Cannellini Beans, 1 × 400 g/14 oz tin Chopped Tomatoes, 50 g/2oz Carrots, chopped 50 g/2oz Celery, chopped 1 tsp Garam Masala, 1 teasp Ground Cardamom, 1/2 teasp Ground Cayenne Pepper, 1/2 teasp Ground Cumin. Heat the olive oil in a large saucepan, add the onions, garlic, and ginger and sauté gently for 5 minutes. Add the remaining ingredients, bring to a boil for a few minutes then reduce the heat and simmer for 1–1/2 hours until the lentils are soft. Allow to cool a little then transfer half the soup to a food processor or liquidizer and process until smooth. Return the pureed soup to the remaining mixture in the pan, mix well and simmer until heated through. It is served as hot soup. Tunisians also have a passion for dhals but often add mint to give it a local characteristic flavour.

Turkey has had significant economic, political and social changes over the past twenty years. The increase in economic potential has seen the country become one of the major lentil exporters producing 500,000 tonnes of lentils which is approximately 150% more than it consumes (De Graaf, 2004). While traditionally, Turkey was one of the leading lentil consuming countries in the world, the increase in affluence over the past 10 years particularly in urban centers has resulted in changes in food towards other products, such as other grains and meat and a consequential decline in lentil consumption. This is also due to migration from rural to urban centers and an increase in lentil prices (De Graf, 2004). Turks have their own version of dhal or red lentil soup known as Kirmizi Mercimek Corbasi which is again quite different from the more lively South Asian and North African dishes. The ingredients are 1 tbsp. butter, 2 onions, chopped; 1½ cups red lentils; 8 cups veal stock; salt; 1 tsp. paprika; 4 tbsp. minced parsley; ¼ cup wine vinegar; 1 tbsp. flour (sifted); 1 tbsp. butter; 3 egg yolks; 1 cup single cream. The cooking instructions are to melt the butter and fry the onions for 2 minutes and add the cleaned lentils with 2 cups of water and boil until the lentils are tender. Add stock, salt and paprika, bring to boil, and then remove from heat. The cooked lentils and then forced through a sieve or liquidized. The mixture is then returned to the pan and kept hot. The rest of the butter and the flour is used to make a roux and cream then added very gradually away from the heat, stirring all the time. Finally, the egg yolks are well-beaten and combine with this mixture and then added to the purée. The dish is served immediately. Garnish with bread croutons that have been rolled in minced parsley and then spoon the wine vinegar over the dish at the table.

Lentil is mainly grown in Ethiopia for its matured seeds that are consumed in different forms. The seeds of lentil are boiled (Nufro) and salted and consumed as snacks. Split (kik) seeds and powdered flour (shiro) are the two major forms used for making a sauce 'kik watt' and 'shiro watt' that are eaten with Injera (flattened bread made of any cereal such as tef, wheat, barley, sorghum, maize, millets). Whole seeds or powdered lentil are used as soup which is particularly popular in towns (Yetneberk and Wondimu, 1994). It is some times mixed with other food legumes for making different sauces. Whole lentil seeds are boiled and mixed with onion and green chili (pepper) and then put inside wheat flour paste to bake and make 'Sambusa'. Cooked and mashed lentil seeds mixed with green chili and onion 'Azifa' is one of the popular dishes particularly in the north-west part of the country. Elbet (paste from flour) is also commonly used in some parts of Ethiopia. Boiled seeds are sometimes mixed with salad or boiled pealed potatoes to garnish. After separating seeds the straws are used as animal feed while the hard root parts are used as firewood (Bejiga, 2006). One particularly popular recipe among Ethiopians appears to be Amhari Mesir Wat- Ethiopian Lentil Bowl, a recipe which is common to the Ethiopian Jews (Phalashi). The ingredients are 1/2 kg Red lentils (*L. culinaris*), 2 large Onions, 1/2 cup Oil, 3 tbsb Tomato paste, 1/2 tsp Paprika; sweet or hot 1 clove Garlic, 1/2 tsp Ground ginger, 1/4 tsp Black pepper, 1 tsp Salt, and 3 cups Water. Lentils should be sorted and washed first and then soaked for 30 minutes before rinsing and draining. While soaking, peel and finely chop the onions and mash the garlic. Heat the oil in large pan and sauté the onion until golden. Add the tomato paste and the paprika and mix. Add half the water and the garlic, ginger pepper and salt. Stir well and then add the rest of the water, stir again, cover and bring to boil. When the water boils, add the lentils, lower the flame and cook 20–30 minutes, until the lentils soften.

Another Jewish recipe with lentil surrounds tradition at Purim when seeds, nuts and pulses are eaten. This arose from the story of Queen Esther who was living in the King's palace in Persia but as a Jew was not allowed to eat non-kosher food at Royal banquets and so lived on a diet of seeds, nuts and pulses. The recipe uses small green lentils that come from central France. These tiny lentils have a distinctive nutty flavor and texture and keep their shape and colour during cooking. The ingredients are 2 tbsp olive oil; 1 leek (trimmed and diced); 2 sweet potatoes and 3 carrots (peeled & diced), 250 g green (Puy) lentils, 2.5 cm piece fresh ginger – peeled and grated, 750 ml hot vegetable stock, 3 oranges – peeled and cut into segments, 200 g spinach leaves and Salt and freshly ground black pepper – to taste. The garnish consists of 2 tablespoons whole toasted almonds; Sea salt and freshly ground black pepper and 1 tsp dried coriander. Heat the olive oil in a large frying pan and sauté the leek, sweet potatoes, carrots and ginger together for 10 minutes. Stir in the lentils and add the hot vegetable stock. Cover and leave to simmer for 40 minutes. When the lentils are completely cooked, stir in the spinach leaves and orange segments and remove from the heat. Toast the almonds in a separate pan with no oil. Season with sea salt, freshly ground black pepper and dried coriander. The required cooking time is about one hour for this dish.

A good example of how lentils can be used to enhance a meat dish as is most often their use in Europe is in the dish known as Potage Saint Hubert which makes a good spicy Christmas soup. Ingredients are 1 lb/450 g soaked brown lentils, 1 onion; 1 leek; thyme; bay leaf; salt and pepper; 1 pheasant; 4 fl oz/100 ml cream. The soaked lentils are cooked in salted water with the onion, the white of the leek, thyme, bay leaf and seasoning. The pheasant is roasted and when cooked the meat is carved from the bone and the best fillets set aside and diced. The rest of the meat is pounded in a mortar. The lentils are strained (reserving the stock) and are added to the meat which is sieved or liquidize and return to the saucepan. This is then thinned with the lentil stock, continuing to add it until the soup is the desired consistency. When it is hot stir in the cream and the diced pheasant. Any cold game could be used for this dish.

Lentil soup in parts of the middle-east is made with bone marrow, but in Lebanon there is a version that is renowned in the region and known as Makhlouta or Lentil Soup with Beans and Rice. The ingredients for this dish are 1 cup lentils; 1 cup chickpeas; $^{1}/_{2}$ cup dried black beans; 1 cup rice; $^{1}/_{2}$ cup olive oil; $^{1}/_{2}$ cup minced onions; $1^{1}/_{2}$ tsp. salt; $^{1}/_{2}$ tsp caraway seeds; 5 cups water. Beans, lentils and chickpeas are washed and then soaked together overnight. Drain, cook with water and salt for 15 minutes under pressure (Pressure cooker). The rice should be boiled separately before adding to the lentil/bean mixture. Onions should be fried in oil until slightly browned and then added with the salt and caraway seeds to the beans and lentils and this is simmered uncovered until well blended. A similar version is also made in Lebanon and Syria called Shawrabat Adas (Lentil Soup) which is made with lentils, Swiss chard leaves or spinach, fresh coriander leaves or celery, garlic cloves, lemon juice, tomato purée and flour. An Iranian equivalent uses ground beef and cinnamon.

Lentil is an ancient crop. It is straight forward to cultivate in many different agro-ecosystems and with such a variety of ways of eating the seeds and all the evidence to show how important it is nutritionally it will continue to be of huge importance across the world for the foreseeable future. The availability of modern technologies and even the prospects of genetic enhancement of lentils may help enhance production to levels unimaginable today but the lentil is above all likely to continue to provide a major high nutrient value – low cost ingredient in the diet of the worlds less well off.

REFERENCES

Abdelgawad, A.S. 1993. Effect of Domestic Processing on Oligosaccharide Content of Some Dry Legume Seeds, *Food Chemistry* 46 (1): 25–31.

Adsule, R.N., S.S. Kadam, and H.K. Leung. 1989. Lentil. In: CRC hand book of world food legumes (eds. D.K. Salunkehe and S.S. Kadam). Boca Raton, Florida, USA: CRC Press.

Anon, 2006 http://www.agr.gc.ca/mad-dam/index_e.php?s1=pubs&s2=bi&s3=php&page=bulletin_17_08_1_ 2004-05-28&PHPSESSID=1539ae40619abe5e07149cdfc6631797 (opened 16 March 2007)

Araki, R., Maruyama, C., Igarashi, S., Yoshida, M., Maruyama, T., Satoh, T., Yoshida, M., Umegaki, K. 2006. Effects of short-term folic acid and/or riboflavin supplementation on serum folate and plasma total homocysteine concentrations in young Japanese male subjects, *European Journal of Clinical Nutrition*, 60: 573–579.

Araya H, Contreras P, Alvina M, Vera G, Pak N 2002 Comparison between an in vitro method to determine carbohydrate digestion rate and the glycemic response in young men. *European Journal of Clinical Nutrition*. 56 (8): 735–739.

Bhattacharya, S., Narasimha, H.V., Bhattacharya, S. 2005. The moisture dependent physical and mechanical properties of whole lentil pulse and split cotyledon *International Journal of Food Science & Technology*, 40, 213–221.

Brand-Miller, J.C., Thomas, M., Swan, V., Ahmad, Z.I., Petocz, P., Colagiuri, S. 2003 Physiological validation of the concept of glycemic load in lean young adults. Journal Of Nutrition 133 (9): 2728–2732.

De Graf, J. 2004. http://ats.agr.ca/europe/3864_e.htm (opened 16th March 2007)

Duke, J.A. 1981. Handbook of legumes of world economic importance. Plenum Press, New York. p. 52–57.

Flight, I., and Clifton, P. 2006. Cereal grains and legumes in the prevention of coronary heart disease and stroke: a review of the literature. *European Journal Of Clinical Nutrition,* 60 (10): 1145–1159.

Bejiga, G. 2006: *Lens Culinaris* Medik. In: Plant Resources of Tropical Africa. 1, Cereals and Pulses (Brink,M & G.Belay eds.) Prota Foundation, Wageningen, Netherlands / Backhuys Publishers, Leiden, CTA, Wageningen, Netherlands pp 91–96.

Giugliano, D., Esposito, K. 2005. Mediterranean diet and cardiovascular health Natural Products and Molecular Therapy. *Annals of the New York Academy Of Sciences*, 1056: 253–260.

Hardacre, A.K., Clark, S.M., Riviere, S., Monro, J.A., Hawkins, A.J. 2006. Some textural, sensory and nutritional properties of expanded snack food wafers made from corn, lentil and other ingredients. *Journal of Texture Studies*, 37 (1): 94–111.

Huisman, J. and van der Poel, A.F.B. 1994. Aspects of the nutritional quality and use of cool season food legumes in anmal feed. p. 53–76. In: F.J. Muehlbauer and W.J. Kaiser (eds.), Expanding the Production and Use of Cool Season Food Legumes. Kluwer Academic Publishers. Dordrecht, The Netherlands.

Hulse, J.H. 1994. Nature, composition and utilization of grain legumes. p. 11–27. In: ICRISAT. Uses of tropical grain legumes: Proceedings of a consultants' meeting, 27–30 Mar, 1989. ICRISAT Center, India, Patancheru, A.P. 502 324. India:ICRISAT.

Iliadis, C. 2001. Effects of harvesting procedure, storage time and climatic conditions on cooking time of lentils (*Lens culinaris* Medikus). Journal of the Science of Food and Agriculture 81 (6): 590–593.

Iqbal, A., Khalil, I.A., Ateeq, N., Khan, M.S., 2006. Nutritional quality of important food legumes. *Food Chemistry,* 97 (2): 331–335.

Jumbunathan, R., Blain, H.L., Dhindsa, K.S., Hussein, L.A., Kogure, K., Li-Juan, L. and Youseef, M.M. 1994. Diversifying use of cool season food legumes through processing. pp. 98–112. In: F.J. Muehlbauer and W.J. Kaiser (eds.) Expanding the Production and Use of Cool Season Food Legumes. Kluwer Academic Publishers. Dordrecht, The Netherlands.

Kay, D. 1979. Food legumes. Tropical Development and Research Institute (TPI). TPI Crop and Product Digest No. 3. p. 48–71. UK.

McGovern, P.E., Glusker, D.L., Moreau, R.A., Nunez, A., Beck, C.W., Simpson, E., Butrym, E.D., Exner, L.J. and Stout, E.C. 1999. A funerary feast fit for King Midas. *Nature*, 402 (6764): 863–864.

Muehlbauer, F.J., Cubero, J.I. and Summerfield, R.J. 1985. Lentil (*Lens culinaris* Medic.). p. 266–311. In: R.J. Summerfield and E.H. Roberts (eds.), Grain Legume Crops. Collins, 8 Grafton Street, London, UK.

Muehlbauer F.J. and Tullu A. 1997. http://www.hort.purdue.edu/newcrop/cropfactsheets/lentil.html (accessed 16 March 2007).

Rani, B. and Kawatra, A. 1994 Fibre Constituents of Some Foods. *Plant Foods for Human Nutrition.* 45: 343–347.

Roberts, D.C.K., Truswell, A.S., Bencke, A., Dewar, H.M. and Farmakalidis, E. 1994. The Cholesterol-Lowering Effect of a Breakfast Cereal Containing Psyllium Fibre. Medical Journal of Australia 161 (11–12): 660–664.

Sarker, A., and Erskine, W. 2006 Recent progress in the ancient lentil *Journal of Agricultural Science* 144: 19–29.

Singh, U. 1999. Cooking quality of pulses. *Journal of Food Science and Technology-Mysore* 36 (1): 1–14.

Singh, U. and Jambunathan, R. 1982. Changes in starch, oligosaccharides and soluble sugars in developing pod wall and seed of chickpea *Phytochemistry*, 21, 297–299.

Van der Maesen, L.J.G. 1972. *Cicer* L., a monograph of the Genus, with special reference to the chickpea (*Cicer arietinum* L.), its ecology and cultivation, *Commun. Agric.* University. Wageningen

Viadel, B., Barbera, R., Farre, R. 2006. Uptake and retention of calcium, iron, and zinc from raw legumes and the effect of cooking on lentils in Caco-2 cells. *Nutrition Research.* 26 (11): 591–596.

Wang, N., and Daun, J.K. 2006. Effects of variety and crude protein content on nutrients and anti-nutrients in lentils (*Lens culinaris*) *Food Chemistry,* 95: 493–502.

Williams, P.C. and U. Singh. 1988. Quality screening and evaluation in pulse breeding. p. 445–457. In: R.J. Summerfield (ed.), World Crops: Cool Season Food Legumes. Kluwer Academic Publishers, Dordrecht The Netherlands.

Williams, P.C., R.S. Bhatty, S.S. Deshpande, L.A. Hussein and G.P. Savage. 1994. Improving nutritional quality of cool season food legumes. p. 113–129. In: F.J. Muehlbauer and W.J. Kaiser (eds.), Expanding the Production and Use of Cool Season Food Legumes. Kluwer Academic Publishers, Dordrecht, The Netherlands.

Yetneberk, S, and Wondimu, A. 1994 Utilization of cool season food legumes in Ethiopia *In* Asfaw Telaye Geletu Bejiga, Saxena,M. C. and Solh, M.B.(*eds*). Cool –season Food Legumes of Ethiopia. Proceedings the First National Cool – season Food legumes Review Conference,16–20 December,1993, Addis Ababa, Ethiopia. ICARDA/Institute of Agricultural Research, ICARDA, Aleppo, Syria. 60–75.

Zapata, L., Pena-Chocarro, L., Perez-Jorda, G., Stika, H.P. 2004. Early Neolithic agriculture in the Iberian Peninsula Journal of World Prehistory, 18 (4): 283–325.

WEBSITES PROVIDING USEFUL ADDITIONAL READING ON CONSUMPTION AND USE OF LENTILS

http://www.icarda.org/Publications/Cook/12/12.html
http://www.puritan.com/vf/healthnotes/HN75_English/Recipe/Lentil_Turkey_Soup.htm#Ingredient-List
http://www.ivu.org/recipes/african/shurit-j.html
http://www.ivu.org/recipes/african/nile-j.html
http://www.ivu.org/recipes/african/amhari-j.html
http://www.icarda.org/Publications/Cook/3/3.html
http://www.natashascafe.com/html/menu.html
http://www.foodreference.com/html/russian-red-lentil.html
http://www.jewishcookery.com
http://www.hort.purdue.edu/newcrop/cropfactsheets/lentil.html

CHAPTER 5

NUTRITIONAL VALUE

GLORIA URBANO[1], JESÚS M. PORRES[1], JUANA FRíAS[2], AND CONCEPCIÓN VIDAL-VALVERDE[2]

[1] *Departamento de Fisiología, Instituto de Nutrición, Universidad de Granada. Campus Universitario de Cartuja s/n, Granada 18071, Spain.*

[2] *Instituto de Fermentaciones Industriales, Consejo Superior de Investigaciones Científicas, Juan de la Cierva 3, Madrid 28006, Spain*

Abstract: The importance of lentils as important dietary sources of macro and micronutrients essential for human welfare has been recognized since ancient times. Lentils provide sufficient amounts of most essential amino acids to meet the nutrient requirements, although they are deficient in sulfur-containing amino acids like most legumes. Lentils also contain fair amounts of other essential nutrients like minerals, vitamins and complex carbohydrates. In contrast, lentils exhibit a considerable amount of non-nutritional compounds like trypsin inhibitors, tannins or phytic acid that are able to interfere with the availability of several nutrients. Different processing conditions that range from the traditional soaking/cooking to germination, fermentation, or several thermal treatments, are usually employed to improve the organoleptic properties of lentil seed and its nutritional value through reducing the negative effect of the above mentioned non-nutritional components. In addition, technological treatments may significantly enhance the functional and beneficial health properties of the processed lentil food products, making consumption of this legume an appealing alternative for today's world

1. INTRODUCTION

Lentil (*Lens culinaris Medik*) is an important dietary source of energy, protein, carbohydrates, fiber, minerals, vitamins and antioxidant compounds, as well as diverse non-nutritional components like protease inhibitors, tannins, α-galactoside oligosaccharides and phytic acid. Lentil is smaller in size than other pulses like the bean, faba bean or chickpea, with a mean diameter and thickness of 4.45–6.82 and 2.36–2.55 mm, respectively and a volume of 56 mm^3 (Fasina et al., 2001; Bhattacharya et al., 2005). Whole lentils have a low weight (13 to 30 seeds to per gram) (Bhatty, 1984; Bhattacharya et al., 2005) and lower seed bulk density than

S.S. Yadav et al. (eds.), Lentil: An Ancient Crop for Modern Times, 47–93.

other legumes such as kidney bean, green peas, black beans or pinto beans (Fasina et al., 2001). In contrast, the kernel density is higher in lentils than all these legumes except for pinto bean.

2. CHEMICAL COMPOSITION OF RAW LENTIL

According with the bibliography the range of values for nutrient composition of lentils is compiled in Table 1.

2.1. Energy

The energy provided by lentil (Table 1) is similar to that of cereals (e.g. wheat) and other pulses (e.g. faba beans, peas or beans). However, it is lower than the crude energy provided by legumes with a higher fat content (e.g. soybeans or lupins).

2.2. Nitrogen

Lentils have an average total nitrogen content of 4.25 g/100 g DM (Table 1), of which nearly 15% is present as soluble non-protein nitrogen mainly composed of free amino acids and small peptides that can be detected by the method of

Table 1. Chemical composition of raw lentil (per 100 grams of dry matter)

	Range	Bibliography
Energy (kJ)	1483–2010	1–3,14,17,30,31,32
Total Nitrogen (g)	3.72–4.88	3,4,10,13,16,18,21,27,30,31
Protein (N × 6.25) (g)	20.6–31.4	1–20, 28–29,30,31,32,33,34,35,36,37
Non-Protein Nitrogen (g)	0.49–1.049	4,21–24
Fat (g)	0.7–4.3	3,11–13,16–19, 21,23,25–27, 29,30,31,33,34,36,37
Carbohydrates (g)	43.4–69.9	17,23,37
Fiber (g)	5.0–26.9	17,23,35
Ash (g)	2.2–4.2	1,11–12,16–18, 21–23,25–27, 28–30,33,34,36,37

1. Kavas and Nehir (1992); 2. Combe et al. (1991); 3. Porres et al. (2002); 4. Sanz et al. (2001); 5. Fasina et al. (2001); 6. Bamdad et al. (2006); 7. Meiners et al. (1976) J Agric Food Chem 24, 1122–1126; 8. Manan et al. (1987); 9. Rehman and Shah (2005); 10. Monsoor and Yusuf (2002); 11. Iqbal et al. (2006); 12. Wang and Daun (2006); 13. Jood et al. (1998) Nahrung 42, 71–74; 14. Solanki et al. (1999); 15. Bhatty (1984); 16. Kylen and McReady (1975); 17. Ereifeg and Haddad (2001); 18. Candela et al. (1997); 19. Urbano et al. (1995); 20. Lombardi-Boccia et al. (2003); 21. Shekib et al. (1985); 22. El-Mahdy et al. (1985); 23. El-Adawy et al. (2003); 24. Periago et al. (1996) Food Res Int 29, 489–494; 25. De Almeida Costa et al. (2006); 26. Valente Mesquita and Reis (2005) Nutr Food Sci 35, 264–270; 27. Danisová et al. (1994); 28. Elhardallou et al. (1999) Food Chem 67, 113–121; 29. Miller McCurdy et al. (1978); 30. Khan et al. (1987); 31. Sika et al. (1995); 32. Solanki et al. (1999); 33. El-Tinay et al. (1989); 34. Perez-Hidalgo et al. (1997) J Food Comp Anal 10, 66–72; 35. Carbonaro et al. (1996); 36. Bartolomé et al. (1995); 37. Cai et al. (2002) J Food Sci 67, 1725–1730; 38. Zhao et al. (2005)

Table 2. Distribution of chemical constituents in the different anatomical parts of lentil seed*.

Component	Proportion to whole seed	Protein (g/100 g DM)	Fat (g/100 g DM)	Crude Fiber (g/100 g DM)	Ash (g/100 g DM)	Nitrogen free extract (g/100 g DM)
Seed Coat	8.0–20	14.3	0.6	29.4	1.94	53.7
Cotyledon	80–90.0	26.5–30.1	3.0	1.0	2.45	63.4
Embryo	2.0	71.1	8.2	2.4	3.94	14.4
Whole seed	100.0	29.6	3.1	3.2	2.40	61.7

* Data taken from Adsule et al. (1989) and Cuadrado et al. (2002a)

Lowry in addition to nucleic acids, puric and pyrimidinic bases, and alkaloids. The remaining nitrogen present in lentils is usually classified as protein nitrogen. A minor proportion of nitrogen from legumes (e.g. peas, lupins) and probably lentils also is present in an insoluble form, (Urbano et al. 2005a,b; Porres et al. 2003, 2005) that arises from non–covalent interactions or disulfide bonds between different proteins and remains associated with the insoluble dietary fiber fraction (Martín-Cabrejas et al., 2003). This insoluble nitrogen is related to the appearance of high-molecular-weight protein aggregates that hardly migrate into the resolving gel when the legume proteins are separated by polyacrylamide gel electrophoresis (Urbano et al., 2005a,b; Porres et al., 2003, 2005). Nevertheless, the insoluble nitrogen of legumes did not seem to have a major effect on protein digestibility assessed by *in vitro* or *in vivo* methods (Porres et al., 2005, 2006), although it could play an important role in technological processes like the preparation of protein isolates where protein solubility is a limiting factor to obtain optimal yields of the extraction process.

Protein content of lentils may change in response to genetic factors and different edaphic and environmental conditions. Table 2 shows that most of protein in lentils is located in the cotyledons (Adsule et al., 1989; Cuadrado et al., 2002a), mainly as storage proteins soluble in salt solution, with a lower proportion as enzymatic proteins soluble in water. It should be noted that the percentage of sulfur-containing amino acids is higher in the enzymatic protein fraction when compared to the storage protein fraction (Bhatty, 1982).

2.2.1. *Amino acid composition and nutritional properties*

Amino acid composition of lentils is described in Tables 3a and 3b. When compared to the nutrient requirements for the 1-year-old infant (FAO/WHO, 1991), the amino acid profile of lentils protein is deficient in sulfur-containing amino acids methionine and cystine, and in tryptophan. The deficiency is greater for growing rats (NRC, 1995), but less severe for the 2–5 year-old infant or the human adult (FAO/WHO, 1991). The low levels of sulfur containing amino acids in legume proteins and their low digestibility (Sarwar and Peace, 1986; Wu et al., 1996; Porres et al., 2002) has lead FAO/WHO to recommend use of the Protein Digestibility-Corrected Amino Acid Score (PDCAAS) as an index to evaluate the nutritional

Table 3a. Non-essential amino acid composition of lentils (g/16 g N)

Asp	Glu	Ser	Gly	Ala	Arg	Pro	Bibliography
9.29	14.45	4.88	4.82	4.82	7.68	3.52	Shekib et al. (1986)
10.75	15.20	4.80	4.25	4.30	9.25	3.65	Combe et al. (1991)
11.56	14.58	4.38	3.36	3.47	8.68	1.19	Kavas and Nehir (1992)
9.94	16.3	2.86	4.33	2.37	3.90	2.58	Urbano et al. (1995)
13.24	16.70	5.19	4.01	4.52	10.12	4.46	Carbonaro et al. (1997)
10.66	18.51	6.02	5.60	4.69	8.95	7.20	Porres et al. (2002)
11.3–14.1	13.2–16.4	5.5–6.4	3.6–4.3	3.8–4.4	6.5–9.5	3.7–5.9	Canadian Grain Commission (2004)
10.1–15.9	12.8–17.3	5.0–6.3	3.3–4.4	3.6–4.3	6.2–11.1	3.5–6.1	Canadian Grain Commission (2004)
10.6–13.8	14.4–18.3	3.8–5.3	3.7–4.8	3.9–5.0	6.7–7.8	3.6–4.6	Wang and Daun (2006)

quality of a protein for humans. The use of true digestibility of the limiting amino acid ($AAS_{(ASAA)}$) suggested by Wu et al. (1996) may be an even better index. Different lentil varieties have a wide range for protein amino acid content (Tables 3a and 3b). Thus selection of varieties should be possible to improve sulfur-containing amino acids and tryptophan content.

The Leu/Ile and Leu/Lys ratios are useful nutitional quality indices. An excess of leucine impairs the utilization of isoleucine and lysine (Harper et al., 1955; Petterson et al., 1997). The Leu/Ile and Leu/Lys ratios ranges for lentils are, 1.24–1.98 and 1.08–2.03, respectively. These values are indicative of good protein quality and similar to other legumes (Fernandez et al., 1996; Nestares et al., 1996, 2001; Martinez-Villaluenga et al., 2006).

2.2.2. *Non protein amino acid composition*

Lentils possess significant amounts of non-protein amino acids (Table 4; taurine, GABA, γ-hidroxyarginine, γ-hidroxyornithine and trigonelline) with biological effects (Rozan et al., 2001; Kuo et al., 2004).

2.3. Fat

The fat content of lentils is low (0.7–3.6%) (Table 1), and similar to other pulses and cereals (peas, beans, wheat) but less than others (chickpea, lupin, soybean). Unsaturated fatty acids (linoleic, oleic, and linolenic) dominate the lentil profile (Table 5) with small amounts of saturated fatty acids, mainly palmitic acid. The lentil fatty acid profile is similar to that of other legumes, (pea, chickpea, soybean; Canadian Grain Commission, 2004; Kumar et al., 2006).

Table 3b. Essential amino acid composition of lentils (g/16 g N)

His	Ile	Leu	Lys	Thr	Val	Tyr	Phe	Met	Cys	Trp	Bibliography
3.36	4.96	7.28	7.09	3.78	4.92	3.23	4.72	0.94	0.96	0.72	Shekib et al. (1986)
2.45	4.05	6.65	6.55	3.40	4.55	3.55	3.60	0.90	1.40	–	Combe et al. (1991)
–	3.44	6.13	6.35	2.69	4.19	2.24	4.34	0.22	1.12	0.74	Kavas and Nehir (1992)
1.33	4.04	7.24	4.27	2.49	3.04	1.13	5.52	1.10	1.46	–	Urbano et al. (1995)
2.56	4.48	7.79	6.84	4.03	5.29	3.06	4.43	0.83	1.04	–	Carbonaro et al. (1997)
1.33	4.31	7.72	5.79	4.23	5.09	1.13	5.77	1.25	0.94	–	Porres et al. (2002)
1.8–3.0	2.6–4.2	5.7–7.1	4.4–7.0	3.6–4.9	3.3–4.9	1.18–3.4	3.7–5.0	0.8–1.2	0.7–1.4	0.9–2.6	Canadian Grain Commission (2004)
2.0–2.8	2.6–4.0	5.8–7.1	4.0–6.7	3.8–4.7	3.4–4.7	1.6–3.3	3.7–4.8	0.8–1.1	0.8–1.4	0.6–0.9	Canadian Grain Commission (2004)
2.6–3.3	4.4–5.5	6.8–8.7	6.3–8.2	3.4–4.4	4.7–6.1	7.4–9.4*		2.2–4.2¥		0.6–0.9	Wang and Daun (2006)
Recommended intakes											
2.6	4.6	9.3	6.6	4.3	5.5	7.2*		4.2¥		1.7	FAO/WHO 1-year-old
1.9	2.8	6.6	5.8	3.4	3.5	6.3*		2.5¥		1.1	FAO/WHO Pre-school (2–5 years old)
1.9	4.1	7.1	6.1	4.1	4.9	6.8*		6.5¥		1.3	NRC (1995) growing rat

*Sum of Phe + Tyr
¥Sum of Met + Cys

Table 4. Free non protein amino acids and trigonelline in lentils (mg/100 g DM)*

Amino acids	L. culinaris	L. ervoides	L. nigricans	L. odemensis	L. orientalis
γ-hydroxy-arginine	2.52–1.58	trace	2.04	0.75	5.73
γ-hydroxy-ornitine	0.01-ND	0.01	0.04	0.02	0.04
α-aminobutyric acid	0.01-ND	0.01	0.02	0.02	0.03
Taurine	0.46–0.03	0.01	0.21	0.019	0.05
Homoarginine	Trace	0.06	0.01	trace	ND
Trigonelline	11.85–0.40	ND	8.31	1.80	18.15

* Results taken from Rozan et al. (2001) and Kuo et al. (2004)

Table 5. Fatty acid composition of lentils

Fatty acid (% in oil)	Lens culinaris cv. Trebisowska♧	Canadian green lentils*	Canadian red lentils*	Australian lentils*
Lauric (C12:0)		ND	ND	–
Myristic (C14:0)	0.40	0.00–0.93	0.42–0.73	0.60–0.60
Palmitic (C16:0)	17.90	10.79–15.36	13.25–15.77	12.70–13.70
Palmitoleic (C16:1)		0.00–0.61	0.00–0.33	–
Stearic (C18:0)	2.00	1.27–1.82	1.34–1.65	1.80–2.10
Oleic (C18:1)	20.10	17.04–25.63	17.05–22.17	22.70–28.00
Linoleic (C18:2)	37.60	40.97–46.14	42.91–45.23	41.90–57.14
Linolenic (C18:3)	6.90	11.93–16.23	12.68–14.66	11.60–12.70
Arachidic (C20:0)		0.77–1.11	0.80–0.92	–
Gadoleic (C20:1)		1.21–1.58	1.12–1.24	1.20–1.30
Eicosadienoic (C20:2)		0.00–0.22	0.00–1.86	–
Behemic (C22:0)		0.81–1.13	0.81–0.91	–
Erucic (C22:1)		0.36–0.74	0.80–0.92	–
Lignoceric (C24:0)		0.47–1.99	0.56–0.70	–
Nervonic (C24:1)		ND	ND	–
Other¥	15.10			

♧Grela and Günter (1995); * Canadian Grain Commission (2004)
¥ Includes C10:0, C12:0, C20:0, C20:1, C20:2, C22:0, C22:1.

2.4. Carbohydrates

Carbohydrate concentration in whole lentils ranges from 43 to 70% (Table 1). The available carbohydrate fraction includes individual soluble sugars such as fructose, glucose and sucrose (1 to 2.5%), galacto-oligosaccharides such as raffinose, ciceritol, stachyose and verbascose (2 to 8%), non-starch polysaccharides (~ 20%), and starch (35 to 63%) with an amylose content of 20–45.5% (Table 6).

Table 6. Carbohydrate composition of lentils

	(g/100 g DM)	References
Total available soluble sugars	1.1–3.2	1–7
Fructose	0.01–0.17	1, 2, 3, 5, 6
Sucrose	1.09–2.97	1, 2, 3, 5, 6, 8
Total α-galactosides	1.8–6.8	1–3, 5–7
Raffinose	0.16–1.49	1–3, 5–10
Ciceritol	0.24–1.99	2, 3, 5, 6
Stachyose	1.1–3.1	1–3, 5–10
Verbascose	ND-1.35	1–3, 5–9
Starch	34.7–65	1, 3, 4, 7–13

1. Frias et al. (1994); 2. Frias et al. (1996a), 3; Frias et al. (1995a); 4. Vidal-Valverde and Frias (1992); 5. Vidal-Valverde et al. (1993a); 6. Vidal-Valverde et al. (1993b); 7. Reddy et al. (1984); 8. Wang and Daun (2006); 9. Fasina et al. (2001); 10. El-Adawy et al. (2003); 11. Jood et al. (1998) Nahrung 42, 71–74; 12. Cai et al. (2002) J Food Sci 67, 1725–1730; 13. Sotomayor et al. (1999)

2.5. Dietary Fiber

Lentil dietary fiber ranges from 9.7 to 24.1% (Table 1). It is mainly cellulose, hemicellulose, pectic substances and lignin (Table 7) (Vidal-Valverde and Frias, 1991; Vidal-Valverde et al., 1992a; Ramulu and Udayasekhara Rao, 1997; Urbano et al., 1999; De Almeida Costa et al., 2006).

2.6. Minerals

Lentils are an important source of dietary essential minerals. These include macronutrients (K, P, Ca, Mg, Na), micronutrients (Fe, Zn, Cu, Mn) and trace elements (Al, Cr, Ni, Pb, Co, Se, Mo) (Tables 8a and 8b). However, the bioavailability of minerals from lentils may be low due to the presence of non-nutritional compounds (phytic acid, tannins, oxalate) that interfere with their nutritive utilization. Most of

Table 7. Dietary fiber in lentils (in grams per 100 g DM)

	Range	Bibliography
Dietary fiber (as NDF)	16.2–21.3	1–3
Cellulose	4.1–5.33	1, 2, 4
Hemicellulose	6.0–15.74	1, 2, 4
Total dietary fiber	11.0–26.9	5–7
Soluble dietary fiber	1.2–6.7	5–10
Insoluble dietary fiber	8.8–31.4	5–10

1. Vidal-Valverde and Frias (1991); 2. Vidal-Valverde et al. (1992a); 3. Arntfield et al. (2001); 4. Reddy et al. (1984); 5. Bartolomé et al. (1995); 6. Perez-Hidalgo et al. (1997) J Food Comp Anal 10, 66–72; 7. Ramulu and Udayasekhara Rao (1997); 8. Elhardallou and Walker (1999) Food Chem 67, 113–121; 9. De Almeida Costa et al. (2006); 10. Candela et al. (1997)

Table 8a. Mineral composition of lentils (mg/100 g DM)

Ca	Total-P	Phytate-P	Inorg.-P	*K*	Mg	Na	Zn	Fe	Bibliography
33							4.6	12.8	Kylen and McReady (1975)
	305.2-409.7	123.5-175.9							Gad et al. (1982)
70	450			1160	100	40			Bhatty (1984)*
40-160	280-630			880-1440	80-140	20-180			
40; 40	153; 247	120; 195				6.9; 4.3	2.5; 3.0	8.9; 11.0	El-Mahdy et al. (1985)
42	380			753		4.61	5.11	7.2	Shekib et al. (1985)
190.1	282.9	46.8					6.1	9.6	Khan et al. (1987)
	331; 467; 389	151; 203; 168							Manan et al. (1987)
60	360	179.8		970	100				Bhatty (1989)
90	380			900	100	40	2.9	12.6	Combe et al. (1991)
							3.7	7.9	Kavas and Nehir (1992)
78.6	239.3							3.7	Danisová et al. (1994)
81.2; 102	282; 308					55.8	3.7; 3.3	7.7; 14.6	Sika et al. (1995)
115-165	250-460							8.0-9.2	Solanki et al. (1999)
128	373	184.3							Urbano et al. (1999)
77							3.57	7.65	Sebastiá et al. (2001)
42.3; 97.9; 60.6	458.5; 315; 441.6			548	129; 119	78.6; 30.4; 68.3	6.2; 3.7; 4.9	13.3; 11.9; 9.2	Ereifeg and Haddad (2001)
	338						4.2	8.2	Koplík et al. (2002)
76	372	338		240	48.5	11			El-Adawy et al. (2003)
							3.4	6.8	Lombardi-Boccia et al. (2003)
53.4							4.1	9.3	Sahuquillo et al. (2003)
69.1	387.6	205.7	57.9		105.1				Porres et al. (2003, 2004)
210	340			2370	220		3.8	6.5	Demirbas (2005)
							2.5	8.2	Erdogan et al. (2006)
120	294			874		79	4.4	9.9	Iqbal et al. (2006)
79.7	509.4	205.5–313.9		1055	138		4.0	7.9	Wang and Daun (2006)*
48.4–107	344–725			550–1268	121–167		2.9–5.9	6.6–9.8	

* The means and range of values have been taken from the bibliographic reference and included in the table.

Table 8b. Mineral composition of lentils (per 100 grams of dry matter)

	Khan et al. (1987)	Combe et al. (1991)	Sika et al. (1995)	Ereifeg and Haddad (2001)	Koplík et al. (2002)	El-Adawy et al. (2003)	Demirbas (2005)	Iqbal et al. (2006)	Erdogan et al. (2006)
S (mg/100 g)							100		
Mn (mg/100 g)	3.4	1.4		1.3; 1.48	1.43	1.8	5.38	1.6	1.17
Cu (mg/100 g)	2.3	3.2		1.1; 0.88	0.85		1.8	3.1	1.01
B (mg/100 g)		0.86	0.992				1.05		
Al (μ g/100 g)			10						
Cr (μ g/100 g)			27.2						
Ni (μ g/100 g)	300		115		189				
Pb (μ g/100 g)			37						
Co (μ g/100 g)					7.1				
Se (μ g/100 g)					103				
Mo (mg/100 g)					1.27		0.73		

the minerals are located in the cotyledons, except for Ca and Fe that are present in a considerable proportion in the lentil seed coat (Adsule et al., 1989). As for other legumes (e.g. peas, lupin, beans; Porres et al., 2003, 2005; Urbano et al., 2006), potassium is the mineral present in quantitatively highest levels, followed by phosphorus, calcium, magnesium and sodium. A high proportion of phosphorus is found in phytic acid with potentially low availability for simple-stomached animals. Less is present as free inorganic phosphate (5.6–14.9%, El-Mahdy et al., 1985; Porres et al., 2004) or other organic phosphorus components (8.6–15.4%, El-Mahdy et al., 1985).

2.7. Vitamins

Lentils is a rich source of water soluble vitamins (Table 9) and, as with most species of legumes contains only small amounts of vitamin C, carotene and retinol.

2.8. Non-nutritional Components

The composition of non-nutritional components found in *Lens culinaris* and their physiological effects are described in Tables 10 and 11. Protease (trypsin and chimotripsin) inhibitors found in lentil can decrease the effectiveness of pancreatic enzymes and interfere with the digestive utilization of protein, causing an enlargement of pancreas. Lentils contain amylase inhibitors in lower values than *Phaseolus* which may contribute to the lower nutritive value of uncooked seeds. While lentils contain lectins Grant et al. (1983) have included lentil in the group of pulses with low reactivity to the erythrocyte agglutination test and low toxicity

Table 9. Vitamin content of lentils

	(mg/100 g DM)	References
Thiamin	0.2–0.72	2, 5
Riboflavin	0.03–0.41	1, 2, 5
Niacin	1.24–1.29	2, 7
B_6	0.55–0.60	3, 5
Retinol	17–112*	4, 5
β-carotene	0.10	5
Biotin	0.132	4
Folic acid	0.03–1.5	4, 5, 8
Pantothenic acid	1,4–1.8	4, 5
Vitamin C	Nd-7	5, 6

*International Units (I.U.).

1. Vidal-Valverde and Frias (1993b); 2. Frias et al. (1995a); 3. Sierra and Vidal-Valverde (1997); 4. Savage (1988) Nutr Abs Rev 5, 319–343; 5. Souci et al. (2000) Food Composition and Nutrition Tables, CRC Press; 6. Frias et al. (2002); 7. Urbano et al. (1995); 8. Han and Tyler (2003) J Agric Food Chem 51, 5315–5318

Table 10. Non-nutritional content of lentils (expressed in dry matter)

	Range	it References
Trypsin and chimotrypsin inhibitors (U/mg)	2.7–6.1	1–9,36
α-amylase inhibitors (U/g)	2–18	10
Lectins (U/mg)	0.20–7.7	1,6,11,12
Tannins (mg/g)	< 0.5–10.9	1,5,8,9,11,12,13,14,15
Oxalate (g/Kg)	1.18–5.4	11,16,17
Phytic acid (g/100 g)	0.15–2.34	1,9,12,13,15,18–32
α-galactosides (g/100 g)	1.8–7.5	9,32
Saponins (mg/100 g)	40–127	33–35

1. El-Mahdy et al. (1985); 2. Vidal-Valverde et al. (1994); 3. Urbano et al. (1995); 4. Frias et al. (1995b); 5. Tabera et al. (1995); 6. Hernández-Infante et al. (1998); 7. Fasina et al. (2001); 8. Porres et al. (2003); 9. Wang and Daun (2006); 10. Jaffé et al. (1973) Nutr Rep Int 7, 169–173; 11. Savage (1988) Nutr Abs Rev 5, 319–343; 12. El-Adawy et al. (2003); 13. Khan et al. (1987); 14. Carbonaro et al. (1996); 15. Rehman and Shah (2005); 16. Massey et al. (2001); 17. Quinteros et al. (2003) Int J Food Sci Nutr 54, 373–377; 18. Bhatty (1989); 19. Bhatty (1989) Can Inst Food Sci Technol J 22, 137–142; 20. El-Tinay et al. (1989); 21. Lombardi-Boccia et al. (1991); 22. Ravindran et al. (1994) Food Chem50, 133–136; 23. Bhatty (1995); 24. Sharma et al. (1996) Nahrung/Food 40, 182–184; 25. Morris and Hill (1996) J Food Comp Anal 9, 2–12; 26. Agte et al. (1998) J Food Sci Technol Mys 35, 330–332; 27. Urbano et al. (1999); 28. Arntfield et al. (2001); 29. Egli et al. (2002); 30. Vidal-Valverde et al. (2002a); 31. Frias et al. (2003b); 32. Porres et al. (2004); 33. Frenwick and Oakenfull (1983) J Sci Food Agric 34, 186–191; 34. Ruiz et al. (1996) J Agric Food Chem 44, 1526–1530; 35. Ruiz et al. (1997); 36. Zhao et al. (2005) J Food Sci 70, S371–S376

when compared to other highly reactive legumes like *Phaseolus vulgaris*, *Phaseolus coccineus* or *Phaseolus acutifolius*. Cuadrado et al. (2002a) purified a lentil lectin extract and fed it to rats at a 5-fold higher dose than the level in lentils without observing any detrimental effect on growth, digestion or metabolism of protein compared to controls. It appears lentil lectin, like other mannose/glucose specific lectins, has a limited effect on the metabolism of experimental animals.

The detrimental effect of phytic acid arises from it forming complexes with protein and minerals in the small intestine making these nutrients poorly available for absorption (Cheryan, 1980). Tannins are polyphenolic compounds that form complexes with salivary and dietary protein and other food components, interfering with their availability and reducing the nutritive utilization of foodstuffs (Bartolomé et al., 1995; Carbonaro et al., 1996).

α-Galactosides are non-nutritional constituents of legumes formed mainly by raffinose, ciceritol, stachyose and verbascose. They are an important contributor to flatus (Granito et al., 2005). Reddy et al. (1984) reported that mung beans and green lentils are less flatulent than navy, kidney, red kidney, chickpea and peas. α-Galactosides are not digested by monogastric animals because the intestinal mucosa lack the hydrolytic enzyme α-galactosidase. These sugars are unable to pass through the intestinal wall (Rackis, 1975). The microflora in the lower intestinal-tract ferments them and produces large amounts of carbon dioxide, hydrogen and small quantities of methane and short chain fatty acids, thus lowering the

Table 11. Physiological effects of non-nutritional components

	Food Intake	Growth	Pancreas	Digestive utilization of protein	Mineral and vitamin bioavailability	Other effects
Trypsin and chimotrypsin inhibitors	ND	↓ (1, 2, 3, 4)	Hyperplasia (1, 2); pancreatic tumors (3, 4) ↑ pancreatic secretion (5)	↓ *In vitro* and *In vivo* (6, 7)	ND	Anti cancer
Lectins	↓ (8)	↓ (2)	Hypertrophy (4, 9, 10)	↓ *In vitro and In vivo* (11, 12)	ND	Intestinal Hypertrophy, Toxicity, Decreased activity of brush border enzymes
Tannins	↓ (14)	↓ (14, 15)		↓ nutritive utilization of protein and other nutrients (16–18)	↓	↑ production and secretion of salivary proteins (24) Antioxidant effects

Phytic acid		↓ *In vitro* (25,26)	↓ Fe; Zn; Ca; Mg; Cu; P; Phytate/Zn molar ratios (27–31)	↓ Glycemic index, Antioxidant, Anti cancer activity, Hypocholes-terolemic (32)
α-galactosides	↓ (33,34) flatus production			
Saponins				Hypocholes terolemic Anti cancer (35,36)

1. Pusztai et al. (1992) Br J Nutr 68, 783–791; 2. Grant et al. (1993) J Nutr 123, 2207–2215; 3. Grant et al (1995) Br J Nutr 73, 17–29; 4. Oliveira et al. (2000) Food Chem 70, 185–191; 5. Laporte and Tremolieres (1973) Nutr Metab 15, 192–206; 6. Leterme et al. (1990) Anim Feed Sci Technol 29, 45–55; 7. Rani et al. (1996) Nahrung/Food 40, 145–146; 8. Grant and Van Driessche (1993) Recent Advances of Research in Antinutritional Factors in Legume Seeds, Wageningen Pers; 9. De Oliveira et al. (1988) Nutr Res 8, 943–947; 10. Bardocz et al. (1989) Med Sci Res 17, 309–311; 11. Jaffé and Brucher (1972) Arch Latinoam Nutr 22, 267–281; 12. Thompson et al. (1986) J Food Sci 51, 150–153; 13. Deglaire et al. (2006) J Agric Food Chem 54, 5197–5202; 14. Mole et al. (1993) Biochem Systemat Ecol 21, 667–677; 15. Jambunatham and Mertz (1973) J Agric Food Chem 21, 692–696; 16. Jansman et al. (1995) J Anim Sci 73, 118–127, 17. Yoneda and Nakatsubo (1998); 18. Mateus et al. (2004) Anal Chim Acta 513, 135–140; 19. Brown et al. (1990) Nutr Res 10, 343–353; 20. Matuschek et al. (2001) J Agric Food Chem 49, 5630–5638; 21. Matuschek and Svanberg (2002) J Food Sci 67, 420–424; 22. Matuschek and Svanberg (2005) Food Chem 90, 765–771; 23. Towo et al. (2006) Food Chem 94, 369–376; 24. Bacon and Rhodes (2000) J Agric Food Chem 48, 838–843; 25. Singh and Krikorian (1982); 26. Kies et al. (2006) J Agric Food Chem 54, 1753–1758; 27. Cheryan (1980); 28. Morris and Ellis (1980) J Nutr 110, 1037–1045; 29. Fordyce et al. (1987) J Food Sci 52, 440–444; 30. Zhou et al. (1992) J Nutr 122, 2466–2473; 31. Porres et al. (2005); 32. Rickard and Thompson (1977) ACS Symp Ser 662, 294–312; 33. Fleming (1981); 34. Granito et al. (2005); 35. Sidhu and Oakenfull (1986); 36. Konoshima et al. (1992)

pH (Rackis, 1975). Fleming (1981) reported a significant positive correlation between intestinal gas production and the content of α-galactosides in legume seeds. However, α-galactosides are considered as prebiotics since as they resist hydrolysis by digestive enzymes and are not absorbed in the upper part of the gastrointestinal tract, they pass into the large bowel and promote the growth of *Bifidobacterium* and *Lactobacillus* (Roberfroid, 2002; Martínez-Villaluenga et al., 2007).

Lentil has a lower saponin content (75–127mg/100g) than soybean or kidney bean (650 and 350 mg/100g, respectively) but a higher level than lupin (38–74 mg/100g) (Ruiz et al., 1997). Saponins have a bitter taste, foam in aqueous solutions and haemolyze red blood cells. However, they also have a benefit in lowering plasma cholesterol levels in humans (Sidhu and Oakenfull, 1986) and have anticancer activity (Konoshima et al., 1992).

3. EFFECT OF DIFFERENT PROCESSING CONDITIONS ON THE CHEMICAL COMPOSITION OF *LENS CULINARIS*

To improve nutritional quality and utilization of lentils as food products, seeds are processed by several methods to remove the undesirable compounds. Physical and chemical methods are used (e.g. soaking, cooking, germination, fermentation, selective extraction, membrane filtration, irradiation and enzyme treatments). These treatments lead to a significant reduction or total elimination of non-nutritional compounds. Breeding programs have also been used, however, progress is slow as there are many different non-nutritional compounds at high levels in raw seeds (Table 10 and 11).

Germination and fermentation processes lead to catabolism of the lentil seed components whereas other processes (e.g. cooking) may cause thermal degradation or may involve extraction of non-nutritional components.

3.1. Soaking and Cooking Processes

Soaking and cooking are employed at both household and industry levels. They improve palatability and destroy heat-labile and heat-resistant non-nutritional components that are able to interfere with the nutritive utilization of protein (Vidal-Valverde et al., 1994). These processes may also reduce total nitrogen, non-protein nitrogen, and minerals, due to leaching into the soaking or cooking solution (Shekib et al., 1985; Rehman and Shah, 2005) (Table 12). Other authors did not observed any substantial loss and have described a considerable increment in total nitrogen (Manan et al., 1987; De Almeida Costa et al., 2006) and ash (Candela et al., 1997) content of lentils after the soaking/cooking process. Candela et al. (1997) studied the effect on chemical composition due to holding cooked lentils at 65 ° C for 3 hours. They observed no considerable changes in the nutrient composition with the exception of certain amino acids like Hys, Leu, Lys, Phe, Thr and Val. Hernández-Infante et al. (1998) reported that both traditional and microwave cooking caused a significant reduction in the content of available lysine

Table 12. Effect of thermal processes on the nutrient composition of lentils (expressed in dry matter)

Reference	Shekib et al. (1985) Decorticated/cooked 30 min		Candela et al. (1997) Soak 12 h/cook 3 h			Urbano et al. (1995, 1999) Dry heat		Rehman and Shah (2005) Normal/Autoclave cooking			De Almeida Costa et al. (2006)	
	Raw	Cooked	Raw	Cooked	Held-warm	Raw	Dry heat	Raw	Norm.	Auto.	Raw	Soaked/cooked
TN (%)	4.88	4.52	4.27	4.05	4.22	4.05	4.01	3.70	3063	3.54–3.63	3.81	4.34
NPN (%)	0.58	0.45										
Starch (%)	67.0	69.5‡	33.8	27.2	30.7	48.7	38.6	42.8	42.4	41.0–42.3	56.4♧	61.8♧
Fat (%)	1.23	0.86	3.59	10.11	10.7	1.3	1.3				2.15	2.36
Fiber (%)	2.45	2.33	35.11@	24.38	26.04	20.6*	15.5				19.0#	21.4
			3.75	7.65	10.21	4.8	11.7				1.44	1.37
			31.36	16.63	15.83	1.3	3.1					
Ash (%)	2.46	2.41	3.21	6.12	5.34							

♧Carbohydrate content.
TN = Total Nitrogen,
NPN = Non-Protein Nitrogen
‡ Nitrogen-free extract
@ Total; soluble; insoluble.
* neutral detergent fiber (NDF), acid detergent fiber (ADF), Lignin # insoluble dietary fiber (IDF) and soluble dietary fiber (SDF)

from lentils. This was in spite of a considerably reduced amount of trypsin inhibitor activity and lectins. Pirman et al. (2001) observed an increase in the content of all essential and non essential amino acids with the exception of Thr as a result of cooking lentils (Table 13).

Carbohydrates.- Soaking and cooking decrease the content of soluble available sugars and α-galactoside oligosaccharides from lentils (Vidal-Valverde et al., 1993a; Vidal-Valverde et al., 2002b). The degree of changes after soaking depend upon several variables including temperature, length, illumination, lentil/water ratio, elimination of the soaking liquid and milling of the lentil seeds. After soaking of lentils for 9 hours in water, citric acid, or bicarbonate solution, Vidal-Valverde et al. (1992b) and Frias et al. (1995a) reported metabolic changes similar to those taking place during germination that led to a significant decrease in α-galactoside oligosaccharides and increments in fructose and glucose, but not sucrose. These authors also reported an increase in the content of riboflavin and the ratio of available to total starch, observations that could imply the higher nutritional value of soaked lentils. The presence of light during the soaking process had a significant effect on the changes of some components like sucrose, raffinose, ciceritol and stachyose, but no effect on total and digestible starch, hydrosoluble vitamins, or dietary fiber content. Sugars such as fructose, glucose and sucrose decreased significantly during cooking. Autoclaving caused a slight but significant decrease in total starch content of cooked lentils (Rehman and Shah, 2005) and improved significantly the *in vitro* digestibility.

Table 13. Effect of cooking process on the essential and non essential amino acid composition of lentils (g/16 g N)*

Amino Acids	Raw Lentil	Cooked Lentil
Non Essential		
Ala	3.65	3.90
Arg	6.36	7.68
Asp	9.46	10.00
Glu	12.92	13.40
Gly	3.59	3.83
His	2.09	2.20
Pro	3.50	3.75
Ser	3.90	4.01
Essential		
Cys	0.07	0.09
Ile	3.64	4.08
Leu	6.57	7.10
Lys	5.84	6.34
Met	0.59	1.24
Phe	4.67	5.06
Thr	3.33	2.95
Tyr	1.40	2.73
Val	4.02	4.53

*Taken from Pirman et al. (2001)

Frias et al. (2003a) established the optimal soaking conditions of lentils for an adequate action of endo and exo α-galactosidase activity and studied the kinetics of soluble carbohydrates as a result of the activity of these enzymes. Soaking led to a significant reduction in the content of sucrose and α-galactoside oligosaccharides matched by an increment in the levels of fructose, glucose and galactose. Therefore, soaking could result in the development of lentil-derived flours with low flatulence and high functionality.

Fat.- Cooking can significantly reduce total fat and fatty acid content of lentils (Shekib et al., 1985; Pirman and Stibilj, 2003) (Tables 12 and 14). The reported losses can be of greater nutritional importance if lentils are consumed after discarding the soaking/cooking solutions. However, other normal additives during cooking (e.g. olive oil replacing the fat content, Candela et al., 1997) may overcome processing losses (Table 12).

Fiber.- Vidal-Valverde and Frias (1991) have reported a significant decrease in the levels of Neutral Detergent Fibre (NDF) and hemicellulose and no major effect of normal and pressure cooking on the Acid Detergent Fiber (ADF), cellulose or lignin content of lentils when expressed on a dry matter basis. Vidal-Valverde et al. (1992a), found soaking lentils in water prior to cooking did not result in any major alteration in the NDF or ADF fractions of lentils, whereas soaking in 0.1% citric acid or 0.07% sodium bicarbonate caused a significant increase in NDF and hemicellulose, with hardly any effect in ADF, cellulose or lignin content. In

Table 14. Effect of cooking process on the fatty acid composition of lentils*

Fatty acid (mg/100 g DM)	Raw lentil	Cooked lentil
14:0	6.3	2.5
15:0	3.8	1.6
16:0	343	161
16:1, n-7	1.5	0.7
17:0	3.4	1.7
17:1, n-7	2.1	0.5
18:0	34.3	17.9
18:1, n-9	628	324
18:2, n-6	1172	619
18:3, n-3	382	198
20:0	10.6	5.9
20:1, n-9	21.3	12.1
22:0	–	–
Total fatty acids	2608	1344
Total saturated	401	191
MUFA	653	337
PUFA	1554	816
Proportion n-6: n-3	1:0.33	1:0.32
Total fat (g/100 g DM)	3.26	1.68

*Taken from Pirman and Stibilj (2003)

contrast, cooking after either one of the soaking processes led to a significant decrease in the levels of NDF and hemicellulose, and significantly enhanced the content of ADF, cellulose and lignin. Ramulu and Udayasekhara Rao (1997) have reported a considerable reduction in total and insoluble dietary fiber content of lentils as a result of dehusking, and a considerable increase in the afore mentioned dietary fiber fractions as a consequence of cooking, whereas none of the processing techniques mentioned caused any major effect on soluble dietary fiber content of lentils. In contrast, Candela et al. (1997) reported a considerable increase in soluble dietary fiber and a significant reduction in the insoluble dietary fiber content of lentils as a result of cooking and warm-holding.

Vitamins.- The content of hydrosoluble vitamins in lentils tended to be decreased or did not change after a short soaking process with the exception of the levels of vitamin B_2 and available niacin under some experimental conditions with ground lentils (Vidal-Valverde et al., 2002b). The effect of different pH of the soaking solution and cooking process on the content of thiamin, riboflavin and available niacin of lentils was studied by Prodanov et al. (2004). These authors did not observe any effect of the different soaking solutions on thiamin content, whereas the levels of riboflavin increase in soaked lentils irrespective of the solution pH and available niacin levels decreased more pronouncedly in water-soaked than in citric acid or bicarbonate-soaked lentils. Cooking caused a significant reduction in thiamin, riboflavin and available niacin of soaked lentils with the exception of the latter vitamin in citric acid or bicarbonate-soaked lentils.

Minerals.- Shekib et al. (1985) have observed a considerable loss of crude ether extract, Ca, P, Fe, Na, K, and Zn content after decorticating and cooking lentils in boiling water for 30 min, whereas hardly any change was observed in the total and non protein nitrogen, crude fiber, ash or nitrogen-free extract content. El-Tinay et al. (1989) studied the effect of soaking and cooking on the nutrient composition of lentil (Table 15). Soaking led to leaching losses of all the minerals studied that

Table 15. Effect of soaking and cooking processes on mineral retention from lentils

	Shekib et al. (1985) mg/100 g DM		El-Tinay et al. (1989) % retention	
	Raw	Cooked	Soaking	Cooking
Ca	42	15.4	78	56
P	380	204	75	48
Fe	7.2	2.4	78	58
Na	4.6	3.8	61	45
K	753	213	75	42
Zn	5.11	3.13	84	42
Cu			91	64
Mg			82	56
Mn			80	60
Phytate			100	67

were further aggravated by cooking. Retention of Cu after the soaking/cooking tended to be higher than that of other minerals while Na showed the greatest lossess. Soaking did not cause a major loss of phytic acid, whereas cooking did result in a considerable loss of this non-nutritional component. Similar results regarding the effect of cooking on phytic acid content of lentils have been reported by other authors (Gad et al., 1982;. Manan et al., 1987; Vidal-Valverde et al., 1994). Loss of phytic acid with soaking and cooking has also been reported in bean and faba bean (Chang et al., 1977; Fernández et al., 1997). Frias et al. (2003b) studied the phytase-catalyzed inositol phosphate degradation in lentils, and found processing conditions which caused a marked reduction in inositol hexa- and pentaphosphate, but had no major effect on inositol tetra- and triphosphate content. Candela et al. (1997) observed a significant increase in the ash content of lentils ascribed to contamination with minerals from the equipment/water used.

3.2. Canning

Martin-Belloso Llanos-Barriobero (2001) reported that proximate composition of canned lentils did not vary significantly with processing time and location where the processing was done with the exception of Na, Ca, (due to brine addition), K (due to soaking in potassium metabisulfite) and ascorbic acid content. There was considerable retention of most nutrients (protein, carbohydrates, fiber, minerals and vitamins) during the canning process (Table 16). Vitamin B_2 was the most sensitive soluble vitamin to canning followed by vitamin B_1, vitamin C, and vitamin B_6.

3.3. Cooking Quality of Lentils

Cooking time is an important factor used to define cooking quality. Others include texture, flavor and appearance of lentil-based foods. Cooking time is affected by genetic, storage, environmental and sowing factors. Long storage periods at relatively high moisture/temperature conditions resulted in increased cooking times. Treatments like soaking overnight, irradiation or tempering can improve cooking characteristics (Khan et al., 1987; Arntfield et al., 2001; Çelik et al., 2004). Iliadis (2001; 2003) found cooking time can be affected by lentil genotype, harvesting process (early harvesting was more desirable for shorter cooking time and lower seed loss due to pods shattering), climatic conditions (rainy years induced longer cooking times than dry years), and by storage conditions, with an increase in the cooking time of stored lentils that was more pronounced in the late harvested lentils. Wet conditions during storage tended to further lengthen the cooking time, especially of those lentils genetically prone to need a longer cooking time. Finally, the influence of genotype was stronger than the influence of soil type for cooking quality of lentils.

The texture of micronization-cooked legumes softened as tempering moisture increased. This was related to increased starch gelatinization and decreased protein solubility during micronization. A reduction in phytic acid and NDF was observed, and the strength of cell walls was also affected. Changes in pectic substances did not appear to influence texture. Addition of different salts to the tempering solution (2% sodium tripolyphosphate, 150 ppm sodium EDTA, mixture of 1% citric acid/2% ascorbic acid) reduced cooking time of lentil when tempering at 40% moisture was carried out, but not when the moisture content was 20%. Arntfield et al. (2001) studied the effect of internal temperatures reached inside the lentil seed during the micronizing process. If micronizing temperature is too high (170 °C), an increase in cooking time and darkening of the seed is also observed relative to a lower micronization temperature (138 °C).

Bhatty (1984) studied the relationship between physical and chemical characteristics and different environmental conditions on the cooking quality of lentils. Location and season of growth and cultivar all had a major influence on the cooking quality of lentil. Correlations were obtained between cooking quality and $Ca^{2+} + Mg^{2+}/P$ ratios. This ratio will depend upon the fertilization, availability and uptake of soil P by the lentil plants. The implications of the correlations are not clear but may relate to some role that P, particularly phytic acid-P, may play in determining the cooking quality of lentil. Higher peak and set-back viscosities of good cooking lentil meals indicated some role of starch and/or other components in affecting the cooking quality of lentil. Erskine et al. (1985) found a positive genetic correlation (0.919) between cooking time and seed size suggesting seed size can be used to predict cooking quality.

Table 16. Effect of canning process on the nutrient composition of lentils*

	Nutrient composition of canned lentils	% nutrient retention and uptake
Moisture (%)	79.4 ± 1.9	–
Protein (%)	6.4 ± 0.7	103
Carbohydrate (%)	10.4 ± 1.0	85
Fat (%)	0.20 ± 0.02	–
Fiber (%)	2.4 ± 0.3	101
Ash (%)	1.26 ± 0.16	232
Na (%)	0.58 ± 0.14	1030
K (%)	0.15 ± 0.002	142
Fe (mg/100 g)	1.24 ± 0.15	101
P (mg/100 g)	58 ± 6	–
Ca (mg/100 g)	17 ± 3	220
Mg (mg/100 g)	14.3 ± 2.2	101
Vitamin B_1 (mg/kg)	0.92 ± 0.13	71
Vitamin B_2 (mg/kg)	0.892 ± 0.009	68
Vitamin B_6 (mg/kg)	0.95 ± 0.12	82
Vitamin C (mg/kg)	10.79 ± 0.5	74

*Taken from Martin-Belloso and Llanos-Barriobero (2001)

Bhatty (1989), concluded that phytic acid was the major determinant of shear force and, hence, cooking quality of lentil. There seems to be a strong positive relationship between phytic acid expressed either as a percentage of the seed weight or seed P, and cooking quality of lentils grown under diverse conditions. Therefore, any condition that lowers the phytic acid content of lentil would likely result in a decrease of their cooking quality. Soaking of poor cooking lentils in phytic acid or EDTA solution made them extremely soft, probably by removing Ca^{2+} from the middle lamellar region during the cooking process. The poor cooking condition persisted in dehulled lentil, suggesting its origin in the cotyledon where Phytic acid is located rather than in the seed coat. The cell wall pectin may play a complementary role in the poor-cooking condition of lentil.

Lack of cell separation and cell wall dissolution is typical of poor cooking lentils (Bhatty, 1990) that showed some evidence of a "lignification-like" mechanism at the cell junctions in the middle lamella. On the other hand, the cell wall was isolated in similar yields from good and poor cooking lentils, and showed a similar chemical composition. Means for hydration coefficient, solids lost, or water absorption were greater for the good-cooking lentil than for the poor-cooking lentil (Bhatty, 1995). The seed coat retarded water uptake and prevented solid loss in the poor-cooking lentils, thus increasing their shear force. However, when the seed coat composition was studied, no major differences were found in the chemical composition in terms of lignin, starch, protein or non-starch polysaccharides, although differences were found in the structure of the seed coat that would impede water uptake during cooking of the poor-cooking lentils. In apparent contrast to the previously described results, Ereifej and Shibli (1995) reported that physical and chemical properties and chemical composition were not related to differences in cookability of three newly developed lentil cultivars or a landrace cultivar (35 min to cook at 100 °C without previous soaking).

3.4. Dry-heat and Irradiation Treatments

With dry-heating processes lentil seeds have no direct contact with an external source of water. In these circumstances total nitrogen content was unaffected (Fasina et al., 2001; Urbano et al., 1995; Carbonaro et al., 1997; Porres et al., 2003), although protein solubility decreased due to thermal denaturation (Fasina et al., 2001; Sanz et al., 2001). Autoclaving significantly increases available carbohydrates like fructose or sucrose and decreases the total starch content (Table 12), not affecting the content of α-galactoside oligosaccharides or hydrosoluble vitamins thiamin, riboflavin and niacin (Urbano et al., 1995). Dry-heating caused a significant reduction in trypsin inhibitor activity in lentil seeds. Infrared heating of lentils did not cause any major effect on total or gelatinized starch or Trypsin inhibitor activity, but did significantly reduce the amount of α-galactoside oligosaccharides (Fasina et al., 2001).

Gamma irradiation is a food preservation technique that offers the potential to protect cereal grains and legumes from insect infestation and microbial contamination during storage. Furthermore, irradiation can significantly alter the functional properties and increase the nutritional value of legume seeds due to inactivation of non nutritional components with antinutritional effect. Gamma irradiation of lentil seeds led to slight differences in the SDS-PAGE patterns of salt-insoluble proteins, causing a slight reduction in band density that could be indicative of minor radiolytic breakdown (Çelik et al., 2004). A general tendency to have higher hydration and swelling index values and lower cooking times were observed that were radiation dose-dependent.

3.5. Germination

Germination mobilizes reserve nutrients required for growth and contributes to the removal of some of the non-nutritive seed compounds such as phytic acid and α-galactosides. At the same time, the content of some other nutrients like minerals and vitamins such as B group and ascorbic acid as well as sugars (fructose, glucose and sucrose, Vidal-Valverde and Frias, 1992; Frias et al., 1996a) can be enhanced. Furthermore, seed sprouts present better organoleptic properties than the dry seeds. During germination, seed enzymatic systems activate and protease activity increases (Urbano et al., 2005a,b) and, consequently, protein nitrogen content decreases whilst non-protein nitrogen, peptides and free protein and non-protein amino acids increase (El-Mahdy et al., 1985; El-Adawy et al., 2003; Rozan et al., 2000; Kuo et al., 2004; Urbano et al., 2005a, b) (Table 17).

This protein pre-fractioning carried out in legume seed can be used into the preparation of dietetic products for population with gastro-intestinal dysfunctions. Hydrolysis and utilization of carbohydrates also takes place during germination and starch digestibility rises in lentil sprouts further increasing digestibility (Vidal-Valverde and Frias, 1992; Frias et al., 1998). Germination also alters the starch granule surface resulting in a higher resistance to temperature changes (Frias et al., 1998). Hemicellulose content is reduced during germination whilst cellulose and lignin increase (Vidal-Valverde and Frias, 1992). Loss of major constituents of the seed (protein and starch) during germination, gives an apparent increase on a dry weight basis of some seed constituents (Urbano et al., 1995; Kavas and Nehir, 1992; El-Adawy et al., 2003; El-Mahdy et al., 1985) (Tables 17 and 18).

Phytic acid decreases during lentil germination since naturally occurring phytases are activated and phytate is progressively degraded (Egli et al., 2002) leading to a decrease in its mineral-binding strength (Sandström and Sandberg, 1992). Vidal-Valverde et al. (2002a) have reported phytate degradation to lower inositol phosphate forms (penta- to triphosphates) using germination periods of 2, 4, and 6 days with or without light. These authors conclude that in order to maximize the beneficial effects of germination on the content of nutrients and non-nutritional components, the germination process should be carried out in the presence of light for a period of 6 days. Germination also seems to reduce hemagglutinating activity possibly due to proteolysis of lectins during germination (Chen et al., 1977).

Table 17. Effect of germination on the chemical composition of lentils

Reference	Kylen and McReady (1975)		El-Mahdy et al. (1985)		Kavas and Nehir (1992)		Danisová et al. (1994)		El-Adawy et al. (2003)		
	Raw	G72h	Raw	G96h	Raw	G96h	Raw	G48–96h	Raw	G72h	G120h
(kJ per 100 g Dry Matter)											
Energy	1572	1592			1492	1647					
(Percent Dry Matter basis %)											
Total N			4.07–4.72	4.63–5.14	4.8	5.33	4.03	4.06	5.02	4.69	4.54
Protein N			3.58–4.09	3.16–3.55							
Non Protein N			0.49–0.63	1.47–1.59					0.54	0.93	1.37
Fat	1.77	1.09					1.56	0.73	1.15	1.09	0.93
Starch									46.2	44.5	43.5
Fiber	5.1	4.0					4.33	6.24	6.75	8.14	9.88
Ash	2.87	2.93	2.62–3.18	2.99–3.84	2.62	2.85	3.73	3.30	4.16	5.87	6.97
(mg mineral per 100 g Dry Matter)											
Na			4.30–6.90	4.60–7.10					11.0	11.6	11.7
K			143–157	180–200					240	280	300
Ca	36.5	44.0	40	45.0–50.0			78.6	73.1	76	86	94
Mg									48.5	49.2	49.7
Total-P			153–247	146–239			239	238	372	383	392
Phytate-P			120–195	60.4–96.8							
Inorganic-P			13.9–19.9	63.0–85.0							
Organic P; non phytate			13.1–38.1	22.6–57.0							
Fe	14.2	11.0	8.9–11.0	8.10–10.0	7.9	9.03	3.73	2.38	8.50	8.70	8.80
Mn									1.80	2.00	2.10
Zn	5.09	5.49	2.50–3.00	2.90–3.60	3.68	4.77					

Table 18. Effect of germination on the essential and non essential amino acid content of lentils (g/16 g N)

Amino Acids	Kavas and Nehir (1992)	Urbano et al. (1995)	Danisová et al. (1994)
Non Essential	Raw G96h	Raw G144h	
Ala	3.47 ± 0.16 3.67 ± 0.54	2.37 3.55	
Arg	8.68 ± 1.75 7.64 ± 0.98	3.90 3.82	
Asp	11.56 ± 0.54 17.47 ± 1.86	9.94 11.97	
Glu	14.58 ± 0.00 12.73 ± 1.86	16.30 16.40	
Gly	3.36 ± 0.00 3.06 ± 0.52	4.43 4.56	
His	–	1.33 1.53	
Pro	1.19 ± 0.59 2.29 ± 0.03	2.58 2.72	
Ser	4.38 ± 0.16 5.01 ± 0.67	2.86 2.75	
Essential	Raw G96h	Raw G144h	Raw G48–96h
Cys	1.12 ± 0.00 1.28 ± 0.02	1.46 1.41	–
Ile	3.44 ± 2.70 4.09 ± 0.43	4.04 4.41	3.07 3.02
Leu	6.13 ± 0.74 7.30 ± 1.06	7.24 8.39	6.13 6.25
Lys	6.35 ± 0.52 7.15 ± 1.20	4.27 5.20	6.91 6.93
Met	0.22 ± 0.00 0.86 ± 0.03	1.10 1.30	0.10 0.35
Phe	4.34 ± 0.29 4.90 ± 0.50	5.52 6.34	4.61 4.65
Thr	2.69 ± 0.64 3.60 ± 0.57	2.49 2.46	3.68 3.71
Tyr	2.24 ± 0.00 3.25 ± 0.25	1.13 1.11	–
Val	4.19 ± 0.20 4.39 ± 0.64	3.04 3.39	4.62 4.38

The reduction of enzyme inhibitor activities during seed germination have been widely reported. Frías et al. (1995) found a progressive but slow decrease on trypsin inhibitor activity (TIA) after 6 days of lentil germination (12% reduction) followed by a faster decrease (up to 45% of control seed value) after 10 days of germination. Similar results were obtained by El-Mahdy et al. (1985). A considerable reduction in chymotrypsin and α-amylase inhibitor activity has been reported after 5-days germination by Sathe et al. (1983).

Vitamin content is also modified as consequence of germination process and while thiamine seems to decrease slightly, riboflavin and available niacin increases sharply as a consequence of germination (Kavas and Nehir, 1992; Urbano et al., 1995; Prodanov et al., 1997; Vidal-Valverde et al., 2002a). Similarly, it has been reported that lentil sprouts are a good source of vitamin C and E (Frias et al., 2002)

3.6. Fermentation

Fermentation is one of the oldest and most economical methods of food production and preservation known. Fermentation can be spontaneously initiated with the microbiota naturally present in the legume or controlled by the use of specific cultures or starters from a batch of previously fermented product. Legumes are fermented to improve their sensory characteristics such as flavor and taste, and to enhance their nutritive value by improving the density and availability of nutrients (Vidal-Valverde et al., 1993b; Svanberg and Lorri, 1997). This can be achieved by degradation of antinutritional factors, by the pre-digestion of certain food components and by the synthesis of promoters for absorption (Frias et al., 1996b). In addition, by increasing the titratable acidity and reducing the pH of the food to levels below 4.5, fermentation precludes the proliferation of contaminating acid-intolerant species of bacteria and fungi.

Changes occurring during the fermentation process are mainly due to endogenous enzymes of the seed and the enzymatic activity of the microflora present in the legume. Differences in the final nutritional value of the fermented product depend on whether the whole seed or flour suspensions with different ratios of flour to water (Kozlowska et al., 1996) are used. Fermentation may be combined with other culinary (soaking and cooking) and technological (germination prior to the fermentation process) treatments to further improve the nutritional value of the fermented product.

Tabera et al. (1995) have studied the effect of natural fermentation for periods of 0–4 days at 42 °C and 79–221 g/L on the nutrient and antinutritional factor content of lentils. The athors reported a slight increase in the total nitrogen content of fermented lentils (Table 19) and a considerable reduction in the trypsin inhibitor activity (62.5%) and tannin/catechin ratio. Changes that were reflected in only a slight increase or no effect on the *in vitro* protein digestibility of the fermented product. Using similar fermentation conditions, Cuadrado et al. (1996, 2002b) obtained a considerable decrease in the levels of lectins and inositol phosphates present in lentils, thus increasing their nutritive potential.

After natural fermentation of lentils for 4 days at 30 °C, Vidal-Valverde et al. (1993b) found a reduction in pH, and in the content of trypsin inhibitor

Table 19. Effect of fermentation on the chemical composition of lentils

Reference	Vidal-Valverde et al. (1993b)		Tabera et al. (1995)		Vidal-Valverde et al. (1997)	
	Raw	4days	Raw	4days	Raw	4days
pH	6.0	3.8				
Titratable acidity	1.08	7.38				
Total Nitrogen (%)			4.12	4.78		
Total starch (g/100 g)	52.3	31.5				
Available starch (g/100 g)	44.8	28.6				
Fiber (%)	19.2	14.1				
Thiamin					0.387	0.370
Riboflavin	0.03	0.09			0.086	0.151
Available niacin					1.30	2.36
TIA (TIU/mg DM)	2.66	1.57	5.09	3.63		
Tannins (mg/g DM)			3.16	6.26		
Catechins (mg/g DM)			0.12	6.67		
α-galactosides (sum)	3.16	ND				

ND = Not detectable; TIA = Trypsin inhibitor activity

activity, sucrose, raffinose, ciceritol, stachyose, total and available starch, NDF, cellulose, and hemicellulose, whereas a considerable increase was found in the titratable acidity and the content of riboflavin, fructose, lignin, and the ratio of available to total starch. The effect of natural fermentation of lentils on the content of fructose and α-galactoside oligosaccharides was further corroborated by Frias et al. (1996b) using different temperatures and flour/water ratios. These authors also found a considerable increase in glucose content of fermented lentils. Using the same fermented samples, Kozlowska et al. (1996) obtained up to 70–75% reduction in the inositol phosphate content. Sotomayor et al. (1999) have reported a considerable decrease in starch content at 0 hours that was further reduced after 96 hours of natural fermentation. Endocorrosion seemed to be the main reaction pattern observed in the lentil starch granule as a consequence of the fermentation process.

Vidal-Valverde et al. (1997) studied the influence of natural fermentation of lentils on the kinetics of thiamin, riboflavin and niacin. Natural fermentation significantly enhanced the content of riboflavin, total, and available niacin content of lentils, but had no effect or caused a slight reduction in thiamin content depending on the different fermentation conditions employed.

3.7. Hydroalcoholic Extraction of Non-nutritional Components

Sanz et al. (2001) developed an extraction procedure to reduce non nutritional components of lentil. They used 80% ethanol at room temperature or 50 °C for different periods of time (1–3 hours). Non-protein nitrogen content (12.9% of total nitrogen) was significantly affected by the different extraction temperatures (increasing at room temperature but falling at 50 °C Total nitrogen increased under all processing conditions. However, nitrogen solubility at pH 3, 5 or 7 was decreased as a result of the extraction process; a reduction that was further aggravated by extraction at 50 °C when compared to extraction at room temperature. The extraction process did not significantly affect the content of amino acids Lys, Hys or Tyr.

3.8. Aqueous Extraction of Protein for the Preparation of Isolates and Films

Monsoor and Yusuf (2002) have described the preparation of protein isolates from lentils, chickpeas or lathyrus beans using and alkaline (pH 11) extraction process followed by isoelectric precipitation of protein at pH 5.4. The authors found a 79.3% protein yield (extractability) for lentil, the highest of the three legumes studied, and a 90.7% protein content in the isolate with an *in vitro* protein digestibility of 95.2% that was very similar to a highly-digestible reference protein like casein and higher than the values found in the protein isolates from chickpeas or lathyrus beans. The thermal treatment of lentil protein isolate at 100 °C for 5 minutes led to a slight improvement in the *in vitro* protein digestibility of the final product.

Lombardi-Boccia et al. (2003) have established processing conditions intended for isolation of legume storage proteins (globulins G1 and G2); using this protocol, the authors obtained a lentil protein isolate with a protein, Fe, and Zn content of 89%, 9.0 and 9.4 mg/100g DM, respectively, in comparison to the 22.6%, 6.8, and 3.4 mg/100g DM originally present in the raw lentil flour. In addition, Fe dialyzability experienced a considerable improvement in the protein isolate when compared to the raw lentil flour in contrast to Zn dialyzability, which suffered a significant decrease as a result of the protein isolation process.

Bamdad et al. (2006) have applied a lentil protein isolate to the preparation of proteinous films oriented to packaging food products with the aim of ensuring a longer preservation process. The authors mixed an aqueous solution of lentil protein (5%) with glycerin to obtain a protein film in which they measured several physicochemical properties like color, tensil strength and percentage elongation, thickness, puncture strength, water vapor permeability, water content or soluble matter. The proteineus film shared similar properties with other soy, pea or whey proteineus films, and could be a feasible way to enhance the industrial use of lentil-based protein isolates.

3.9. Functional Properties

Lentil protein solubility is susceptible to a high degree of variation depending on the pH of the extraction solution used. In a similar way to other legumes, lentil protein exhibits a relatively wide isoelectric pH range of 4.0–6.0 in which the lowest solubility is observed. Protein solubility is sharply increased at both sides of the isoelectric pH range due to its high proportion of acidic and basic amino acids, reaching values that are close to 100% when the pH of the extraction solution is in the range of 10–12 (Carbonaro et al., 1997; Fasina et al., 2001; Sanz et al., 2001). Protein solubility from lentils and other pulses can be negatively affected by several technological treatments like thermal treatments or organic solvent extraction (Carbonaro et al., 1997; Fasina et al., 2001; Sanz et al., 2001), or positively by dehulling or germination processes (Ghavidel and Prakash, 2006). Sadowska et al. (1999) have reported the enhancing effect of fermentation process on protein solubility of lentils in the acidic pH range (3–5), whereas protein solubility was decreased in the neutral to basic pH range when compared to the raw lentil without processing. The fermentation process also produced modifications in other functional properties of lentil flour like the emulsifying capacity and the water or oil absorption capacity, and the functional properties of lentil-wheat flour mixes, affecting their bread-making properties.

Due to the influence of protein solubility on other important functional properties like the emulsifying capacity or the foam or gel forming capacity, the solubility profile of lentil protein is of outmost importance for lentil inclusion in the preparation of different food products. Succinylation of lentil globulins has been shown to modify the isoelectric pH from 4.5 in the native globulins to 3.5 in succinylated proteins due to a higher negative charge of the later proteins (Bora, 2002); succinylation also improved protein solubility in the pH range of 4.5–8, but decreased it below pH 4. The functional properties of lentil globulins were affected by the succinylation process that improved the water absorption capacity, viscosity, emulsion activity and emulsion stability, decreasing the oil absorption capacity and not exerting any significant effect on foaming capacity or foaming stability. No effects on the functional properties of lentil globulins were observed as a result of different extents of succinylation.

Fernandez-Orozco et al. (2002, 2003) have studied the Superoxide Dismutase like-activity, peroxyl radical-trapping capacity, trolox-equivalent antioxidant capacity and the content of soluble protein and several low molecular weight antioxidants like carotenoids, ascorbic acid, polyphenols and reduced glutathione in different lentil varieties, as well as the effect on these parameters of two processing technologies like cooking and germination. In general, SOD-like activity varied within different lentil cultivars and was not affected by the germination process, although it was considerably reduced by the cooking process. The SOD-like activity corresponded mainly to enzymatic SOD, and only minor proportions of the total activity corresponded to the total phenol contents, albumin protein, ascorbic acid or reduced glutathione. With regard to the other lentil components studied, the amount of soluble protein increased slightly with the germination process and was significantly reduced by cooking. The content of phenolic compounds was slightly decreased by both processing conditions, whereas considerable losses were observed in tocopherol and reduced glutathione. Ascorbic acid was not present in raw or cooked lentils, whereas significant amounts of this component were detected in germinated lentils. The antioxidant capacity varied among the different lentil varieties tested, and decreased significantly as a result of germination, with only slight changes being found in cooked lentils when compared to raw lentils. The authors observed a markedly higher molar percentage contribution of phenolic compounds to the antioxidant capacity in raw or cooked lentils when compared to the other low molecular weight antioxidants or soluble proteins. In contrast, the contribution of tocopherols, ascorbic acid and reduced glutathione in germinated lentils was considerably higher.

Cari et al. (2002) prepared protein curds from 6 different legumes and studied the textural properties of the products obtained in relationship to their protein constituents. According to these authors, curds produced from soybeans, chickpeas, and faba beans exhibited a better texture and higher texture scores for hardness, springiness and cohesiveness than curds produced from lentils, smooth peas or mung beans due to the higher amount of 11S globulins with superior levels of sulfur-containing amino acids in the former three legumes. Zhao et al. (2005) prepared spaghetti with semolina containing 5–30% milled flours of green pea, yellow pea or lentil and evaluated physical-chemical properties, descriptive sensory and consumer acceptance characteristics. Lentil fortification decreased the lightness, shiny appearance, elasticity and overall quality of spaghetti products and there was a trend for the optimal cooking time to increase and significant increments in the cooking loss and firmness of the product. With regard to consumer acceptance, panelists liked the control spaghetti (100% wheat semolina) the most; spaghetti containing lentil flour had a lower color score than the products containing other 3 legume flours, but had higher flavor acceptability than spaghetti containing chickpea and yellow pea flours. General comments of the panelists indicated that legume spaghetti had a beany off flavor that could be related to the activity of lipoxygenase activity prior to drying and cooking of pasta product.

Banerjee et al. (2003) have designed different lentil-based extruded expanded products aimed at the preparation of snack foods, and studied the effect of temperature and moisture on process optimization. Incorporation of lentil to the snack food significantly enhanced the protein content of the functional product, and the extrudate properties could be significantly modified by different temperature and moisture conditions used in the preparation process designed by response surface methodology.

Lentil has been pin-milled and air classified into starch and protein fractions for incorporation into food and feed products and for industrial utilization (Gonzalez and Perez, 2002). The isolated lentil starch was microwave-irradiated and extruded in order to improve its functional and rheological characteristics. Both treatments caused internal granule compaction that was reflected in higher density and lower water absorption and swelling power of the modified starches. The restricted swelling and solubilization of processed starches could be responsible for the reduction of their amylographic viscosity. Moisture, crude protein and fiber contents were also reduced by the thermal treatments, which also reduced the retrogradation tendency of starch; that could be interesting from an industrial point of view, since starch retrogradation has been one of the limiting factors for a wider industrial use of legume-derived starches.

Dalgetty and Baik (2006) have prepared breads with wheat flour fortified with 3, 5, and 7% legume hulls or insoluble fiber, or with 1, 3, and 5% soluble cotyledon fibers isolated from pea, lentil, or chickpea flour. Inclusion of legume hulls or insoluble fiber led to increases in dough water absorption, mixing time and loaf weight, but decreased loaf volume when compared to control bread without legume hulls or insoluble fiber. Color and crumb uniformity was better in breads containing soluble fiber when compared to breads containing either hulls or insoluble fiber. In general, legume fiber or hull fortification of bread led to higher moisture content compared to the control bread.

4. BIOLOGICAL EVALUATION OF RAW AND PROCESSED *LENS CULINARIS*

4.1. Effect of Lentil Composition on Daily Food Intake

Legumes are seldom eaten raw, but are usually processed as previously described. Daily food intake of a raw lentil diet with a protein content in the range of 23–24.5% by growing rats was inferior to a casein-methionine control diet (Urbano et al., 1999; Porres et al., 2003). Lower food intake can be attributed to a range of factors (Mercer et al., 1989) as experimental animals respond with a voluntary suppression of food intake to a nutrient imbalanced diet. A range of possible causes are discussed below but the cause or causes are not yet confirmed.

Carbohydrates.- The content of palatable soluble sugars in lentils is usually low (Frias et al., 1996a,b) and could lower intake. However, the increase in soluble sugars in response to germination and fermentation processes could improve palatability and thus increase food intake. Excessive germination time (5–6 days or more), may hamper intake due to changes in food texture or food chemistry (Rozan et al., 2000). The high content of dietary fiber present in lentil may have contributed to the lower daily food intake obtained, due to its satiating effect (Slavin, 2005).

Protein level and protein quality of the diet.- Increased amounts of dietary protein can reduce food intake (Johnson and Anderson, 1982). However, the reduced food intake derived from lentil consumption appears to be related to protein quality rather than to the levels of protein in the diet. Dietary amino acid imbalances can decrease food ingestion (Harper et al., 1970), and specific relationships between amino acid ratios and daily food intake indicate these food constituents are active participants in the regulation of food consumption.

Minerals and vitamins.- Mineral and vitamin deficiency can lead to an important reduction in the nutritive utilization of food, and experimental animals show a depression in food intake as one of the initial symptoms of mineral-vitamin deficiency (O'Dell and Reeves, 1989). Lentils are generally deficient in certain essential minerals like calcium or iodine, and in liposoluble vitamins, choline, and niacine. Moreover, a sufficient amount of minerals and vitamins to meet the nutrient requirements does not ensure the availability of these nutrients for the animal. Supplementation of a mineral-vitamin premix to a raw lentil diet led to a significant increase in daily food intake by growing rats (Porres et al., 2003).

Fat.- The quantity and nutritional quality of fat present in the diet is known to have a significant influence on daily food intake (Le Magnen, 1983). Lipid peroxidation processes may cause a reduction in the organoleptic properties of food (Villaume et al., 1993). Unsaturated fatty acids are by nature more prone to become rancid by the action of lypoxygenase enzyme present in lentil (Maccarrone et al., 1997). However, the low levels of fat present in lentils suggest they are an unlikely cause of suppressed palatability

Non-nutritional components.- The presence of α-galactoside oligosaccharides in legumes has a negative effect on daily food intake due to the flatus-producing potential of these compounds. However, thermal treatment reducing α-galactoside of lentils has no effect on dietary intake (Urbano et al., 1995). Possibly these compounds are not present in sufficient quantities in lentils to play a major role in food intake. Dietary levels of α-galactoside oligosaccharides up to 3% do not exhibit any measurable flatulence effect, but do show important prebiotic benefits. Moreover, Maillard products formed during the thermal treatment can negatively affect the palatability of heat-treated foods (Sarriá et al. 2001) and might have affected the palatability of lentils, thus hindering the beneficial effects of removing α-galactoside oligosaccharides. While the astringent sensation produced by tannins can lead to a significant reduction of daily food intake levels found in lentils do not appear to be high enough to exert any significant effect.

4.2. Nutritive Utilization of Protein from Raw and Processed Lentils

4.2.1. Digestive Utilization

Digestive utilization of lentil protein determined *in vivo* using the growing rat as experimental model ranges from 62.3–92.0% (Table 20), using *in vitro* methodologies it varies from 37.6–49.6% and from 82.3–92.5% depending on the different *in vitro* experimental conditions used. The *in vivo* digestive utilization of lentil protein is higher than that found for the common bean (Nestares et al., 2001), similar to faba beans and chickpeas (Fernandez et al., 1996; Nestares et al., 1996), and lower than what has been reported for peas and lupin (Urbano et al., 2003; Porres et al., 2006).

Table 20. Digestive utilization of protein (%) from lentils and effect of technological processing methods

	Raw	Heated/ Cooked	Germination	Fermentation	Ext. EtOH
In vivo	75[1];75.6[7]; 79.0[8]; 62.3[10]; 79.0[11]; 65.7[13]; 76.1[15]; 80.7–84.2[16]	71.5[7]; 79.0[8]; 89.5[13]; 72.1–78.7[15] 81.7–85.1[16]	72.1[7]		
In vitro					
pH-stat	87.7[2]; 82.5[5]; 92.4[6]	81.66[5]		77.4–92.6[2]	91.4–92.8[6]
pH-drop	84.20[12]; 82.3–84.8[17]		85.15–87.53[12]		
Pepsin digestion	37.6[3] 41.5[4]	67.8–72.7[3], 47.3[4]			
Pancreatin digestion	25.0[4]	38.3[4]			
Pepsin + Pancreatin digestion	49.6[4]; 95.2[9] 74.8[14] 69.78[14]	63.9[4]; 97.9[9]	80.5[14] 92.5[14]		

1. Combe et al. (1991); 2. Tabera et al. (1995); 3. Rehman and Shah (2005); 4. Shekib et al. (1986); 5. Carbonaro et al. (1997); 6. Sanz et al. (2001); 7. Urbano et al. (1995); 8. Porres et al. (2003); 9. Monsoor and Yusuf (2002); 10. Cuadrado et al. (2002a); 11. Sarwar and Peace (1986); 12. El-Adawy et al. (2003); 13. Manan et al. (1987); 14. El-Mahdy et al. (1985); 15. Hernández-Infante et al. (1998); 16. Savage and Scott (1989); 17. Carbonaro et al. (1996)

The poorer digestive utilization of plant proteins when compared to animal proteins can be attributed to the specific structure of plant proteins that makes them more resistant to the attack by digestive proteinases, and to the presence in legumes of certain non-nutritional components that interfere with the activity of digestive enzymes and induce a higher secretion of endogenous nitrogen to the digestive tract. One of the main factors responsible for the higher losses of endogenous nitrogen derived from the consumption of legume-based diets appears to be the presence of lectins in legumes. Lectins bind to glycoproteins of the brush border membrane and increase the turnover of intestinal mucosal cells, thus inducing a higher nitrogen loss in feces as well as other metabolic effects. However, lectins present in lentils have been described to be of the mannose/glucose specific type (Rubio et al., 1995), and did not exert any particular effect on the digestive utilization of protein due to their low level in lentils and their low degree of reactivity (Grant et al., 1983; Hernández-Infante et al., 1998; Cuadrado et al., 2002a). Instead, the low digestibility of lentil proteins could be attributed to the reduced digestibility of lentil globulins and/or the inhibitory effect of the lentil seed coat (Cuadrado et al., 2002a).

Protease inhibitors (trypsin and chimotrypsin) present in lentils can interfere with the digestive utilization of protein by binding and blocking the activity of these digestive proteinases. This interfering action will lead to an incomplete digestion of dietary protein, and to a higher fecal excretion of endogenous nitrogen. In addition, the experimental animals have a higher requirement for sulfur containing amino acids in order to synthesize new digestive proteinases; amino acids that are limiting in legume seeds (Green and Lyman, 1972).

Tannins also exist in lentils which may interact with dietary protein and decrease its susceptibility to the attack by digestive proteinases. Furthermore, tannins themselves can interact with digestive enzymes interfering with their activity. Higher tannin/catechin ratios imply a higher degree of polymerization of tannins and, therefore, a lower digestive utilization of protein (Yoneda and Nakatsubo, 1998).

The role of phytic acid on digestive utilization of protein is not clear in view of the differences reported by several authors that either point out to a reduction in the activity of trypsin and chimotrypsin caused by phytic acid, or else do not find any relationship between a reduction in phytic acid content and improvements in the digestive utilization of protein *in vitro* or *in vivo* (Singh and Krikorian, 1982; Manan et al., 1987; Knuckles et al., 1989; Porres et al., 2005, 2006).

4.2.1.1. Effect of processing on the digestive utilization of protein

Thermal treatments.- Dry heating or autoclave treatment of non-soaked lentil seeds caused a reduction in the content of TIA, tannins and phytic acid, and increased the tannin/catechin ratio, without significantly affecting the digestive utilization of protein assessed by *in vitro* (Monsoor and Yusuf, 2002) or *in vivo* (Urbano et al., 1995; Porres et al., 2003) methodologies. This lack of improvement in digestive utilization of protein can be attributed to the higher degree of tannin polymerization and protein denaturation that results from the thermal treatment and is able to counteract the benefits of eliminating heat-labile non-nutritional components like TIA. On the other hand, soaking of lentil seeds followed by a traditional or autoclave cooking process, caused a significant decrease in phytic acid and tannin content (Rehman and Shah, 2005), and led to a significant increase in the *in vitro* and *in vivo* protein digestibility (Savage and Scott, 1989; Rehman and Shah, 2005), and the *in vitro* starch digestibility (Rehman and Shah, 2005). In contrast, other authors have found no appreciable differences or even a reduction in the digestive utilization of lentil protein as a result of different soaking and cooking processes. Carbonaro et al. (1997) did not find a significant difference in protein digestibility of lentils measured by an *in vitro* ph-stat methodology between raw and cooked lentils, Aparna et al. (2000) found a reduction in the *in vitro* protein digestibility of lentils after cooking in a salt, bicarbonate, tartaric or citric acid solution, and Hernández-Infante et al. (1998) reported that a reduction in the content of TIA and hemaglutinins caused by microwave treatment of raw or soaked lentils or by cooking previously soaked lentils was not reflected in any improvement of protein digestibility.

Germination.- The process of germinating *Lens culinaris* seeds has been shown to decrease the levels of TIA, tannins and phytic acid (Vidal-Valverde et al., 1994; Urbano et al., 1995; El-Adawy et al., 2003) and improve the digestibility of lentil protein after short term germination period of up to 75 hours (El-Adawy et al., 2003). In contrast, Urbano et al. (1995) have reported a noticeable reduction in the digestive utilization of lentil protein after 6-days germination in spite of the above mentioned decrease in the content of non-nutritional components. Changes in the food matrix and seed chemical composition that take place during germination could be partially responsible for the enhancing or inhibitory effect of this technological process on protein digestibility.

Fermentation.- Shekib (1994) have studied the effect of natural fermentation of *Lens culinaris* for a period of 4 days on the content of different nitrogen fractions and the nutritive utilization of protein, finding a considerable increase in the levels of non-protein nitrogen and a slight increment in crude protein content, whereas no significant change was observed in the amount of protein nitrogen. With regard to the nutritive utilization of protein, fermentation process led to a significant improvement in the *in vitro* protein digestibility of the fermented product.

Tabera et al. (1995) have reported that natural fermentation of lentils using temperatures of 28–35 °C, and lentil flour concentrations of 79–150 g/L during 0–4 days decreased the levels of TIA by 17.1–62.5% and the tannin/catechin index, increasing the tannin content. The changes in tannin content and tannin/catechin index were associated to the soaking process of lentils carried out prior to fermentation. However, very slight changes in the *in vitro* protein digestibility of lentils were observed by these authors as a result of natural fermentation process.

Ethanol Extraction.- Lentil flour extraction with 70% ethanol during 1–3 hours at room temperature or 50 °C led to a reduction in protein solubility at different pH, but did not cause any major alteration in *in vitro* protein digestibility assessed by a pH-stat methodology (Sanz et al., 2001). Similar results have been reported by Porres et al. (2005, 2006) for lupin in which the amount of insoluble nitrogen was increased as a consequence of α-galactoside-oligosaccharide extraction using 50% ethanol at 40 °C without any major effect on protein digestibility assessed by *in vitro* or *in vivo* methodologies.

Protein amino acid digestibility.- The pattern of fecal amino acid composition in rats that consumed lentil diets was characterized by the excretion of endogenous protein that contains high amounts of methionine, cystine and lysine (Sarwar and Peace, 1986; Porres et al., 2002). This high excretion justifies the lower fecal digestibility of these amino acids when compared to the rest of dietary amino acids and crude protein. In addition, sulfur-containing amino acids of lower digestibility are deficient in legume-based diets, where the nutritional demands for these amino acids are increased for the *de novo* synthesis of pancreatic enzymes due to the effect of non-nutritional components like TIA or tannins, all that contributes to the lower nutritive utilization of lentil protein.

The chemical score of lentil protein is within 61–88%, whereas the Protein Digestibility-Corrected Amino Acid Score (PDCAAS) ranges between 52–71%, and the chemical score corrected by the digestibility of the limiting amino acid ($AAS_{(ASAA)}$), usually the sulfur-containing amino acids in lentils, ranges between 41–57% (Sarwar and Peace, 1986; Porres et al., 2002; Iqbal et al., 2006). The thermal treatment in autoclave alters the nutritive utilization of lentil as indicated by the lower PDCAAS and $AAS_{(ASAA)}$ indices, and by the lower true digestibilities of methionine and tyrosine (Porres et al., 2002). However, chemical score and protein digestibility were not affected by the autoclave treatment.

In conclusion, it is difficult to assess the influence of any individual non-nutritional component on protein digestibility based on the effect of different technological treatments, given that multiple changes in chemical composition and food matrix are known to take place during the course of processing that may consistently affect the nutritive utilization of protein. The effects of processing on protein digestibility should be ascribed not only to changes in a single food component but rather to the combined effect of several food constituents.

4.2.2. *Metabolic utilization*

The nutritional quality of raw lentil protein assessed by nutritional indices like the percentage of retained to absorbed nitrogen, weight gain or Protein Efficiency Ratio

Table 21. Metabolic utilization of protein from lentils

PER	0.30[1]; 0.64[2]; 0.83[3]
Weight gain (g/day)	0.67[1]; 1.44[2]; 6.7[4]; 0–0.23[5]
Nitrogen retention (mg/day)	51.3[1]; 74.9[2]
% Retained to Absorbed Nitrogen	19.1[1]; 26.4[2]; 71.8[4]; 29.9[5]; 41.0[6]

1. Urbano et al. (1995); 2. Porres et al. (2003); 3. Hernández-Infante et al. (1998); 4. Combe et al. (1991) ; 5. Cuadrado et al. (2002a); 6. Manan et al. (1987)

(PER) is usually low (Table 21) and inferior to other legumes like the pea, faba bean or lupin (Fernandez et al., 1996; Urbano et al., 2003; Porres et al., 2006), although similar to beans and chickpeas (Nestares et al., 1996; Nestares et al., 2001). The deficiency in sulfur-containing amino acids and certain essential minerals and vitamins together with the low availability of some of these nutrients from lentils are responsible, at least in part, for the low nutritional value of lentil protein. In fact, supplementation of raw lentil flour with methionine or a mineral-vitamin premix (Combe et al., 1991; Urbano et al., 1995; Porres et al., 2003) led to a significant improvement in the metabolic utilization of protein. The beneficial effects of methionine supplementation to a raw or cooked lentil diet have been also reported by Savage and Scott (1989), who, nevertheless, found a higher biological value in cooked lentils than in raw lentils even after methionine supplementation. With regard to the mineral-vitamin supplementation of raw lentil flour, the improvement in nutritive utilization of dietary carbohydrates may have contributed to the improved nutritive utilization of protein, given that protein can be then oriented to plastic rather than to energetic needs.

Miller McCurdy et al. (1978) and Davis et al. (1984) have studied the nutritive utilization of lentil protein using an experimental model based on the protozoan *Tetrahymena pyriformis. This* protozoan requires the same amino acids essential for weanling rats and can degrade intact proteins; the growth period requires only 4 days and maintenance costs are relatively inexpensive. The growth responses of *Tetrahymena* to lentil were significantly lower than those obtained fro casein or yellow pea (Miller McCurdy et al., 1978; Davis et al., 1984). The albumin fraction of lentils was of superior quality to promote *Tetrahymena* growth than the whole lentil seed protein, whereas the globulin fraction was similar or slightly inferior. Methionine was the first limiting amino acid in lentils and growth responses of *Tetrahymena* improved significantly in response to supplementation of this amino acid to lentils. Nevertheless, even when lentil powders were supplemented with essential amino acids to simulate casein, they supported less than half as much growth of *Tetrahymena* than the intact casein standard. Lentil protein concentrate had a superior nutritional value when compared to the unprocessed legume flour and this nutritional value was improved by the individual supplementation of Met or Cys, but not by the combination of Met + Cys or Met + Cys + Lys.

Combe et al. (2004) have studied the differential effect of lentil feeding on proteosynthesis rate in different tissues when compared to a casein control diet. According to these authors, feeding of cooked lentils induced a partitioning of the flux for protein synthesis, preferentially to the benefit of intestinal tissues at the expense of liver and muscle tissue. The reduced amino acid flux to the liver in lentil-fed animals was not related to a change in real digestibility and could be attributed to a higher intestinal amino acid uptake. Thus, there would be an important effect of the alimentary tract on amino acid availability for the support of other physiologic processes like growth. The trophic effect of lentils in small and large intestine can be produced by products derived from gut fermentation of complex carbohydrates or proteins that are not digested in the small intestine.

4.2.2.1. Effect of technological treatments on the metabolic utilization of lentil protein

Thermal processing.- The dry heat or thermal treatment in autoclave of lentils did not improve the metabolic utilization of protein assessed by the percentage of retained to absorbed nitrogen or the Protein Efficiency Ratio when compared to the raw lentil (Urbano et al., 1995; Porres et al., 2003). Furthermore, Hernández-Infante et al. (1998) found a significant reduction in PER as a result of traditional cooking of lentils, although they did not find any significant effect caused by autoclave cooking. In contrast, Manan et al. (1987) have reported a significant improvement in the NET Protein Utilization (NPU) and Biological Value of lentils after soaking in water at room temperature for 4 hours and cooking for 40 minutes followed by drying at 105 °C. Finally, Miller McCurdy et al. (1978) have reported that lentils cooked for the shortest period of time retained a higher nutritional value of protein assessed by the *Tetrahymena pyriformis* experimental model.

Germination.- Germination of lentils for a 6-day period caused a significant reduction in the metabolic utilization of protein assessed by the percentage of retained to absorbed nitrogen, but did not affect the Protein Efficiency Ratio (Urbano et al., 1995). Similar undesirable effects of a long germination period have been reported by other authors in soyabean and pea (Bau et al., 2000; Urbano et al., 2005a,b). It appears that germination for longer periods has a detrimental effect on the organoleptic and nutritional properties of sprouted legume seeds, whereas germination for shorter periods of 48–96 hours have been shown to significantly improve the metabolic utilization of peas (Urbano et al., 2005a,b).

4.3. Nutritive Utilization of Minerals

4.3.1. Effect of Processing Conditions

Lentils are good dietary sources of essential minerals like P, Mg, Ca, K, Zn, Fe, and Mn. Nevertheless, mineral bioavailability from legumes is usually poor as a result of their high dietary fiber content or the presence in legumes of non-nutritional components like phytic acid, oxalate, or polyphenols that may interfere with mineral absorption. Dietary fiber is capable of causing significant losses of Mg and calcium by means of a solvent drag mechanism or ionic interactions with calcium and other food components in case of magnesium, or by adsorption of calcium to cellulose and interaction with other dietary fiber components like lignin (Hardwick et al., 1991; Luccia and Kunkel, 2002). Massey et al. (2001) have found appreciable quantities of oxalate in cooked lentils (1.18 mg/g or 100 mg/serving). Even though these values are lower than those reported for the soyabean and some of its products, the common bean or peanut butter, they are still over the recommendations made for patients with kidney stone (10 mg per serving or 50–60 mg/day).

Lentils have a substantial proportion of phosphorus in the form of phytic acid and low contents of free inorganic phosphorus (El-Mahdy et al., 1985; Porres et al., 2004). Phytic acid phosphorus has been described as being poorly available for monogastrics (Reddy et al., 1982). Furthermore, at the basic pH found in the small intestine of monogastrics, phytic acid may form an insoluble complex with divalent and trivalent cations, which renders them unavailable for absorption (Cheryan, 1980). Nevertheless, when low levels of Ca are present in the diet and the dietary source of P is mainly phytate, it has been shown that part of phytate-phosphorus from lentils and faba beans was available for absorption by growing rats in an attempt to maximize the absorption of this mineral and thus meet the nutritional requirements of the animal. The degree of phytic acid degradation can be modulated by endogenous phytase present in the seed (although phytase activity is usually negligible in non-germinated pulses), by the presence in the brush border membrane of a phosphatase with phytase activity, or by the ability of phytase to absorb the phytic acid molecule. In addition, phytic acid that escapes degradation in the small intestine will be efficiently hydrolyzed by the microbiota present in the large intestine of the animals where the absorption of phosphorus could take a especial relevance under conditions of decreased dietary P supply or unpaired absorption in the small intestine.

Calcium digestibility from lentils has been assessed by an *in vitro* methodology by Sahuquillo et al. (2003). These authors found that digestibility (46.6%) was higher in lentils when compared to other legumes like the bean or the chickpea. Using an *in vivo* experimental model, the digestive utilization of calcium assessed by the Apparent Digestibility Coefficient ranged between 54.3–59.8% (Urbano et al., 1999; Porres et al., 2003). These values are lower than the digestive utilization of calcium from a casein-methionine control diet with similar protein content tested by Urbano et al. (1999). A finding that may be attributed to dietary factors like the lower protein quality of lentil protein, the presence of low levels of vitamin D, or the influence of dietary fiber and non-nutritional components like phytic acid or oxalate.

Due to the low amount of calcium provided by lentil diets, experimental animals made an extremely efficient metabolic use of this mineral, which was reflected in the high percentages of retained to absorbed calcium found in the animals that ingested raw or processed lentils. The low amount of calcium provided by lentils caused a reduction in the amount of this mineral accumulated in the *Longissimus dorsi* muscle, but not in the femur, probably due to the short experimental period of 10 days that was used.

Dahl et al. (1995) found no significant effect of dietary lentils on the final calcium balance of nine human volunteers, although they did report a reduction in the urinary calcium and sodium excretion and increments in the amount of potassium excreted. Homeostasis of P is closely associated with that of calcium. Therefore, in view of the low levels of calcium provided by lentil diets, metabolic utilization of phosphorus was low in spite of a reasonable net absorption, and a large amount of this mineral was excreted in the urine by the experimental animals (Urbano et al., 1999; Porres et al., 2003, 2004).

Digestive utilization of Mg from lentils (45.9%, Porres et al., 2004) was inferior to a casein-methionine control diet (Urbano et al., 2006), possibly as a consequence of the low vitamin D content of lentils or the high amount of phytic acid and magnesium provided by lentil diets. Nevertheless, net Mg absorption was still sufficient to meet the nutrient requirements of experimental animals. Metabolic utilization of magnesium from lentils was low due to the high intake and net absorption of this mineral, although it was higher than what has been reported for other legumes like the chickpea or the common bean by Nestares et al. (1997, 2003) who obtained null balances of this mineral.

Regarding the digestive utilization of zinc, Porres et al. (1996) have reported null absorption values of this mineral from lentils using the growing rat as experimental model, and associated this low absorption to the phytic acid content of lentil flour studied. Due to the low absorption of zinc from lentils by the experimental animals, the balances of this mineral in growing rats that ingested lentil diets were very reduced or null in spite of a low urinary excretion (Porres et al., 1996). In contrast, Sebastiá et al. (2001) found a considerable amount of dialyzable zinc in lentils and other raw legumes like chickpeas and common beans. Similar results have been reported by Sahuquillo et al. (2003) using an *in vitro* model based on the amount of soluble mineral after simulated gastric and intestinal digestions, and by Lombardi-Boccia et al. (2003), who described a percentage of dialyzable zinc close to 25% in raw lentil flour, similar to chickpeas and beans and higher than lupins. Zinc dialyzability decreased slightly in isolated globulins when compared to the raw lentil flour.

Iron bioavailability from legumes, and specifically lentils, determined in humans by an extrinsic tag methodology was found to be low (Gillooly et al., 1983; Lynch et al., 1984). These low bioavailability values are in agreement with the low *in vitro* availability of this mineral from lentils reported by Hazell and Johnson (1987), Sebastiá et al. (2001), Sahuquillo et al. (2003), and Lombardi-Boccia et al. (1991, 2003), although the latter authors found a significant improvement in Fe dialyzability after isolating the globulin protein fraction of lentils (up to 10% dialyzable Fe), which was higher than Fe dialyzability from globulins isolated from other legumes like lupin, common bean are chickpea. The low iron absorption found using *in vitro* or *in vivo* models can be attributed to several factors e.g. 1) the presence of seed components that act as inhibitors of iron absorption (inositol phosphates, tannins, oxalate or structural components), 2) the different Fe storage in lentil and/or differing behaviour of lentil ferritin with regard to iron absorption, compared to iron associated with soybean ferritin which is reasonably absorbed (Davila-Hicks et al., 2004), 3) the low levels of seed components with an enhancing effect on iron absorption, such as organic acids (Teucher et al., 2004) and sulfur-containing amino acids (Layrisse et al., 1984), and 4) the possible endogenous iron losses in the gastrointestinal tract that would be increased by lentil flour experimental diets.

4.3.1.1. Effect of processing on mineral availability

Mineral and vitamin supplementation.- Supplementation of raw lentil flour with a mineral-vitamin premix gave rise to a significant increase in the nutritive utilization of calcium, magnesium and zinc that matched a significant improvement of different parameters related to growth of the animals and nutritive utilization of the lentil meal. Nevertheless, due to the greater amount of these minerals provided by the mineral-supplemented when compared to the unsupplemented lentil diet, a reduction was observed in the digestive utilization of total phosphorus and the percentage of phytate-phosphorus absorbed by the animals (Porres et al., 1996, 2003, 2004).

Thermal treatment.- Dry heating of lentils led to a significant increase in the digestive utilization of Ca and P by growing rats (Urbano et al., 1999). A finding that the authors related to the reduction in the levels of phytate and hemicellulose present in heated lentils. In contrast, no significant effect on Ca, P or Mg digestibility by growing rats was found in response to autoclaving of lentils seeds when compared to the raw lentil flour (Porres et al., 2003, 2004). Traditional or autoclave cooking of lentils resulted in a significant reduction in calcium and zinc dialyzability, but no changes in Fe dialyzability when compared to raw lentil (Sebastiá et al., 2001). Nevertheless, dialyzability of the three minerals studied was significantly improved in a commercial ready-to-eat lentil product as a result of the addition of preserving agents like EDTA or citric acid to the commercial product that are well known enhancers of mineral solubility and dialyzability (Porres et al., 2001; Yeung et al., 2002). Similar results have been reported by Viadel et al. (2006) using a Caco-2 cell line *in vitro* model. Quinteros et al. (2001) have studied the effect of the same technological processes described by Sebastiá et al. (2001) on the amount of soluble Fe from lentils and its speciation, reporting a decrease in the content of total and soluble Fe as a result of traditional or autoclave cooking process, whereas an increase in the amount of soluble Fe was found in the ready-to-eat commercial lentil product. In all the legume samples studied by these authors, the amount of soluble Fe (III) present in legume seeds was significantly higher than the amount of soluble Fe (II) with a potentially higher bioavailability.

Lombardi-Boccia et al. (1991) have reported a considerable reduction in Fe dialyzability as a result of the extrusion process, and a considerable improvement in the dialyzability of this mineral after enzymatic dephytinization of lentils. Similar results have been reported in pea by Urbano et al. (2006), who found a significant improvement in iron absorption after treatment of the pea flour with exogenous microbial phytase which led to inositol hexaphosphate levels that were below those reported by Sandberg and Svanberg (1991) to be detrimental for Fe availability (above $0.5\,\mu mol/g$), and in lupins or whole wheat bread by Porres et al. (2001, 2005), who observed a higher dialyzability of Fe and other minerals like P, Zn or Mn alter phytase enzyme treatment. On the other hand, Hazell and Johnson (1987) have observed a significant decrease in the percentage of diffusible ^{59}Fe after addition of phytic acid to different plant foods.

Germination.- Germination of *Lens culinaris* for a period of 6 days led to a significant increase in the digestive utilization of Ca, P, and Zn by growing rats assessed by the Apparent Digestibility Coefficient (Porres et al., 1996; Urbano et al., 1999).

This improvement in mineral availability can be attributed to the reduction in the levels of non-nutritional components like phytic acid and tannins that are able to interfere with the availability of the above mentioned cations, to changes in the content of the different dietary fiber fractions (Vidal-Valverde and Frias, 1992; Urbano et al., 1999; Urbano et al., 2006), and to the synthesis of different ligands like free amino acids or organic acids with an enhancing effect on mineral absorption (Sripriya et al., 1997).

REFERENCES

Adsule R, N.; Kadam, S. S.; Leung, H. K. Lentil in *Handbook of World Food Legumes: Nutritional Chemistry, Processing Technology, and Utilization*. Salunkhe, D. K., Kadam, S. S., eds. CRC Press, Inc, Boca Raton, Florida, Vol II, 1989, pp 131–152.

Aparna, K.; Khatoon, N.; Prakash, J. Cooking quality and *in vitro* digestibility of legumes cooked in different media. *J. Food. Sci. Technol. Mys.* **2000**, *37*, 169–173.

Arntfield, S. D.; Scanlon, M. G.; Malcolmson, L. J.; Watts, B. M.; Cenkowski, S.; Ryland, D.; Savoie, V. Reduction in lentil cooking time using micronization: Comparison of 2 micronization temperatures. *J. Food. Sci.* **2001**, *66*, 500–505.

Bamdad, F.; Goli, A. H.; Kadivar, M. Preparation and characterization of proteinous film from lentil (*Lens culinaris*) edible film from lentil (*Lens culinaris*). *Food. Res. Int.* **2006**, *39*, 106–111.

Banerjee, S.; Ghosh, A.; Chakraborty, P. Characteristics of effect of temperature and moisture on lentil based extruded expanded product and process optimization. *J. Food Sci. Technol. Mys.* **2003**, *40*, 597–605.

Bartolomé, B.; Jiménez-Ramsey, L. M.; Butler, L. G. Nature of the condensed tannins present in the dietary fibre fractions in foods. *Food Chem.* **1995**, *53*, 357–362.

Bau, H. M. ; Villaume, C. H.; Méjean, L. Effects of soybean (*Glycine max*) germination on biologically active components, nutritional values of seeds, and biological characteristics in rats. *Nahrung-Food.* **2000**, *44*, 2–6.

Bhattacharya, S.; Narasimb, H. V.; Bhattacharya, S. The moisture dependent physical and mechanical properties of whole lentil pulse and split cotyledon. *Int. J. Food Sci. Technol.* **2005**, *40*, 213–221.

Bhatty, R. S. Albumin proteins of eight edible grain legume species: Electrophoretic patterns and amino acid composition. *J. Agric. Food Chem.* **1982**, *30*, 620–622.

Bhatty, R. S. Comparisons of good- and poor-cooking lentils. *J. Sci. Food. Agric.* **1995**, *68*, 489–496.

Bhatty, R. S. Cooking quality and losses of phytic acid, calcium, magnesium and potassium of lentils soaked in different solutions. *Can. Inst. Food Sci. Technol. J.* **1989**, *22*, 450–455.

Bhatty, R. S. Cooking quality of lentils: The role of structure and composition of cell walls. *J. Agric. Food Chem.* **1990**, *38*, 376–383.

Bhatty, R. S. Protein subunits and amino acid composition of wild lentil. *Phytochemistry.* **1986**, *25*, 641–644.

Bhatty, R. S. Relationship between physical and chemical characters and cooking quality in lentil. *J. Agric. Food Chem.* **1984**, *32*, 1161–1166.

Bora, P. S. Functional properties of native and succinylated lentil (*Lens culinaris*) globulins. *Food Chem.* **2002**, *77*, 171–176.

Canadian Grain Commission. Lentil *(Lens culinaris)*. In *The Chemical Composition and Nutritive Value of Canadian Pulses.* **2004**, Canadian Grain Commission. http://www.pulsecanada.com/pdf/crn/Reports/Table (CND **pulses** vs Aust).pdf

Candela, M.; Astiasaran, I.; Bello, J. Cooking and warm-holding: Effect on general composition and amino acids of kidney beans (*Phaseolus vulgaris*), chickpeas (*Cicer arietinum*), and lentils (*Lens culinaris*). *J. Agric. Food Chem.* **1997**, *45*, 4763–4767.

Carbonaro, M.; Cappelloni, M.; Nicoli, S.; Lucarini, M.; Carnovale, E. Solubility-digestibility relationship of legume proteins. *J. Agric. Food Chem.* **1997**, *45*, 3387–3394.

Carbonaro, M.; Virgili, F.; Carnovale, R. Evidence for protein–tannin interaction in legumes: Implications in the antioxidant properties of faba bean tannins. *Lebens. Wiss. Technol.* **1996**, *29*, 743–750.

Çelik, S.; Yalçin, E.; Basman, A.; Köksel, H. Effect of irradiation on protein electrophoretic properties, water absorption and cooking quality of lentils. *Int. J. Food. Sci. Nutr.* **2004**, *55*, 641–648.

Chang, R.; Schwinner, S.; Burr, H. Phytate removal from whole dry beans by enzymatic hydrolysis and diffusion. *J. Food Sci.* **1977**, *42*, 1098–1110.

Chen, L. H.; Thacker, R. R.; Pan, S. H. Effect of germination on hemagglutinating activity of pea and bean seeds. *J. Food. Sci.* **1977**, *42*, 1666.

Cheryan M. Phytic acid interactions in food systems. *CRC. Crit. Rev. Food Sci. Nutr.* **1980**, *13*, 297–335.

Combe, E.; Achi, T.; Pion, R. Compared metabolic and digestive utilization of faba bean, lentil and chickpea. *Reprod. Nutr. Dev.* **1991**, *31*, 631–646.

Combe, E.; Pirman, T.; Stekar, J.; Houlier, M. L.; Mirand, P. P. Differential effect of lentil feeding on proteosynthesis rates in the large intestine, liver and muscle of rats. *J. Nutr. Biochem.* **2004**, *15*, 12–17.

Cuadrado, C.; Ayet, G.; Robredo, L. M.; Tabera, J.; Villa, R.; Pedrosa, M. M.; Burbano, C.; Muzquiz, M. Effect of natural fermentation on the content of inositol phosphates in lentils. *Zeitsch. Lebensm. Unters. Forsch.* **1996**, *203*, 268–271.

Cuadrado, C.; Grant, G.; Rubio, L. A.; Muzquiz, M.; Bardocz, S.; Pusztai, A. Nutritional utilization by the rat of diets based on lentil (*Lens culinaris*) seed meal or its fractions. *J. Agric. Food Chem.* **2002a**, *50*, 4371–4376.

Cuadrado, C.; Hajos, G.; Burbano, C.; Pedrosa, M. M.; Ayet, G.; Muzquiz, M.; Pusztai, A.; Gelencser, E. Effect of natural fermentation on the lectin content of lentils measured by immunological methods. *Food. Agric. Immunol.* **2002b**, *14*, 41–49.

Dahl, W. J.; Whiting, S. J.; Stephen, A. M. Dietary lentils and calcium balance in adult men. *Nutr. Res.* **1995**, *15*, 1587–1598.

Dalgetty, D. D.; Baik, B. K. Fortification of bread with hulls and cotyledon fibers isolated from peas, lentils, and chickpeas. *Cereal. Chem.* **2006**, *83*, 269–274.

Danisová, C.; Holotnáková, E.; Hozová, B.; Buchtová, V. Effect of germination on a range of nutrients of selected grain and legumes. *Acta Alimentaria.* **1994**, *23*, 287–298.

Davila-Hicks, P.; Theil, E. C.; Lönnerdal, B. Iron in ferritin or in salts (ferrous sulfate) is equally bioavailable in nonanemic women. *Am. J. Clin. Nutr.* **2004**, *80*, 936–940.

Davis, K. R.; Costello, M. J.; Mattern, V.; Schroeder, C. Effect of age of sample and of amino acid supplementation on the *Tetrahymena*-relative nutritive value of lentils, green and yellow split peas, and their processed forms. *Cereal. Chem.* **1984**, *61*, 311–315.

De Almeida Costa, G. E.; da Silva Queiroz-Monici, K.; Pissini Machado Reis, S. M.; Costa de Oliveira, A. Chemical composition, dietary fibre and resistant starch contents of raw and cooked pea, common bean, chickpea and lentil legumes. *Food. Chem.* **2006**, *94*, 327–330.

Demirbas, A. ß-glucan and mineral nutrient contents of cereals grown in Turkey. *Food Chem.* **2005**, *90*, 773–777.

Egli, I.; Davidsson, L.; Juillerat, M. A.; Barclay, D.; Hurrell, R. F. The influence of soaking and germination on the phytase activity and phytic acid content of grains and seeds potentially useful for complementary feeding. *J. Food. Sci.* **2002**, *67*, 3484–3488.

El-Adawy, T. A.; Rahma, E. H.; El-Bedawey, A. A.; El-Beltagy, A. E. Nutritional potential and functional properties of germinated mung bean, pea and lentil seeds. *Plant. Food. Hum. Nutr.* **2003**, *58*, 1–13.

El-Mahdy, A. R.; Moharram, Y. G.; Abou-Samaha, O. R. Influence of germination on the nutritional quality of lentil seeds. *Z. Lebensm. Unters. Forsch.* **1985**, *181*, 318–320.

El-Tinay, A. H.; Mahgoub, S. O.; Mohamed, B. E.; Hamad, M. A. Proximate composition of mineral and phytate contents of legumes grown in Sudan. *J. Food. Comp. Anal.* **1989**, *2*, 69–78.

Erdogan, S.; Erdemoglu, S. B.; Kaya, S. Optimization of microwave digestion for determination of Fe, Zn, Mn and Cu in various legumes by flame atomic absorption spectrometry. *J. Sci. Food. Agric.* **2006**, *86*, 226–232.

Ereifej, K. I.; Haddad, S. G. Chemical composition of selected Jordanian cereals and legumes as compared with the FAO, Moroccan, East Asian and Latin American tables for use in the Middle East.

Trends. Food Sci. Technol. **2001**, *11*, 374–378.

Fernández, M.; Aranda, P.; López-Jurado, M.; García-Fuentes, M.; Urbano, G. Bioavailability of phytic acid phosphorus in processed *Vicia faba* L. var major. *J. Agric. Food Chem.* **1997**, *45*, 4367–4371.

Fernandez, M.; Lopez-Jurado, M.; Aranda, P.; Urbano, G. Nutritional assessment of raw and processed faba bean (*Vicia faba* L.) cultivar major in growing rats. *J. Agric. Food Chem.* **1996**, *44*, 2766–2772.

Fernandez-Orozco, R.; Zielinski, H.; Frias, J.; Vidal-Valverde, C.; Piscula, M. K. Superoxide dismutase-like activity of raw, cooked and germinated lentils. *Pol. J. Food. Nutr. Sci.* **2002**, *11/52*, 39–44.

Fernandez-Orozco, R.; Zielinski, H.; Piskula, M. Contribution of low-molecular-weight antioxidants to the antioxidant capacity of raw and processed lentil seeds. *Nahrung/Food.* **2003**, *47*, 291–299.

Fleming, S. E. A study of relationships between flatus potential and carbohydrate distribution in legume seeds. *J. Food Sci.* **1981**, *46*, 794.

Frias, J.; Bakhsh, A.; Jones, D.; Arthur, A. E.; Vidal-Valverde, C.; Rhodes, M. J. C.; Hedley, C. L. Genetic analysis of the raffinose oligosaccharide pathway in lentil seeds. *J. Exp. Bot.* **1999**, *50*, 469–476.

Frias, J.; Diaz-Pollan, C.; Hedley, C. L.; Vidal-Valverde, C. Evolution and kinetics of monosaccharides, disaccharides and α-galactosides during germination of lentils. *Z. Lebensm. Unters. Forsch.* **1996a**, *202*, 35–39.

Frias, J.; Doblado, R.; Vidal-Valverde, C. Kinetics of soluble carbohydrates by action of endo/exo α-galactosidase enzyme in lentils and peas. *Eur. Food Res. Technol.* **2003a**, *216*, 199–203.

Frias, J.; Doblado, R.; Antezana, J. R.; Vidal-Valverde, C. Inositol phosphate degradation by the action of phytase enzyme in legume seeds. *Food Chem.* **2003b**, *81*, 233–239.

Frias, J.; Fernandez-Orozco, R.; Zielinski, H.; Kozlowska, H.; Vidal-Valverde, C. Effect of germination on the content of vitamin C and E in lentils. *Pol. J. Food Nutr. Sci.* **2002**, *11/52*, 76–78.

Frias, J.; Fornal, J.; Ring, S. G.; Vidal-Valverde, C. Effect of germination on physico-chemical properties of lentil starch and its components. *Lebensm. Wiss. Technol.* **1998**, *31*, 228–236.

Frias, J.; Prodanov, M.; Sierra, I.; Vidal-Valverde, C. Effect of light on carbohydrates and hydrosoluble vitamins of lentils during soaking. *J. Food. Protec.* **1995a**, *58*, 692–695.

Frias, J.; Vidal-Valverde, C. Basks, A.; Arthur, A. E.; Hedley, C. An assessment of variation for nutritional and non-nutritional carbohydrates in lentil seeds (Lens culinaris). *Plant Breeding.* **1994**, *113*, 170–173.

Frias, J.; Vidal-Valverde, C.; Kozlowska, H.; Tabera, J.; Honke, J.; Hedley, C. L. Natural fermentation of lentils. Influence of time, flour concentration, and temperature on the kinetics of monosaccharides, disaccharides and alpha-galactosides. *J. Agric. Food Chem.* **1996b**, *44*, 579–584.

Gad, S. S.; Mohamed, M. S.; El-Zalaki, M. E.; Mohasseb, S. Z. Effect of processing on phosphorus and phytic acid contents of some egyption varieties of legumes. *Food Chem.* **1982**, *8*, 11–19.

Ghavidel, R. A.; Prakash, J. Effect of germination and dehulling on functional properties of legume flours. *J. Sci. Food. Agric.* **2006**, *86*, 1189–1195.

Gillooly, M.; Bothwell, T. H.; Torrance, J. D.; MacPhail, A. P.; Derman, D. P.; Bezwoda, W. R.; Mills, W.; Charlton, R. W.; Mayet, F. The effects of organic acids, phytates and polyphenols on the absorption of iron from vegetables. *Br. J. Nutr.* **1983**, *49*, 331–342.

Granito, M.; Torres, A.; Frías, J.; Guerra, M.; Vidal-Valverde, C. Influence of fermentation on the nutritional value of two varieties of *Vigna sinensis*. *Eur. Food Res. Technol.* **2005**, *220*, 176–181.

Grant, G.; More, L. J.; McKenzie, N. H.; Stewart, J. C.; Pusztai, A. A survey of the nutritional and haemagglutination properties of legume seeds generally available in the UK. *Br. J. Nutr.* **1983**, *50*, 207–214.

Green, G. M.; Lyman, R. L. Feedback regulation of pancreatic enzyme secretion as a mechanism for trypsin inhibitor induced hypersecretion in rats. *Proc. Soc. Exp. Biol. Med.* **1972**, *140*, 6–12.

Grela, E. R.; Günter, K. D. Fatty acid composition and tocopherol content of some legume seeds. *Anim. Feed Sci. Technol.* **1995**, *52*, 325–331.

Hardwick, L. L.; Jones, M. R.; Brautbar, N.; Lee, D. B. N. Magnesium absorption: Mechanisms and the influence of vitamin D, calcium and phosphate. *J. Nutr.* **1991**, *121*, 13–23.

Harper, A. E.; Benevenga, N. J.; Wohlueter, R. M. Effect of ingestion of disproportionate amounts of amino acids. *Physiol. Rev.* **1970**, *50*, 428–558.

Harper, A. E.; Benton, D. A.; Elvbejhem, C. A. L-Leucine, an isoleucine antagonist in the rat. *Arch. Biochem. Biophys.* **1955**, *57*, 1–12.

Hazell, T.; Johnson, I. T. *In vitro* estimation of iron availability from a range of plant foods. Influence of phytate, ascorbate and citrate. *Br. J. Nutr.* **1987**, *57*, 223–233.

Hernández-Infante, M.; Sousa, V.; Montalvo, I.; Tena, E. Impact of microwave heating on hemagglutinins, trypsin inhibitors and protein quality of selected legume seeds. *Plant. Food. Hum. Nutr.* **1998**, *52*, 199–208.

Iliadis, C. Effect of harvesting procedure, storage time and climatic conditions on cooking time of lentils (*Lens culinaris* Medikus). *J. Sci. Food. Agric.* **2001**, *81*, 590–593.

Iliadis, C. Influence of genotype and soil type on cooking time in lentil (*Lens culinaris* Medikus). *Int. J. Food Sci. Technol.* **2003**, *38*, 89–93.

Iqbal, A.; Khalil, I. A.; Ateeq, N.; Khan, M. S. Nutritional quality of important food legumes. *Food Chem*, **2006**, *97*, 331–335.

Johnson, D. J.; Anderson, G. H. The prediction of plasma amino acid concentration from diet amino acid content. *Physiol. Rev.* **1982**, *43*, R99–R103.

Kavas, A.; Nehir, S. Changes in nutritive value of lentils and mung beans during germination. *Chem. Mikrobiol. Technol. Lebensm.* **1992**, *14*, 3–9.

Khan, M. A.; Rana, I. A.; Ullah, I.; Jaffery, S. Physicochemical characters and nutrient composition of some improved lines of lentils grown in Pakistan. *J. Food. Comp. Anal.* **1987**, *1*, 65–70.

Knuckles, B. E.; Kuzmicky, D. D.; Gumbmann, M. R.; Betschaart, A. A. Effect of myo-inositol phosphate esters on in vitro and in vivo digestion of protein. *J. Food. Sci.* **1989**, *54*, 1348–1350.

Konoshima, T.; Kokumai, M.; Kozuka, M. Antitumor promoting activities of afromosin and soyasaponin I isolated from *Wistaria brachybotrys*. *J. Nat. Prod.* **1992**, *55*, 1776–1778.

Koplík, R.; Borková, M.; Mestek, O.; Komínková, J.; Suchanek, M. Application of size-exclusion chromatography-inductively coupled mass spectrometry for fractionation of element species in seeds of legumes. *J. Chromat. B.* **2002**, *775*, 179–187.

Kozlowska, H.; Honke, J.; Sadowaska, J.; Frias, J.; Vidal-Valverde, C. Natural fermentation of lentils. Influence of time, concentration and temperature on the kinetics of hydrolysis of inositol phosphates. *J. Sci. Food.* Agric. **1996**, *71*, 367–375.

Kumar, V.; Rani, A.; Solanki, S.; Hussain, S. M. Influence of growing environment on the biochemical composition and physical characteristics of soybean seed. *J. Food. Comp. Anal.* **2006**, *19*, 188–195.

Kuo, Y. H.; Rozan, P.; Lambein, F.; Frías, J.; Vidal-Valverde, C. Effects of different germination conditions on the contents of free protein and non-protein amino acids of commercial legumes. *Food Chem.* **2004**, *86*, 537–545.

Kylen, A. M.; McReady, R. M. Nutrients in seeds and sprouts of alfalfa, lentils, mung beans and soybeans. *J. Food Sci.* **1975**, *40*, 1008–1009.

Layrisse, M.; Martínez-Torres, C.; Leets, I.; Taylor, P.; Ramírez, J. Effect of histidine, cysteine, glutathione or beef on iron absorption in humans. *J. Nutr.* **1984**, *114*, 217–223.

Le Magnen, J. Body energy balance and food intake: A neuroendocrine regulatory mechanism. *Physiol. Rev.* **1983**, *63*, 314–386.

Lombardi-Boccia, G.; Dilullo, G.; Carnovale, E. *In vitro* iron dialyzability from legumes: Influence of phytate and extrusion cooking. *J. Sci. Food. Agric.* **1991**, *55*, 599–605.

Lombardi-Boccia, G.; Ruggeri, S.; Aguzzi, A.; Cappelloni, M. Globulins enhance *in vitro* iron but not zinc dialyzability: a study of six legume species. *J. Trace Elem. Med. Biol.* **2003**, *17*, 1–5.

Luccia, B. H. D.; Kunkel, M. E. In vitro availability of calcium from sources of cellulose, methylcellulose, and psyllium. *Food Chem.* **2002**, *77*, 139–146.

Lynch, S. R.; Beard, J. L.; Dassenko, S. A.; Cook, J. D. Iron absorption from legumes in humans. *Am. J. Clin. Nutr.* **1984**, *40*, 42–47.

Maccarrone, M.; Veldink, G. A.; Vliegenthart, J. F. G.; Agro, A, F. Ozone stress modulates amine oxidase and lipoxygenase expression in lentil (*Lens culinaris*) seedlings. *FEBS. Lett.* **1997**, *408*, 241–244.

Manan, F.; Hussain, T.; Alli, I.; Iqbal, P. Effect of cooking on phytic acid content and nutritive value of Pakistani peas and lentils. *Food Chem.* **1987**, *23*, 81–87.

Martin-Belloso, O.; Llanos-Barriobero, E. Proximate composition, minerals and vitamins in selected canned vegetables. *Eur. Food. Res. Technol.* **2001**, *212*, 182–187.

Martín-Cabrejas, M. A.; Ariza, N.; Esteban, R.; Mollá, E.; Waldron, K.; López-Andréu, F. J. Effect of germination on the carbohydrate composition of the dietary fibre of peas (*Pisum sativum* L.). *J. Agric. Food Chem.* **2003**, *51*, 1254–1259.

Martínez-Villaluenga, C.; Frias, J.; Vidal-Valverde, C. Alpha-galactosides: Antinutritional Factors or Functional Ingredients? *Crit. Rev. Food Sci. Nutr.* **2007**, (in press).

Massey, L.; Palmer, R. G.; Horner, H. T. Oxalate content of soybean seeds (*Glycine max*: Leguminosae), soyfoods, and other edible legumes. *J. Agric. Food Chem.* **2001**, *49*, 432–466.

Mercer, L. P.; Dodds, S. J.; Schweisthall, M. R.; Dunn, J. D. Brain histidine and food intake in rats fed diets deficient in single amino acids. *J. Nutr.* **1989**, *119*, 66–74.

Miller McCurdy, S.; Scheier, G. E.; Jacobson, M. Evaluation of protein quality of five varieties of lentils using *Tetrahymena pyriformis* W. *J. Food Sci.* **1978**, *43*, 694–697.

Monsoor, M. A.; Yusuf, H. K. M. *In vitro* protein digestibility of lathyrus pea (*Lathyrus sativus*), lentil (*Lens culinaris*), and chickpea (*Cicer arietinum*). *Int. J. Food Sci. Technol.* **2002**, *37*, 97–99.

National Research Council. Nutrient Requirements of Laboratory Animals. Fourth Revised Edition.; National Academy Press, Washington, D.C. **1995**.

Nestares, T. Barrionuevo, M.; Urbano, G.; López-Frías, M. Nutritional assessment of protein from beans (*Phaseolus vulgaris* L) processed at different pH values, in growing rats. *J. Sci. Food. Agric.* **2001**, *81*, 1522–1529.

Nestares, T.; Barrionuevo, M.; López-Frías, M.; Vidal, C.; Urbano, G. Effect of different soaking solutions on nutritive utilization of minerals (Calcium, Phosphorus, and Magnesium) from cooked beans (*Phaseolus vulgaris*, L) in growing rats. *J. Agric. Food Chem.* **2003**, *51*, 515–520

Nestares, T.; Barrionuevo, M.; Urbano, G.; López-Frías, M. Effect of processing methods on the calcium, phosphorus, and phytic acid contents and nutritive utilization of chickpea (*Cicer arietinum* L.). *J. Agric. Food Chem.* **1999**, *47*, 2807–2812.

Nestares, T.; López-Frías, M.; Barrionuevo, M.; Urbano, G. Nutritional assessment of raw and processed chickpea (*Cicer arietinum* L.) protein in growing rats. *J. Agric. Food Chem.* **1996**, *44*, 2760–2765.

Nestares, T.; Urbano, G.; López-Frías, M.; Barrionuevo, M. Nutritional assessment of magnesium from raw and processed chickpea (*Cicer arietinum* L.) in growing rats. *J. Agric. Food Chem.* **1997**, *45*, 3138–3142.

O'Dell, B. L.; Reeves, P. G. Zinc status and food intake. In *Zinc in Human Biology*. Mills, C. F. Ed. London, International Life Sciences Institute, **1989**, pp. 183–220.

Petterson, D.; Sipsas, S.; McIntosh, J. B. The chemical composition and nutritive value of Australian pulses. Canberra: Grains Research and Development Corporation. **1997**.

Pirman, T.; Stibilj, V. An influence of cooking on fatty acid composition in three varieties of common beans and in lentil. *Eur. Food Res. Technol.* **2003**, *217*, 498–503.

Pirman, T.; Stibilj, V.; Stekar, J. M. A.; Combe, E. Amino acid composition of beans and lentil. *Zb. Bioteh. Fak. Univ. Ljubl., Kmet. Zooteh.* **2001**, *78*, 57–68.

Porres, J. M.; Aranda, P.; López-Jurado, M.; Urbano, G. Effect of Natural and Controlled Fermentation on Chemical Composition and Nutrient Dialyzability from Beans (*Phaseolus vulgaris*, L.). *J. Agric. Food Chem.* **2003**, *51*, 5144–5149.

Porres, J. M.; Aranda, P.; López-Jurado, M.; Urbano, G. Nutritional potential of raw and free α-galactosides lupin (*Lupinus albus* var. *multolupa*) seed flours. Effect of phytase treatment on nitrogen and mineral dialyzability. *J. Agric. Food Chem.* **2005**, *53*, 3088–3094.

Porres, J. M.; Aranda, P.; López-Jurado, M.; Urbano, G. Nutritional evaluation of protein, phosphorus, calcium and magnesium bioavailability from lupin (*Lupinus albus* var. *multolupa*)-based diets in growing rats: effect of α-galactoside oligosaccharide extraction and phytase supplementation. *Brit. J. Nutr.* **2006**, *95*, 1102–1111.

Porres, J. M.; Etcheverry, P.; Miller, D. D.; Lei, X. G. Phytase and citric acid supplementation in whole-wheat bread improves phytate-phosphorus release and iron dialyzability. *J. Food Sci.* **2001**, *66*, 614–619.

Porres, J. M.; López-Jurado, M.; Aranda, P.; Urbano, G. Bioavailability of phytic acid-phosphorus and magnesium from lentils (*Lens culinaris* M.) in growing rats: Influence of thermal treatment and vitamin-mineral supplementation. *Nutrition*. **2004**, *20*, 794–799.

Porres, J. M.; López-Jurado, M.; Aranda, P.; Urbano, G. Effect of heat treatment and mineral vitamin supplementation on the nutritive use of protein and calcium from lentils (*Lens culinaris* M.) in growing rats. *Nutrition*. **2003**, *19*, 451–456.

Porres, J. M.; López-Jurado, M.; Aranda, P.; Urbano, G. Effect of processing on the bioavailability of phytic acid and zinc in lentils. In *COST 98 Effect of Antinutrients on the Nutritional Value of Legume Diets*, Bardocz, S., Muzquiz, M., Pusztai, A, Eds. European Commission, Directorate General XII, Science, Research and Development, 1996, Volume 4, pp. 28–31.

Porres, J. M.; Urbano, G.; Fernández-Fígares, I.; Prieto, C.; Pérez, L.; Aguilera, J. F. Digestive utilisation of protein and amino acids from raw and heated lentils by growing rats. *J. Sci. Food. Agric.* **2002**, *82*, 1740–1747.

Prodanov, M.; Sierra, I.; Vidal-Valverde, C. Effect of germination on the thiamine, riboflavin and niacin contents in legumes. *Z. Lebensm. Unters. Forsch.* **1997**, *205*, 48–52.

Prodanov, M.; Sierra, I.; Vidal-Valverde, C. Influence of soaking and cooking on the thiamin, riboflavin and niacin contents of legumes. *Food Chem.* **2004**, *84*, 271–277.

Quinteros, A.; Farré, R.; Lagarda, M. J. Optimization of iron speciation (soluble, ferrous and ferric) in beans, chickpeas and lentils. *Food. Chem.* **2001**, *75*, 365–370.

Rackis, J. J. Oligosaccharides of food legumes: α-galactoside activity and flatus problems. In *Physiological Effects of Food Carbohydrates*. Allen J., Heilge, J. Eds. American Chemical Society, Washington, DC, **1975**, pp. 207.

Ramulu, P.; Udayasekhara Rao, P. Effects of processing on dietary fiber content of cereals and pulses. *Plant. Food. Hum. Nutr.* **1997**, *50*, 249–257.

Reddy, N. R.; Pierson, M. D.; Sathe, S. K.; Salunkhe, D. K. Chemical, nutritional and physiological aspects of dry bean carbohydrates: A review. *Food Chem.* **1984**, *13*, 25–68.

Reddy, N. R.; Sathe, S. K.; Salunkhe, D. K. Phytates in legumes and cereals. *Adv. Food. Res.* **1982**, *28*, 1–92.

Rehman, Z.; Shah, W. H. Thermal heat processing effects on antinutrients, protein and starch digestibility of food legumes. *Food. Chem.* **2005**, *91*, 327–331.

Roberfroid, M. Functional food concept and its application to prebiotics. *Dig. Liver Dis.* **2002**, *34*, S105.

Rozan, P.; Kuo, Y. H.; Lambein F. Amino acids in seed and seedling of the genus Lens. *Phytochemistry*. **2001**, *58*, 281–289.

Rozan, P.; Kuo, Y.; Lambein, F. Free amino acids present in commercially available seedlings sold for human consumption. A potential hazard for consumers. *J. Agric. Food Chem.* **2000**, *48*, 716–723.

Rubio, L. A.; Grant, G.; Scislowski, P. W. O.; Brown, D.; Bardocz, S.; Pusztai, A. The utilization of lupin (*Lupinus angustifolius*) and faba bean (*Vicia faba*) globulins by rats is poorer than of soybean globulins and lactalbumin but the nutritional value of lupin seed meal is lower only than that of lactalbumin. *J. Nutr.* **1995**, *125*, 2145–2155.

Ruiz, R. G.; Price, K. R.; Rose, M. E.; Fenwick, G. R. Effect of seed size and testa color on saponin content of Spanish lentil seed. *Food Chem.* **1997**, *3*, 223–226.

Sadowska, J.; Fornal, J.; Vidal-Valverde, C.; Frias, J. Natural fermentation of lentils. Functional properties and potential in breadmaking of fermented lentil flour. *Nahrung*, **1999**, *43*, 396–401.

Sahuquillo, A.; Barberá, R.; Farré, R. Bioaccesibility of calcium, iron, and zinc from three legume samples. *Nahrung/Food*. **2003**, *47*, 438–441.

Sandberg, A. S.; Svanberg, U. Phytate hydrolysis by phytase in cereals. Effects on *in vitro* estimation of iron availability. *J. Food Sci.* **1991**, *56*, 1330–1333.

Sandström, B.; Sandberg, A. S. Inhibitory effects isolated inositol phosphates on zinc absorption in humans. *J. Trace. Elem. Electrolytes. Health. Dis.* **1992**, *6*, 99–103.

Sanz, M. A.; Blázquez, I.; Sierra, I.; Medrano, M. A.; Frias, J.; Vidal-Valverde, C.; Hernández, A. Nutritional evaluation of ethanol-extracted lentil flours. *J. Agric. Food Chem.* **2001**, *49*, 1854–1860.

Sarriá, B.; López-Fandiño, R.; Vaquero, P. Does processing of a powder or in-bottle-sterilized liquid infant formula affect calcium bioavailability? *Nutrition*, **2001**, *17*, 326–331.

Sarwar, G.; Peace, R. W. Comparisons between true digestibility of total nitrogen and limiting amino acids in vegetable proteins fed to rats. *J. Nutr.* **1986**, *116*, 1172–1184.

Sathe, S. K.; Deshpande, S. S.; Reddy, N. R.; Goll, D. E.; Salunkhe, D. K. Effect of germination on proteins, raffinose oligosaccharides and antinutritional factors in the Great Northern beans (*Phaseolus vulgaris* L.). *J. Food Sci.* **1983**, *48*, 1796–1800.

Savage, G. P.; Scott, S. K. Effect of cooking and amino acid supplementation on the nutritive value of lentils (*Lens culinaris* M.). In *Recent Advances of Research in Antinutritional Factors in Legume Seeds. Proceedings of the 1st International Workshop on Antinutritional Factors in Legume Seeds.* Huisman, J., Van der Poel, F. B., Liener, I. E., Eds. Pudoc Public, Wageningen, The Netherlands, **1989**, pp. 243–248.

Sebastiá, V.; Barberá, R.; Farré, R.; Lagarda, M. J. Effects of legume processing on calcium, iron and zinc contents and dialysabilities. *J. Sci. Food. Agric.* **2001**, *81*, 1180–1185.

Shekib, L. A. E.; Zouil, M. E.; Youssef, M. M.; Mohammed, M. S. Effect of cooking on the chemical composition of lentils, rice and their blend (Koshary). *Food Chem.* **1985**, *18*, 163–168.

Shekib, L. A. H.; Zoueil, M. E.; Youssef, M. M.; Mohamed, M. S. Amino acid composition and *in vitro* digestibility of lentil and rice proteins and their mixture (*Koshary*). *Food. Chem.* **1986**, *20*, 61–67.

Shekib, L. A. Nutritional improvement of lentils, chickpea, rice and wheat by natural fermentation. *Plant. Food. Hum. Nutr.* **1994**, *46*, 201–205.

Sidhu, G. S.; Oakenfull, D. G. A mechanism for the hypocholesterolaemic activity of saponins. *Br. J. Nutr.* **1986**, *55*, 643–649.

Sika, M.; Terrab, A.; Swan, P. B.; Hegarty, P. V. J. Composition of selected Moroccan cereals and legumes: Comparison with the FAO table for use in Africa. *J. Food. Comp. Anal.* **1995**, *8*, 62–70.

Singh, M.; Krikorian, A. D. Inhibition of trypsin activity by phytate. *J. Agric. Food Chem.* **1982**, *30*, 799–800.

Slavin, J. L. Dietary fiber and body weight. *Nutrition*. **2005**, *21*, 411–418.

Solanki, I. S.; Kapoor, A. C.; Singh, U. Nutritional parameters and yield evaluation of newly developed genotypes of lentils (*Lens culinaris* Medik.). *Plant. Food. Hum. Nutr.* **1999**, *54*, 79–87.

Sotomayor C.; Frias, J.; Fornal, J.; Sadowska, J.; Urbano, G.; Vidal-Valverde, C. Lentil starch content and its microscopical structure as influenced by natural fermentation. *Starch/Stärke*. **1999**, *51*, 152–156.

Sripriya, G.; Antony, U.; Chandra, T. S. Change in carbohydrate, free amino acids, organic acids, phytate, and HCl extractability of minerals during germination and fermentation of finger millet (*Eleusine coracana*). *Food Chem.* **1997**, *58*, 345–350.

Svanberg, U.; Lorri, W. Fermentation and nutrient availability. *Food Control*. **1997**, *8*, 319–327.

Tabera, J.; Frías, J.; Estrella, I.; Villa, R.; Vidal-Valverde, C. Natural fermentation of lentils. Influence of time, concentration and temperature on protein content, Trypsin inhibitor activity and phenolic compound content. *Z. Lebensm. Unters. Forsch.* **1995**, *201*, 587–591.

Teucher, B.; Olivares, M.; Cori, H. Enhancers of iron absorption: ascorbic acid and other organic acids. *Int. J. Vitam. Nutr. Res.* **2004**, *74*, 403–419.

Urbano, G.; Aranda, P.; Gómez-Villalva, E.; Frejnagel, S.; Porres, J. M.; Frías, J.; Vidal-Valverde, C.; López-Jurado, M. Nutritional evaluation of pea (*Pisum sativum* L.) protein diets alter mild hydrothermal treatment and with and without added phytase. *J. Agric. Food Chem.* **2003**, *51*, 2415–2420.

Urbano, G.; Aranda, P.; Vílchez, A.; Aranda, C.; Cabrera, L.; Porres, J. M.; López-Jurado, M. Effects of germination on the composition and nutritive value of proteins in *Pisum sativum* L. *Food Chem.* **2005a**, *93*, 671–679.

Urbano, G.; López-Jurado, M.; Aranda, C.; Vilchez, A.; Cabrera, L.; Porres, J. M.; Aranda, P. Evaluation of zinc and magnesium bioavailability from pea (*Pisum sativum*, L.) sprouts. Effect of illumination and different germination periods. *Int. J. Food Sci. Technol.* **2006**, *41*, 618–626.

Urbano, G.; López-Jurado, M.; Fernández, M.; Moreu, M. C.; Porres-Foulquie, J.; Frias, J.; Vidal-Valverde, C. Ca and P bioavailability of processed lentils as affected by dietary fiber and phytic acid content. *Nutr. Res.* **1999**, *19*, 49–64.

Urbano, G.; López-Jurado, M.; Frejnagel, S.; Gómez-Villalva, E.; Porres, J. M.; Frías, J. Vidal-Valverde, C.; Aranda, P. Nutritional assessment of raw and germinated pea (*Pisum sativum* L.) protein and carbohydrate by in vitro and in vivo techniques. *Nutrition.* **2005b**, *21*, 230–239.

Urbano, G.; Lopez-Jurado, M.; Hernandez, J.; Fernández, M.; Moreu, M. C.; Frias, J.; Diaz-Pollan, C.; Prodanov, M.; Vidal-Valverde, C. Nutritional assessment of raw, heated, and germinated lentils. *J. Agric. Food Chem.* **1995**, *43*, 1871–1877.

Urbano, G.; Porres, J. M.; Frejnagel, S.; López-Jurado, M.; Gómez-Villalva, E.; Vidal-Valverde, C.; Aranda, P. Improvement of iron availability from phytase-treated Pisum sativum, L flour. *Food Chem.* **2006**, *103*, 389–395.

Viadel, B.; Barberá, R.; Farré, R. Effect of cooking and legume species upon calcium, iron and zinc uptake by Caco-2 cells. *J. Trace. Elem. Medic. Biol.* **2006**, *20*, 115–120.

Vidal-Valverde, C.; Frias, J. Changes in carbohydrates during germination of lentils. *Z. Lebensm. Unters. Forsch.* **1992**, *194*, 461–464.

Vidal-Valverde, C.; Frias, J. Legume processing effects on dietary fiber components. *J. Food. Sci.* **1991**, *56*, 1350–1352.

Vidal-Valverde, C.; Frias, J.; Esteban, R. Dietary fiber in processed lentils. *J. Food. Sci.* **1992a**, *57*, 1161–1163.

Vidal-Valverde, C.; Frias, J.; Estrella, I.; Gorospe, M. J.; Ruiz, R.; Bacon, J. Effect of processing on some antinutritional factors of lentils. *J. Agric. Food Chem.* **1994**, *42*, 2291–2295.

Vidal-Valverde, C.; Frias, J.; Prodanov, M.; Tabera, J.; Ruiz, R.; Bacon, J. Effect of natural fermentation on carbohydrates, riboflavin and trypsin inhibitor activity of lentils. *Z. Lebensm. Unters. Forsch.* **1993b**, *197*, 449–452.

Vidal-Valverde, C.; Frias, J.; Sierra, I.; Blazquez, I.; Lambein, F.; Kuo, Y. New functional legume foods by germination: effect on the nutritive value of beans, lentils and peas. *Eur. Food Res. Technol.* **2002a**, *215*, 472–477.

Vidal-Valverde, C.; Frias, J.; Valverde, S. Changes in the carbohydrate composition of legumes after soaking and cooking. *J. Am. Diet. Assoc.* **1993a**, *93*, 547–550.

Vidal-Valverde, C.; Frias, J.; Valverde, S. Effect of processing on the soluble carbohydrate content of lentils. *J. Food. Protect.* **1992b**, *55*, 301–304.

Vidal-Valverde, C.; Prodanov, M.; Sierra, I. Natural fermentation of lentils. Influence of time, temperature and flour concentration on the kinetics of thiamin, riboflavin and niacin. *Z. Lebensm. Unters. Forsch.* **1997**, *205*, 464–469.

Vidal-Valverde, C.; Sierra, I.; Frias, J.; Prodanov, M.; Sotomayor, C.; Hedley, C.; Urbano, G. Nutritional evaluation of lentil flours obtained after short-time soaking process. *Eur. Food Res. Technol.* **2002b**, *215*, 138–144.

Villaume, C.; Chandrasiri, V.; Bau, H. M.; Nicolas, J. P.; Méjean, L. Effet du traitement technologique de préparation des protéines de soja sur la prise alimentaire du rat en croissance. *Sci. Alim.* **1993**, *13*, 377–383.

Wang, N.; Daun, J. K. Effects of variety and crude protein content on nutrients and anti-nutrients in lentils (*Lens culinaris*). *Food Chem.* **2006**, *95*, 493–502.

Wu, W.; Williams, W.; Kunkel, M.; Acton, J.; Huang, Y.; Wardlaw, F.; Grimes, L. Amino acid availability and availability-corrected amino acid score of red kidney beans (*Phaseolus vulgaris,* L.). *J. Agric. Food Chem.* **1996**, *44*, 1296–1301.

Yeung, A. C.; Glahn, R. P.; Miller, D. D. Comparison of the availability of various iron fortificants in bread and milk using an *in vitro* digestion Caco-2 cell culture method. *J. Food Sci.* **2002**, *67*, 2357–2361.

Yoneda, S.; Nakatsubo, F. Effects of the hydroxylation patterns and degrees of polymerization of condensed tannins on their metal-chelating capacity. *J. Wood Chem. Technol.* **1998**, *18*, 193–205.

Zhao, Y. H.; Manthey, F. A.; Chang, S. K. C.; Hou, H. J.; Yuan, S. H. Quality characteristics of spaghetti as affected by green and yellow pea, lentil, and chickpea flours. *J. Food Sci.* **2005**, *70*, S371–S376.

CHAPTER 6

GLOBAL PRODUCTION AND WORLD TRADE

DAVID L. McNEIL[1], GEORGE D. HILL[2], MICHAEL MATERNE[3], AND BRUCE A. McKENZIE[2]

[1] *School of Agricultural Science, University of Tasmania, Private Bag 54, Hobart, Tasmania 7001, Australia*

[2] *Agriculture Group, Agriculture and Life Science Division., PO Box 84, Lincoln University, Canterbury, New Zealand*

[3] *Grains Innovation Park, The Department of Primary Industries, Private Bag 260, Horsham, Victoria 3401, Australia*

E-mail: david.mcneil@utas.edu.au

Abstract: Lentils are a major international pulse crop (4 million Ha harvested in 2005). However, they fall well behind the major cereal and oilseed crops in planted area as well as behind the other pulse crops of peas, chickpeas and beans. Yields tend to be low (global mean of approximately 0.8 t/ha over the last 16 years) with 95% of the crop raingrown. There are three major areas of production N America, the Indian sub continent and Turkey. There are other areas of production in Australia, Iran, Syria and China. Between them these areas account for over 90% of global production. There are two major groups red (70–80%) and green lentils with Canada being the largest global producer of green lentils. Lentil production in the developing world is relatively static while the population in South Asia, where most lentils are consumed, has been rapidly increasing. This has left countries such as India with a very low supply per head of population. This deficit has to be made up by increases in world trade. The major world player in lentil exports is Canada which in 2005 exported 576,000 t. Other major exporters in the same year were Turkey (118,000 t), Australia (108,000 t) the United States of America (160,000 t). Most importing countries import relatively small quantities from a number of countries. In 2004 the largest lentil importers were Bangladesh (110,000 t), Sri Lanka (93,000 t), Egypt (89,000 t) and Colombia (63,000 t). A recent nine month ban by India on lentil exports has lead to a sharp increase in their price on the world market. In the past some countries, such as Turkey, imported lentils from Canada, processed them, and then re-exported them

1. INTRODUCTION

Traditionally lentils have been consumed where they are grown as a peasant crop. Approximately 70% of world lentil production is consumed in the country where they are grown (Agriculture and Agri-Food Canada, 2002c). Lentils are grown world wide

S.S. Yadav et al. (eds.), Lentil: An Ancient Crop for Modern Times, 95–105.

as a dryland crop with relatively little grown under irrigation as they respond poorly to irrigation and the high inputs that are characteristic of the system. For example the The Southeastern Anatolia Project, in Turkey planned to have 7.5% of the area (envisaged at 1.7 million hectares at completion) sown to lentils wheras the actual amount sown is only 1.5% (Anonymous 2007). Lentils are not irrigated in Canada, USA or Australia. Some irrigated cropping occurs in various parts of Asia. However, it tends to be very small with less than 0.2% of the irrigated area in Vaishali under lentils (Reddy, 2006) and less than 10% of the total pulse crop being irrigated in India (Gupta 2003). Thus all statistics given are dominated by dry land production.

In countries such as India in the last 40 years average lentil yield has hardly changed compared with increases in cereal yields such as rice and wheat. In 1961 lentil yield in India was 453 kg ha^{-1} by 2004 it had only risen to 760 kg ha^{-1} (FAOSTAT 2007). Over the same period the population of India rose from 439 to 1,029 million (Registrar General of India 2007). Bangladesh, in the same region, has had considerable population growth and is now home to more than 147 million people (CIA 2007) while the population of Pakistan went from 40 million to 136 million by 1995 and is predicted to reach 357 million by 2050 (IIASA 2007). Overall IIASA (2007) predicts that most world population growth will be in Asia and among countries where lentil is a common item of diet. Population growth will be high in India, Pakistan, and Bangladesh. Given all of these countries have limited available land, and water, to further increase pulse grain production, the shortfall in production will have to come from significantly increased imports from developed countries such as Canada and Australia.

2. GLOBAL PRODUCTION SITUATION

Lentils fall into several cartegories based primarilty on cotyledon and seed coat colour. Green and red lentils are the predominat lentil types grown, consumed and traded internationally. Green lentils have a yellow cotyledon and pale green seed coat and red lentils have an orange cotyledon and usually a dark seed coat, although the dominant seed coat colour varies between countries. Green lentils are typically cooked and consumed whole and red lentils split for use in products such as soups and dhal. Red lentils constitute 70–80% of world production (Patterson 2006). These two groups may be further subdivided based on size (small, medium and large). Generally the green lentils are also larger sized than red lentils, however, there are small green and medium-large red lentils. In addition there are a range of minor niche varieties (low tannin, black, dark green, speckled and brown) which may be locally important For example, the French have traditionally grown and prefer the DuPuy type lentil that has a mottled green and blue seed coat and yellow cotyledon and a brown dotted lentil with yellow cotyledon is consumed in Spain. Internationally the minor varieties only represent a small component of trade, constitute less than 3% of total Canadian production (Skrypetz 2000), less than 1% of production in Australia (Materne pers. comm.). In the USA medium green lentils and the Spanish brown type variety Pardina are grown with 20,000 t of Pardina exported to Spain anuually. FAO figures group all the lentils types into a single category (FAOSTAT, 2007). The type of lentil produced and preferred in traditional lentil growing countries tends to be the type grown for many centuries. Therefore, being a relatively newly traded commoditiy, export orientated countries are targeting the production of types traditionally grown in importing countries.

Production in Australia, and the Indian subcontinent region is dominated by red lentil production (Materne, pers comm 2007). Turkey also produces mainly red lentils but also approximately 6% green lentils (DeGraaf 2004) a situation similar to Syria. Alternatively Canadian production has shifted from 13% red lentils in 1998–1999 (Skrypetz 2000) to the present situation of approximately 56% red lentil production (Patterson 2006, Skrypetz 2006). Largely due to a decrese in green lentil production as a result of low prices. The type of lentil imported and consumed in regions depends on local preferences, for example, Spain preferring brown lentils and France the dark green speckled lentils and Algeria the large green lentils (Anonymous 2005).

While locally important and a valuable pulse crop in many regions lentils are a relatively minor crop on a global scale. In 2004 total global lentil production was approximately 4×10^6 metric tonnes (Figure 1) compared with 23 for beans, 12 for dry peas 8.5 for chickpeas 206 for soy beans and 626 for wheat (FAOSTAT 2007). Lentils are grown in many countries but production is dominated by India, Canada and Turkey with annual production of 2,125,000 tonnes or 61.4% of world production in 2006. A positive trend in lentil production is largely due to increased area rather than increasing yields (Figure 1). If demand countries to rise, due to population and economic growth, then the current production trend increase (2.5% per year) is likely to continue and production will increase to 5,000,000 tonnes by 2020. Increased production will primarily come from the benefits of research and new varieties (chapter 12), but also from increased area, predominantly in current lentil growing regions. For example, an increase in production in Nepal has come from replacing fallow with lentil and Canada could potentially return to its 2005 production area of 900,000 ha.

Lentils are also predominantly a subsistence crop with 33% of the global crop traded internationally between 2000/1 and 2004/5 years (FAOSTAT 2007). Canada has dominated world trade for this period accounting for 39% of global exports with Turkey next at 14%, Australia 15%, India 11%, USA 8% and Syria 3%. In total these 6 countries constitute over 90% of global exports. Imports are more widely distributed and more variable with 30 separate countries accounting for only 85% of imports. Major importers have been Bangladesh, Sri Lanka and Egypt each of which has imported over 100,000 mt in at least 1 year between 2000 and 2005. Import volumes can be variable depending on local crop production and political (eg tarif) influences. In 2000 Turkey imported 141,000 mt of lentils compared with only 6,000 in 2004 and 64,000 in 2005. However, it should be remembered that countries such as Turkey, Egypt and Dubai may import lentils for processing and then export.

Lentil production varies widely between years in response to domestic weather conditions, carry over stocks, world prices and farm policies (Figures 2–4). For example Australian lentil production and yields have fluctuated wildly in response to climatic conditions over the last 5 years (Table 1 taken from data supplied by Pulse Australia, 2007). Alternatively, yields in India have been more reliable but are low.

Canadian production has responded strongly to market oversupply of green lentils. Total lentil production in 2005/6 rose to 1.3×10^6 metric tonnes from 0.3×10^6

Table 1. Australian lentil production statistics from 2002 to 2006

Year	Harvested Yield 000 mt	Planted Area 000 ha	Average yield t/ha
2002	45	165	0.27
2003	160	131	1.22
2004	93	131	0.7
2005	209	113	1.85
2006	38	152	0.25

Compiled from http://www.pulseaus.com.au/statistics_and_market_overview/crop_production/

metric tonnes in 2002/3. In the same period area seeded rose from 600–900 kha. This resulted in carry over stocks of 0.6×10^6 metric tonnes of predominantly green lentils going into the 2006/7 season (Skrypetz 2007). The grower response was a drop in planted area to less than 600 kha in 2006 a shift towards lower yielding red lentils and an expected halving of total production (Figures 2–4).

Government policies also have an impact on production and trade. For example, it is no coincidence that lentil production has increased since the introduction of the U.S. Farm Bill during 2002 which provides Loan Deficiency Payment for US lentil producers if prices fall below a predetermined level. World lentil production has generally been trending upwards from increases in both yields and area harvested (Figure 1). While fluctuating widely most large producers have been increasing production (Figure 2) with the exception of Turkey where the lentil harvested area fell by 50% between 1990 and 2004. This was partially off set by increasing yields (up from 0.9 to 1.2 t/ha; Figure 3).

3. GLOBAL PRODUCTION FIGURES

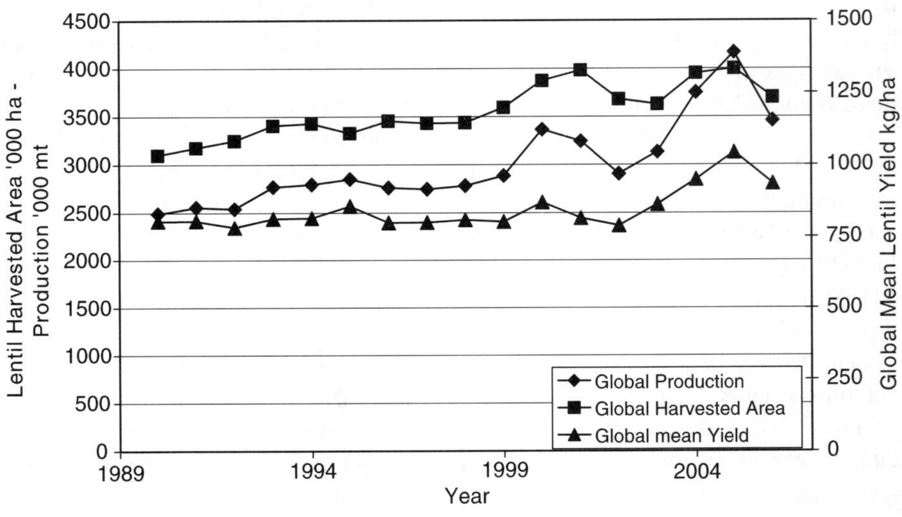

Figure 1. Global production, harvested areas and mean yields for lentils from 1990 to 2006. Data for 2006 are estimates. Data compiled from FAOSTAT 2007, Skrypetz 2006 and Pulse Australia 2006

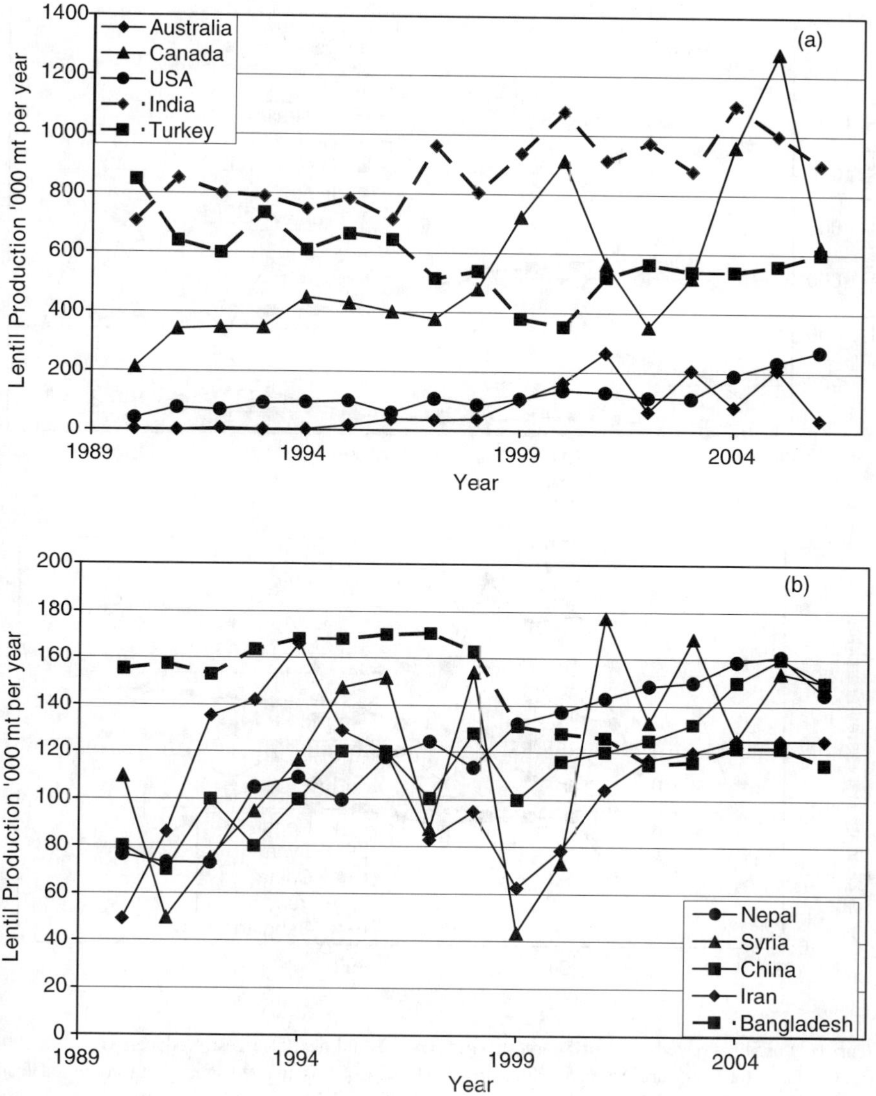

Figure 2. Lentil production for 5 largest producers (a), and next 5 largest producers (b) from 1990 to 2006. Data for 2006 are estimates. Data compiled from FAOSTAT 2007, Skrypetz 2006 and Pulse Australia 2006

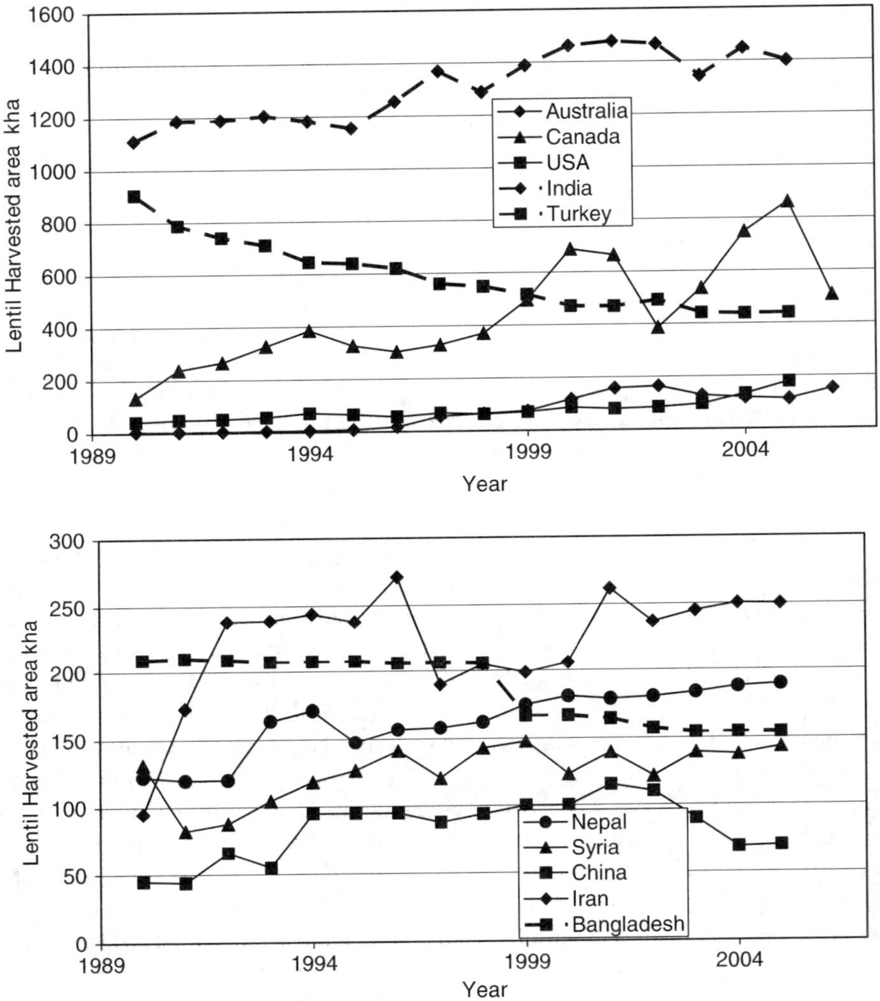

Figure 3. Lentil harvested area for 5 largest producers (a), and next 5 largest producers (b) from 1990 to 2006. Data for 2006 are estimates, area for Australia 2006 is planted area. Data compiled from FAOSTAT 2007, Skrypetz 2006 and Pulse Australia 2006

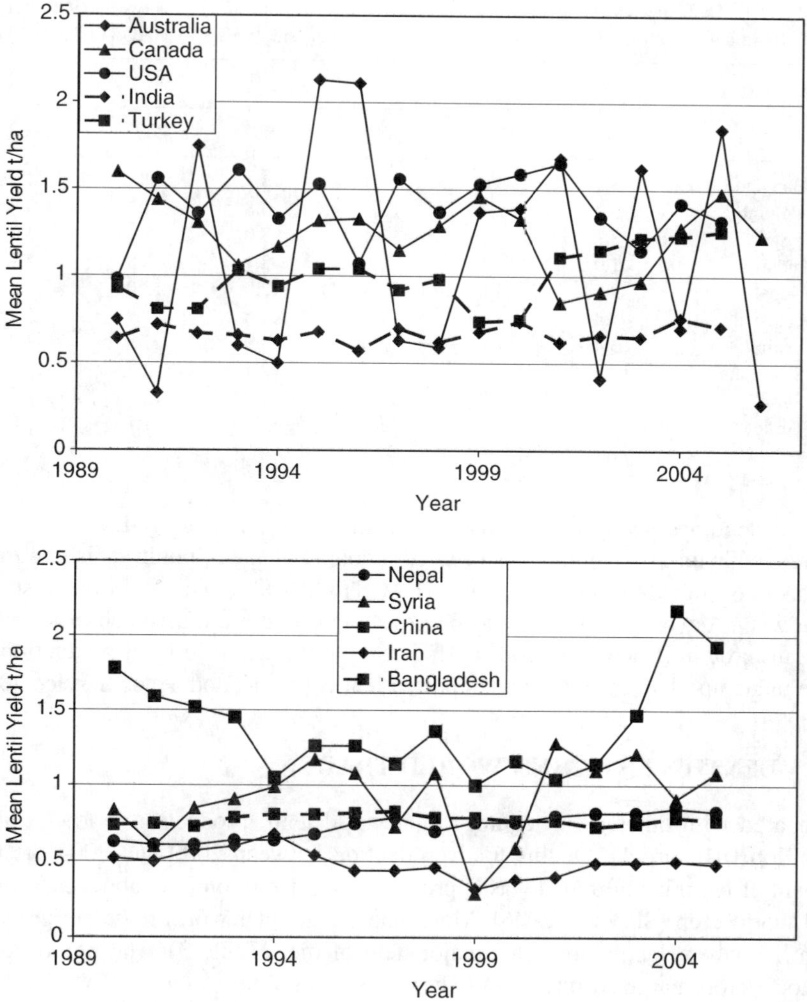

Figure 4. Lentil mean yields per ha of harvested area for 5 largest producers (a), and next 5 largest producers (b) from 1990 to 2006. Data for 2006 are estimates, yield for Australia 2006 is relative to planted area. Data compiled from FAOSTAT 2007, Skrypetz 2006 and Pulse Australia 2006

4. LENTIL AVAILABILITY

The substantial increases in south Asian populations outlined above have had a negative effect on the availability of lentil seed for human consumption (Table 2).

In countries such as Pakistan and India lentil availability is little more than 1 to 2 g day^{-1} (FAOSTAT 2007). Lentil producing countries, where lentils are regularly consumed, such as Nepal, Syria and Turkey have a higher daily consumption of 7 to 12 g. Spain is a European country where there is a long tradition of consumption

Table 2. Daily per capita availability of lentil for human consumption between 1999 and 2004 in countries where lentils are a major item of human diet (g day^{-1}) (FAOSTAT, 2007)

Country	Year					
	1999	2000	2001	2002	2003	2004
Bangladesh	3.1	3.0	3.1	3.1	4.2	4.0
Colombia	3.3	3.4	3.5	3.8	4.2	3.8
Egypt	3.2	3.1	3.1	3.0	3.1	3.0
India	1.9	2.1	2.0	2.1	1.8	2.1
Nepal	9.8	14.0	12.9	12.0	12.0	12.5
Pakistan	1.4	1.3	1.7	1.6	1.9	1.2
Spain	3.4	4.2	3.9	3.9	3.8	3.9
Sri Lanka	10.7	8.5	13.6	12.1	12.7	13.3
Syria	6.5	8.3	10.2	11.8	12.4	7.0
Turkey	8.9	10.9	11.9	11.9	7.1	8.0

of grain legumes. Daily lentil consumption in Spain is about 4 g day^{-1} (Table 2). However, Spain is a country where large amounts of meat, poultry, fish and eggs are also consumed as protein sources (FAOSTAT 2007). This is in contrast with India where devout Hindus tend to be vegetarians and are thus much more reliant on plant protein (Price et al. 2003). Therefore any shortfall in local production has to be made up by the purchase of imported lentils (Agri-Food Trade Service 2005).

5. LENTILS ENTERING WORLD TRADE

Compared with other grain legumes total world lentil production is low at about 4 million t (Figure 1). Of this total production between 2000 and 2004 the total amount of lentil being traded was approximately 1.1 million t i.e. about 27% of the total world crop (Skrypetz 2006). Most lentils that enter world trade are grown in countries where lentils are not a major item of diet (Table 3). The proportion of the local crop marketed internationally varies from almost none (CRNIndia 2007) in India to a large proportion of the crop in Canada (Skrypetz, 2006).

In June 2006 India put an embargo on lentil, and other pulse, exports. This was initially to 26 December 2006 but was later extended to 31 March 2007 (Subramani, 2006; Statpub 2007). The move drew negative responses from both Sri Lanka and Bangladesh both of which had imported significant quantities of lentil from India (Subramani, 2006). It also lead to an increase in the price of lentils on the world market.

6. MAJOR LENTIL EXPORTERS

CRNIndia (2007) list, in order of decreasing quantity exported, Canada, Turkey, Australia, India, the United States, Syria, China, UAE, Nepal and Belgium as the principal lentil exporters. However, the first seven of these countries account for nearly 90% of total world lentil exports. Table 3 gives the exports of these countries from 1999 to 2005.

Over the period reported exports have fluctuated quite widely in major producing countries. Canadian exports varied from 352,000 t in 2002 to 519,000 t in 2000. Statistics Canada gives its export data for 2005/6 as 669,000 t and forecasts exports of 730,000 in 2006/7, and 580,000 t in 2007/8 (Skrypetz 2007) these values are being highly influenced by carryover stocks which rose from 55,000 t in 2002/3 to 590,000 t in 2005/6. Over the same period Australian exports ranged from 25,000 to 242,000 t and those of the United States from 76,000 to 160,000 t. As the vast majority of lentils are grown as a rain fed crop (Chapter 11) these wide year to year variations in exports volumes were probably related to varying climatic conditions in the major exporting countries.

7. MAJOR LENTIL IMPORTERS

Table 4 lists those fifteen countries that, on average, import more than 20,000 t of lentil in 2004 (The last year for which complete export figures are available in FAOSTAT). A further 13 countries imported between 18,000 and 6,000 t. The rest of the world accounted for approximately a further 180,000 t. Generally importing countries import relatively small amounts of lentils.

The largest single lentil importer is Bangladesh which in recent years has imported more than 100,000 t each year. In terms of import volumes it is followed by Sri Lanka (93,000 t), Egypt (89,000 t), Colombia (63,000 t), Spain (41,000 t) Algeria (39,000 t) and Pakistan (36,000 t). All other importing countries import less than 32,000 t year^{-1} (Skrypetz 2006).

8. LENTIL RE-EXPORTS

A number of countries have imported lentils and then re-exported them depending on local supply and demand. In the Middle East The United Arab Emirates is

Table 3. Exports of lentils by major exporting countries from 1999 to 2005 (1,000 t) (FAOSTAT 2007, Skrypetz 2006)

Country	Year						
	1999	2000	2001	2002	2003	2004	2005
Australia	25	134	218	242	85	150	108
Canada	417	519	491	351	371	374	576
China	22	18	15	21	33	38	34
India	147	191	106	86	83	137	*n/a*
Turkey	105	100	159	119	217	171	118
Syria	40	17	12	11	70	71	*n/a*
United States	76	80	99	103	97	88	160

Table 4. Major lentil importers 2000 to 2005 (1,000 t) (Skrypetz, 2006)

Country	Year					
	2000	2001	2002	2003	2004	2005
Algeria	72	47	63	67	39	86
Bangladesh	37	47	63	123	110	n/a
Colombia	67	50	65	53	63	67
Egypt	77	113	100	61	89	n/a
France	36	32	31	32	27	33
Germany	37	26	21	21	24	20
India	21	87	67	38	27	n/a
Italy	28	28	27	31	27	28
Mexico	26	31	29	29	31	30
Pakistan	37	68	67	81	36	n/a
Peru	25	28	27	20	25	n/a
Saudi Arabia	15	25	21	24	26	n/a
Spain	50	47	47	47	41	54
Sri Lanka	80	91	107	91	93	n/a
Sudan	22	14	20	14	32	n/a

responsible 80% of total pulse grain imports into the region. However, it is estimated that 60% of these imports are processed in some way, repackaged, and sold to India and other local countries (Kizirian and Taha 2007). Turkey both imports and exports lentils (Sarigedik 2006). Turkish imports are mainly green lentil and exports mainly red lentil. Formerly, Turkey used to import large quantities of lentils from Canada which were re-exported. However, since 2000 changes in policy in Turkey have made it more difficult to import and re-export large quantities of lentil (Agriculture and Agri-Food Canada (2002a).

India also imports lentils and after processing them re-exports some of them to Sri Lanka and Pakistan (Agriculture and Agri-Food Canada (2002c). Statpub (1998) reported that only very small amounts of Chinese lentils were re-exported. Thus although a proportion of world lentils that enter trade is subject to re-exportation it would appear that at present the total amount involved is relatively small.

REFERENCES

Agriculture and Agri-Food Canada (2002a) Pulse crops in the Middle East and North Africa. Bi-weekly Bulletin vol 15 edn 5 pp 1–4.

Agriculture and Agri-Food Canada (2002b) Pulse crops in South Asia. Bi-weekly Bulletin vol 15 edn 10 pp 1–4.

Agriculture and Agri-Food Canada (2002c) Lentils/Fababeans. Bi-weekly Bulletin vol 15 edn 11 pp 1–6.

Agri-Food Trade Service (2005) Market information Asia Pacific, Agri-Food Country Profile, India September 2005. http://its.agr.ca/asia/3664_e.htm

CIA (2007) The World Factbook – Bangladesh, pp 1–11. https://www.cia.gov/cia/publications/factbook/print/bg.html.

Anonymous (2007) Factors affecting cropping patterns in irrigated areas and preventative measures necessary in the southeastern Anatolia project (GAP). http://www.gap.gov.tr/English/Dergi/D7121999/onlem.html

Anonymous (2005) Canada's lentil industry, Agriculture and Agri-food Canada, Special crops profile, http://www.agr.gc.ca/misb/spec/index_e.php?s1 = len&s2 = &page = intro&type = prof

CRNIndia (2007) Commodity lentil (masur). http://www.crnindia.co/commodity/masur.html

DeGraaf, J. (2004) Turkey lentil market report. Canadian Agrifood Trade Service. http://ats.agr.ca/europe/384_e.htm

FAOSTAT (2007) Food and Agriculture Organisation of the United Nations Rome. http://faostat.fao.org.

Gupta, R.N. (2003) Market Information, Asia Pacific, Pulses – Brief – India, August 2003. Canadian Agrifood Trade Service. http://ats.agr.ca/asia/3825_e.htm IIASA (2007) World population: Major trends, pp 1–7. http://www.iiasa.ac.at/Research/LUC/Paper/gkh1/chap1.htm.

Kizirian H, Taha M. (2007) GCC-5 Pulses market: Worth more than a hill of beans. pp 1–4, FASonline. http://www.fas.usda.gov/info/agexporter/1997/November%201997/gcc5.html.

Patterson, G. (2006) Shift from green to red lentils: the right thing to do. Sakatchewan Pulse Growers Journal, PulsePoint, October 2006, page 30.

Price G K, Landes R, Govindan A. (2003). India's pulse sector: Results of field research. Electronic Outlook Report from the Economic Research Service, USDA. www.ers.usda.gov.

Pulse Australia (2007) Crop production market overviews, http://www.pulseaus.com.au/statistics_and_market_overview/crop_production/

Reddy 2006 ICAR research complex for eastern region, Patna, Bihar, India: Site Presentation. http://www.gecafs.org/meetings_gecafs/2006_06_27/Dr%20ReddyPatna.ppt.

Registrar General of India (2007) Population of India (1951–2001), S-117. http:/indiabudget.nic.in.

Sarigedik U. (2006) Turkey grain and feed annual report 2006. USDA Foreign Agricultural Service GAIN Report No TU6010, pp 1–24.

Skrypetz, S. (2000) Lentils Situation and Outlook. Bi-weekly Bulletin, Agriculture and Agri-food Canada, December 15, 13(21). pp 4.

Skrypetz, S. (2006) Lentils Situation and Outlook. Bi-weekly Bulletin, Agriculture and Agri-food Canada, May 12, 19(7). pp 4.

Skrypetz, S. (2007) Canada: Pulse and special crops outlook Agriculture and Agri-food Canada, Feb. 6, 1 pp 4.

Statpub (1998) China expects 85,000 lentil crop. http://www.statpub.com/stat/open/0rajjpj.html

Statpub (2006) India extends pulse export ban. http://www.statpub.com/open/203055.phtml

Subramani M. (2006) Centre notified ban on pulses export. http://www.thehindubusinessline.com/2006/06/29stories/2006062904261200.htm

CHAPTER 7

LENTIL-BASED CROPPING SYSTEMS

H. S. SEKHON, GURIQBAL SINGH, AND HARI RAM

Department of Plant Breeding, Genetics and Biotechnology, Punjab Agricultural University, Ludhiana, Punjab, India

E-mail: sekhonhsd@yahoo.com

Abstract: Lentil is a protein-rich winter season pulse crop. Its cultivation is concentrated mostly in semi-arid regions in the Indian sub-continent and dry areas of Middle East. It can be grown under conserved moisture conditions after monsoon rains. The sowing of lentil is popular as mono and sequential cropping, intercropping, mixed cropping, relay cropping and multistorey cropping in various countries. In India, Pakistan, Bangladesh and Nepal, rice-lentil system is more common but its cultivation is also done after maize, cotton, sorghum and pearlmillet. It can be intercropped successfully in wheat, barley, mustard and linseed. For mixed/intercropping optimum seeding and planting configuration is very important to achieve higher total productivity. In various experiments lentil + wheat (30%), lentil + mustard 5 : 1 row ratio and lentil + linseed 5 : 1 row ratio showed highest land equivalent ratio than sole lentil. In autumn sugarcane, lentil intercropping revealed higher cane equivalent yield than sole sugarcane. In eastern India, the broadcasting of lentil seed in standing rice about 15 days before the harvest gave significantly higher grain yield than lentil sown after the harvest of rice. The inclusion of lentil in various cropping systems improves physical properties of soil and increases the yield of succeeding cereal crop due to biological nitrogen fixation and other rotational effects. Proper agronomic management, use of bio-fertilizers and mechanical cultivation may not only improve productivity but also help to bring large area under lentil in various cropping systems

1. INTRODUCTION

Pulses play an important role in agriculture and lentil is one of the important pulse crops. Like other pulse crops, lentil is known for biological nitrogen fixation (Unkovich and Pate, 2000) and thereby improves soil health, particularly in poorer areas, as well as human health by providing protein-rich grains. Due to demand of land for other uses cropped land area in the world is not going to increase. However, human population is increasing day by day, thereby demanding more food production including from pulses such as lentil. Therefore, there is a need to increase cropping intensity by including lentil in different cropping systems.

S.S. Yadav et al. (eds.), Lentil: An Ancient Crop for Modern Times, 107–126.

Lentil growing is concentrated mostly in the semi-arid regions in the Indian subcontinent and dry areas of the middle east. Globally, in 2004 it occupies 3.95 m ha area and has 3.75 mt production and 948 kg/ha average yield (FAOSTAT, 2004). In South Asia it is primarily grown in India, Nepal, Bangladesh, Pakistan and Myanmar. The productivity of lentil in Canada, Australia and U.S.A. is 1304, 1571 and 1274 kg/ha, respectively (Tickoo *et al.* 2005). At present the developed countries contribute 38% of the global output. In contrast, Asia has recorded decline in its share from 78% in 1991 to 60% in 2001. Out of 51 countries where lentil is grown, recently Canada has emerged the largest producer followed by India, Turkey, Australia, Nepal, Bangladesh, Syria and Iran (Ali and Kumar, 2005). In India, lentil is grown on 1.45 m ha area and the average yield is 759 kg/ha. The major areas are Bundelkhand region of Uttar Pradesh and Madhya Pradesh and Tal (low lying) area of Bihar. Among the cultivated species there are two races i.e. microsperma (small and round seeded) and macrosperma (large and flat seeded). Countries of South Asia mainly grow microsperma type whereas the macrosperma are predominant in Southern Europe, North Africa and Latin America. In India, 40–45% of the total area is occupied by macrosperma and their cultivation is concentrated in Madhya Pradesh and Jhansi division of Uttar Pradesh (Bahl *et al.*, 1993).

Agro-climatically, the lentil growing areas of South Asia can be classified into North Western Plains (province of Pakistan and states of Haryana, Punjab and Western Uttar Pradesh of India); North Eastern Plains (Myanmar, Bangladesh and Eastern Uttar Pradesh, Bihar and West Bengal states of India); Central Highlands/Plateau (plateau of Madhya Pradesh, Uttar Pradesh and Maharashtra states of India) and *Terai* Region (Himalayan foothills of Pakistan, India and Nepal). In South Asia, lentil is mostly grown during the winter season (October–March) under conserved soil moisture during the preceding monsoon months.

A diverse range of farm practices for growing lentils are followed in different regions of the world. Lentil can be grown as mono and sequential cropping, mixed cropping, intercropping, relay cropping and multi-storey cropping. There are many factors, including agro-climatic, technological and socio-economic, which influence the success and adoption of any crop/cropping system in any area. Lentil is successfully grown in various cropping systems in different parts of the world. This chapter discusses briefly the various aspects of lentil-based cropping systems.

2. MONO AND SEQUENTIAL CROPPING

Monocropping (monoculture – 100% cropping intensity) refers to a system in which same crop is grown year after year in the same field. It is an age-old practice, mainly under rainfed conditions. The sequence includes fallow-lentil. Traditionally the system was sustainable because the population pressure on land was low and the subsistence farming was in vogue. However, the system of monoculture is risky, unstable over different seasons, provides low returns and also causes the build up of diseases and nematodes. These problems may be solved, to some extent, by ideal crop rotations (sequential cropping). The sequential cropping may include two or more crops in a year over the same field. The double cropping (200% cropping intensity) system is very popular in the Uttar Pradesh, Bihar, West Bengal states of India and in Nepal (Ali *et al.*, 1993). In India, sowing of lentil after monsoon rains fallows is popular in Madhya Pradesh and adjoining areas of Uttar Pradesh where generally large seeded lentils are grown. The same practice is more common in Eastern parts of Indian subcontinent where lentil is sown after monsoon/flood water recedes. In the *Tal* areas (*diara* land) where lentils are sown after receding flood water, grain yields are quite high as soils are fertile because of silt deposits and have good soil moisture. In Bangladesh, practice of sowing lentil in October – November in fields kept fallow after the harvest of *aus* rice is more common. The late sowing of lentil in November – December is done after the harvest of *aman* (late) rice. The other practice is deep-water rice-lentil. In Punjab, Haryana, Rajasthan and Maharashtra states of India and in parts of Pakistan the lentil crop is taken after the harvest of *kharif* (rainy season) crops such as rice, maize, cotton, pearlmillet and sorghum. In these areas the lentil crop is benefited from irrigation. India is divided into 15 agro-climatic zones and the important lentil-based cropping systems in these zones are presented in Table 1. Important sequential cropping, intercropping, mixed cropping relay cropping and multi-storey cropping system involving lentil in various countries are given in Table 2.

Sequential cropping is virtually the only form of cropping of lentil in Canada, Turkey, USA, Australia and Spain that together comprise 48% of global lentil production (FAOSTAT, 2004). Rotational crops include cereals, oilseeds, pastures and fallows (e.g. Peoples *et al.*, 2001).

3. MIXED OR INTERCROPPING

Mixed cropping denotes a system of growing two or more crops (or varieties) simultaneously with no row arrangement on the same piece of land. Crop mixtures may be of a cereal crop, a pulse crop and an oilseed crop in different proportions and the mixture is generally sown by a broadcast method. Mixed cropping varies from one area to another and even differs among farmers within a single location. Indeed, each system tends to reflect farmers' needs, resources, economic considerations and convenience, marketing feasibility and labour availability (Aiyer, 1949). The traditional practice of mixed cropping has gained popularity in recent years in the form of intercropping with a suitable modification in planting pattern. Scientifically, intercropping (polyculture) is defined as growing of two or more dissimilar crops simultaneously on the same piece of land in a distinct row arrangement. This may occur by using a base crop to which is added rows of an additional component crop. The recommended optimum plant population of the base crop is suitably combined with appropriate additional plant density of the component crop. In fact, intercropping is the space and time dependent form of multiple cropping.

Intercropping provides significant advantages in land use efficiency, crop productivity and monetary returns as compared with sole cropping under diverse agroecologial situations. It also results in more efficient use of solar energy and

Table 1. Important lentil-based cropping systems in different agro-climatic zones of India

Sr. No.	Zone	States represented	Annual rainfall (mm)	Cropping systems
1	Western Himalayan Region	Jammu & Kashmir, Himachal Pradesh, Uttar Pradesh	1650–2000	Rice-lentil Ragi-lentil
2	Eastern Himalayan Region	Assam, West Bengal, Manipur, Meghalya Nagaland, Arunachal Pradesh	1840–2030	Maize-lentil Jute-urdbean-lentil Jute-winter rice-lentil
3	Low Gangetic Plains Region	West Bengal	1300–1600	Maize-lentil Rice-lentil Rice-chickpea+lentil
4	Middle Gangetic Plains Region	Uttar Pradesh, Bihar	1200–1470	Rice-lentil
5	Uttar Pradesh Upper Gangetic Plains Region	Uttar Pradesh	780–900	Rice-lentil
6	Trans Gangetic Plains Region	Punjab, Haryana	360–890	Rice-lentil Maize – lentil Sorghum(fodder)-lentil Cotton-lentil
7	Eastern Plateau and Hills Region	Madhya Pradesh, Maharashtra, Orissa, West Bengal	1270–1430	Early rice- lentil Rice-lentil + linseed/ mustard/barley
8	Central Plateau and Hills Region	Madhya Pradesh, Rajasthan, Uttar Pradesh	490–1570	Rice-lentil Maize-lentil Cotton-lentil
9	Western Plateau and Hills Region	Maharashtra, Madhya Pradesh, Rajasthan	600–1040	Soybean – lentil Sorghum – lentil Pearlmillet – lentil
10	Southern Plateau and Hills Region	Andhra Pradesh Tamil Naidu Karnataka	680–1000	Sorghum – lentil Pearl millet - lentil
11	East Coast Plains and Hills Region	Orissa, Andhra Pradesh, Tamil Naidu, Pandicherry	780–1290	Maize – lentil
12	West Coast Plains and Hills Region	Tamil Naidu, Kerala, Goa, Karnataka, Maharashtra	2230–3640	–

13	Gujarat Plains and Hills Region	Gujarat	340–1790	–
14	Western Dry Region	Rajasthan	400	Sorghum – lentil Pearlmillet – lentil
15	Island Region	Andeman and Nikobar Island, Lakshdeep Islands	1500–3090	–

inputs. However, selection of crop/variety and agronomic requirement aspect is very important. As a principle component crops should have contrasting maturities to reduce competition for the same resources at the same time, variable rooting pattern for better utilization of moisture and nutrients from different plant depths, different plant height for efficient use of light. Erect and compact plant type of component crops is the main considerations in intercropping. Indeed, the intercrop should have either synergistic or complementary effect relative to the base crop. There should be no mutual inhibition.

3.1. Wheat-Lentil Intercropping

Intercropping of lentil in wheat with optimum seeding and planting configuration may provide appreciable increase in total productivity. In Nepal, Pathak and Malla (1981) observed that wheat and lentil in 1:1 ratio was beneficial in land utilization as this combination showed high land equivalent ratio (LER). In Bangladesh, Ahmed *et al.* (1987) in an experiment on mixed cropping of wheat and lentil, found that 2:1 was the most compatible, promising and economically profitable seeding ratios for wheat and lentil. Rahman and Shamusuddin (1981) obtained the highest total productivity, LER and net return when 30% (by weight relative to lentil sead) of wheat seed was sown in between lentil rows spaced 30 cm apart. Miah and Rahman (1993) found maximum LER (1.45) in lentil (100%) + wheat (30%) in 10:3 seed rate ratio (Table 3). Mixed cropping of lentil (80%) + wheat (20%) or lentil (70%) – wheat (30%) provided high LER (Ciftci and Ulker, 2005). Volunteer wheat interference does not influence lentil growth for the first 80 days after crop emergence, but afterwards, the lentil height, straw and grain yield is affected by the presence of wheat (Ghosheh and El-Shatnawi, 2006). Further the semi-tall cultivar 'Hourani' had more adverse effect on lentil growth and yield than the semi-dwarf cultivar, 'ACSADES'. With intercropping of lentil, in wheat, though the height of lentil plants was increased, yet the number of branches/plant, number of pods/plant and grain yield of lentil were reduced (Khan *et al.*, 2005).

3.2. Mustard-Lentil Intercropping

Mustard and lentil are important winter season oilseed and pulse crops of North Western parts of India. The success of mustard + lentil cropping system depends

Table 2. Important lentil-based sequential cropping, intercropping, mixed cropping, relay cropping multi-storey cropping systems

Cropping system	Area	References
Mixed Cropping		
Wheat + lentil	Mymensingh, Bangladesh	Akter *et al.* (2004)
Linseed + lentil	Uttar Pradesh, India	Mishra and Ali (2002)
Lentil + wheat	Van, Turkey	Ciftci and Ulker (2005)
Lentil + barley	Van, Turkey	Ciftci and Ulker (2005)
Sequential cropping		
Lentil-finger millet	Uttar Pradesh, India	Prakash *et al.* (1991)
Lentil-wheat	Kermanbshah, Iran	Sayadian and Taliee (2001)
Rice-lentil	West Bengal, India	Brahmachari *et al.* (2004)
Rice-lentil	Uttar Pradesh, India	Singh *et al.* (2000a)
Rice-lentil	Uttar Pradesh, India	Singh *et al.* (2001a & 2001b)
Soybean-lentil		Bhattacharya *et al.* (2006)
Lentil-wheat		Badarneh (2005)
Soybean-lentil	Madhya Pradesh, India	Gupta *et al.* (2003)
Wheat-lentil	Syria	Harmsen *et al.* (2001)
Rice-lentil		Sarkar *et al.* (2003)
Rice-lentil	Uttaranchal, India	Prakash *et al.* (2002)
Lentil-canola	Saskatchewan, Canada	Miller *et al.* (2003)
Lentil-mustard	Saskatchewan, Canada	Miller *et al.* (2003)
Lentil-durum wheat	Saskatchewan, Canada	Gan *et al.* (2003)
Intercropping		
Sugarcane (Autumn) + lentil	Uttaranchal, India	Rana *et al.* (2006)
	West Bengal, India	Suman *et al.* (2006)
		Giri (2005)
Wheat + lentil	Dakota	Carr *et al.* (1995)
Wheat + lentil	El-Gemmeiza, Egypt	Mahrous *et al.* (1998)
	Dera Ismailkhan, Pakistan	Khan *et al.* (2005)
Lentil + linseed	West Bengal, India	Sarkar *et al.* (2004)
Lentil + mustard	Bihar, India	Singh *et al.* (2000b)
Chickpea + lentil	Punjab, Pakistan	Ali *et al.* (2005a & 2005b)
Rapeseed + lentil	West Bengal, India	Sarkar and Pal (2005)
Wheat + lentil	Mymensingh, Bangladesh	Akter *et al.* (2004)
Barley + lentil	Central Europe	Schmidtke *et al.* (2004)
Maize + lentil	Uttar Pradesh, India	Misra *et al.* (2001)
Lentil + linseed	Uttar Pradesh, India	Singh *et al.* (1998)
Lentil + Indian mustard	Uttar Pradesh, India	Singh *et al.* (1998)

Utera/Relay cropping		
Rice + lentil	Madhya Pradesh, India	Dwivedi and Sharma (2005)
Rice + lentil	Faizabad, Pakistan	Jabbar *et al.* (2005)
Multi-storey cropping		
Shisham (*Dalbergia sissoo*) + lentil	Haryana, India	Nandal and Singh (2001)
Eucalyptus + lentil		Kumar and Nandal (2004)

Table 3. Grain yield of lentil and wheat and LER as influenced by different seed ratios

Treatment	Ratio	Grain yield (t/ha)		LER
		Lentil	Wheat	
Lentil sole	1:0	1.52	–	1.00
Lentil + 10% wheat	10:1	1.51	0.32	1.21
Lentil + 20% wheat	5:1	1.48	0.48	1.30
Lentil + 30% wheat	10:3	1.47	0.70	1.45
Lentil + 50% wheat	2:1	1.46	0.58	1.31
Wheat sole		–	1.46	1.00

Source : Miah and Rahman (1993).

on the selection of variety, soil type, cultivation conditions, planting pattern etc. Iqbal (1989) evaluated the performance of mustard and lentil grown in different mixed and intercrops combinations under variable seed rate ratios and plant geometry. The yield advantage was the highest (20%) in the mixed cropping of mustard and lentil with seed ratio of mustard (75%) + lentil (25%) followed by the yield advantage of 15% from the intercropping of mustard and lentil with (2 paired rows) + lentil broadcast. Singh and Rajput (1996) obtained the highest lentil – equivalent yield and net profit when lentil and mustard (var. Karanti) were sown in 6:1 row ratio. Mustard variety ND 8501 showed the lowest lentil equivalent yield and net returns. The row ratio of 6:1 during both the years of study gave higher grain yield than 4:1 row ratio in all the varieties. This might be attributed to lesser competition for moisture, nutrients, space and light. Intercropping of lentil + Indian mustard (*Brassica juncea*) in 6:1 (Singh *et al.*, 1998) and in 5:1 (Kumar *et al.*, 2001) row arrangement produced higher grain yield compared to other cropping systems. Singh *et al.* (2000b) revealed maximum lentil equivalent yield and land equivalent ratio in lentil (100%) + wheat (25%) plant density at 5:1 row ratio (Table 4).

3.3. Linseed – Lentil Intercropping

In numerous studies, the intercropping of lentil with linseed at different sowing configurations showed the advantage of intercropping over a sole crop. Lentil + linseed in 3:2 pattern had maximum (1.61) LER (Miah and Rahman, 1993).

Table 4. Seed yield, lentil equivalent yield, land equivalent ratio, net return and benefit : cost ratio of lentil + mustard intercropping under different plant densities and row arrangements in Bihar, India

Treatment		Row ratio	Seed yield (t/ha)		Lentil equivalent yield (t/ha)	Land-equivalent ratio	Net return (Rs/ha)	Benefit cost ratio
Plant density (%)								
Lentil	Mustard		Lentil	Mustard				
100	25	2:1	0.86	0.60	1.48	1.24	14914	2.93
100	50	2:1	0.65	0.72	1.40	1.17	13391	2.45
100	25	3:1	0.75	0.46	1.23	1.02	11469	2.25
100	50	3:1	0.68	0.78	1.49	1.25	14638	2.68
100	25	5:1	1.12	0.70	1.85	1.54	19824	3.89
100	50	5:1	1.00	0.68	1.70	1.42	17558	3.21
67	33	2:1	0.70	0.62	1.34	1.12	13338	2.78
75	25	3:1	0.85	0.45	1.32	1.09	13003	2.72
83	17	5:1	1.10	0.35	1.46	1.20	14989	3.15
Sole lentil			1.25	–	1.25	1.00	12234	2.63
Sole mustard			–	1.10	1.14	1.00	10477	2.13
CD (P = 0.05)			0.16	0.12	0.21		2502	0.53

Source: Singh *et al.* (2000b).

In intercropping system genotypes of lentil do differ in productivity (Tomar *et al.*, 2000). In Uttar Pradesh (India), in lentil + linseed system, lentil genotype 'L 4076' produced higher grain yield than 'DPL 62' (Mishra and Ali, 2002), thereby indicating its better compatibility in lentil–linseed intercropping. Further, lentil equivalent yield and LER and area-time equivalent ratio were the highest in 'L 4076' lentil + linseed system under a 6:2 arrangement indicating that this combination was more efficient in utilizing time and area. Contrarily Neupane and Bharati (1993) in Nepal did not observe beneficial effect of lentil + linseed intercropping. Sarkar *et al.* (2004) showed highest lentil equivalent yield in the case of lentil (100%) + linseed (25%) plant density at a row ratio of 5:1 (Table 5).

3.4. Sugarcane-Lentil Intercropping

Sugarcane is a long-duration commercial crop sown at wide row spacing and its initial growth is slow. Therefore, a short-duration pulse crop can be intercropped in autumn planted sugarcane. In the North Eastern plains lentil can be successfully intercropped with autumn sugarcane (Srivastava, 1975). Panwar *et al.* (1990) observed higher cane yield sown during autumn with lentil intercrop. Rana *et al.* (2006) observed higher cane equivalent yield in sugarcane + lentil system than sugarcane alone. The heaviest cane growth was recorded in crop intercropped with lentil, attributed to no shading effect and nitrogen fixation (Table 6).

4. RELAY CROPPING

Relay cropping refers to sowing of succeeding crop after flowering and before the harvest of a standing crop. It is analogous to a relay race where one crop hands the baton to the next crop in quick succession. Farmers having assured irrigation can take to relay cropping system to augment their income. This system is usually practiced in the low lands of Eastern India where lentil is sown in a standing crop of rice as the latter reaches physiological maturity. This system is also called *paira* or *utera* cropping. The seeds of lentil are broadcast in a standing rice crop after the excess water is drained about two weeks before its harvest. It is generally followed to make the best use of residual moisture of rice fields. The practice enables better and early establishment of lentil seedlings due to an adequate availability of soil moisture which otherwise is lost quickly once the rice is harvested.

In a study at Berhampur (West Bengal), Chakraborty *et al.* (1976) obtained high productivity of lentil under *paira* cropping with *aman* rice (late rice). At Dholi in Bihar, Roy Sharma *et al.* (1997) found that *paira* cropping of lentil produced higher (1.27 t/ha) when sown after the harvest of rice than its seeding on well prepared land after the harvest of paddy (0.65 t/ha). The cost-benefit ratio in the case of *paira* and late sown crop was 4.26 and 1.26, respectively. The higher yield from *paira* cropping over late sown was due to better growth because of increased growth duration. Late sown crop is also subjected to low temperatures (7 to 8° C) during germination and early growth

Table 5. Seed yield, lentil equivalent yield, land equivalent ratio, net return and benefit : cost ratio of lentil and linseed intercropping under different plant densities and row arrangements in West Bengal, India

Treatment		Row ratio	Seed yield (t/ha)		Lentil equivalent yield (t/ha)	Land-equivalent ratio	Net return (Rs/ha)	Benefit cost ratio
Plant density (%)								
Lentil	Linseed		Lentil	Linseed				
100	25	2:1	0.93	0.74	1.57	1.41	15082	2.48
100	50	2:1	0.87	0.85	1.61	1.46	15324	2.45
100	25	3:1	0.95	0.64	1.51	1.37	13830	2.33
100	50	3:1	0.84	0.83	1.56	1.43	14396	2.35
100	25	5:1	1.05	0.82	1.76	1.58	17632	2.65
100	50	5:1	1.00	0.84	1.73	1.56	16890	2.55
67	33	2:1	0.82	0.94	1.64	1.52	15872	2.52
75	25	3:1	0.95	0.83	1.67	1.54	15998	2.47
83	17	5:1	1.12	0.61	1.65	1.44	15448	2.40
Sole lentil			1.30	–	1.30	1.00	10600	2.03
Sole linseed			–	1.05	0.91	1.00	4204	1.40
CD (P = 0.05)			0.01	0.02	0.02			

Source : Sarkar *et al.* (2004).

Table 6. Effect of lentil and *rajmash* intercropping in sugarcane

Treatment	Shoot height (cm)	Cane equivalent yield (t/ha)
Sugarcane sole	95	131.5
Sugarcane + lentil	101	138.0 (1.07)*
Sugarcane + *rajmash*	93	133.7 (0.85)

*Values in parentheses are intercrop yields
Source : Rana *et al.* (2006).

stages thus producing low biomass. Studies conducted by various workers indicated that sowing at the optimum time for relay cropping is very important.

Roy Sharma *et al.* (1984) reported that lentil should be broadcast in the standing rice about 3–4 weeks before the harvest of rice. Experiments conducted at Khumaltar in Nepal indicated that the best time for relay cropping of lentil in rice was 15–20 days prior to the harvesting of rice (NGLRP, 1990). Field trials laid out at Berhampur (West Bengal) under the All India Coordinated Research Project (AICRP) showed that relay crop of lentil sown in rice 15 days before its harvest gave significantly higher yield than that sown 7 days before the harvest of rice (AICRP, 2004 & 2006) (Table 7).

Panwar *et al.* (1981) and Roy Sharma *et al.* (1984) reported beneficial effect of *Rhizobium* inoculation in *paira* cropping. Roy Sharma *et al.* (1984) suggested to use 70 kg/ha seed rate of lentil for *paira* cropping. Similar results were found in AICRP trial (AICRP, 2006) undertaken at Berhampur. This study also showed higher yields with seed soaking in KH_2PO_4 2% solution for 6 h or with the use of sprouted seeds for sowing (AICRP 2004 & 2006) (Table 7). The grain yield of

Table 7. Grain yield of lentil as affected by seed soaking, seed rate and time of sowing of lentil in rice *utera* system at Berhampore West Bengal, India

Treatment	Grain yield (t/ha) (2003–04)*	Treatment	Grain yield (t/ha) (2005–06)**
Time of sowing		**Seed rate (kg/ha)**	
7 days before rice harvest	1.16	40	0.77
15 days before rice harvest	1.23	50	0.79
C.D. 5%	0.04	60	0.96
		70	1.05
		C.D. 5%	0.04
Seed soaking		**Seed soaking**	
No soaking (control)	1.01	No soaking (control)	0.79
Soaking in water for 6 h	1.17	Soaking in water for 6 h	0.86
Soaking in KH_2PO_4 2% solution for 6 h	1.26	Soaking in KH_2PO_4 2% solution for 6 h	0.93
Sowing of sprouted seeds	1.33	Sowing of sprouted seeds	0.98
C.D. 5%	0.03	C.D. 5%	0.04

Source : *AICRP (2004), **AICRP (2006).

Table 8. Nutrient requirement of lentil in rice-lentil production system

Nutrients (kg/ha)	Grain yield of lentil (t/ha)	
	No-tillage	Deep tillage (10 cm)
30 : 20 P_2O_5 and S	1.16	1.28
60 : 30 P_2O_5 and S	1.30	1.41
30 : 20 : 15 P_2O_5, S and Zn	1.39	1.51
60 : 30 : 15 P_2O_5, S and Zn	1.43	1.57
30 : 20 : 15 : 5 P_2O_5, S, Zn and B	1.39	1.60
60 : 20 : 15 : 5 P_2O_5, S, Zn and B	1.58	1.71

C.D. 5% Tillage × Nutrients = 0.17
Source : IIPR, (2004).

lentil in the case of control, inoculation, inoculation + 20 kg N/ha, inoculation + 20 kg N + 30 kg P_2O_5/ha was 0.92, 1.1, 1.4 and 1.59 t/ha, respectively. However, the increment in yield under inoculation, N and P top dressing was less in late sown crop. Studies conducted at the Indian Institute of Pulses Research, Kanpur showed that deep tillage was essential for obtaining higher yield of lentil. The deep tillage treatment with 30:20:15:5 kg/ha P_2O_5, S, Zn and B gave grain yield almost equivalent to 60:20:15:5 kg/ha P_2O_5, S, Zn and B without tillage (IIPR, 2004) (Table 8).

5. MULTI-STOREY CROPPING

Multi-storey or multitier cropping is a system of growing together crops of different heights at the same time on the same piece of land and thus using land, water, nutrients, space and light most efficiently and economically i.e., papaya + citrus + lentil or coconut + papaya + lentil.

Tall trees are generally planted at a wider spacing. Therefore, tree canopies let through most of light and it is possible to grow other crops underneath the tall trees. The low-growing crop should be more mesophytic, less light demanding and with horizontal leaves as these can intercept more light. On the other hand, the top of the canopy (dominant species) should consist of a high light requirement and should have erect and narrow leaves, so that it may cause less shade. Kumar and Nandal (2004) revealed that the yield of lentil, *berseem*, wheat, potato and mustard under two and a half year old eucalyptus planted at 6 m × 2 m decreased the grain yield to the tune of 16.3, 52.1, 62.3, 80.8 and 82.4%, respectively. Thus, lentil recorded the least reduction in yield. Semwal *et al.* (2002) suggested that in tree-crop mixed cropping lopping of trees could help in increasing the productivity of crops as compared to no lopping.

6. EFFECT OF LENTIL ON YIELD OF SUCCEEDING CROP

In lentil-based cropping systems, grain/seed yields of crops are generally higher after lentil as compared to after non-legumes including cereals, oilseeds, etc. Higher yields after lentil could be due to increased soil fertility owing to biological nitrogen fixation or due to other rotational effects. Prasad *et al.* (1990) in a field experiment conducted at the Indian Agricultural Research Institute, New Delhi showed that rice grain yield was 0.3–0.5 t/ha more when grown after lentil than after fallow. Grain yield of rice could be as much as 23.4% higher after lentil as compared to after wheat (Prakash *et al.*, 2002). Furthermore, lentil is as good as chickpea in increasing the productivity of succeeding rice crop. Fingermillet (*Eleusine coracana* L.) yielded higher after lentil (2.19 t/ha) than after wheat (1.75 t/ha) (Prakash *et al.*, 1991).

7. NITROGEN BENEFITS TO THE SUCCEEDING CROP

Lentil, being a legume, has the ability to fix atmospheric nitrogen in the soil in association with specific rhizobia. Lentil could derive two-third of its nitrogen needs from the atmosphere (Badarneh, 2005). At maturity, the level of symbiotically fixed nitrogen in two years has been reported to be 154 and 117 kg N/ha in monocropped lentil and 95 and 41 kg N/ha in lentil intercropped with barley (Schmidtke *et al.*, 2004). In Pakistan, in lentil, nitrogen derived from nitrogen fixation ranged from 50 to 87% (mean 73%), crop N fixed was 42–85 kg N/ha (mean 68 kg N/ha) and average nitrogen balanced was + 27 kg N/ha (with residues) and + 16 kg N/ha (without residues) (Shah *et al.*, 2003). Use of bio-inoculants such as *Rhizobium* spp. and *Azospirillum brasilense* helps in enhancing nodulation in lentil (Tiwari and Misra, 2000; Chandra and Pareek, 2002; Kumar and Chandra, 2005) and thereby nitrogen fixation. Therefore, these inoculants may need to be used to ensure greater inputs of nitrogen fixation in lentil-based cropping systems.

Lentil, when grown as sole or intercropped with wheat, produced yield advantage on succeeding fingermillet (*Eleusine coracana* L.) equivalent to 19.81 and 10.38 kg N/ha, respectively (Prakash *et al.*, 1991).

8. EFFECT ON PHYSICAL PROPERTIES OF SOIL

Inclusion of lentil in a cropping system helps in improving physical properties of soil. Soil from a Soybean – lentil rotation had lower bulk density than soybean – wheat rotation (Prakash *et al.*, 2004; Bhattacharyya *et al.*, 2006). Furthermore, under zero tillage, soybean – lentil rotation resulted in better soil water retention and transmission properties than soybean – wheat.

9. WEED MANAGEMENT

Season-long weed competition could cause as high as 73% yield reduction as compared to weed free plots (Phogat *et al.*, 2003). Herbicides need to be carefully selected for controlling weeds in inter/mixed cropping systems, as these should be safe to both the crops (main and inter crops.). Some herbicides applied in a crop may have long life and thereby cause some adverse effect on the succeeding sensitive crop. On the other hand, such herbicide could provide effective weed control in the succeeding crop as well. Therefore, in lentil-based cropping systems, it is essential to know whether the previous crop had received any herbicides application which could be detrimental to the lentil crop. Atrazine or pendimethalin application at the recommended dose in maize does not leave any residues to affect succeeding lentil crop (Reddy and Tyagi, 2005). Application of anilofos + 2, 4-DEE, anilofos, butachlor or pendimethalin followed by one hand weeding at 40 days after sowing (DAS) in rice recorded the lower weed population and higher yield of succeeding lentil, indicating these herbicides to be safe to lentil as well as their increased carryover effect for effective weed management (Singh *et al.*, 2000a).

Removal of weeds in lentil sown by *utera* system is very important factor. One hand weeding about 30 DAS gave as good yield as in two hand weedings done at 30 and 45 DAS. The post-emergence application of oxyflourfen at 0.05 kg/ha produced 10% higher yield than the unweeded treatment (Table 9).

Intercropping systems may have smothering effect on weeds. Intercropping lentil with wheat is very effective in reducing vegetative production by weeds compared with sole-cropped lentil (Carr *et al.*, 1995).

10. NUTRIENT MANAGEMENT

All crops need nutrients to provide high yields. However, fertilizer recommendations should not be based on a single crop, rather these should be based on a cropping system. In rice + lentil (*Utera*) cropping system application of fertilizers to both the crops provide higher income than application to rice only (Dwivedi and Sharma, 2005). However, in wheat – lentil cropping system, lentil benefits greatly from phosphorus fertilizer applied to the preceding wheat crop (Harmsen *et al.*, 2001), and thereby resulting in reduction in the cost of lentil production without significantly reducing lentil yields. In rice – lentil cropping sequence, grain yields of lentil are higher with organic sources either alone or combined with inorganic nutrients (Sarkar *et al.*, 2003). Full recommended dose of nitrogen application to rice significantly increases the grain yield and nutrient uptake of succeeding lentil (Singh *et al.*, 2001b).

Table 9. Grain yield of lentil as influenced by weed management under *utera* conditions at Dholi, India

Weed management	Grain yield (t/ha)
Unweedy check	1.81
Weed free	2.15
One hand weeding 30 DAS	2.08
One hand weeding 45 DAS	1.87
Two hand weedings 30 + 45 DAS	2.07
Oxyfluorfen 0.05 kg/ha	1.99
C.D. (P = 0.05)	0.26

Source : AICRP (1997).

11. DISEASE INCIDENCE

Lentil – cowpea – mungbean cropping sequence reduces nematode population more than chickpea – okra – chilli, mustard – mungbean – tomato and tomato – fallow – okra (Wani, 2005). Infestations by some pathogens, such as ascochyta and sclerotinia fungi, may appear more extensive in intercropped lentil with wheat than sole cropping (Carr *et al.*, 1995), probably because the microclimate created by the wheat canopy could favour disease development. However, these researchers reported disease incidence in one out of three years experimentation, which clearly indicates that moist conditions favorable for disease development in intercropping system may not appear very frequently to discourage intercropping.

12. ECONOMICS

Economics is one of the important factors influencing adoption of any crop/cropping system by the farmers. Intercropping of mustard, common bean, rapeseed and lentil with sugarcane provided net return of Rs. 62104, 65067, 67138 and 69040 against Rs.71145 in sole cropping of sugarcane (Rana *et al.*, 2006). In another study, in Bangladesh, sugarcane + lentil intercropping was found not profitable because it produced less net benefit than sole sugarcane (Hossain *et al.*, 2004), whereas in West Bengal, India, sugarcane + lentil was found to provide the highest benefit : cost ratio (3.11) (Giri, 2005). In Lebanon, lentils were found to provide higher net revenue ($ 160 per ha) than chickpea ($ 97 per ha) and barley ($ 27 per ha) (Yau, 2004). In Bangladesh, mixed cropping of lentil (100%) and wheat (40%) recorded the highest land equivalent ratio (1.52), monetary advantage (63%) and benefit : cost ratio (1.84) (Akter *et al.*, 2004). In West Bengal, India, an intercropping of 100% lentil + 25% linseed in 5:1 row ratio provided maximum net return and benefit : cost ratio (Sarkar *et al.*, 2004) (Table 5). Similarly in Bihar, India, an intercropping of 100% lentil + 25% mustard in 5:1 row ratio provided maximum net return and net benefit : cost ratio (Singh *et al.*, 2000b) (Table 4) The lentil + Indian mustard as well as lentil + linseed intercropping in 6:1 row ratios gave higher net profit and benefit : cost ratio than other row ratios involving lower numbers of lentil rows (Singh *et al.*, 1998).

13. THRUST AREAS

Lentil is an important pulse for human consumption. Bold seeded lentils are preferred in the market and these may respond to high levels of seed rate and phosphorus (Singh *et al.*, 2003). Therefore, there is a need to develop high yielding, short-duration, disease resistant, bold seeded varieties and study their agronomy to realize their full yield potential. Many diseases such as Fusarium wilt, rust, Ascochyta blight, botrytis gray mould, stemphylium blight, stem rot, collar rot, root rot, powdery mildew, downy mildew, etc. reduce the grain yields of lentil. International organizations should come forward for providing disease resistant material to the lentil growing countries. For example, International Centre for Agricultural Research in the Dry Areas (ICARDA) has a rich collection of over 10000 accessions of cultivated lentil and more than 500 accessions of lentil wild relatives, which could be used as base materials for sources of resistance (Sarker *et al.*, 2005). Development of tall varieties, suitable for mechanical harvesting, could help in bringing large area under lentil in various cropping systems.

There is a need to intensify research on the use of bio-fertilizers, nutrients and weed control particularly in relay cropping systems of lentil. The effective strains of rhizobia under different agroclimatic conditions should be identified. Improvement should be made through improved techniques for efficient utilization of residual soil moisture for better crop establishment. There is also a dire need to work out the optimum seed ratios, proper planting geometry, etc. with new varieties for increased productivity and economic returns.

REFERENCES

Ahmed, A., Rahman, M.A. and Kalley, T.G. 1987. Study on the mixed cropping of wheat and lentil at varying seeding ratios under different levels of fertility. *Bangladesh Journal of Agricultural Research* 12(1), 53–59.

AICRP (All India Coordinated Research Project). 1997. All India Coordinated Research Project on Improvement of MULLaRP. Indian Institute of Pulses Research, Kanpur.

AICRP (All India Coordinated Research Project). 2004. All India Coordinated Research Project on Improvement of MULLaRP. Indian Institute of Pulses Research, Kanpur.

AICRP (All India Coordinated Research Project). 2006. All India Coordinated Research Project on Improvement of MULLaRP. Indian Institute of Pulses Research, Kanpur.

Aiyer, A.K.Y.N. 1949. Mixed cropping in India. *Indian Journal of Agricultural Sciences*. 19, 439–543.

Akter, N., Alim, M. A., Islam, M. M., Naher, Z., Rahman, M. and Hossain, A. S. M. I. 2004. Evaluation of mixed cropping and intercropping of lentil and wheat. *Journal of Agronomy* 3 (1), 48–51.

Ali, H., Khan, M. A. and Ahmad, S. 2005a. A case study of competition functions in chickpea based intercropping. *Indus Journal of Biological Sciences* 2 (3), 357–361.

Ali, H., Khan, M. A. and Ahmad, S. 2005b. Quantitative response of chickpea grown in association with different intercrops. *Indus Journal of Biological Sciences* 2 (3), 371–374.

Ali, M. and Kumar, S. 2005. Current status and prospects of pulses production. In : *Pulses* (eds. Singh, G., Sekhon, H.S. and Kular, J.S.) Agrotech Publishing Academy, Udaipur, pp. 9–31.

Ali, M., Saraf, C.S., Singh, P.P., Rewari, R.B. and Ahlawat, I.P.S. 1993. Agronomy of lentil in India. In : Lentil in South Asia (eds. Erskine, W. and Saxena, M.C.). International Center for Agricultural Research in the Dry Areas, ICARDA, Aleppo, Syria. pp. 103–127.

Badarneh, D. M. D. 2005. Crop nitrogen uptake in a legume-wheat rotation using 15N methodology. *Dirasat Agricultural Sciences* 32 (2), 229–238.

Bahl, P.N., Lal, S. and Sharma, B.M. 1993. An overview of the production and problems of lentil in South Asia. In : *Lentil in South Asia* (eds. Erskine, W. and Saxena, M.C.). International Center for Agricultural Research in the Dry Areas, ICARDA, Aleppo, Syria. pp. 1–10.

Bhattacharyya, R., Prakash, V., Kundu, S. and Gupta, H. S. 2006. Effect of tillage and crop rotations on pore size distribution and soil hydraulic conductivity in sandy clay loam soil of the Indian Himalayas. *Soil & Tillage Research* 86 (2), 129–140.

Brahmachari, K. , Somnath Pal and Mondal, N. N. 2004. Preparation of fishmeal (well decomposed) and its effects on different crops under rice-based cropping system in Coastal Saline Zone of West Bengal. *Indian Agriculturist* 48 (3/4), 203–205.

Carr, P.M., Gardner, J.C., Schatz, B.G., Zwinger, S.W. and Guldan, S.J. 1995. Grain yield and weed biomass of a wheat – lentil intercrop. *Agronomy Journal* 87, 574–579.

Chakraborty, L.N., Sen, S.N., Mandal, S.K., Gupta, S.K. and Mukherjee, D. 1976. Possibility of utilizing rice fallows in West Bengal. *Field Crop Abstracts* 29, 2382.

Chandra, R. and Pareek, R.P. 2002. Effect of rhizobacteria in urdbean and lentil. *Indian Journal of Pulses Research* 15(2), 152–155.

Ciftci, V. and Ülker, M. 2005. Economic benefits of mixed cropping of lentil (*Lens culinaris*) with wheat (*Triticum aestivum*) and barley (*Hordeum vulgare*) at different seeding ratios. *Indian Journal of Agricultural Sciences* 75 (2), 100–102.

Cuberi, J.I. 1981. Origin, taxonomy and domestication. In : Lentil (eds. Walb, C. and Hawtin, G.) *Commonwealth Agricultural Bureaux*, England, pp. 15–38.

Dwivedi, R. K. and Sharma, R. S. 2005. Nutrient management of utera lentil (*Lens culinaris*) under rice-based cropping system. *Crop Research* (Hisar) 29 (2), 179–181.

FAOSTAT 2004. http://faostat.fao.org. Food and Agriculture Organization of the United Nations, Rome.

Gan, Y. T., Miller, P. R., McConkey, B. G., Zentner, R. P., Stevenson, F. C. and McDonald, C. L. 2003. Influence of diverse cropping sequences on durum wheat yield and protein in the semiarid Northern Great Plains. *Agronomy Journal* 95 (2), 245–252.

Ghosheh, H. Z. and El-Shatnawi, M. K. 2006. Influence of volunteer durum wheat (*Triticum durum*) cultivars and density on lentils (*Lens culinaris*). *Acta Agronomica Hungarica* 54 (1), 101–108.

Giri, S. 2005. Influence of different intercrops on the incidence of borer pests, productivity and profitability of autumn planted sugarcane in Gangetic alluvial zone of West Bengal. *Indian Sugar* 55 (2), 105–108.

Gupta, S. K., Tiwari, R. K. and Khan, R. A. 2003. Economic evaluation of soybean-based cropping system under rainfed conditions of Kymore plateau, Madhya Pradesh. *JNKVV Research Journal* 37 (1), 102–103.

Harmsen, K., Matar, A. E., Saxena, M. C. and Silim, S. N. 2001. Yield response to phosphorus fertilizer in a wheat-lentil rotation in a Mediterranean environment. NJAS – Wageningen *Journal of Life Sciences* 49 (4), 385–404.

Hossain, G. M. A., Haque, M. A., Mahmud, K., Haque, M. I. and Anam, M. R. 2004. Feasibility study of different intercrops with sugarcane at Chuadanga Region. *Journal of Agriculture & Rural Development* (Gazipur) 2 (1), 115–120.

Iqbal, M.T. 1989. Performance of mustard and lentil grown in different mixed and intrcrop combinations. M.Sc. Agronomy. Dissertation BAU, Mymensingh.

Jabbar, A., Ahmad, R., Ehsanullah and Nazir, M. S. 2005. Agro-economic performance of diversified rice-based relay cropping systems at zero and conventional tillage under strip plantation. *Pakistan Journal of Agricultural Sciences* 42 (1/2), 18–21.

Khan, M., Khan, R. U., Muhammad, D. and Rashid, A. 2005. Mutual effect of legume and cereal intercropping under rodkohi rainfed conditions of D.I. Khan. *Sarhad Journal of Agriculture* 21 (4), 629–632.

Kumar, A. and Nandal, D. P. S. 2004. Performance of winter crops under Eucalyptus tereticornis based agrisilviculture system. *Indian Journal of Agroforestry* 6 (2), 97–98.

Kumar, R. and Chandra, R. 2005. Effect of adhesives on survival of inoculated *Rhizobium leguminosarum* on seed and symbiotic performance in lentil under field conditions. *Indian Journal of Pulses Research* 18(2), 206–210.

Kumar, R., Prakash, O. and Singh, B. P. 2001. Effect of lentil based intercropping on root growth, crop yield, protein and oil production under drylands. *Indian Journal of Pulses Research* 14 (1), 48–51.

Mahrous, M. A., Eisa, M. S. and Abd-alla, A. A. 1998. Effect of intercropping wheat with lentil at varying nitrogen fertilization rates on yield and their components. *Annals of Agricultural Science*, Moshtohor 36 (1), 61–69.

Miah, A.A. and Rahman, M.M. 1993. Agronomy of lentil in Bangladesh. In : *Lentil in South Asia* (eds. Erskine, W. and Saxena, M.C.). International Center for Agricultural Research in the Dry Areas, ICARDA, Aleppo, Syria, pp. 128–138.

Miller, P. R., Gan, Y., McConkey, B. G. and McDonald, C. L 2003. Pulse crops for the Northern Great Plains: II. Cropping sequence effects on cereal, oilseed, and pulse crops. *Agronomy Journal* 95 (4), 980–986.

Mishra, J. P. and Ali, M. 2002. Effect of intercropping patterns on biological and economic sustainability of linseed (*Linum usitatissimum*) with lentil (*Lens culinaris*) genotypes. *Indian Journal of Agricultural Sciences* 72 (10), 577–580.

Misra, B. N., Singh, B. and Rajput, A. L. 2001. Yield, quality and economics as influenced by winter maize (*Zea mays*)-based intercropping system in eastern Uttar Pradesh. *Indian Journal of Agronomy* 46 (3), 425–431.

Nandal, D. P. S. and Singh, R. R. 2001. Productivity of different cropping sequences in *Dalbergia sissoo* Roxb. based agro-silviculture system. *Indian Journal of Forestry* 24 (4), 433–436.

Neupane, R.K. and Bharati, M.P. 1993. Agronomy of lentil in Bangladesh. In : *Lentil in South Asia* (eds. Erskine, W. and Saxena, M.C.) International Center for Agricultural Research in the Dry Areas, ICARDA, Aleppo, Syria, pp. 139–145.

NGLRP (National Grain Legume Research) 1990. Report on winter grain legumes research in Nepal. Paper presented in Winter Crops Workshop, NWDP, 10–14 September, 1990. Bhairahawa, Nepal.

Panwar, K.S., Faroda, A.S., Malik, D.S., Ahlawat, I.P.S., Dhingra, K.K., Rao, M.R., Singh, R.C., Dahiya, D.R., Singh, atar and Sharma, R.P. 1981. A Review on Quarter Century of Research Work on Grain Legumes Agronomy in India. 1955–1980. Published by *Indian Society of Agronomy*, New Delhi. pp. 176–202.

Pathik, D.S. and Malla, M. 1981. Wheat based intercropping trial. Report presented at Winter Crops Workshop, NWDP, September 1990, Bhairahwa, Nepal.

Peoples, M.B., Bowman, A.M., Gault, R.R., Herridge, D.F., McCallum, K.M., McCormick, M.H., Norton, R.M., Rochester, I.J., Scammell, G.J. and Schwenke, G.D. 2001. Factors regulating the contributions of fixed nitrogen by pasture and crop legumes to different farming systems of eastern Australia. *Plant and Soil* 228, 29–41.

Phogat, S.B., Kumar, S., Sangwan, N. and Hooda, R.S. 2003. Effect of herbicides and cultural practices on weed flora in lentil. *Indian Journal of Pulses Research* 16(2), 119–121.

Prakash, V., Bhattacharya, R. and Srivastva, A. K. 2004. Effect of tillage management on yield and soil properties under soybean (*Glycine max*)-based cropping system in mid-hills of north-western Himalayas. *Indian Journal of Agricultural Sciences* 74 (11), 573–577.

Prakash, V., Ghosh, B. N., Pandey, A. K. and Gupta, H. S. 2002. Effects of preceding winter legumes and nitrogen rates on N uptake, yield attributes and yield of rice, and monetary returns from rotation. *Annals of Agricultural Research* 23 (3), 402–406.

Prakash, V., Kumar, S. and Bhatnagar, V.K. 1991. Yield and N management in finger millet (*Eleusine coracana*) as affected by preceding winter legumes. *Indian Journal of Pulses Research* 4(2), 173–176.

Prasad, R., John, P.S., George, M., Singh, S. and Sharma, S.N. 1990. Effect of lentil residue management on the productivity and NPK removal by lentil-rice double cropping. *LENS Newsletter*. 17 (1), 5–8.

Rahman, M.A. and Shamusuddin 1981. Intercropping of lentil and wheat. *Bangladesh Journal of Agricultural Research* 6(2), 27–31.

Rana, N.S., Kumar, S., Saini, S.K. and Panwar, G.S. 2006. Production potential and profitability of autumn sugarcane-based intercropping systems as influenced by intercrops and row spacing. *Indian Journal of Agronomy* 51(1), 31–33.

Reddy, D. A. and Tyagi, S. K. 2005. Integrated weed management in maize-groundnut sequential cropping system – a review. *Agricultural Reviews* 26 (4), 235–248.

Roy Sharma, R.P., Thakur, H.C., Sharma, H.M., Mishra, S.S. and Thakur, S.S. 1984. Effect of fertilization and inoculation of *paira* and late sown pure lentil. *Indian Journal of Agronomy* 29(4), 459–462.

Roy Sharma, R.P., Sharma, H.M., and Mishra, S.S. 1997. Pulses in non-traditional areas and cropping systems – Possibilities and future research needs. In : *Recent Advances in Pulses Research* (eds. Asthana, A.N. and Ali, M.) Kanpur : Directorate of Pulses Research. pp. 547–566.

Sarkar, B. and Pal, A. K. 2005. Water use and yield of rapeseed (*Brassica campestris* var yellow sarson) and lentil (*Lens culinaris* Medik.) grown as sole crop and as inter crop. *Journal of Interacademicia* 9 (1), 10–15.

Sarkar, R. K., Malik, G. C. and Pal, P. K. 2004. Effect of intercropping lentil (*Lens culinaris*) and linseed (*Linum usitatissimum*) under varying plant density and row arrangement on productivity and advantages in system under rainfed upland. *Indian Journal of Agronomy* 49 (4), 241–243.

Sarkar, S., Singh, S. R. and Singh, R. P. 2003. The effect of organic and inorganic fertilizers on soil physical condition and the productivity of a rice-lentil cropping sequence in India. *Journal of Agricultural Science* 140 (4), 419–425.

Sarker, A., Bayaa, B. and Erskine, W. 2005. Combating lentil diseases through host-plant resistance. In *Proceedings of the 1st International Edible Legume Conference in conjunction with the IVth World Cowpea Congress,* Durban, South Africa, 17–21 April 2005, pp. 1–5.

Sayadian, K. and Taliee, A. A. 2001. Investigation to determine a suitable rotation for rain-fed wheat in Kermanshah. *Seed and Plant* 16 (4), 495–508.

Schmidtke, K., Neumann, A., Hof, C. and Rauber, R. 2004. Soil and atmospheric nitrogen uptake by lentil (*Lens culinaris* Medik.) and barley (*Hordeum vulgare* ssp. *nudum* L.) as monocrops and intercrops. *Field Crops Research* 87 (2/3), 245–256.

Semwal, R. L., Maikhuri, R. K., Rao, K. S., Singh, K. and Saxena, K. G. 2002. Crop productivity under differently lopped canopies of multipurpose trees in Central Himalaya, India. *Agroforestry Systems* 56 (1), 57–63.

Shah, Z., Shah, S. H., Peoples, M. B., Schwenke, G. D. and Herridge, D. F. 2003. Crop residue and fertilizer N effects on nitrogen fixation and yields of legume-cereal rotations and soil organic fertility. *Field Crops Research* 83 (1), 1–11.

Singh, D.P. and Rajput, A.L. 1996. Evaluation of Indian mustard (*Brassica juncea*) varieties in association with lentil (*Lens culinaris*). *Indian Journal of Agronomy*. 41(1), 27–29.

Singh, D.P., Rajput, A.L. and Singh, S.K. 1998. Productivity and economics of lentil (*Lens culinaris*) – based cropping system. *Indian Journal of Agronomy* 43(4), 588–590.

Singh, G. Wade, L.J., Singh, B.B., Singh, R.K. and Singh, V.P. 2001a. Nutrient management in semi-deep water (30–50 cm) rice (*Oryza sativa*) and its effect on succeeding lentil (*Lens culinaris*) crop. *Indian Journal of Agronomy* 46(1), 12–16.

Singh, G., Singh, B.B., Agarwal, R.L. and Nayak, R. 2000a. Effect of herbicidal weed management in direct–seeded rice (*Oryza sativa*) and its residual effect on succeeding lentil (*Lens culinaris*). *Indian Journal of Agronomy* 45(3), 470–476

Singh, M.K., Thakur, R., Pal, S.K., Verma, U.N. and Upasani, R.R. 2000b. Plant density and row arrangement of lentil (*Lens culinaris*) and mustard (*Brassica juncea*) intercropping for higher productivity under Bihar plateau. *Indian Journal of Agronomy* 45(2), 284–287.

Singh, O.N., Sharma, M. and Dash, R. 2003. Effect of seed rate, phosphorus and FYM application on growth and yield of bold seeded lentil. *Indian Journal of Pulses Research* 16(2), 116–118.

Singh, S.K., Varma, S.C. and Singh, R.P. 2001b. Effect of integrated nutrient management on yield, nutrient uptake and changes in soil fertility under rice (*Oryza sativa*) – lentil (*Lens culinaris*) cropping system. *Indian Journal of Agronomy* 46(2), 191–197.

Srivastava, S.C. 1975. Performance of legumes as intercrops in sugarcane. *Indian Journal of Genetics* 35, 269–270.

Suman, A., Lal, M., Singh, A. K. and Gaur, A. 2006. Microbial biomass turnover in Indian subtropical soils under different sugarcane intercropping systems. *Agronomy Journal* 98 (3), 698–704.

Tickoo, J.L., Sharma, B., Mishra, S.K. and Dikshit, H.K. 2005. Lentil (*Lens culinaris*) in India : Present status and future perspectives. *Indian Journal of Agricultural Sciences*. 75 (9), 532–62.

Tiwari, V.N. and Misra, S.K. 2000. Studies on nitrogen and bio-inoculants on biological nitrogen fixation and productivity of lentil. *Indian Journal of Pulses Research* 13(2), 39–42.

Tomar, S.K., Tripathi, P. and Rajput, A.L. 2000. Effect of genotype, seeding method and ammonium phosphate on yield and protein and nutrient uptake by lentil (*Lens culinaris*). *Indian Journal of Agronomy* 45(1), 148–152.

Unkovich, M.J. and Pate, J.S. 2000. An appraisal of recent field measurements of symbiotic N_2 fixation by annual legumes. *Field Crops Research* 65, 211–228.

Wani, A. H. 2005. Effect of cropping sequences and ploughing on plant parasitic nematodes and plant growth in field. *Indian Journal of Nematology* 35 (1), 63–67.

Yau, S. K. 2004. Safflower agronomic characters, yield and economic revenue in comparison with other rain-fed crops in a high-elevation, semi-arid Mediterranean environment. *Experimental Agriculture* 40 (4), 453–462.

CHAPTER 8

RHIZOBIUM MANAGEMENT AND NITROGEN FIXATION

DAVID L. McNEIL[1] AND MICHAEL MATERNE[2]

[1]*School of Agricultural Science, University of Tasmania, Private Bag 54, Hobart, Tasmania 7001, Australia*

[2]*Grains Innovation Park, The Department of Primary Industries, Private Bag 260, Horsham, Victoria 3401, Australia*

E-mail: david.mcneil@utas.edu.au

Abstract: Through fixing their own nitrogen, growing lentils offers a substantial saving in the need to use fossil fuels to generate fixed nitrogen for agricultural production. Much of the nitrogen fixed by the lentil crop may then be available for subsequent crops in the rotation as crop residues break down. Estimates for the contribution of N to soils by lentils are generally in the order of 20 kg N ha^{-1} yr^{-1}. However, the level of N fixed by lentils varies considerably, spatially and temporally in response to a host of environmental and ecological factors. Lentils require effective infection by *Rhizobium leguminosarum* in order to fix nitrogen. This infection process could fail due to a number of reasons including a lack of or inappropriate strains of rhizobia, failure of the plant to invest in the symbiosis, or through altered metabolism. While conditions that suit better growth of the lentil crop will normally enhance the nitrogen fixation of the crop there may also be specific situations in which the fixation process is more sensitive and fixation limits the growth of the crop

1. INTRODUCTION

Although the earth's atmosphere is 80% di-nitrogen (N_2) gas, nitrogen availability often limits agricultural production as this form of nitrogen cannot be utilised by plants. Plants require their nitrogen in a fixed form (e.g. ammonia, nitrate or organic compounds) primarily for use in synthesising proteins and nucleic acids. This fixed nitrogen limitation on global productivity is likely to increase with increasing global demand for food. Nitrogen fixation occurs both biologically and non-biologically. Burns and Hardy (1975) gave an early estimate of global (biological and non biological) fixed nitrogen of about 175 Tg N yr^{-1} which has been updated and increased by Cleveland et al (1999) who used over100 pre-existing published estimates of BNF to generate global-level estimates of biological N fixation. Their

S.S. Yadav et al. (eds.), Lentil: An Ancient Crop for Modern Times, 127–143.

best estimate of nitrogen fixation by biological ecosystems is 195 Tg N yr^{-1}, with a range of 100–290 Tg N yr^{-1}. This compares with 82 Tg N yr^{-1} from N fertilizer production in 1998 (primarily by the Haber-Bosch process and up from 3.5 Tg N yr^{-1} in 1950; Mosier, 2002). There are also lesser non biological inputs from combustion and lightning (Nesbitt et al., 2000).

Biological nitrogen fixation depends on bacterial enzymic reduction of N_2 via nitrogenase. This reduction can occur through the actions of free living, associative or symbiotic bacteria. The reduction to ammonia requires large amounts of energy and reductant. At least 16 molecules of ATP are consumed during the reduction of each molecule of di-nitrogen. However, reduction of NO_3^- by plants also requires energy (or direct use of photosynthetic electrons) at a similar level (Atkins, 1982). Legume root nodules consist of a symbiotic association between bacteria and plants and are a major source of fixed N for crops. Together with other crop systems (e.g. rice) this source of fixed N resulting from human crop production activities provides about 40 Tg N yr^{-1} equivalent to half of the total N applied annually as industrially produced fertilizers (Vitousek et al., 1997). They also point out that both biological and non biological anthropogenic (i.e. resulting from human activity) N fixation has been increasing. While this has greatly increased global food productivity it is also associated with increased environmental damage. The high cost and negative environmental impacts of artificial nitrogen fertilizers gives N fixing legume crops (e.g. lentils) a competitive advantage of being independent of soil nitrogen. Through fixing their own nitrogen, growing lentils offers a substantial saving in the need to use fossil fuels to generate fixed nitrogen for agricultural production. Much of the nitrogen fixed by the lentil crop may then be available for subsequent crops in the rotation as crop residues break down. Estimates for the contribution of N to soils by lentils are generally in the order of 20 kg N ha^{-1} yr^{-1} (Peel, 1998). However, the levels of N fixed by legumes (including lentils) varies considerably, spatially and temporally in response to a host of environmental and ecological factors (e.g. soil N level, water, P, pathogens etc.; Vitousek et al., 2002). For example, lentil plant residues after the harvest of seed varied from 1.0 T/ha to 5.8 T/ha across 3 years and 4 sowing times in Australia (Materne 2003).

2. PHYSIOLOGY OF N FIXATION IN LENTIL

Lentils form a symbiosis with *Rhizobium leguminosarum* (Humphrey et al., 2001) in which the bacteria are enclosed within root nodules in a specialized form known as bacteroids. Formation of this symbiosis is host specific through the interaction of both plant and bacterial genes (Begum et al., 2001, Smit et al., 1992). The complex structure of nodules exists to enable the functioning of nitrogenase enzymes that are highly sensitive to oxygen damage and the rhizobia which are obligate aerobes. In nodules, internal oxygen is regulated to 3–30 nM and oxygen diffusion is facilitated through intercellular air spaces and an oxygen binding haemoprotein (leghaemoglobin). The bacteroids are enclosed within a host-derived peribacteroid membrane that regulates flows between the bacteroids and their environment. The plant provides organic acids as a carbon source for the bacteria and the bacteroids export fixed N in the form of ammonia. The lentil nodules then modify the ammonia to export amides to the above ground plant parts in the xylem (Chopra et al., 2002). Detailed energy balances have been carried out for a number of symbioses but not lentil (Atkins, 1984). Rates of fixation by the bacteria are regulated by access to oxygen which may be controlled by the plant and subject to influences from a range of external factors (McNeil et al., 1984). These processes have not been well studied in lentil, however, processes appear to be similar to those in other amide exporting *Rhizobium leguminosarum* infected grain legume species (Begum et al., 2001). The structure and morphology of lentil nodules has been well documented in a light and electron microscopy study by Haswell et al., (2001) where nodules were examined every 1–2 weeks throughout the lifecycle of greenhouse grown plants.

3. TOTAL FIELD NITROGEN FIXATION

Peoples et al., (1995) summarised publications for % total plant nitrogen gained from fixation (%Ndfa) and total nitrogen from fixation for a range of crops, trees and pastures. They provide a range of values for lentil of 39–87% and 10–192 kg N ha^{-1}. While estimates are not as numerous for lentil as for some other grain legume crops (eg peas) it is possible to get a general consensus of likely N fixation outcomes for lentils (Table 1). Generally %Ndfa is high and similar to other species under

Table 1. Examples of trials from different regions around the world determining lentil fixation percentages and absolute rates either alone or relative to other pulses

Source	location	Lentil		Situation & plant parts analysed		Other pulses*	
		%N from fixation	Total fixed kg N ha^{-1}			%N from fixation	Total fixed kg N ha^{-1}
Peoples et al., 2001	SE wheat belt Australia	79	90	D#	Above ground	75–89	128–160
Shah et al., 2003	NWFP Pakistan	73	68	D	Above ground	75	112
Shah et al., 2002	Peshawa Pakistan	82–96	42–91	D	Above ground	–	–
Rennie & Dubetz, 1986	Alberta Canada	67	84	I	Above ground	79–85	176–216
van Kessel, 1994	Saskatchewan Canada	92	127	I	Above ground	–	–
Moawad et al., 1998	Egypt	53	127	I	Above ground	–	–

D = Dryland, I = Irrigated,

* these differed among experiments but included, mung, lupin, chickpea, pea, faba bean.

both dryland and rainfed conditions in eastern and western hemispheres. However, total N fixed tends to be lower than from other pulse crops in all situations and is likely to reflect the generally lower total biomass yields of lentils compared to other pulses. A survey by Unkovich and Pate (2000) of the quantities of N_2 fixed concluded that the principal grain crop legumes were ranked in the following descending order: soyabean, lupin, field pea, faba bean, common bean (*Phaseolus vulgaris*), lentil and chickpea but total nitrogen fixation will still be dependent on where and how well the crop is grown, and in species comparative studies the relative adaptation of species to the local testing environment. Of particular note also, however, is the paucity of experiments that have also included the below ground fixed N in their calculations. While often referred to, it is generally not measured and there do not seem to be any good values for lentils in the literature. Peoples et al., (2001) assumed an increase in total plant N of 50% (that is 33% of total plant N is below ground at harvest) if below ground N is included across a range of temperate pulses. Use of this figure would substantially increase the rates of N fixed ha^{-1}. Pate et al., (1979) give an estimate of 38% of N in below ground parts of a 70 day old lupin plant. Hood et al., (1990) give a very similar estimate of 37% of total plant N is either lost to the soil or held in roots and nodules in a chickpea plant at harvest. Khan et al., (2002) using four grain legumes, not including lentils, generally produced estimates of between 30 and 50% of total N held in the roots at harvest. This below ground plant N will greatly increase the calculated carryover of N to the next crop assuming 50–80% of the above ground N is removed in harvested product.

Using averaged data from table 1 and FAOSTAT data for 2004 it is possible to make some calculations of global lentil N fixation and carry over to subsequent crops. FAOSTAT indicated a global harvested area of 3.200 million ha of lentils, global production of 3.7 million tonnes and thus average yields of about 1.16 tonnes ha^{-1}. Use of these data and an average from the dryland experiments reported in table 1 of approximately 73 kg N fixed ha^{-1} yr^{-1} or 110 kg N fixed $ha^{-1}yr^{-1}$including below ground parts, means lentils fix in the vicinity of 0.23 to 0.35 Tg N yr^{-1} with and without below ground N included. However, with an average removal of approximately 65 kg N ha^{-1} yr^{-1} in harvested grain, global carryover of additional fixed N to following crops is relatively low at 8 kg N fixed ha^{-1} yr^{-1} or 45 kg N fixed $ha^{-1}yr^{-1}$ with and without below ground N included. This total net N balance of fixed N $ha^{-1}yr^{-1}$ returned by lentils is relatively low compared to the values of 92–126 kg fixed N $ha^{-1}yr^{-1}$ reported for pea and lupin by Peoples et al., (2001). However, it is consistent with the ranking for fixation among grain legume crops reported by Unkovich and Pate (2000).

4. NITROGEN ROTATION BENEFITS

Failure to account for below ground fixed nitrogen can partly explain many claims in the literature (e.g. Buddenhagen, 1990) that harvested grain legume crops are likely to have little if any net return to the cropping system. Incorporating the below ground fixed N estimates a general expectation of nitrogen carryover benefits for a wheat crop following a harvested lentil crop is possible using the total net N balance of 45 kg fixed N $ha^{-1}yr^{-1}$ returned by lentils reported above. It is also important to account for multiyear effects of the nitrogen carryover such as Strong et al., (1986) found when growing wheat for two years after a range of cereal and grain legume crops. In the first year after lentil rather than cereal crops they had an increased yield of 510 kg ha^{-1}(50% increase) with 19 additional kg of N in the wheat crop (105% increase). In the second year there was also a benefit to the wheat crop of 500 kg ha^{-1}(19% increase) with 12 additional kg of N in the wheat crop (27% increase). Soil N analyses allowed them to attribute the additional benefits of the rotation system almost entirely to different levels of available nitrogen. Strong et al., (1986) found a first year yield increase of 19% for wheat following oilseeds and a second year increase of 5%. Taken over the two seasons this indicated the non fixed nitrogen benefit of a broadleaf crop in the rotation accounted for 1/3rd of the total benefit. Similarly, in a more recent series of rotation trials in Canada across 3 years and 2 locations, Miller et al., (2003) found yield increases of 19% for wheat following mustard and 29% (605 kg ha^{-1}) for wheat following lentil with little difference in mean grain protein (< 1% averaged) among the treatments. Expectations would be much higher using lentil as a green manure crop where the full 110 kg N fixed $ha^{-1}yr^{-1}$ would be returned to the soil. Of course these numbers would be modified by the particular yield and fixation levels of individual lentil crops as well as the rate of breakdown of the stubble, degree of N limitation of the following crop and the presence of other limiting factors in the following crop such as water stress or disease. For example Kirkegaard et al., (2004) reported residual water amounts of 59 mm following wheat against 74 mm following lentils. The benefits of the extra water could easily be confused as a nitrogen benefit. Equally situations have been reported where there is no yield or economic benefit from including lentils in a cereal rotation (e.g. for lentil wheat rotation in Jordan by Badarneh, 2005). In such circumstances other limiting factors may need to be overcome before the benefits of fixed N is realised. Research by ICARDA (1980) in Syria found a 70% increase in wheat yields following lentils but only if the lentils were fertilized with 50 kg P ha^{-1}. In many areas fixed N is supplemented with N fertilisers to achieve an optimum yield, and in these circumstances N is not limiting if the target yield is not reached, primarily due to low water availability, disease or abiotic stresses. It is thus worth looking more generally to see how well this prediction for a nitrogen based rotation response with lentils/cereal is borne out by experimental results.

Numerous experiments around the world have reported responses of following cereals after lentils rather than 0 N treated cereals. Prakash et al., (2002) reported a 23.4% increase in rice yields (460 kg $ha^{-1}yr^{-1}$) following lentil rather than wheat in a rabi crop grown in India. Such increases have not been confined to the developing world with Guy and Gareau, (1998) producing a 7% (200 kg $ha^{-1}yr^{-1}$) increase in Idaho, USA. This crop was of interest because the comparison crop to lentil in the rotation was mustard which would have been expected to give other rotational benefits exclusive of the N effect. Thus the yield increase is likely to be more directly attributable to N rather than other compounding effects. Miller et al., (2003) also found in Canada that adjusting fertilizer rates for additional N input when comparing mustard and lentil prior to wheat in a rotation effectively eliminated the additional benefit of lentils. Shah et al., (2003) reported 49% gains in maize and sorghum yields in NWFP, Pakistan following lentils. Yau et al., (2003) reported 44% increases in barley yields in Lebanon. Miller et al., (2002) reported 21% increases in wheat yield in Canada. Importantly they also found a 5 kg N ha^{-1}increase in soil N of the lentil stubble in the 0–120 cm depths prior to sowing the wheat as well as an 8% increase in grain protein. This allowed them to directly link yield increases with increased N availability. An even higher increase in soil N (40 kg ha^{-1}) following lentil has been reported from Thailand (Patwary et al., 1989). In other experiments in Canada wheat lentil rotations were found to be the most profitable (including increased wheat protein levels) as well as having a 19% reduction in CO_2 emissions (primarily due to reduced N fertilizer inputs) making the option the most sustainable of those trialled in the experiments (Zentner, 2002). These data also indicate there are other potential benefits in rotation systems from the N fixation including reduced fertilizer costs, increased wheat quality and reduced greenhouse gas emissions.

Other, in crop, differences may also confound the nitrogen benefit. The general health of the crop and productivity may increase or decrease growth and thereby alter the level of fixation and benefits of the lentil nitrogen fixation in the rotation. A comparison of the dryland (mean of 73 kg N fixed $ha^{-1}yr^{-1}$) and irrigated (mean of 113 kg N fixed $ha^{-1}yr^{-1}$) fixation rates in table 1 indicates a substantially greater fixation in the irrigated crops (60 kg N fixed $ha^{-1}yr^{-1}$ if below ground fixation is included).

There is a need however to be reasonable in the expectations of benefits from nitrogen fixed by lentils. As Unkovich and Pate (2000) have indicated lentils are relatively low on the scale compared with most other grain legume crops. However, compared to many pasture systems lentils are very low. Peoples et al., (2001) found on average 30% to 80% higher levels of N fixation in southern Australian vetch, lucerne and subterranean clover pastures when compared to lentils. In southern Lebanon Yau et al., (2003) found vetch in a rotation had almost twice the benefit of lentils on barley production. Looking globally Peoples et al., (1995) found reports of a maximum of 192 kg N fixed $ha^{-1}yr^{-1}$for lentil compared with maximum annual pasture fixation rates of 386 (lucerne), 373 (red clover), 380 (desmodium) and 291 kg N fixed $ha^{-1}yr^{-1}$ (white clover). Vigil and Nielsen (1998) compared the economic value of replacing a clean fallow with a lentil green manure crop in a water limited environment in Colorado, USA. They found yield reductions due to water limitations of up to 1050 kg ha^{-1}. Their conclusion was thus, “At current fertilizer costs, legume N (in this system) was too expensive to be considered a reasonable alternative to chemical fertilizer”. Pikul et al., (1997) also point to another potential problem with green manuring lentils. They found yield for spring wheat grown in Montana USA, was 25% less following a lentil green manuring treatment. This was in spite of no differences in water availability in the treatments and no differences in most soil parameters. The exception was soil NO_3^- N was 35% less in green manured treatments in spite of the potential for a 66% increase due to the green manure. They concluded slow breakdown of the residues was preventing the N from becoming available. This situation is quite different from the data of Strong et al., (1986) in near tropical Queensland and Evans et al., (2003) in the SW slopes of NSW, Australia where breakdown in the intervening season is high. Evans et al., (2003) found wheat yield increases of up to 930 kg ha^{-1} in the first season after a green manure and 370 kg ha^{-1} in the second season. While lentils were not in their experiments they achieved wheat yield increases following grain pea of up to 520 kg ha^{-1}. The slow release of N resulting in greater available N late in the season can also be beneficial for increasing protein concentration in cereal grain. As an alternative, late applications of N fertilisers have also been used in cereals to increase protein (Woodard and Bly, 1998).

5. RHIZOBIAL LIMITATIONS ON LENTIL N FIXATION

Lentils require effective infection by *Rhizobium leguminosarum* (Humphrey et al., 2001) in order to fix nitrogen. This infection process could fail due to a number of reasons including; no or few rhizobia in the soil, inability of the rhizobia to survive in the soil due to intrinsic limitations, inappropriate strains of rhizobia in the soil, efficient rhizobia being out competed by less efficient/effective soil rhizobia, failure of the plant to invest in the symbiosis either through lack (eg early in the seedling stage) of time or resources, or through altered metabolism (e.g. towards nitrate reduction in soils high in available nitrogen). In any of these situations fixation may be reduced and the grain yield and or rotational value of the lentil will be reduced.

Attempts to overcome rhizobial limitations of fixation have been conducted for a considerable period (e.g. McNeil et al., 1981) and have typically been of the following form. Potential inoculating bacteria are selected for increased growth and performance, relative to uninoculated controls, of target legume lines or species in sterile media with or without additional limitations (e.g. acidity). The performance of the bacteria (survival, growth e.g. Slattery and Pearce, 2002) may also be tested in soils or media with intrinsic limitations such as high pH or high salinity. Responses to inoculation are then attributed to the nitrogen fixation capacity of the inoculating bacteria. Typically these bacteria are then also screened in the field for real effects on lentil production by addition of the inoculating bacteria (with or without other inputs such as *Azospirillum* { El-Komy and Wahab, 1998} or N and P {Shah et al., 2000}) to field plots. Yield and various fixation parameters (e.g. nodule mass/number or various methods for directly measuring fixation{McNeil, 1982b, Tewari et al., 2004, Hardarson and Danso, 1993}) are then measured and increases attributed to fixation. Care needs to be taken in interpreting these results for a number of reasons. Firstly, strong agreement between the greenhouse and field experiments are not always evident (Slattery and Pearce, 2002). Second, many of the negative results may not be reported. Additionally, however, several other observations are needed to gain good interpretation of these trials. These are; the level of rhizobia initially in the soil (Slattery and Pearce, 2002), the percentage occupancy of nodules by the inoculating bacteria (McNeil et al., 1983) and the survival of the inoculating bacteria in the soil (Slattery and Coventry, 1999).

Many soils will not have grown lentil crops previously or recently in many production areas. Particularly if the soils also suffer from a constraint such as acidity, waterlogging or high salinity *Rhizobium leguminosarum* populations may be low and there may be significant responses in the field to inoculation with appropriate strains. In low pH (4.4–5.0) soils in the field in Victoria Slattery and Pearce, (2002) identified low populations of *Rhizobium leguminosarum* and using acidity tolerant bacteria as inoculants achieved nodule number increases of 8 fold, grain yield increases of almost 50% and 25% increases dry matter for 13 test strains relative to uninoculated field controls. The three parameters were reasonably well correlated. However, some strains were very ineffective and gave no or negative responses. Dilworth et al., (2001) have reviewed the selection of rhizobia for acid soil tolerance. Rai and Singh (1999) selected salt tolerant rhizobia and found them successful in increasing yield in salt tolerant lentil genotypes on sodic soils. Similarly Shah et al. (2000) found in a Pakistan soil low in N, P and rhizobia that lentil yield increases of 77% (393 kg ha^{-1}) were achieved via inoculation. They also found 71–95% occupancy of the nodules by the inoculant bacteria confirming the direct effect of the inoculation. In a rhizobium deficient soil in Chile Herrera and Longeri (1985) achieved 85 fold increases in nodulation, plant dry weight increases of 341%, seed yield increased by 114% from 530 kg ha^{-1} to 1150 kg ha^{-1} and seed protein yield by 132%. In situations where soil rhizobia are more common yield increases are often not as great. Khurana and Sharma (1995) evaluated several lentil cultivar by strain treatments in the Punjab in India. They found interactions among the treatments with an average yield response to inoculation of 16%, however, they do not provide data on the nodule occupancy by strains. In other trials in India Dhingra et al., (1988) only found significant yield increases in 3 out of 5 years. In South Eastern Australia no increase in yield was found after inoculating lentils grown on fertile, alkaline soils that had a long history of host crops and naturally high levels of rhizobia (Slattery and Pearce, 2002).

The interaction between the competitive ability of the inoculant strains in the soil and their ability to enhance yields is well demonstrated by Moawad et al., (1998) in trials in Egypt. They found less than 10% of field isolates to be high in their effectiveness. In field trials they obtained an average of 17% occupancy by inoculant lines with a non significant average 5% yield increase. This represents a 2.9% increase in yield for each 10% increase in nodule occupancy. However, their data for berseem clover for which the field isolates were equally ineffective showed the potential of inoculation. They achieved an average of 66% occupancy by inoculant strains and an average 38% increase in yield. This represents a 5.8% increase in yield for each 10% increase in nodule occupancy. Shah et al., (1996) also found correlations in Pakistan between nodule occupancy and response to inoculation. They found higher native populations of *Rhizobium leguminosarum* in soils which had recently or frequently grown lentils. Kumar and Chandra (2005) found differences in application methods resulted in different levels of nodule occupancy. When nodule occupancy by the introduced strains was high they also found significant yield increases. In a meta analysis of soybean field inoculation trials which had determined nodule occupancy done up to 1984 (McNeil, pers. comm. NifTAL, Hawaii, 1984) there was a significant regression between nodule occupancy by inoculant bacteria and yield. For each 10% increase in nodule occupancy by the inoculant strain (USDA 110) soybean yields increased by 1%. The data from lentil for nodule occupancy would seem to support a similar, but potentially higher, level of response.

A number of authors have also looked at improving fixation by combined addition of rhizobia and mycorrhiza or soil bacteria. The latter in an attempt to increase P availability which is in high demand by legume crops. Badr El-Din and Moawad (1988) found a synergistic response to combined inoculation in a soil low in both mycorrhiza and rhizobia consistent with responses of N fixation for improved P availability. However, successful commercial scale establishment of mycorrhiza on plants in the field is extremely difficult. El-Komy and Wahab, (1998) found increases in nodulation, yield and fixation for lentil simultaneously inoculated with *Azospirillum* and rhizobia. At this stage there do not appear to be commercial applications of these dual inoculation technologies, however, there may be some applications in the future if they prove reliable in achieving infection and/ or high populations of both inoculants in the field.

The general indications from these experiments are therefore that responses to inoculation are common in lentils but that a primary limitation is the ability to out compete existing, less efficient, bacteria in the soil. It is thus only when suitable efficient bacteria can achieve high levels of nodule occupancy that responses are achieved.

6. OTHER ENVIRONMENTAL LIMITATIONS ON N FIXATION

While it has already been stressed that conditions that suit better growth of the lentil crop will normally enhance the nitrogen fixation of the crop (e.g. irrigation in Table 1) there may also be specific situations in which the fixation process is more sensitive and fixation limits the growth or %Ndfa of the crop. Some of these that have been suggested are: high soil available N, micronutrient limitations (B, Mb, Co, Fe), Phosphorous limitation, limited N availability at specific stages, and water limitations (either drought or waterlogging). However, in a fixation limited crop, growth will also be limited and it is difficult to separate the specific effects on fixation from effects on growth limitation. Numerous studies exist in which various combinations of minor and micro nutrients have been added to lentils in deficient soils and responses have occurred (e.g. Sinha, 1994), however, these will be dealt with elsewhere. There are, however, some situations where it is specifically possible to separate the effects. For example, significant levels of Mo and Co are only required by the fixation system, and deficiencies would inhibit N fixation without affecting growth in soils where N was non limiting (Lucinski et al., 2002) or where a transient stress destroys N fixation capability (nodule death) and the plant either cannot re-establish nodules (eg a near senescent plant {McNeil and LaRue, 1984} or must significantly reinvest in their structure). Alternatively small amounts of added N have been suggested in the form of starter N to assist in establishing nodules, or as leaf applied N to delay the onset of nodule senescence (McNeil and LaRue, 1984).

6.1. Micronutrient Limitations

Co, Mo, and Fe are all involved in N fixation and may be required specifically by the fixation process in greater amounts than needed by the plant. Cu and Mn are involved in nitrogen metabolism. Mo limitations on legumes has been widely observed and generally overcome in southern Australian soils (Donald and Prescott, 1975). Specific responses (11% yield increase) to Mo seed application by lentils have also been observed (Golubev and Lugovskikh, 1974) and fertilisation is generally needed to overcome the problem. More recently work by Gahoonia et al., (2005) has indicated that lentil lines exist with different root morphologies that may be better able to scavenge micronutrients from the soil and give 10–20% yield increases. Lentils have been shown to respond to Co in pot experiments (Sarada and Polasa, 1992) but levels needed are very low 0.15 ppm. Lentil genotypes have been shown to differ in iron absorption and translocation and fixation has responded positively to greater absorption (Rai et al., 1984). However, evidence does not exist for a specific fixation dependant Fe effect in lentil.

6.2. Phosphorus Limitations

Phosphorus has been suggested to be more limiting for N fixers, particularly agricultural grain legumes and pasture species, than other species (Vitousek, 2002, Robson, 1983). Thus it seems reasonable that P limitation may lead to reductions in fixation greater than simply reductions in plant growth. Badr El-Din and Moawad, (1988) have shown responses of fixation to mycorrhizal inoculation presumably due to enhanced P uptake. There are numerous reports of yield increases and N fixation stimulation with application of P to lentils (e.g. ICARDA, 1980, Shah et al., 2000). Badarneh (1995) reported that %Ndfa increased in lentils grown under water stress when P was added. This suggests that the P was specifically benefiting the fixation process. It is clear that lentils need P for both fixation and growth and probably that the effects on fixation are direct as well as by reduction of overall plant growth.

6.2.1. Nitrogen interactions

Research on the use of starter nitrogen (small amounts at planting to establish the fixation process) have generally found advantages in greenhouse experiments but no or little advantage in the field (Turay et al., 1991). This is consistent with most reports for other grain legume crops. McNeil and LaRue, (1984) reported that small doses of applied nitrogen could inhibit fixation and lead to loss of fixation capacity and yield reductions or stimulate fixation when applied to leaves late in the plant life cycle and stimulate yield in soybeans. Lentils like all other legumes have their fixation rates suppressed by available soil nitrogen (Bremer et al., 1988). The ecological basis for this is simply to prevent surrounding competing vegetation gaining access to the N and hence suppressing legume growth. Nitrate use and N fixation are energetically similar in their requirements. However, when nitrate is directly assimilated in the leaf it may directly access photosynthetic electrons and thus require less carbohydrates. In addition if the Rhizobium are inefficient (e.g. Hup- strains) energy costs of fixation may increase (Atkins, 1982) relative to nitrate assimilation. Typically fertilization of lentils with nitrogen does not make any yield benefits (Dutta, 1985, Islam and Afandi, 1980) provided the crop is adequately inoculated either artificially or naturally. However, the %Ndfa may fall substantially under elevated soil N. Selection with or without mutation may produce symbioses effectively uncoupled from nitrate inhibition (Carroll et al., 1985) though this has not yet been done for lentils. Minor gains are possible by selecting rhizobia which perform better in the presence of elevated soil nitrate (Saxena et al., 1996).

6.2.2. *Salinity and sodicity interactions*

Maher et al., (2003) have demonstrated it is possible to select salt tolerant lentils and that their growth and fixation is enhanced. Rai and Singh (1999) found interactions exist between rhizobium strain, lentil cultivar, yield and nitrogen fixation in lentil. They suggested that the salt stress was directly affecting the performance of the symbiosis and a combined selection for salt tolerant genotypes and bacteria could increase lentil yields (up to 28% above salt tolerant genotypes alone, substantially more relative to salt intolerant genotypes) fixation and fixation enzyme activities.

6.2.3. *Water availability*

Potentially waterlogging can cause reductions in fixation and nodulation and if severe enough lead to loss of nodules. McNeil and Slatter (pers comm.. Nepal 2003) observed severe loss of nodules after a waterlogging event in some, but not all, lentil lines growing in a field trial in Nepal. This suggests the possibility that part of the effect of waterlogging on lentils could be the effect on N fixation and the subsequent energy cost of re-establishing nodules. Drought stress may also act in this way. Athar (1998) demonstrated differences in rhizobium survival in droughted soils in Pakistan and this would interact with nodulation and fixation under those conditions. Badarneh (1995) found that %Ndfa was lower under drought conditions than well watered conditions suggesting more adverse effects of drought on lentil fixation than total productivity. Bremer et al., (1988) also found using ^{15}N approaches that the proportion of N assimilated from the atmosphere declined with increasing soil nitrate levels and increasing drought stress. This may mean that the symbiosis was preferentially sensitive to drought stress or may simply be consistent with the concept that soil N is preferentially used first. Thus under drought stress or any other stress the reduced growth means it takes longer to overcome the soil available N limitation on nitrogen fixation.

6.3. Potential to Breed for Increased N Fixation

The concept of breeding both rhizobia and grain legumes (e.g. by reducing fixation sensitivity to soil N, of rhizobia {McNeil, 1982a} or soybean plants {Carroll et al., 1985}) for increased nitrogen fixation both to increase pulse crop yields and increase the rotation benefits described above has a long history with a breeding program in place in CIAT in the early 1980's (Graham, 1981). The area has been well reviewed by Herridge and Danso (1995). Their conclusion was that only a relatively few cultivars have been released with a specifically improved capacity for nitrogen fixation. For lentil, which are typically grown on low N soils with high %Ndfa (Table 1) it is likely that selection for plant growth (and indirectly plant yield), particularly under adverse conditions, will produce better adapted, higher yielding, higher fixing plants. This is particularly the case when selecting for tolerance to conditions that specifically limit N fixation more than overall plant growth (e.g. soil N level, available P) indicated in section 6 above. Hobson et al., (2004) have produced substantial increases in lentil photosynthesis rate and green matter accumulation, and thus N fixation using boron tolerant lines in high boron soils.

It would be expected that increasing the total levels of biomass through agronomy or breeding would also increase levels of N fixation. In lentil, earlier sowing increased total biomass and seed yield in Australia but improvements in disease resistance are necessary to enable early sowing without risk of yield and quality losses (Materne 2003). High biomass lentil varieties have also been released from ICARDA for the developing world (Sarker, et al., 2004). However an increase in harvest index will continue to be a focus of breeding programs to increase seed yield and thus result in a greater removal of fixed N from the system. Lentil can effectively convert biomass to seed, achieving harvest indices of up to 50% (Materne 2003).

Levels of nitrogenase activity have been shown to decrease less rapidly under soil sodicity in an improved (PL-406) lentil line than in a local cultivar (Singh et al., 1993) and this has been suggested as a reason for the superiority of Pl-406. However, the cultivar was not specifically bred for increased N fixation. A second breeding option is the specific selection for N fixation by plants under defined conditions. This has been done in soybeans for tolerance of high soil N (e.g. Carroll et al., 1985) as well as selections for high nodulation in chickpeas (Rao et al., 1993). However, there does not appear to be any successful releases of lentil lines bred in this manner. A third option is, the selection for, and better inclusion in the symbiosis, of more efficient (e.g Hup-, see Atkins, 1982), and more effective rhizobia. There have been limited gains also in this area which are covered in the earlier section (5) on rhizobia.

7. CONCLUSIONS

Lentils occupy a minor position in global N fixation relative to other major crops such as soybean and world N fertilizer production. They generally have fixation rates towards the lower end of those achieved by pasture and grain legume species. However, where lentils are grown for profitability in their own right their N fixation can have significant beneficial effects on their own yield and that of non N fixing rotation crops. Lentil fixation is generally highest under conditions that promote good growth of the crop in total. However, in situations where field populations of rhizobia are low or where inoculant bacteria can achieve high nodule occupancies they can achieve significant responses to inoculation. There are also specific environmental conditions which are not favourable to fixation that can limit the %Ndfa and total fixation by the crop. While economic data indicate that lentil fixed N is not a commercially competitive fertilizer in most cropping situations, when N carryover from lentil crops results as a by-product of growing the crop it can have significant advantages, especially in improving cereal quality. Thus when growing lentils as a crop it is worthwhile managing the crop to achieve maximum fixation and thereby maximum total productivity and profitability.

REFERENCES

Atkins, C.A., 1982. Efficiencies and inefficiencies in the legume/Rhizobium symbiosis—A review. *Plant and Soil*. 82: 273–284.

Badarneh, D.M.D. 1995. Magnitude of nitrogen fixation by lentil at different rates of phosphorus using 15N technique. *Journal of Agronomy and Crop Science*. 175: 7–14.

Badarneh, D.M.D. 2005. Crop nitrogen uptake in a legume-wheat rotation using 15N methodology. *Dirasat. Agricultural Sciences*. 32: 229–238.

Badr el-Din, S.M.S. and Moawad, H. 1988. Enhancement of nitrogen fixation in lentil, faba bean, and soybean by dual inoculation with rhizobia and mycorrhizae. *Plant and Soil*. 108: 112–124.

Begum, A.A., Leibovitch, L., Migner, P and Zhang, F. 2001. Specific flavonoids induced *nod* gene expression and pre-activated *nod* genes of *Rhizobium leguminosarum* increased pea (*Pisum sativum* L.) and lentil (*Lens culinaris* L.) nodulation in controlled growth chamber environments. *Journal of Experimental Botany*. 52: 1537–1543.

Bremer, E.; Rennie, R. J.; Rennie, D. A. 1988. Dinitrogen fixation of lentil, field pea and faba bean under dryland conditions. *Canadian Journal of Soil Science*. 68: 553–562.

Buddenhagen, I.W. 1990. Legumes in farming systems in Mediterranean climates. In *'The role of legumes in the farming systems of the Mediterranean areas.'* (Eds AE Osman, MH Ibrahim,MA Jones) pp 3–29. (Kluwer Academic Publishers: Dordecht, The Netherlands).

Burns, R.C. and R.W.F. Hardy. 1975. *Nitrogen Fixation in Bacteria and Higher Plants*. New York: Springer-Verlag.

Carroll, B.J., McNeil, D.L. and Gresshoff, P.M. 1985. Isolation and properties of novel soybean (*Glycine max*. L. Merr.) mutants that nodulate in the presence of high nitrate concentrations. *Proceedings National Academy Sciences USA*. 82: 4162–4166.

Chopra, J., Kaur, N., and Gupta A.K.A. 2002. Comparative developmental pattern of enzymes of carbon metabolism and pentose phosphate pathway in mungbean and lentil nodules. *Acta Physiologiae Plantarum.* 24: 67–72.

Cleveland, C.C., Townsend, A.R., Schimel, D.S., Fisher, H., Howarth, R.W., Hedin, L.O., Perakis, S.S., Latty, E.F., von Fischer, J.C., Elseroad, A. and Wasson, M.F., 1999. Global patterns of terrestrial biological nitrogen (N_2) fixation in natural ecosystems, *Global Biogeochemical Cycles.* 13: 623–645.

Dhingra, K.K., Sekhon, H.S,; Sandhu, P.S. and Bhandari, S.C. 1988. Phosphorus-Rhizobium interaction studies on biological nitrogen fixation and yield of lentil. *Journal of Agricultural Science, UK* 110: 141–144.

Dilworth, M.J., Howieson, J.G., Reeve, W.G., Tiwari, R.P. and *Glenn, A.R.* 2001. Acid tolerance in legume root nodule bacteria and selecting for it. *Australian Journal of Experimental Agriculture.* 41: 435–446.

Donald, C.N. and Prescott, J.A. 1975. Trace elements in Australian crop and pasture production, 1924–1974. In: *Trace elements in teh soil-plant-animal continuum Nicholas, D.J.D. and Egan, A.R. (eds). Academic Press, New York. pp 7–37.*

Dutta, R. K. 1985. Productivity of lentil in relation to N-availability and population density. *Lens Newsletter.* 12: 21–24.

El-Komy, H.M. and Wahab, A.M.A. 1998. Effect of simultaneous inoculation of *Azospirillum* and *Rhizobium* spp. on growth, nodulation and nitrogen fixation of two legumes using the 15N-isotope dilution technique (IDT) and the difference method (DM). *Acta Microbiologica Polonica* 47: 283–296.

Evans J, Scott G, Lemerle D, Kaiser A, Orchard B, Murray GM, Armstrong EL. 2003. Impact of legume 'break' crops on the yield and grain quality of wheat and relationship with soil mineral N and crop N content. *Australian Journal of Agricultural Research.* 54: 777–788.

Gahoonia, T.S. multi-location grain yield and benefit-cost ratio of two lentil (*Lens culinaris*, Medikus.) varieties. *Plant and Soil.* 272: 153–161.

Golubev, V.D. and Lugovskikh, M.A. 1974. Pre-sowing treatment of lentil seeds with ammonium molybdate. *Khimiya v Sel'skom Khozyaistve* 12: 24–25.

Graham, P.H. 1981. Some problems of nodulation and nitrogen fixation in Phaseolus vulgaris L.: a review. *Field Crops Research.* 4: 93–112.

Guy, S.O. and Gareau, R.M. 1998. Crop rotation, residue durability, and nitrogen fertilizer effects on winter wheat production. *Journal of Production Agriculture.* 11: 457–461.

Hardarson, G., and Danso, S.K.A. 1993. Methods for measuring biological nitrogen fixation in grain legumes. *Plant and Soil.* 152, 19–23.

Haswell, M., Humphry, D.R., Cummings, S.P., and Andrews, M. 2001. Nodule structure and development in lentil (Lens culinaris): a light and electron microscopy study. *Aspects of Applied Biology.* 63: 83–84.

Herrera, A., Longeri, L. 1985. Response of lentil (*Lens culinaris* Medik) to inoculation with *Rhizobium leguminosarum. Ciencia e Investigacion Agraria.* 12: 49–53.

Herridge, D.F. and Danso, S.K.A. 1995. Enhancing crop legume N_2 fixation through selection and breeding. *Plant and Soil.* 174: 51–82.

Hobson, K.B., Armstrong, R.D., Nicolas, M., Connor, D.J. and Michael A. Materne, M.A. 2004. Boron tolerance of lentil – highlights of a research program. *New directions for a diverse planet: Proceedings of the 4th International Crop Science Congress. Brisbane, Australia, 26 Sep–1 Oct 2004. Available online at: www.cropscience.org.au Paper published – I will send reference.*

Hooda, R.S., Sheoran, I.S. and Singh, R. 1990. Partitioning and utilization of carbon and nitrogen in nodulated roots and nodules of chickpea (*Cicer arietinum*) grown at two moisture levels. *Annals of Botany.* 65: 111–120.

Humphrey, D.R., Cummings, S.P. and Andrews, M. (2001). Comparison and tentative identification of Rhizobiaceae isolated from nodules of lentil grown in New Zealand and the United Kingdom. *Aspects of Applied Biology.* 63: 101–110.

International Center for Agricultural Research in the Dry Areas. 1980. Lentil in crop rotation. *News from ICARDA* (No.7): 2–3.

Islam, R. and Afandi, F. 1980. Responses of lentil cultivars to Rhizobium inoculation and nitrogen fertilization. Lens 7: 50–51.

Khan, D.F., Peoples, M.B., Chalk, P.M. and Herridge, D.F. 2002. Quantifying below-ground nitrogen of legumes. 2. A comparison of ^{15}N and non isotopic methods. *Plant and Soil.* 239: 277–289.

Khurana, A.S. and Sharma, P. 1995. Variety and Rhizobium strain interactions in lentil. *Lens Newsletter.* 22: 34–36.

Kirkegaard, J., Christen, O., Krupinsky, J. and Layzell, D. 2004. Break crop benefits in temperate wheat production. *New directions for a diverse planet: Proceedings of the 4th International Crop Science Congress.Brisbane, Australia, 26 Sep–1 Oct 2004. Available online at: www.cropscience.org.au*

Kumar, R. and Chandra, R. 2005. Effect of adhesives on survival of inoculated *Rhizobium leguminosarum* on seed and symbiotic performance in lentil under field conditions. *Indian Journal of Pulses Research* 18: 206–210.

Lucinski, R., Polycn, W., and Ratajczak, L., 2002. Nitrate reduction and nitrogen fixation in symbiotic association *Rhizobium* – legumes. *Acta Biochimica Polonica.* 49: 537–546.

Maher, L., Armstrong, R. and Connor, D. 2003. Salt tolerant lentils – a possibility for the future? *Solutions for a better environment: Proceedings of the 11th Australian Agronomy Conference, Geelong, Victoria, Australia, 2–6 February* 2003: 0–4.

Materne, M. A. 2003. Importance of phenology and other key factors in improving the adaptation of lentil (*Lens culinaris* Medikus) in Australia. Thesis presented for the degree of Doctor of Philosophy at The University of Western Australia, School of Plant Biology and Centre for Legumes in Mediterranean Agriculture (CLIMA), Faculty of Natural and Agricultural Sciences.

McNeil, D.L. 1982a. Variations in the ability of *Rhizobium japonicum* strains to nodulate soybeans and maintain fixation in the presence of nitrate. *Applied and Environmental Microbiology.* 44: 647–652.

McNeil, D.L. 1982b. Quantification of symbiotic nitrogen fixation using ureides: a review. *In*: P.H. Graham, S. Harris, Eds *'Biological nitrogen fixation for tropical agriculture'* CIAT, Cali 609–617.

McNeil, D.L. and LaRue, T.A. 1984. Effect of nitrogen source on ureides in soybeans. *Plant Physiology.* 74: 227–232.

McNeil, D.L., Borton, S., Amara, D. and Vora, M.S. 1983. Use of antibiotic resistant Rhizobium mutants for competition studies with *Cajanus cajan. International Pigeonpea Newsletter.* 2: 71–72.

McNeil, D.L., Carroll, B.J. and Gresshoff, P.M. 1984. The nitrogen fixation capacity of bacteroids extracted from soybean nodules inhibited by nitrate ammonia or dark treatments *In*: *'Symbiotic Nitrogen Fixation (1)'* ed. B.S. Ghai. USG Publishers, Ludhiana, p 79–88.

McNeil, D.L., Croft, L. and Sandhu, T.S. 1981. Response of chickpeas to inoculation with Rhizobium in Hawaii. *International Chickpea Newsletter.* 3: 26–27. 1981.

Miller, P.R., Gan, Y., McConkey, B.G.eand McDonald, C.L. 2003. Pulse crops for the northern Great Plains: II. Cropping sequence effects on cereal, oilseed, and pulse crops. *Agronomy Journal.* 95: 980–986.

Miller, P.R., Waddington, J., McDonald, C.L. and Derksen, D.A. 2002. Cropping sequence affects wheat productivity on the semiarid northern Great Plains. *Canadian Journal of Plant Science.* 82: 307–318.

Moawad, H., Badr El-Din, S.M.S. and Abdel-Aziz R.A. 1998. Improvement of biological nitrogen fixation in Egyptian winter legumes through better management of Rhizobium. *Plant and soil.* 204: 95–106.

Mosier, A.R., 2002. Environmental challenges associated with needed increases in global nitrogen fixation. *Nutrient Cycling in Agroecosystems.* 63: 101–116.

Nesbitt, S.W., Zhang, R. and Orville, R.E., 2000. Seasonal and global NOx production by lightning estimated from the Optical Transient Detector (OTD). *Tellus B.* 52: 1206–1215.

Omar Ali, Sarker, A., Rahman, M.M. and Erskine, W. 2005. Root traits, nutrient uptake, Athar, M. 1998. Drought tolerance by lentil rhizobia (*Rhizobium leguminosarum*) from arid and semiarid areas of Pakistan. *Letters in Applied Microbiology.* 26: 38–42.

Pate, J.S., Layzell, D.B. and McNeil, D.L. 1979. Modeling the Transport and Utilization of Carbon and Nitrogen in a Nodulated Legume. *Plant Physiology.* 63:730–737.

Patwary, S.U., Haque, Q. and Badruddin, M. 1989. Role of legume on nitrogen balance and A-value of soil under different sequential cropping systems. Thai Journal of Agricultural Science 22: 213–221.

Peel, M.D., 1998. Crop Rotations for Increased Productivity. *North Dakota State University extension Bulletin.* 48.

Peoples, M.B., Bowman, A.M., Gault, R.R., Herridge, D.F., McCallum, K.M., McCormick, M.H., Norton, R.M., Rochester, I.J., Scammell, G.J. and Schwenke, G.D. 2001. Factors regulating the contributions of fixed nitrogen by pasture and crop legumes to different farming systems of eastern Australia. *Plant and Soil.* 228: 29–41.

Peoples, M.B., Herridge, D.F. and Ladha, J.K. 1995. Biological nitrogen fixation: an efficient source of nitrogen for sustainable agricultural production? *Plant and Soil.* 174: 3–28.

Pikul, J.L., Aase, J.K. and Cochran, V.L. 1997. Lentil green manure as fallow replacement in the semiarid northern Great Plains. *Agronomy Journal.* 89: 867–874.

Prakash, V., Ghosh, B.N., Pandey, A.K. and Gupta, H.S. 2002. Effects of preceding winter legumes and nitrogen rates on N uptake, yield attributes and yield of rice, and monetary returns from rotation. *Annals of Agricultural Research.* 23: 402–406.

Rai, R. and Singh, R.P. 1999. Effect of salt stress on interaction between lentil (lens culinaris) genotypes and Rhizobium spp. Strains: symbiotic N_2 fixation in normal and sodic soils. *Biol Fertil Soils.* 29: 187–195.

Rai, R., Prasad, V., Choudhury, S.K. and Sinha, N.P. 1984. Iron nutrition and symbiotic N_2-fixation of lentil (*Lens culinaris*) genotypes in calcareous soil. *Journal of Plant Nutrition.* 7: 399–405.

Rao, D.L., Giller, K.E., Yeo, A.R. and Flowers, T.J. 2002. The Effects of Salinity and Sodicity upon Nodulation and Nitrogen Fixation in Chickpea (*Cicer arietinum*). *Annals of Botany.* 89: 563–570.

Rennie, R.J. and Dubetz, S. 1986. Nitrogen-15-determined nitrogen fixation in field-grown chickpea, lentil, fababean, and field pea. *Agronomy Journal.* 78: 654–660.

Robson, A.D. 1983. Mineral nutrition. In: Broughton, W.J. (ed). *Nitrogen Fixation Vol 3 Legumes pp 36–55, Clarendon Press, Oxford.*

Sarada, R.L. and Polasa, H. 1992. Effect of manganese, copper and cobalt on the in vitro growth of *R. leguminosarum*-2001 and on the symbiotic nitrogen fixation in lentil plants. *Indian Journal of Agricultural Research.* 26: 187–194.

Sarker, A., Aydogan, A., Sabaghpour, S.H., Kusmenoglu, I., Sakr, B., Erskine, W. and F.J. Muehlbauer, F.J. 2004. Lentil Improvement for the Benefit of Highland Farmers.*New directions for a diverse planet: Proceedings of the 4th International Crop Science Congress. Brisbane, Australia, 26 Sep 1 Oct 2004. Available online at: www.cropscience.org.au*

Saxena, A.K., Rathi, S.K., Tilak and K.V.B.R. 1996. Selection and evaluation of nitrate-tolerant strains of *Rhizobium leguminosarum* biovar*viceae* specific to the lentil. Biology *and Fertility of Soils.* 22: 126–130.

Shah, M.S., Nawaz, H. and Idris, M. 2002. Nitrogen fixation in farmers fields under rainfed dry conditions. proceedings of 17th WCSS, 14–21 August, Thailand. Poster 150: 1–5.

Shah, N.H., Hafeez, F.Y., Arshad, M. and Malik, K.A. 2000. Response of lentil to *Rhizobium leguminosarum bv. viciae* strains at different levels of nitrogen and phosphorus. *Australian Journal of Experimental Agriculture.* 40: 93–98.

Shah, N.H., Hafeez, F.Y., Hussain, A. and Malik, K.A. 1996. Influence of seasonal variation on the indigenous population of *Rhizobium leguminosarum bv. viceae* and competitive ability of introduced rhizobia in lentil. *Lens Newsletter* 23: 32–37.

Shah, Z., Shah, S.H., Peoples, M.B., Schwenke, G.D., and Herridge, D.F. 2003. Crop residue and fertiliser N effects on nitrogen fixation and yields of legume-cereal rotations and soil organic fertility *Field Crops Research.* 83: 1–11.

Singh, B.B., Tewari, T.N. and Singh, A.K. 1993. Stress studies in lentil (*Lens esculenta* Moench). III. Leaf growth, nitrate reductase activity, nitrogenase activity and nodulation of two lentil genotypes exposed to sodicity. *Journal of Agronomy and Crop Science.* 171: 196–205.

Sinha, A.C., Mandal, B.B. and Jana, P.K. 1994. Yield and water-use efficiency of rainfed lentil (*Lens culinaris*) as influenced by boron, zinc and molybdenum. *Indian Journal of Agricultural Sciences* 64: 863–866.

Slattery, J.F. and Coventry, D.R. 1999. Persistence of introduced strains *of Rhizobium leguminosarium* bv*trifolii* in acidic soils of north-eastern Victoria. *Australian Journal of Experimental Agriculture*. 39: 829–837.

Slattery, J.F. and Pearce, D. 2002. Development of elite inoculant Rhizobium strains in southeastern Australia. In: *Inoculants and nitrogen fixation of legumes in Vietnam. (ed. D Herridge). ACIAR proceedings* 109e. p 86–94.

Smit, G., Swart, S., Lugtenberg, B.J.J, and Kijne, J.W. 1992. Molecular mechanisms of attachment of *Rhizobium* bacteria to plant roots. *Molecular Microbiology*. 6: 2897–2903.

Strong, W.M., Harbison, R.G.H., Nielsen, B.D., Hall, B.D. and Best, E.K. 1986. Nitrogen availability in a Darling Downs soil following cereal, oilseed and grain legume crops 2. effect of residual soil nitrogen and fertiliser nitrogen on subsequent wheat crops. *Australian Journal of Experimental Agriculture*. 26: 353–359.

Tewari, K., Suganuma, T., Fujikake, H., Ohtake, N., Sueyoshi, K., Takahashi, Y. and Ohyama, T. 2004. Effect of Deep Placement of N Fertilizers and Different Inoculation Methods of Bradyrhizobia on Growth, N_2 Fixation Activity and N Absorption Rate of Field-grown Soybean Plants. *Journal of Agronomy and Crop Science* 190; 46–58.

Turay, K.K., Andrews, M. and McKenzie, B.A. 1991. Effects of starter nitrogen on early growth and nodulation of lentil (*Lens culinaris* Medik.). *Proceedings Annual Conference – Agronomy Society of New Zealand*. 21: 61–65.

Unkovich, M.J. and Pate, J.S. 2000. An appraisal of recent field measurements of symbiotic N_2 fixation by annual legumes. *Field Crops Research*. 65: 211–228.

van Kessel, C. 1994. Seasonal accumulation and partitioning of nitrogen by lentil. *Plant and Soil*.164: 69–76.

Vigil, M.F. and Nielsen, D.C. 1998. Winter wheat yield depression from legume green fallow. *Agronomy Journal*. 90: 727–734.

Vitousek, P.M., Aber, J.D., Howarth, R.W., Likens, G.E., Matson, P.A., Schindler, D.W., Schlesinger, W.H. and Tilman, D.G. 1997. Human alteration of the global nitrogen cycle: Causes and consequences. *Ecological Applications*. 7: 737–750.

Vitousek, P.M., Cassman, K., Cleveland, C., Crews, T., Field, C. B. Grimm, N. B. Howarth, R. W., Marino, R., Martinelli, L., Rastetter, E.B. and J. I. Sprent, J.I. 2002. Towards an ecological understanding of biological nitrogen fixation. *Biogeochemistry*. 57&58: 1–45.

Woodard, H.J. and *Bly, A.* 1998. Relationship of nitrogen management to winter wheat yield and grain protein in South Dakota. *Journal of Plant Nutrition*. 21: 217–233.

Yau, S.K., Bounejmate, M., Ryan, J., Baalbaki, R., Nassar, A. and Maacaroun, R. 2003. Barley-legumes rotations for semi-arid areas of Lebanon. *European Journal of Agronomy* 19: 599–610.

Zentner, R.P., Campbell, C.A., Biederbeck, V.O., Miller, P.R., Selles, F. and Fernandez, M.R. 2002. In search of a sustainable cropping system for the semiarid Canadian prairies. *Journal of Sustainable Agriculture*. 18: 117–136.

CHAPTER 9

NUTRIENT AND IRRIGATION MANAGEMENT

B.A. McKENZIE[1], M. ANDREWS[2], AND G.D. HILL[1]

[1] *Agriculture Group, Agriculture and Life Science Division, PO Box 84, Lincoln University, Canterbury, New Zealand*

[2] *School of Sciences, University of Sunderland, Sunderland SR1 3SD, UK*

Email: mckenzie@lincoln.ac.nz

Abstract: Lentils are often grown in difficult edaphic conditions on stored soil water. This usually results in low yields. While water stress is often responsible for these low yields, low levels of soil fertility can also be a contributing factor. Lentils can usually fix enough N for their own requirements, although if sown into soil with extremely low available N or into cold, wet soil they may require a small amount of starter N to ensure adequate early growth to support nodulation and N fixation. While responses to fertilisers are variable, there are many reports that P at 40–60 kg/ha can help provide increased yields. Also, application of K at around 20 kg/ha may be beneficial in sandy or eroded soils. Sulfur is unlikely to be deficient if fertilisers containing S impurities are applied to other crops in the rotation. Of the micronutrients Zn and B are the two most likely to be deficient.While breeding and management can reduce the effects of drought stress lentils will often respond to irrigation with substantial yield increases. However, their high sensitivity to water logging means that irrigation can be overdone and thus decrease yield. Irrigation requirements are best calculated by considering the limiting water use deficit of the crop and irrigating to control soil moisture deficit to a value suited to the variety and soil type

1. INTRODUCTION

In most lentil growing regions the crop is grown on marginal lands on stored soil moisture with no inorganic fertiliser nutrient inputs (Sinha 1977). Under these conditions, lentil yields are quite low e.g. average yields in India are 0.71 t/ha (FAO STAT 2006). However, on high fertility soils with adequate soil moisture in countries such as New Zealand, the United Kingdom and Canada, yields of 3 t/ha can be achieved (Andrews et al. 2001; FAO STAT 2006). Thus even under 'optimal' conditions, lentil is not a highly productive crop when compared to cereals or other grain legumes. In New Zealand, where average lentil yields have been around 2 t/ha

S.S. Yadav et al. (eds.), Lentil: An Ancient Crop for Modern Times, 145–158.

(Jermyn et al. 1981) and many growers attain 3 t of seed/ha on a consistent basis, total dry matter production is seldom greater than 10 t/ha (McKenzie and Hill 1990). Wheat, which in New Zealand frequently yields more than 10 t grain/ha often produces in excess of 20 t DM/ha (White et al. 1999), and peas (*Pisum sativum*) can produce up to 8 t of seed/ha and about 16 t DM/ha (Wilson et al. 1984).

One of the main reasons for low yields in many lentil producing regions is water stress (Sarker et al. 2002; McKenzie and Hill 2004). Drought stress is the major restriction on lentil yield in lentil growing regions, worldwide (Muehlbauer et al. 1995, 2006; McKenzie and Hill 2004). For example, Erskine and El Ashkar (1993) reported that 80% of the variation in lentil seed yield in Mediterranean climates was accounted for by differences in seasonal rainfall. In the Mediterranean regions of West Asia and North Africa, lentil is usually grown in areas of 300–400 mm rain/year (Erskine et al. 1994). In these areas, most rain falls in winter, and from March until crop maturity in May, the crop experiences water and high temperature stress to an extent that restricts yield.

Lack of water restricts cell and tissue expansion, and limits plant ability to utilise available nutrients. In arid regions, nutrient inputs are primarily via animal manure and household wastes (El Dessougi 2006). However, in India with lentils, the traditional practice is to use no fertilizers or manure (Verniau, 2007). Improved practices under irrigation are to apply 18–20 kg of N and 17–22 kg P/ha plus K, Zn and S if required (Verniau, 2007). When grown under more favourable edaphic conditions, lentils are more productive, and will utilise greater amounts of soil nutrients. In the United States, Muehlbauer et al. (1995) recommended, for lentils in the Palouse region: 35 g/ha of Mo,17–22 kg S/ha, 44–66 kg P/ha if soil tests show less than 4 ppm P, and 22 kg K/ha on sandy or eroded soils. Utilising the mineral content of lentils reported by Labuda and Weslowski (1992), 10 t of lentil dry matter would contain 333, 45.5, 209, 18.3 kg/ha of N, P, K and S respectively. Saxena (1981) reported that a 2 t/ha crop of lentil seed would require about 100 kg N/ha, 12 kg P and 65 kg of K/ha. Clearly lentils may require fertiliser inputs to maximise productivity.

Agricultural scientists are trying to improve dryland lentil yields through breeding with the objective of increasing lentil adaptation to environmental stress. The key strategy used to combat drought has been to match crop development with the period of soil moisture availability. This has been achieved in two ways. Firstly, genotypes with early seedling establishment, early and rapid biomass development and early flowering and maturity have been selected (Kusmenoglu and Muehlbauer 1998; Sarker et al. 2002 a, b; Muehlbauer et al. 2006). This has involved the selection of drought tolerant varieties in sites with an average annual rainfall of less than 300 mm (Sarker et al. 2002b). Secondly, seed has been sown earlier in the spring or in the autumn (Kusmenoglu and Muehlbauer 1998). Autumn sowing can increase leaf area duration and hence give greater radiation interception. It can also reduce the period the crop is exposed to water stress due to its earlier maturation. An alternative or complementary strategy to combat drought, in areas where water is available, is irrigation. This has been shown to result in very large consistent yield increases compared with growing dry land lentil crops with inadequate rainfall.

This chapter considers firstly the use of starter N in lentil crops; secondly, the nutrient requirements of lentil other than N and finally strategies for quantifying the relationships between irrigation and yield, and options for managing drought.

2. STARTER NITROGEN

As shown in Chapter 8, lentils can fix adequate N for their own requirements in most situations. However, generally, the legume/rhizobium symbiosis can be affected by many factors and as a consequence is variable in productivity. Ledgard and Giller (1995) reviewed factors that can affect N fixation and their work can be summarised as follows:

1. An effective symbiosis is required for N fixation to occur.
2. Poor soil nutrition can result in poor nodulation or N fixation.
3. Low soil pH can limit N fixation.
4. High soil N levels can inhibit nodulation and N fixation.

Grain legumes, including lentils, can take up and assimilate N from the soil prior to nodulation, to an extent that affects growth (Andrews et al. 1992). In lentil, Turay et al. (1991), showed that both in the glasshouse and in the field early N increased leaf area, and shoot and root dry weight (Table 1). However, three weeks after nodules were first visible, nodule dry weight of lentil was decreased by up to 92% with increased early N. This decline in nodule dry weight was likely to be due to increased uptake of NO_3^- as legume species which are less affected by early N such as field beans take up lower amounts of NO_3^- (Table 1, Andrews et al. 1992).

Table 1. Effects of different concentrations of applied nitrate on early growth of three legume species. Shoot, root, and cotyledon (cot) measurements were taken when nodules were first visible; nodule dry weight was determined three weeks after nodules were first visible (Turay et al. 1991)

Species	Applied nitrate (mol/m^3)	Leaf area (cm^2)	Dry weight (mg/plant)			
			Shoot	Root	Cot	Nodule
Lens culinaris	0	11.1	59	41	–	26
	1	15.3	73	52	–	16
	10	25.0	119	74	–	2
	SEM	1.7	6	4	–	2
Phaseolus vulgaris	0	32.5	194	114	–	154
	1	46.3	200	123	–	131
	10	75.7	289	160	–	0
	SEM	3.1	13	9	–	8
Vicia faba	0	33.0	137	166	81	168
	1	39.1	164	153	87	140
	10	38.1	161	109	127	67
	SEM	2.1	10	7	20	11

Nevertheless, in soils that are very low in available soil N, small amounts of N added at sowing have been shown to increase early growth and N fixation in lentils (Saxena 1981). When lentils were sown early, into cold wet conditions which were not conducive to N fixation, application of 10–25 kg N/ha increased early growth (Saxena 1981). Thus, starter N may be useful under poor edaphic or low soil N conditions, but is probably not of value in regions with moderate soil fertility and good growing conditions.

3. MAJOR NUTRIENTS

3.1. Phosphorus

Since lentils are effective N fixers under most conditions, P may be the most common nutrient limiting lentil growth and yield. Phosphorus plays a major role in many plant processes including storing and transfer of energy; stimulation of root growth, flowering, fruiting and seed formation; nodule development and N fixation (McLaren and Cameron 1996; Ali et al. 1997).

The literature shows a wide variation in response of lentils to the application of P. While soils of many lentil growing areas are low in P (Saxena 1981), Table 2 shows that over a wide range of countries, and even in India the response of lentil to P application is quite variable; ranging from no response to a nearly 60% yield increase. In New Zealand, where soils on cropping farms are highly fertile, and soil P is usually not limiting McKenzie and Hill (unpublished data) have never found a response of lentil yield to the application of P in a range of student experiments over a number of years.

However, Muehlbauer et al. (1995) recommended 44–66 kg P/ha for lentils in the Palouse region, USA. Also, Saxena (1981) reported that the optimum level of fertilizer P application ranges from 17–43 kg of P/ha. Matar (1976) reported that the critical level of available P in the soil was about 4 ppm in years with average rainfall: in dry years lentils responded linearly to available P up to 9 ppm.

Phosphorus can also interact with rhizobium inoculants to provide significant benefits to lentil crops. Sekhon et al. (1986) reported that 8.6 kg of P/ha plus

Table 2. Variability in response of lentil yield to phosphorus application

Treatments (per ha)	Response	Country	Source
0, 12 and 23 kg P	Nil	Sudan	El-Sarrag and Nourai 1983
0 and 16 kg P	Nil	Pakistan	Ali et al. 1988
0 and 50 kg P	Nil	New Zealand	McKenzie et al. 1989
0 and 13 kg P	Nil	India	Sharma et al. 1993
0 kg NPK and 18 kg N, 13 kg P, and 14 kg K	59% inc.	India	Sharma et al. 1993
0, 5, 11 and 16 kg P	41% inc.	India	Rathore et al. 1992
0, 5, 11,16 and 22 kg P	21% inc.	India	Khare et al. 1988

Rhizobium gave similar yields to 17.2 kg of P/ha with no Rhizobium. Phosphorus application increased both the number of nodules and nitrogenase activity.

While response to P is clearly variable, most authors seem to agree that if soil test values indicate that there is less than 4 ppm P in the soil, then P fertiliser applied in the range of about 40 to 60 kg P/ha will be beneficial.

3.2. Potassium

There are very few reports in the literature of lentils responding to K. Jain et al. (1995) showed that 16.6 kg/ha of K significantly increased K contents in both the whole plant and lentil seed, however, there was no increase in net return per hectare. In the same experiment, Sharma et al. (1993) reported that K had no effect on seed yield. On a sandy loam soil in the Punjab, Azad et al. (1995) did find a small significant increase of 19% in grain yield with the addition of 33.2 kg K/ha. Muehlbauer et al. (1995) also reported that adding 18.3 kg /ha of K to sandy or eroded soils increased yield and may improve lentil cooking quality (Wassimi et al. 1978).

3.3. Sulfur

Sulfur is extremely important for grain legumes because it is required for nodulation and N fixation (Oke 1969). It is also needed to help increase the generally low levels of S containing amino acids found in lentil seed. While S deficiencies occur widely throughout the world (Mahler et al. 1988) the S requirements of many crops are generally met through application of superphosphate (McLaren and Cameron 1996). Where superphosphate fertilisers are not used, however, S deficiency can limit lentil production.

In New Delhi Shivakumar et al. (1995) reported that addition of 75 kg S/ha gave a 27% increase in seed yield compared to control plots. Other workers (Gwal et al. 1995; Sharma et al. 1993) found S had no effect on lentil yield.

Published results suggest that if fertilisers containing S impurities are applied to other crops in the rotation, it is likely that there will be adequate amounts in the soil for lentil.

4. MICRONUTRIENTS

According to Mahler et al. (1988) there are eight micronutrients essential for the growth of grain legumes. These are: B, Cl, Co, Cu, Fe, Mn, Mo Zn. In lentils, the micronutrients which are most frequently cited as likely to limit growth and yield are: Zn, B and Mo.

Saxena (1981) reported lentils are very susceptible to Zn deficiency particularly when grown after paddy rice as Zn deficiency is common in these soils. The critical limit for available Zn ranges from about 0.5 ppm to 1.81 ppm dependent upon the extractant used (Saxena 1981). A number of reports in the literature show that lentils respond to Zn. Sharma et al. (1993) and Jain et al. (1995) both reported that in India maximum lentil yields were obtained from fertiliser regimes which contained Zn. In Nepal, Srivastava et al. (1999) found that lentil responded positively to Zn up to 2.0 kg Zn/ha. However, in Syria, Zn at 5 kg/ha had no effect on yield in a pot experiment (Singh and Saxena 1986). Shuknesha (1977) found that P and Zn can interact such that when the P:Zn ratio is greater than 400, Zn may become deficient.

Molybdenum is a key micronutrient especially for legumes. This is because it is found in nitrogenase and nitrate reductase. These enzymes are respectively involved in N fixation and the conversion of nitrate −N to amino-acid N (McLaren and Cameron 1996). While there are few reports of Mo deficiencies in lentils, in Bulgaria application of Mo as a seed treatment increased seed yield (Golubev and Lugovskikh 1974).

Boron deficiency has also been reported in lentil and can cause leaf chlorosis and low yields in Nepal (Srivastava et al. 1999). These authors reported that grain yield of lentils with no B averaged 0.1 t/ha, but with 0.5 kg B/ha yield it increased to 1.4 t/ha.

5. RESPONSE TO IRRIGATION

In the literature, the response of lentils to irrigation has been highly variable. This is due to a number of reasons. The crop will respond more favourably to irrigation in a dry season than in a wet season. The response to irrigation may be varied by other limiting factors such as nutrient deficiency. There may be genetic variation in response to water stress and finally, some of the variability will be due to scientists applying irrigation in various ways and measuring different environmental and crop responses.

This variability of reported responses of lentil seed yield to irrigation is demonstrated in Table 3. As an example, at Tel Hadya, Syria, the maximum yield response to irrigation in a very dry season was a yield increase of 680% (Zhang et al. 2000), while in New Zealand, in a year with mean rainfall 7% above the long term mean of 695 mm, full irrigation gave an 87% decrease in seed yield when compared to the rainfed plots (McKenzie and Hill 1990). Even at a single site, response to irrigation can be quite variable. For example, Zhang et al. (2000) showed that over ten years in Tel Hadya, Syria, rainfed seed yields ranged from 176 kg/ha to 2195 kg/ha. Full irrigation gave seed yield increases ranging form 9.5% to 680% (Table 3). This was primarily due to variability in rainfall.

What is also clear from Table 3 is, that like most crop species, lentil usually responds positively to irrigation. Recently, however, Shrestha et al. (2006) reported that in a glasshouse trial, water deficit at flowering or podding resulted in seed yield increases in a West Asian and a South Asian genotype respectively. This was due to an increased numbers of flowers and to maintenance of pod and seed set after re-watering.

While lentils will usually respond to irrigation in dry seasons, they are extremely sensitive to both excess irrigation or rainfall and flooded soils (Summerfield 1981).

Table 3. Reported variations in response of lentil seed yield to irrigation across a range of seasons and locations

Irrigated yield (kg/ha)	Increase over control (%)	Irrigation method	Location	Source
1703	80	Water applied	Sudan	Ageeb (1975/76)
2460	40	One irrigation at flowering	Pantnegar, India	Lal et al. (1988)
1970	107	Three irrigations (0–2.0) at veg (0–1.0) at reproductive growth	New Delhi, India	Saraf and Baitha (1985)
1382	680	Fully irrigated max. response	Tel Hadya, Syria	Zhang et al. (2000)
2124	9.5	Fully irrigated min. response		
2170	174	Fully irrigated max. response	Tel Hadya, Syria	Oweis et al. (2004)
2020	25	Fully irrigated min. response		
3520	18	Fully irrigated max. response	Canterbury, New Zealand	McKenzie and Hill (1990).
700	−87	Fully irrigated min. response		

McKenzie and Hill (1990) reported yield losses of about 90% when a lentil crop was fully irrigated during a season with slightly more than average rainfall. The crop was not flooded, but rainfall of 113 mm in November (113% more than the November long term mean) resulted in excessive vegetative growth, serious disease issues (mainly due to *Botrytis cinerea*) and a concomitant collapse of both seed yield and harvest index. Flooding can also cause serious yield losses in lentil and there is evidence that, in comparison with a range of other grain legumes including pea, faba bean, chickpea and white lupin, it is more sensitive to water logging (Tang and Thomson 1996). Summerfield (1981) reported that waterlogged soils cause nutrient toxicity or deficiencies, acidity and anoxia, while Ali et al. (1997) reported that poor drainage can inhibit N fixation and increase susceptibility to root diseases.

6. QUANTIFYING THE RESPONSE TO WATER STRESS

As reported above, the response of lentil to irrigation is quite variable. Many researchers have in the past attempted to relate yield to the number of irrigations or to the amount of water applied. This is a very common technique and is particularly useful when growers cannot control the timing of their irrigations. Many reports show good relationships between yield and the number of irrigations. In Tel Hadia, Hamdi et al. (1992) found that two irrigations increased seed yield per plant by

20%. Abdel-Rahman et al. 1980) found that irrigating every 20d gave higher seed yields than irrigating every 40 or 60d. Saraf and Baitha (1985) reported that two irrigations increased seed yield by 64%, while 3 or 4 irrigations increased seed yield over the unirrigated control plants by 107% and 106% respectively. In general, good relationships between yield and the number of irrigations are found in arid regions, where lentils are grown on stored soil water only. However, in temperate regions where rainfall is more consistent there is often little relationship between yield and the number of irrigations (McKenzie et al. 1985).

These variables may provide a reliable indication of potential yield, but this will only be the case if rainfall is consistently absent. Generally, studies which have attempted to quantify drought by measuring drought intensity and its duration have shown a more stable relationship between irrigation and yield (French and Legg 1979; Wilson et al. 1984, McKenzie and Hill 1990).

7. IRRIGATION AT CRITICAL PERIODS OF SENSITIVITY

A close look at the scientific literature suggests that many crops have a period of their life cycle when they are particularly sensitive to drought. Wilson et al. (1984) showed that when averaged over three sowing dates peas performed best when irrigated at flowering and pod fill. Rathore et al. (1992) reported that for lentil, irrigation at branching and podding gave a 51% yield increase. Lal et al. (1988) showed that maximum grain yield was obtained when lentil was irrigated once at flowering, this also gave maximum water use efficiency. On the other hand, McKenzie and Hill (1990), under rain shelters, found no difference in seed yield between treatments which received either no irrigation until flowering and then full irrigation or full irrigation until flowering and then none. This was despite the fact that fully irrigated plants yielded 245 g seed/m^2 compared to 32 g/m^2 in unirrigated plots.

Clearly there is large disparity in the idea that flowering or podding is a critical period of sensitivity to drought in lentil. Wilson et al. (1984) suggested that a possible reason for uncertainty about critical periods at least in temperate sub humid climates like Canterbury is that rain often falls at inopportune times. Indeed, they suggest flowering and podding are not critically sensitive times and that untimely rainfall affected their results. The work of McKenzie and Hill (1990) does suggest that at least for the small seeded cultivar Titore, there is no critical period of sensitivity in lentil.

8. RELATING YIELD TO WATER STRESS

Penman (1962), designed a new approach to examine the effect of water stress on crop yield. This approach was based on relating yield to the maximum water stress which the crop experienced during its life cycle. Water stress or soil moisture deficit (SMD) is calculated as:

The sum of daily potential evapotranspiration (P_{ET}) minus rainfall (R) and irrigation (I)

(1) $SMD = \Sigma\ P_{ET} - (R + I)$

This approach utilises P_{ET} and characterises the growing season. In hot dry seasons with low rainfall and irrigation, SMD can build up to very high levels, while in cool moist years, or if sufficient irrigation is available, the SMD may stay low. There is often a very good relationship between yield and the maximum potential SMD which a crop experiences during the growing season (French and Legg 1979). In very dry conditions, lentil yields are very accurately predicted by this model. McKenzie and Hill (1990) showed that both seed and total dry matter production of lentil grown beneath rain shelters were related to maximum potential SMD (Figure 1).

Figure 1 suggests that on the silt loam cropping soil on which this crop was grown, the deficit at which yield was reduced (the limiting deficit) was about 130 mm and after this deficit, seed yield was lost at the rate of about 0.39% per mm of SMD. The high limiting deficit (D_l) for this lentil crop is one explanation for why the crop is considered drought tolerant. French and Legg (1979) reported D_ls ranging from 80 mm for *Vicia faba* to 140 mm for *Triticum* spp. The D_l is very sensitive to soil type and rooting depth. McKenzie (1987) showed that lentils are relatively deep rooting, with a tap root that can grow to at least 55 cm. This is

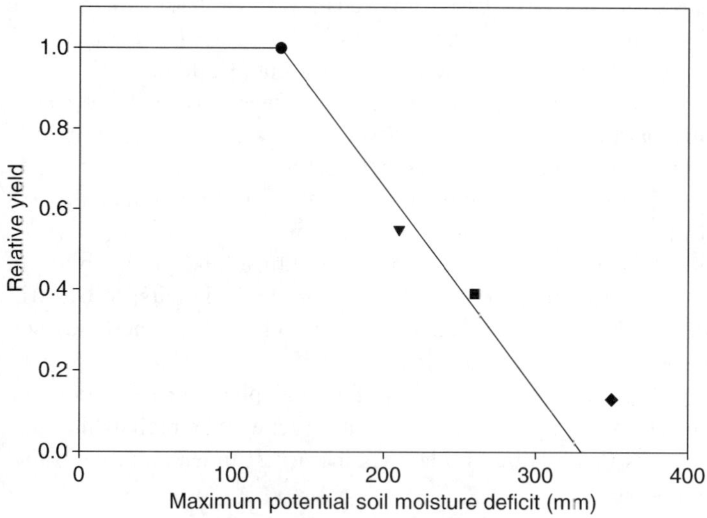

Figure 1. The relationship between relative seed yield (yield of fully irrigated crop/yield of unirrigated crop) and maximum potential soil moisture deficit of lentils grown beneath rain shelters in Canterbury New Zealand. Full irrigation (•), full irrigation until flowering then none (▼), nil irrigation until flowering then full (■) and nil irrigation (♦). The equation of line is $y = 1.45 - 0.0039\times$; $R^2 = 0.962$. (From McKenzie and Hill 1990)

similar to the 60–65 cm root depth reported by Shrestha et al. (2005) and could explain the relatively high D_l shown in Figure 1.

Lentil is drought tolerant in comparison with other temperate grain legumes. For example, the limiting soil moisture deficit for lentil is similar to that for chickpea and substantially greater than that for pea or faba bean (McKenzie and Hill 2004). However, the amount of yield loss experienced for each mm of drought past the limiting soil moisture deficit was greater for lentil than chickpea or faba bean. Differences in limiting soil moisture deficits across species are probably due to variation in their rooting depth and root proliferation (McKenzie and Hill 2004). Lentil roots are capable of extracting water from at least 90 cm depth (Sharma and Prasad 1984; McKenzie 1987). The variation in yield loss per mm of drought above the limiting soil moisture deficit was probably caused by factors which can influence water use, such as leaf size and orientation, stomatal number and orientation, and radiation use efficiency. Differences in osmotic adjustment could also be important (Nielsen 2001).

9. WATER USE EFFICIENCY

Water use efficiency (WUE) of lentil can be calculated in a variety of ways. This section will only compare WUEs based on total crop water use. Water use efficiency is the ratio of yield (seed or total dry matter production) to total crop water use or total ET. A wide range of WUEs have been reported for lentil. In Canterbury, New Zealand, WUE for dry matter production (WUE_{dm}) ranged from 13.2 kg/ha.mm to 28.1 kg/ha.mm (McKenzie and Hill, 1990). In Nepal, WUE_{dm} had a mean value of 18.3 kg/ha, while in Syria it was 13.7 kg/ha mm (Zhang et al. 2000). Values for WUE_{dm} were about 30 kg/ha mm in Southwestern Australia (Siddique et al. 1998). Water use efficiencies for grain yield are considerably lower than for dry matter production. McKenzie and Hill (1990) reported WUE_{seed} values ranging from 4.8–7.76 kg/ha.mm. Other values reported include: 6.3, Shrestha et al. (2005); 3.8, Zhang et al. (2000) and 11 kg/ha.mm Siddique et al. (1998). In India Lal et al. 1988 reported WUE_seed ranged from 7.6–10.4 kg seed/ha.mm.

Water use efficiency is markedly affected by both climate and productivity. In areas with very hot temperatures and high vapour pressure deficits, WUE will be lower than under cooler conditions. Any climatic factor which increases soil evaporation as a proportion of Et, will also reduce WUE.

There may be scope for improving WUE of lentil through plant breeders. Siddique et al. (1990) found that WUE_{seed} of modern wheat varieties was higher than that of old varieties. This was mainly due to increases in harvest index and more rapid phenological development.

10. COPING WITH DROUGHT

In most lentil growing countries, the crop is grown on stored soil moisture only. It is well adapted for this purpose. McKenzie and Hill (1990) showed that a lentil crop could survive and produce a yield of 320 kg/ha when grown under rain shelters from 1 July 1985 until final harvest 6 months later. While the unirrigated crop produced only 320 kg/ha, the fully irrigated crop produced 2450 kg seed/ha.

While lentil is well suited as a dry land crop, yield potential is low. There are, however, a number of strategies which can be utilised to improve lentil productivity in arid regions. Silim et al. (1993) found that early maturity was correlated with seed yield. These authors suggested that selecting and sowing early flowering cultivars will be beneficial in drought prone areas. In dry east coast New Zealand, early sowing of both peas and lentils can help these crops escape the usual summer droughts.

The indeterminate growth habit of lentils may also provide a benefit in dry seasons. Shrestha et al. (2006), found that water stress, during reproductive growth, affects both vegetative and reproductive development. This has a range of effects on the numbers of flowers, pods and seeds. Indeed, in two varieties, upon rewatering, they increased flower production and pod and seed retention. This increased yield after water stress. An indeterminate growth habit clearly makes recovery after drought possible.

11. CONCLUSIONS

Lentils do not require large amounts of fertiliser to produce high yielding crops. They produce enough N through N fixation for their own requirements. However, in very low N soil, or when sown into cold wet conditions they may require from 10–25 kg of starter N. Crop P requirements should be determined through use of a soil test. If soil P is less than about 4 ppm, P at 20–60 kg/ha will probably benefit yield. Responses to K are quite variable and it is probably only required on sandy or eroded soils at about 20 kg/ha. Sulfur is unlikely to be needed if fertilisers with S impurities are included in the farm rotation. Among micronutrients, Zn at 2 kg/ha and B at 0.5 kg/ha have proved beneficial to yields in some circumstances.

Lentils are well adapted to dry land farming, but under severe water stress the crop responds favourably to irrigation. Response to irrigation can be variable, with some researchers reporting that the reproductive phase is critically sensitive to water stress. Other researchers have reported that yield is linearly related to the amount of water stress the crop receives. Water use efficiency is variable, but it may be possible to improve crop WUE through breeding. The most useful strategy for improving yields in drought prone areas is through the use of early flowering and maturing varieties.

REFERENCES

Abdel-Rahman KA, Shalaby EM, Abdallah MM (1980) Seed and quality of lentil as affected by different sowing dates and irrigation frequency. Research Bulletin No. 1234. Faculty of Agriculture, Ain Shams University, Cairo, Egypt.

Ageeb OAA (1975/76) Annual Report, Hudeiba Research Station cited by: Ali M, Dahan R, Mishra JP, Saxena NP (1997) Towards the more efficient use of water and nutrients in food legume cropping. In R Knight (ed) Linking Research and Marketing Opportunities for Pulses in the 21 st Century.

Proceedings of the Third International F ood Legume Research Conference, Adelaide, S. Australia 22–26 September, 1997 355–368.

Ali A, Khan BR, Keatinge JDH (1988) Effects of inoculation and phosphate fertilizer on lentil under rainfed conditions in upland Baluchistan. Lens Newsletter **15:** 29–33.

Ali M, Dakan JP, Saxena NP (1997) Towards the more efficient use of water and nutrients in food legume cropping. In: Knight R (ed) Linking Research and Marketing Opportunities for pulses in the 21st Century. Proceedings of the third International Ford Legumes Research Conference. Adelaide, S. Australia Sept. 22–26, 1997, 355–368.

Andrews M, Hill GD, Raven JA, Sprent JI (1992) Nitrate effects on leaf growth of grain legumes prior to nodulation: species differences relate to nitrate uptake. Proceedings of the 1st European Conference on Grain Legumes: 139–140.

Andrews M, McKenzie BA, Joyce A, Andrews ME (2001) The potential of lentil (*Lens culinaris*) as a grain legume crop in the UK: an assessment based on a crop growth model. Annals of Applied Biology 139: 293–300

Azad AS, Manchanda JS, Bains SS, Gill AS (1995) Phosphorus and potassium fertilizer interactions increase grain yield of lentil on sandy loam soil of Punjab, India. Lens Newsletter **22:** 16–18.

El Dessougi H (2006) Restoring soil fertility. ICARDA Caravan **23**: 23–25.

El-Sarraq G, Nourai AH 1983. A review of research on lentils (*Lens culinaris*) in the Sudan. Lens Newsletter **10,** 1–12.

Erskine W, El Ashkar F (1993) Rainfall and temperature effects on lentil (*Lens culinaris*) seed yield in the Mediterranean environment. Journal of Agricultural Science, Cambridge 121: 347–354

Erskine W, Tufail M, Russell A, Tyagi MC, Rahman MM, Saxena MC (1994) Current and future strategies in breeding lentil for resistance to biotic and abiotic stresses. Euphytica 73: 127–135

French BK, Legg BJ (1979) Rothamsted irrigation 1964-76. Journal of Agricultural Science, Cambridge **92,** 15–37.

Golubev VD, Lugovskikh MA (1974) [Pre-sowing treatment of lentil seeds with ammonium molybdate] Khimiya v Selskom Khozyaisve **12:** 24–25.

Gwal HB, Tiwari RJ, Gupta DK (1995) Fertilizer management of lentil under rain-fed conditions in Madhya Pradesh. Lens Newsletter **22:** 11–12.

Hamdi A, Erskine W, Gates P (1992) Adaptation of lentil seed yield to varying moisture supply. Crop Science **32:** 987–990.

Jain RC, Tiwari RJ, Nema DP (1995) Integrated nutrient management for lentil under rain-fed conditions in Madhya Pradesh II. Nodulation. Nutrient content and economies. Lens Newsletter **22:** 13–15.

Jermyn WD, Goulden DS, Lancaster IM, Banfield RA (1981) Lentil evaluation in New Zealand. Proceedings Agronomy Society of New Zealand **11:** 77–81.

Khare JP, Tomar GS, Tiwari UK, Sharma HL (1988) Response of lentil to nitrogen and phosphorus under rainfed conditions in central India. Lens Newsletter **15:** 12–14.

Kusmenoglu I, Muehlbauer FJ (1998) Genetic variation for biomass and residue production in lentil (*Lens culinaris* Medik.). ll. Factors determining seed and straw yield. Crop Science 38: 911–915

Labuda S, Weslowski M (1992) Chemical and mineral composition of lentil in flowering stage as affected by herbicide application in the field. Lens Newsletter **19,** (2): 39–42.

Lal M, Gupta PC, Pandey RK (1988) Response of lentil to different irrigation schedules. Lens Newsletter **15,** 20–23.

Ledgard SF, and Giller KE (1995) Atmospheric N_2 fixationas an alternative N source. In: Bacon PE (ed) Nitrogen fertilisation in the environment. Marcel Dekker Inc, New York, 443–486.

Mahler RL, Saxena MC, Aeschrimann J (1988) Soil fertility requirements of pea lentil, chickpea and faba bean. In: Summerfield RJ (ed) World Crops: Cool Season Food Legumes. Proceedings of the International Food Legume Research Conference on pea, lentil, faba bean and chickpeas, Spokane, Washington 6–11 July 1986, 277–289.

Matar AE (1976) Correlation between $NaHC0_3$ extractable P in soil and yield of wheat and lentil grown under dry farming conditions. ACSAD, Damascus, Soil Science P-2, 16 pp.

McKenzie BA, Sherrell C, Gallagher JN, Hill GD (1985) Response of lentil to irrigation and sowing date. Proceedings of the Agronomy Society of New Zealand. **15,** 47–50.

McKenzie BA (1987) The Growth, Development and Water Use of Lentils (*Lens culinaris* Medik.). PhD. Thesis. Lincoln College, University of Canterbury 223 pp.

McKenzie BA, Hill GD (1990) Growth, yield and water use of lentils (*Lens culinaris*) in Canterbury, New Zealand. Journal of Agricultural Science, Cambridge **114**: 309–320.

McKenzie BA, Hill GD (2004) Water use in grain legumes. In: Proceedings of the 5th European Conference on Grain Legumes/2nd International Conference on Legume Genomics and Genetics, AEP – l'Association Européenne de Recherche sur les Protéagineuse, Paris, France, 61–62.

McKenzie BA, Miller ME, Hill GD (1989) The relationship between lentil crop production and weed biomass production in Canterbury. Proceedings of the Agronomy Society of New Zealand **19:** 11–16.

Muehlbauer FJ, Kaiser WJ, Clement SL, Summerfield RJ (1995) Production and breeding of lentil. Advances in Agronomy 54: 283–332

Muehlbauer FJ, Cho S, Sarker A, McPhee KE, Coyne CJ, Rajesh PN, Ford R (2006) Application of biotechnology in breeding lentil for resistance to biotic and abiotic stress. Euphytica 147: 149–165

Oke OL (1969) Sulphur nutrition of legumes. Experimental Agriculture **5,** 111–116.

Nielsen DC (2001) Production functions for chickpea, field pea and lentil in the Central Great Plains. Agronomy Journal 93: 563–569.

Oweis T, Hackum A, Pala M (2004) Lentil production under supplemental irrigation in a Mediterranean environment. Agricultural Water Management **68,** 251–265.

Penman HL (1962) Woburn irrigation 1951-59 1. Purpose, design and weather. Journal of Agricultural Science, Cambridge **58,** 343–348.

Rathore RS, Khandwe R, Khandwe N, Singh PP (1992) Effect of irrigation schedules, phosphorus levels and phosphate solubilising organisms on lentil. I. Yield. Lens Newsletter **19,** (1) 17–19.

Saraf CS, Baitha SP (1985) Water use patterns and water requirement of lentil planted on different dates. Lens Newsletter **12,** 12–18.

Sarker A, Aydin N, Aydogan A, Sabaghpour SH, Ketata H, Kusmenoglu I, Erskine W (2002a) Winter lentils promise improved nutrition and income in West Asian Highlands. ICARDA Caravan 16: 14–16

Sarker A, Neupane RK, Sakr B, El Ashkar F, Lutfir A, Erskine W (2002b) More grain from less rain: ICARDA'S strategy to improve lentil for resource-poor farmers in dry areas. ICARDA Caravan 17: www.icarda.org/publications/

Saxena MC (1981) Agronomy of lentils. In: Webb C, Hawtin G (eds) Lentils Commonwealth Agricultural Bureaux, Norwich. pp 111–129.

Sekhon HS, Dhingra KK, Sandhu PS, Bhandari SC (1986) effect of time of sowing, phosphorus and herbicides on the response to Rhizobium inoculation. Lens Newsletter **13:** 11–15.

Sharma AK, Billore SD, Singh RP (1993) Integrated nutrient management for lentil under rainfed conditions. Lens Newsletter **20:** 15–16.

Sharma SN, Prasad R (1984) Effect of soil moisture regimes on the yield and water use of lentil (*Lens culinaris* Medic.). Irrigation Science 5: 285–293.

Shivakumar BG, Saraf CS, Patil RR (1995) Effect of phosphorus and sulphur levels and limited irrigation on the performance of Macrosperma lentils. Lens Newsletter **22:** 19–23.

Shrestha R, Turner NC, Siddique KHM, Turner DW (2006) Physiological and seed yield responses to water deficits among lentil genotypes from diverse origins. Australian Journal of Agricultural Research **57**, 903–915.

Shrestha R, Siddique KHM, Turner NC, Turner DW, Berger JD (2005) Growth and seed yield of lentil (*Lens culinaris* Medikus) genotypes of West Asian and South Asian origin and crossbreds between the two under rainfed conditions in Nepal. Australian Journal of Agricultural Research **56**, 971–981.

Shuknesha A (1977) Influence of zinc on survival of rhizobia, nutrient uptake and nitrogen fixation by lentil (*Lens esculenta* Moevch). Pantnagar Journal of Research **2:** 259.

Siddique KHM, Loss SP, Pritchard DL, Regan L, Tennant D, Jettner RL, Wilkinson D (1998) Adaptation of lentil (*Lens culinaris* Medic.) to Mediterranean type environments: effect of time of sowing on growth, yield and water use. Australian Journal of Agricultural Research **49,** 613–626.

Siddique KHM, Tennant D, Perry MW, Belford RK (1990). Water use and water use efficiency of old and modern wheat cultivars in a Mediterranean-type environment. Australian Journal of Agricultural Research **41**, 431–437.

Silim SN, Saxena MC, Erskine W (1993) Adaptation of lentil to Mediterranean environment. I. Factors affecting yield under drought conditions. Experimental Agriculture **29,** 9–19.

Singh NP, Saxena MC (1986) Response of lentil to phosphorus and zinc application. Lens Newsletter **13:** 227–28.

Sinha, SK (1977) Food legumes: distribution, adaptability and biology of yield. FAO Plant Production and Protection. Paper No. 3 124 pp.

Srivastava SP, Joshi M, Johansen C, Rego TJ (1999) Boron deficiency of lentil in Nepal. Lens Newsletter **26:** 22–24.

Summerfield, RJ (1981) Adaptation to environments. In. C. Webb and G Kawtin, eds Lentils 91–110. Commonwealth Agricultural Bureaux.

Tang C, Thomson BD (1996) Effects of solution pH and bicarbonate on the growth and nodulation of grain legume species. Plant and Soil 186: 321–330.

Turay KK, Andrews M, McKenzie BA (1991) Effects of starter nitrogen on early growth and nodulation of lentil (*Lens culinaris* Medik). Proceedings of the Agronomy Society of New Zealand **21,** 61–65.

Verniau, S. 2007. Lentil. In The World Fertilzer Use Manual Taken from http://www.fertilizer.org/ifa/publicat/html/pubman/lentil.htm downloaded 5 February, 2007.

Wassimi NS, Abu-Shakra R, Tannous R, Hallab AH (1978) Effect of mineral nutrition on cooking quality of lentils. Canadian Journal of Plant Science **58**: 165–168.

White J, Millner J, Moot DJ (1999) Cereals. In White J and Hodgson J (eds) New Zealand Pasture and Crop Science. Oxford University Press, Auckland 213–234.

Wilson D, Jamieson P, Hanson R (1984) Analysis of response of field peas to irrigation and sowing date 1. Conventional methods. Proceedings of the Agronomy Society of New Zealand **14:** 71–74.

Wilson DR, Jamieson PD, Jermyn WA, Hanson R (1984) Models of growth and water use of field peas (*Pisum sativum* L.) in P.D. Hebblethwaite, M.C. Heath and T.C.K. Dawkins (eds). The Pea Crop. A Basis for Improvement. London: Butterworths; 139–151.

Zhang H, Pala M, Oweis T, Harris H (2000) Water use and water-use efficiency of chickpea and lentil in a Mediterranean environment. Australian Journal of Agricultural Research **51,** 295–304.

CHAPTER 10

WEED MANAGEMENT

JASON BRAND[1], N. T. YADURAJU[2], B. G. SHIVAKUMAR[3], AND LARN MCMURRAY[4]

[1] *Department of Primary Industries, Victoria PB 260, Horsham, Victoria 3401, Australia*
[2] *National Agricultural Innovation Programme, KAB II, Pusa, New Delhi 110012, India*
[3] *Division of Agronomy, Indian Agricultural Research Institute, Pusa, New Delhi 110012, India*
[4] *South Australian Research and Development Institute, PO Box 822, Clare, South Australia 5453, Australia*
E-mail: jason.brand@dpi.vic.gov.au

Abstract: Lentils compete poorly with weeds and yield reductions in excess of 80% due to competition have been recorded. In lentil, due to poor early vigour and short height it is critical to minimise weed populations throughout the whole lifecycle. To manage weeds a range of cultural, physical and chemical practices can been employed, both 'in crop' and throughout the farming system. Cultural practices include land preparation, seed preparation, sowing and crop establishment method, nutritional management, management of insects and disease and irrigation scheduling. While physical removal or killing of weeds generally occurs either manually or with the tools designed for interculturing. These methods integrated with chemicals applied pre-sowing, post-sowing pre emergence and post emergence, provide a sustainable and profitable long term strategy to minimise weeds and maximise grain yield, particularly when implemented with crop rotations throughout the farming system. This chapter provides an overview and discusses principles behind these strategies in lentils across different production regions

1. INTRODUCTION

Lentil (*Lens culinaris* Medikus) is an important grain legume throughout the world. In 2005, it was cultivated over an area of 4 million hectares producing 4 million tonnes of grain. Although lentil is cultivated in more than 50 countries, about 10 countries account for more than 95% of this production (Chapter 6). It is grown primarily for grain production as its seed is rich in protein (approximately 22%). It is a major component of the diet in many countries. In addition, the straw can be used as a high quality feed for livestock.

S.S. Yadav et al. (eds.), Lentil: An Ancient Crop for Modern Times, 159–172.

Lentil is grown in a range of cropping systems (sequentially or intercropping) worldwide both for its rotational (eg. biological nitrogen fixation, disease break) and economic benefits. Due to its relatively short height and slow early growth, it competes poorly with weeds (Basler *et al.*, 1981). Yield reductions of up to 84% due to weed competition have been recorded in the literature (Al Thahabi *et al.*, 1994; Boerboom and Young, 1995; Kumar and Kolar, 1989; Mohamed *et al.*, 1997; Salkini and Nygaard, 1983; Saxena and Wassimi, 1980). Weeds compete with crops for nutrients, moisture and space as well as harbouring insect pests and pathogens that may adversely affect the lentil crop. In addition to yield loss, several of the important weeds (i.e. *Lathyrus aphaca, Vicia sativa* and *V. hirsuta* produce seed similar in shape and size to that of lentil and separation from the lentil seed is difficult, posing serious problems in seed production and processing.

In order to manage weeds in a crop of lentils a range of cultural, physical and chemical practices have been employed worldwide. The specific practices and combination of practices, varies widely from country to country and is dependent on social, economic and environmental issues. This chapter provides an overview of strategies that have been used to prevent weed infestation and subsequent grain yield loss in lentils across different production regions.

2. WEED FLORA AND COMPOSITION

Weed flora observed in lentil crops varies considerably between countries owing to variability in climatic conditions, soil types and cropping systems. In general, weeds that cause significant yield loss in lentils can be broken into 4 major groups: annual grasses, annual broadleafs, perennial and biennial grasses and broadleafs, and parasitic (Table 1). Within each country and cropping region there are many production manuals for growers that identify many of the key weed species and their importance (Day *et al.*, 2006; Hntowich, 2000; Holding and Bowcher, 2004; Moorthy *et al.*, 2002; Singh *et al.*, 2001; Wilding *et al.* 1998).

3. WEED COMPETITION

Lentil competes poorly with weeds due to its poor seedling vigour and short stature (Basler, 1981) which is further compounded when growing season temperatures are low. Critical periods for weed competition in lentil vary and are dependent on region, climate and weed species present. To ensure a profitable lentil crop it is essential that weeds are adequately controlled. It is important that in each cropping system, the growth and development patterns of key weeds is understood for effective control. Further, the degree of yield loss depends on the nature of weeds and the stage and duration of weed crop competition (Day *et al.*, 2006; Radosevich *et al.*, 1997). This information enables the development of successful, cost-effective and sustainable weed management practices.

Generally in lentil it is important to minimise weeds populations throughout the whole lifecycle, including pre-sowing, emergence, vegetative, reproductive and

Table 1. Common annual, perrenial and parasitic weed flora of lentil crops

Common name	Scientific name	Family
Annual grasses		
Little seed canary grass	*Phalaris minor* (L.) Retz.	Poaceae
Wild oat	*Avena Spp.*	Poaceae
Annual blue grass	*Poa annua* L.	Poaceae
–	*Polypogon monspeliensis* (L.) Desf.	Poaceae
Brome grass	*Bromus spp.*	Poaceae
Silver grass	*Vulpia bromoides*	Poaceae
Barley grass	*Critesion murinum*	Poaceae
Rye grass	*Lolium Spp.*	Poaceae
Green foxtail	*Setaria viridis*	Poaceae
Broad leaf weeds		
Common lambs quarters	*Chenopodium album* L.	Chenopodiaceae
Bifora	*Bifora testiculata*	Apiaceae
Milk thistle	*Sonchus oleraceus*	Asteraceae
Prickly lettuce(Whip thistle)	*Lactuca serriola*	Asteraceae
Musk weed	*Myagrum perfoliatum*	Brassicaceae
Bedstraw	*Galium aparine*	Rubiaceae
Medic	*Medicago Spp.*	Fabaceae
Wild mustard	*Sinapis arvensis & Sisymbrium Spp.*	Brassicaceae
Wild radish	*Raphanus raphanistrum*	Brassicaceae
Wild turnip	*Brassica tournefotii & Rapistrum rugosum*	Brassicaceae
Sheep weed (White iron weed)	*Buglosoides arvensis*	Boraginaceae
White sweet clover	*Melilotus alba* Medicus	Leguminosae
Corn spurry	*Spergula arvensis* L.	Caryophyllaceae
Wild safflower	*Carthamus oxyacantha* Bieb.	Asteraceae
Cudweed	*Gnaphalium indicum* L.	Asteraceae
Arrow weed	*Pluchea lanceolata* Oliv.	Asteraceae
–	*Launia nudicaulis* H.K.	Asteraceae
Yellow sweet clover	*Melilotus indica* (L.) All.	Leguminosae
Wild pea	*Lathyrus aphaca* L.	Leguminosae
Field bindweed	*Convolvulus arvensis* L.	Convolvulaceae
Scarlet pimpernel	*Anagallis arvensis* L.	Primulaceae
Wild onion	*Asphodelus tenuifolius* Cav.	Liliaceae
Bur clover	*Medicago denticulata*	Leguminosae
Wood sorrel	*Rumex dentatus* L.	Polygonaceae
Fumitory	*Fumaria Spp.*	Papavaraceae
Bedstraw	*Gallium Spp.*	Rubaceae
Common vetch	*Vicia Spp.*	Leguminosae
Wooly pod vetch	*Vicia villosa*	Fabaceae
Swine cress	*Coronopus didimus* (L.) Sm.	Cruciferae
Black nightshade	*Solanum nigrum* L.	Solanaceae
Buckwheat/Wireweed	*Polygonum Spp.*	Polygonaceae

(Continued)

Table 1. (Continued)

Common name	Scientific name	Family
Red Root Pigweed	*Amaranthus palmeri*	Amaranthaceae
Perennials		
Canada thistle	*Cirsium arvense* (L) Scop.	Asteraceae
Nut grass	*Cyperus rotundus* L.	Cyperaceae
Bermuda grass	*Cynodon dactylon* (L.) Pers.	Poaceae
Tiger grass	*Saccharam spontaneum* L.	Poaceae
Mignonette	*Reseda spp.*	Resedaceae
Parasitic weeds		
Dodder	*Cuscuta spp.*	Convolvulaceae
Broomrape	*Orobanche spp*	Orobancheaceae

maturity stages to maximize grain yield. In growing regions, where the crop is grown on stored moisture or where growing season rainfall is limiting, weed control prior to sowing is also critical (Radosevich *et al.*, 1997). Similarly, during the emergence and vegetative stage, minimizing weeds is essential to reduce competition for water, but also during this stage weeds can compete for nutrients and photosynthetically active radiation (Radosevich *et al.*, 1997; Pathan *et al.*, 2006). Weeds that germinate and grow during the flowering and maturity stages are often less competitive for water, nutrients and photosynthetically active radiation, but can cause difficulties during the harvesting process (particularly in mechanized situations) resulting in significant yield losses and reduction in grain quality (Day *et al.*, 2006).

4. WEED MANAGEMENT PRACTICES

Weed control is one of the major limitations to growing lentil worldwide (Al Thahabi *et al.*, 1994; Boerboom and Young, 1995; Mohamed *et al.*, 1997). Many cultural, physical and chemical practices have been employed and often the total elimination of weeds during the entire period of crop growth is not economical or required. The control of weeds during critical phases of crop weed competition for the particular production region will be sufficient with weeds emerging thereafter not significantly affecting yield and quality. Generally best results are achieved when a range of management practices are followed and weeds are adequately controlled throughout the whole farming system, rather than just focusing on control in the lentil crop.

4.1. Cultural

Cultural practices here refer to non-herbicide practices involved in crop management which aid weed prevention and crop competitiveness. These include land preparation, seed preparation, sowing and crop establishment method, nutritional management, management of insects and disease and irrigation scheduling.

Land preparation, in terms of tillage practice and stubble management, is critical for the production of lentil. Lentil has been successfully grown in all systems from conventionally cultivated through to zero till and with and without the presence of stubble (Battikhi and Suleiman, 1999; Day *et al.* 2006; Matus *et al.*, 1997; Pala *et al.*, 2000). Traditionally, due to its short stature lentil has been grown in systems where the soil has been cultivated to create a flat seed bed and stubble removed to allow good establishment and ease of harvest. These conditions are also generally regarded as optimal for herbicide application where applicable (see below). Cultivation of soil can incorporate weed seed into top soil layers allowing for maximum germination when soil is moist and the potential for a knockdown herbicide or follow-up cultivation to control emerging weeds prior to sowing in some situations (Roberts, 1981). Cultivation can also bury weed seed below their optimum emergence depth inducing dormancy and inhibiting germination by reducing the oxygen concentration (Preston, 2007). Future disturbance may reintroduce the seed to an appropriate depth for germination to occur. The cultivation process can also stimulate the germination of some weed species by damaging the seed coat which break seed dormancy (Preston, *pers. comm.*). Stubble from previous crops can be removed by burning or physical incorporation into the soil. Burning can reduce weed numbers of some species. Walsh and Newman (2006) found that all seed of Annual Ryegrass (*Lolium rigidum*) and Wild Radish (*Raphanus raphanistrum L.*) were destroyed by burning, provided burning temperatures were above 400 and 500°C, respectively. Removing stubble and allowing the soil to be solarized can have a significant impact on weed populations. Linke (1994) showed that soil solarization reduced the population in most weed species present, with only a minority increasing in population. Current farming systems in mechanized regions of the world tend to be moving toward, no-till or zero-till practices, where minimal soil disturbance occurs and stubble is maintained. New machinery and technology also allows sowing between standing stubble rows. The minimal soil disturbance ensures that most weed seeds are left on the soil surface. It has been shown that seed survival of weed species such as green foxtail and wild oat (*avena fatua* L.) are reduced 5 fold by leaving seed on the surface in comparison with burying (Banting *et al.*, 1973; Saga and Mortimer, 1976; Thomas *et al.*, 1986). Modelling and field observations appear to confirm trends that preventing seed entry into the surface seed bank will reduce weed populations in the long term (Anderson, 2004; Mohler, 1993; Preston, 2007). In addition, standing stubble rows can provide trellising for the crop resulting in improved harvestability.

Seed preparation and variety selection can be important tools to help maximize the competitiveness of the lentil crop. It is important that high quality (i.e. disease, virus, weed free), viable seed lots are used, selected from fertile areas in the paddock that have not been chemically desiccated to aid with harvesting operations (Day *et al.*, 2006). Lentil varieties vary in growth habit and morphology (Erskine and Goodrich, 1991) and differences in competitive ability could be expected. However, research has been inconclusive with Tepe *et al.* (2005) reporting minor differences in competitiveness between varieties, but indicating that none of the varieties were really effective as a 'stand-alone' component for weed control in lentils. McDonald *et al.* (2007) also showed that differences in early vigour between genotypes were insufficient to affect the competitive ability of lentil.

Seeding depth and rate, sowing time, and row space can influence crop establishment, competitiveness and grain yield. Shallow sowing (3–5 cm) can increase plant emergence, height, plant dry weight and grain yield compared with deep sowing (8–10 cm), but may also result in greater crop injury through pre sowing or post sowing pre emergent herbicides (Brand *et al.*, 2001; van Rees, 1997). Increased sowing rates can reduce weed populations, due to increased crop competition (Ball *et al.*, 1997; Boerboom and Young, 1995; McDonald *et al.*, 2007; Paolini *et al.*, 2003). Delayed sowing can reduce early vigour, crop competitiveness and resultant yield (Holding and Bowcher, 2004; Mishra *et al.*, 1996), however, it can also provide increased opportunity for weed control by mechanical and/or chemical methods prior to sowing (Brenzil *et al.* 2006; Day *et al.*, 2006). Narrow row-spacing theoretically should improve weed competition as the lentil crop will reach canopy closure earlier, however, Chaudhary and Singh (1987) found it to have no effect on weed populations. Alternatively, in wider row spacings, the crop is likely to be less competitive with weeds, but this method allows alternative methods of inter-row weed removal.

Nutrient supply must at least match nutrients removed by the crop, to ensure optimum crop growth and maximize competitiveness against weeds. For example, rhizobial inoculation and N and P fertilization has been shown to increase biomass and yield (Bolland *et al.* 1999; Shah *et al.* 2000) which may increase weed or crop competitiveness depending on which gains the greatest advantage from the improved fertility. Reducing impacts of disease and insects through resistant cultivars and appropriate pesticides will help to maintain growth and competitiveness of the crop. For example, seed applied fungicides increase seedling establishment and reduce disease transmission from seed to seedling which results in improved vigour and grain yield (Bretag, 1989; Morrall and Beauchamp, 1984). Similarly, several in-crop fungicides have been shown to reduce the impacts of diseases such as anthracnose (Chongo *et al.* 1999) and ascochyta blight (Beauchamp *et al.* 1986). In irrigated lentil production systems, irrigation before seed-bed preparation significantly reduced grass and broad-leaved weeds and increased grain yield compared with nil irrigation (Mohamed *et al.*, 1997). Overall, the selection of appropriate cropping systems combined with optimum cultivar, sowing depth and time, plant density and nutritional, disease and insect management will assist in the suppression of weeds and complement other methods of weed control.

4.2. Physical / Mechanical

Physical/mechanical in crop removal of weeds is common in countries where costs of labour are low or where this method is required because of organic status of production, and no herbicides are permitted. It involves the physical removal and killing of weeds either manually or with the tools designed for inter-culturing. Hand weeding is generally as effective as herbicide application with best results achieved with multiple weeding (Dawood 1994; Sekhon *et al.*, 1993; Turk and Tawaha, 2002). However, it is time consuming and often expensive. Inter-cultural practices either with bullock drawn or tractor mounted implements are only possible in situations where the crop is sown in rows wide enough for implements to pass. Snobar and Haddad (1998) showed that mechanical cultivation with chiselling twice during the season was as effective in controlling the weeds as the herbicide and manual treatments, and gave comparable grain yields despite the increased row spacing. In organic systems, tillage post seeding prior to emergence and post emergent in the seedling stage can be effective in controlling weeds (Brenzil *et al.*, 2006). However, most mechanical tillage practices require high energy input which can be a limiting factor in regards to machinery requirements and costs. In addition, thermal methods, such as flaming, have been successfully used for weed control (Holmoy and Netland, 1994). A detailed summary of the various mechanical and thermal methods for weed control can be found in Bond and Grundy (2001). Physical/mechanical methods need to be cautiously used due to the sensitivity of lentil shoots and roots to damage (Stringi *et al.*, 1988). Any potential weed control benefits may be negated by reduced productivity from the physical damage.

4.3. Chemical

Chemical weed control is employed mostly in large scale intensive agricultural systems where labour costs are high and a majority of the farm operation is mechanized. In these situations, chemicals enable efficient and timely weed control over large areas of land with minimal labour input. Chemical weed control is becoming more popular in developing and underdeveloped countries as labour costs increase and mechanization occurs. It is most effective when used as part of an integrated weed management package (see 5. Integrated weed Management) including cultural methods described above and crop rotations, particularly as the repeated use of any one effective herbicide may result in development of resistant weed types. For example, Maurice and Billet (1991) observed that green foxtail (*Setaria viridis*) became resistant to trifluralin and ethalfluralin in southern Manitoba in Canada following their repeated use in lentil-wheat-flax-canola cropping system. Similarly in Australia there is widespread resistance of annual ryegrass (*Lolium rigidum*) to chemicals such as haloxyfop, trifluralin and clethodim (Day *et al.*, 2006).

The feasibility of using herbicides depends on the cropping systems, land preparation methods, soil conditions, cultivar tolerance and anticipated weed problems in the crop (Moyer *et al.*, 1992; Tepe *et al.*, 2004). For example, the use of herbicides is the only cost effective option to control weeds in no-till, stubble retained systems which are popular in mechanized regions. Most herbicides, particularly residual, can cause significant crop damage (Saxena and Wassimi, 1980; Yasin *et al.*, 1995; Nitschke, 2003) and it is important that recommendations found on labels are followed to minimize risks. Application practices, environmental conditions (eg. soil type and moisture, rainfall patterns, sunlight etc) and herbicide chemistry (i.e. solubility and leaching index) can significantly influence both weed efficacy and crop response to the herbicide. For example, for many herbicides applied to the soil (particularly those with residual activity) it is important that they are applied to moist soil and seed is sown deeper to avoid germination directly into the herbicide layer. In some cases with herbicides applied pre-sowing, minimum rainfall requirements are needed between application timing and sowing of lentil. In addition, many of the selective herbicides are applied at concentrations which are toxic to the target weed, but not the lentil crop and there is commonly a narrow margin between toxicity to the weed and toxicity to the crop. For example, metribuzin applied at recommended rates resulted in no crop damage, while when applied at double recommended rates significant crop damage and yield loss occurred (Nitschke, 2003). In this section we will discuss principles behind various chemical weed control methods. Registration and use of chemicals varies widely from country to country and specific recommendations for use can be found in production manuals for growers (Day *et al.*, 2006; Hntowich, 2000; Holding and Bowcher, 2004; Moorthy *et al.*, 2002; Singh *et al.*, 2001).

4.3.1. *Pre-sowing*

Pre-sowing herbicides can be used to kill all weeds growing before the sowing of lentil (ie glyphosate) or selective weeds emerging during the life of the residual chemical (ie trifluralin) and provide a weed free environment for crop establishment, promoting early vigour and improved competition of the crop. Control of weeds prior to sowing also prevents loss of soil water by transpiration, providing the following crop with maximum water supply to optimize growth. Pre-sowing weed control also has a major role in controlling herbicide resistant weeds (ie annual rye grass) or hard to control weeds in crop (ie medic) and often seeding is delayed at the expense of grain yield to allow significant germinations of these weeds. Both knockdown and residual herbicides are used pre-sowing, to control weed populations present and prevent emergence of weeds with the lentil crop. Knockdown herbicides used are generally non-selective and have no residual value (e.g. glyphosate) and may be used multiple times in the period up to sowing, to kill all plant material present.

Herbicides with residual value can be used pre-sowing as “spikes” to knockdowns to increase the success of killing larger and older weeds (e.g. dicamba). In addition, some residual herbicides are applied and incorporated into the soil surface that don’t kill weeds present, rather kill emerging weeds (e.g. trifluralin).

4.3.2. *Post sowing pre-emergence*

Post sowing pre-emergence herbicides are applied to the soil surface immediately after sowing, but before emergence. It is the primary method of weed control in mechanized countries. Both non-residual knockdown (e.g. glyphosate) and residual herbicides (e.g. metribuzin, simazine, propyzamide, prometryn) can be used to control emerging grass and broadleaf weeds. They tend to be used in a targeted manner, where-by the herbicides and rates used are dependent on weed species present (Holding and Bowcher, 2004). These herbicides are essential to minimize weed populations during the early growth phase of the crop, when it is least competitive.

4.3.3. *Post-emergence*

Post emergence herbicides are generally used when other weed control methods have not adequately controlled the emergence of weeds or where there has been emergence of new flushes of weeds at later stages of crop growth. Few broad-leaf selective post – emergence herbicides are suitable for use in lentil (e.g. diflufenican, flumetsulam), however, there are a number of options for grass weed control (e.g. clethodim, haloxyfop, fluzifop-butyl). Many of the post emergent herbicides also affect the crop and thus may cause some yield reduction associated with controlling the weeds. Alternatively, in wide row farming systems knockdown herbicides can be applied inter-row with shielded sprayers, however, will not control weeds germinating in the row.

Several other knockdown herbicide weed control methods are used to control weeds later in crop growth. Wickwiping is a method were a rope wick containing knockdown herbicides, such as glyphosate with paraquat and diquat, is traversed just above crop height and makes contact with all weeds taller than the crop, preventing seed set of the weeds (Holding and Bowcher, 2004). Crop topping is the technique of applying a herbicide to a mature crop to kill any green weeds and sterlize any weed seeds. Desiccation is similar to crop topping, but the herbicide is applied at physiological maturity of the crop and an earlier growth stage of the weed than with crop topping (Holding and Bowcher, 2004). All these methods can aid the harvesting process, reduce grain contamination and assist in control of resistant weeds.

5. INTEGRATED WEED MANAGEMENT

Integrated weed management (IWM) is the synergistic use of a combination of cultural, mechanical, chemical and biological practices to prevent and control weeds throughout the farming system. One of the primary aims of IWM is to improve sustainability of the farming system via concurrent increases in crop productivity and reductions in economic losses, risks to human health and potential damage to flora and fauna (Yaduraju and Mishra, 2005).

As discussed earlier, weed resistance to herbicides, is a major constraint to chemical weed control. It is therefore necessary to use integrated strategies whereby different chemical, cultural and mechanical control methods are used. In addition, to minimize reliance on 'in crop' control methods, so far discussed, it is important to maintain weed control throughout the entire farming system (i.e. in cropping or pasture phases that occur in rotation with lentil). Several studies have indicated the benefits of crop rotation for weed control (Brenzil *et al.*, 2006; Derken *et al.*, 2002; Holding and Bowcher, 2004). Selection of suitable crop rotations can allow sustainable and economical control of problem weeds in lentil to occur in other phases of the rotation. Lentils have a very limited range of herbicides suitable for use in crop (Preston, 2002). However, other pulse crops i.e. peas (*pisum sativum*) have wider range of suitable herbicides giving different options for weed control. In addition, the use of vetch or oats for hay will allow all weed set to be controlled in other phases of a rotation. In some countries, due to the relative high gross margins, there is a tendency to grow lentils every 2 or 3 years in a tight rotation with cereals. This can result in difficulties in adequately controlling weeds which can only be controlled effectively in cereal phases of the rotation (Preston, 2002). Overall, widening rotations with the use of alternative pulse or break crops is important for the long term sustainability of lentil in the farming system.

In addition, in determining the best strategy for integrated control it is also important that the biology of the major weeds which affect production is understood. For example, Matthews *et al.* (1996) found that a delay in sowing, with pre-emergent herbicide application integrated with crop topping in the pulse phase of the rotation and catching weed seed at harvest, reduced rye grass densities and improved crop yields throughout the farming system.

Utilizing integrated strategies is important with weeds that are particularly difficult to control in the lentil crop (i.e. various broadleaf and parasitic weeds) and allows the use of various other control techniques, which would generally not be used within a lentil crop. For example, knife rolling over cover crops has been used to aid weed control (Ashford and Reeves, 2003). Green/brown manuring or cutting a forage crop for hay can also help to minimize weeds throughout the farming system (Holding and Bowcher, 2004). Prevention of weeds in crops leading up to the lentil phase of the rotation is often the best method to minimize expensive chemical, cultural and mechanical inputs into the lentil crop to control weeds and to reduce weed burdens so optimum sowing times can be used to maximize yields.

6. ECONOMIC ANALYSIS

A major factor affecting weed control methods is economic viability, both directly to the lentil crop and indirectly to the whole farming system. Costs of production in lentil and returns from the crop vary widely throughout the world, so specific weed management practices will also vary widely. In determining the optimum weed management strategy, producers need to assess a complex range of factors:

1. Yield potential in weed free situation,
2. Potential returns from the crop,
3. Potential yield loss at different levels of weed control,
4. Costs associated with other crop management activities (e.g. delayed sowing, disease and insect management and harvest),
5. Costs of weed control methods,
6. Effectiveness of weed control methods,
7. Potential economic benefits to following crops.
8. Cost of storage and re-cleaning of weed infested grain.

Table 2. Simplified example of impact of changes in grain yield potential and weed control costs on the profitability of lentil.

Potential yield (t/ha)	2	1.5	1	0.5	0.5	0.5	0.5	0.5
Weed control costs ($A/ha)	100	100	100	100	20	40	60	80
Effectiveness[1]	95%	95%	95%	95%	30%	50%	70%	90%
	$A/ha[5]	$A/ha	$A/ha	$A/ha	$A/ha	$A/ha	$A/ha	$A/ha
Potential net profit (weed free)[2]	800.00	600.00	400.00	200.00	200.00	200.00	200.00	200.00
Potential net profit (weeds uncontrolled)[3]	160.00	120.00	80.00	40.00	40.00	40.00	40.00	40.00
Potential gross profit[4]	570.00	380.00	190.00	0.00	24.00	30.00	28.00	18.00

[1] Effectiveness. Weed control expected when control methods have been applied. i.e. 95% is equivalent to an expected 5% grain yield loss.
[2] Potential net profit (weed free) = Potential yield × Price received for grain ($400/t)
[3] Potential net profit (weeds uncontrolled) = Potential net profit (weed free) × potential yield loss from weeds (80% yield loss used in this example)
[4] Potential gross profit = (Potential net profit (weed free) – weed control costs – general management costs ($100/ha used in this example)) × Effectiveness
[5] $A/ha = Australian dollars

The combination of these factors and a producers inherent attitude to risk will determine the amount spent on controlling weeds. Table 2 illustrates the impact of some of the factors listed above on gross profits. It can be seen that in high yielding situations, more can be spent on managing weeds, while in lower yielding situations the cost of removing a high proportion of weeds may exceed potential returns.

7. SUMMARY

Lentil is one of the most important pulse crops grown in many countries. Weeds cause severe yield and quality losses in all lentil growing areas. Several sustainable and economically viable cultural, mechanical and chemical options to control weeds are available to producers. The best options for weed control vary widely between the cropping regions around the world. However the philosophy of using integrated weed management, a synergistic combination of cultural, mechanical, chemical and biological practices to prevent and control weeds throughout the farming system appears to be the most sustainable and profitable long term strategy.

REFERENCES

Al-Thahabi, S.A., Yasin, J.Z., Abu-Irmaileh, B.E., Haddad, N.I. and Saxena, M.C. 1994. Effect of weed removal on productivity of chickpea (*Cicer arietinum* L.) and lentil (*Lens culinaris Med.*) in a Mediterranean environment. *Journal of Agronomy and Crop Science*. 172(5): 333–341

Anderson, R.L. 2004. Impact of subsurface tillage on weed dynamics in the central great plains. *Weed Technology*. 18: 186–192.

Ashford, D.L. and Reeves, D.W. 2003. Use of a mechanical roller-crimper as an alternative kill method for cover crops. *American Journal of Alternative Agriculture*. 18(1): 37–45.

Ball, D.A., Ogg, A.G. Jr. and Chevalier, P.M. 1997. The influence of seeding rate on weed control in small-red lentil (*Lens culinaris*). *Weed Science*. 45(2): 296–300.

Banting, J.D., Molberg, E.S., and Gephardt, J.P. 1973. Seasonal emergence and persistence of green foxtail. *Canadian Journal of Plant Science*. 53: 369–376.

Basler, F. 1981. Weeds and their control. In. Lentil. Webb, C. and Hatin, G. Commonweatlth Agricultural Bureaux.

Battikhi, A.M. and Suleiman, A.A. 1999. Effect of tillage and plant residue management practices on shrinkage of a vertisol. *Journal of Agronomy and Crop Science*. 182(4): 285–290.

Beauchamp, C.J., Morrall, R.A.A. and Slinkard, A.E. 1986. Effects of scheduling applications of benomyl, captafol and chlorothalonil on Ascochyta blight of lentil. *Canadian Journal of Plant Pathology*. 8(3): 260–268.

Boerboom, C.M. and Young, F.L. 1995. Effect of postplant tillage and crop density on broadleaf weed control in dry pea (*Pisum sativum*) and lentil (*Lens culinaris*). *Weed Technology*. 9(1): 99–106.

Bolland, M.D.A., Siddique K.H.M., Loss, S.P. and Baker, M.J. 1999. Comparing responses of grain legumes, wheat and canola to applications of superphosphate. *Nutrient Cycling in Agroecosystems*. 53(2): 157–175.

Bond, W. and Grundy, A.C. 2001. Non-chemical weed management in organic systems. *Weed Research*. 41: 383–405.

Brand, J.D., Materne, M. and Armstrong, R.A. 2001. Utilising the full yield potential of new pulse cultivars in Victoria through improved agronomy. *In* 'Proceedings of the 10th Australian Agronomy Conference.' Hobart, Tasmania, Australia 28th Jan - 1st Feb.

Brenzil, C., Reckseidler, B., Johnson, E. and Frick, B. 2006. Organic Crop Production: Weed Management. Agriculture and Food, Saskatchewan.

Bretag, T.W. 1989. Evaluation of fungicides for the control of Ascochyta blight in lentils. *Tests of agrochemicals and cultivars*. Apr 1989 10: 44–45.

Chaudhary, M. and Singh, T.P. 1987. Studies on weed control in lentil. *Indian Journal of Agronomy*. 32(3):2 95–297.

Chongo, G., Bernier, C.C. and Buchwaldt, L. 1999. Control of anthracnose in lentil using partial resistance and fungicide applications. *Canadian Journal of Plant Pathology*. 21(1): 16–22.

Cobb, W.T. 1978. Ethalfluralin, a new herbicide for podded crops. *Proceedings of the Western Society of Weed Science*. 31: 98–101.

Dangol, D.R. 1990. Lentil weeds in Rampur, Chitwan Valley, Nepal. *Lens*. 17(1): 11–13.

Dawood, R.A. 1994. Hand weeding in lentil (*Lens culinaris* Med.) grown on beds at various growth stages under different phosphorus levels. *Assiut Journal of Agricultural Sciences*. 25(3): 131–142.

Day, T., Day. H., Hawthorne, W., Mayfield, A., McMurray, L., Rethus, G. and Turner, C. 2006. Grain Legume Handbook. Eds. Lamb, J. and Poddar, A.

Dersen, D.A., Anderson, R.L., Blackshaw, R.E. and Maxwell, B. 2002. Weed dynamics and management strategies for cropping systems in the northern great plains. *Agronomy Journal*. 94: 174–185

Erskine, W. and Goodrich, W.J. 1991. Variability in lentil growth habit. *Crop Science*. 31(4): 1040–1044

Holding, D. and Bowcher, A. 2004. Weeds in Winter Pulses – integrated solutions. CRC for Australian Weed Management Technical series #9.

Hntowich, G. 2000. Saskatchewan Pulse Growers – Pulse Production Manual. http://www.saskpulse.com/growing/index.php

Holmoy, R. and Netland, J. (1994) Band spraying, selective flame weeding and hoeing in late white cabbage, part 1. *Acta Horticulturae* 372, Engineering for Reducing Pesticide Consumption and Operator Hazards, 223–234.

Kumar, K. and Kolar, J.S. 1989. Effect of chemical weed control and Rhizobium inoculation on the yield of lentil. *Journal of Research Punjab Agricultural University*. 26(1): 19–24.

Linke, K.H. 1994. Effect of soil solarization on arable weeds under Mediterranean conditions: control, lack of response or stimulation. *Crop Protection*. 13(2): 115–120.

Matthews, J.M., Llewellyn, R., Powles, S. and Reeves, T. 1996. Integrated weed management for the control of herbicide resistant annual ryegrass. *In* 'Proceedings of the 8th Australian Agronomy Conference.'

Matus, A., Derksen, D.A., Walley, F.L., Loeppky, H.A., Kessel, C. van and Van Kessel, C. 1997. The influence of tillage and crop rotation on nitrogen fixation in lentil and pea. *Canadian Journal of Plant Science*. 77(2): 197–200

McDonald, G.K., Hollaway, K. and McMurray, L. 2007. Weed competition in lentil (*Lens culinaris*) *Australian Journal of Experimental Agriculture*. 47(1): 48–56.

Mishra, J.S., Singh, V.P. and Bhan, V.M. 1996. Response of lentil to date of sowing and weed control in Jabalpur, India. *Lens Newsletter*. 23(1/2): 18–23.

Mohamed, E.S., Nourai, A.H. and Mohamed, G.E., Mohamed, M.I. and Saxena, M.C. 1997. Weeds and weed management in irrigated lentil in northern Sudan. *Weed Research Oxford*. 37(4): 211–218.

Mohler, C.L. 1993. A model of the effects of tillage on emergence of weed seedlings. *Ecological Applications* 3: 53–73.

Moorthy, B.T.S., Mishra, J.S. and Dubey, R.P. 2002. Teaching Manual on Recent Advances in Weed Management. National Research Centre for Weed Science, Jabalpur.

Morrall, R.A.A and Beauchamp CJ (1984) Evaluation of fungicide seed treatments for ascochyta infected lentil seed. *Pesticide Research Report* 285.

Moyer, J.R., Bergen, P. and Schaalje, G.B. 1992. Effect of 2,4-D and dicamba residues on following crops in conservation tillage systems. *Weed Technology*. 6(1): 149–155.

Nitschke, S.J. 2003. Herbicide tolerance within cereal and pulse varieties. *In* South Australian Field Crop Evaluation Program Post Harvest Report 2002/2003 pp 93–104. Eds Wheeler, R.D. and McMurray, L.S., SARDI, Primary Industries and Resources, Adelaide, South Australia, Australia.

Pala, M., Harris, H.C., Ryan, J., Makboul, R. and Dozom, S. 2000. Tillage systems and stubble management in a Mediterranean-type environment in relation to crop yield and soil moisture. *Experimental Agriculture*. 36(2): 223–242.

Paolini, R., Colla, G., Saccardo, F. and Campiglia, E. 2003. The influence of crop plant density on the efficacy of mechanical and reduced-rate chemical weed control in lentil (*Lens culinaris* Medik.). *Italian Journal of Agronomy*. 7(2): 85–94.

Pathan, S., French, R.J. and Hashem, A. 2006. Competitive effects of wild radish (*Raphanus raphanistrum* L. on lupin cultivars (*Lupinus angustifolius* L.). *In* Proceedings of the 13th Australian Agronomy Conference. Perth, Western Australia, Australia 10th–15th Sept.

Preston, C. 2007. Weed biology – the missing link to better weed management. *In* GRDC Grains Research Update, South Australia, Adelaide, South Australia, Australia 7th – 8th Feb.

Preston, C. 2002. Managing an eternal pests – weeds. *In* Lentil Focus 2002 National Conference. Horsham, Victoria, Australia 15th to 17th Oct.

Radosevich, S., Holt, J. and Ghersa, C. 1997. Weed Ecology – Implications for Management 2nd Edn. John Wiley and Sons Pub.

Roberts, H.A. 1981. Seed banks in soils. Advances in Applied Biology. 6: 1–55.

Saga, G.R. and Mortimer, A.M. 1976. An approach to the study of population dynamics of plants with special reference to weeds. *Advances in Applied Biology*. 1: 1–47

Salkini, A.B. and Nygaard, D. 1983. Survey of weeds in lentils in north and north-eastern Syria. *Lens*. 10(2): 17–20.

Saxena, M.C. and Wassimi, N. 1980. Crop weed competition studies in lentils. *Lens*. 7: 55–57.

Sekhon, H.S., Guriqbal Singh and Brar, J.S. 1993. Effect of chemical, mechanical and cultural manipulations on weed growth and grain yield of various pulse crops. Integrated weed management for sustainable agriculture Proceedings of an Indian Society of Weed Science. International Symposium, Hisar, India, 18–20 November 1993. III: 141–146.

Shah, N.H., Hafeez, F.Y., Arshad, M. and Malik, K.A. 2000. Response of lentil to Rhizobium leguminosarum bv. viciae strains at different levels of nitrogen and phosphorus. *Australian Journal of Experimental Agriculture* 40(1): 93–98.

Singh, V.P., Dixit, Anil, Mishra, J.S., Singh, P.K., Raghuwanshi, M.S. and Bhan, V.M. 2001. Cropping System: An approach to integrated weed management.

Snobar, B.A. and Haddad, N.I. 1998. Evaluation of weed control methods in lentil (*Lens culinaris* Med.) in Jordan. *Dirasat Agricultural Sciences*. 25(2): 203–213.

Stringi. L., Amato, G., Gristina, L. and Cibella, R. 1988. Weed control in lentils and first results obtained in a semi-arid environment. *Informatore Agrario*. 44(25): 59–63.

Tepe, I., Erman, M., Yazlik, A., Levent, R. and Ipek, K. 2004. Effect of different control methods on weeds, yield components and nodulation in the spring lentil. *Turkish Journal of Agriculture*. 28: 49–56.

Tepe, I., Erman, M., Yazlik, A., Levent, R. and Ipek, K. 2005. Comparison of some winter lentil cultivars in weed-crop competition. *Crop Protection*. 24(6): 585–589.

Thomas, A.G., Banting, J.D. and Bowes, G. 1986. Longevity of green foxtail seeds in a Canadian praire soil. *Canadian Journal of Plant Science*. 66: 189–192.

Turk, M.A. and Tawaha, A.M. 2002. Lentil (*Lens culinaris* Medik.) response to frequencies of hand weeding. *Indian Journal of Agricultural Research*. 36(2): 137–140.

van Rees, H. 1997. In: Southern Mallee and Northern Wimmera Crop and Pasture Production Manual. (*Birchip Cropping Group*: Birchip, Victoria). p. 55.

Walch, M. and Newman, P. 2006. Burning narrow windrows for weed seed destruction. *In* Proceedings of the 13th Australian Agronomy Conference. Perth, Western Australia, Australia 10th–15th Sept.

Wilding, J.L., Barnett, A.G. and Amor R.L. 1998. Crop Weeds. R.G. and F.J. Richardson Publishers.

Yaduraju, N.T. and Mishra, J.S. 2005. Weed Management in Pulses. In: Guriqbal Singh, H.S. Sekhon and J.S. Kolar (eds.) *Pulses*. Agrotech Publishing Academy, Udaipur. pp. 359–376.

Yasin, J. Z., Al Thahabi, S., Abu Irmaileh, B.E., Saxena, M.C. and Haddad, N.I. 1995. Chemical weed-control in chickpea and lentil. *International Journal of Pest Management*. 41(1): 60–65.

CHAPTER 11

COMMERCIAL CULTIVATION AND PROFITABILITY

MICHAEL MATERNE[1] AND A. AMARENDER REDDY[2]

[1] *Grains Innovation Park, The Department of Primary Industries, Private Bag 260, Horsham, Victoria, 3401, Australia*

[2] *Centre for Poverty and Rural development, Administrative Staff College of India, Hyderabad*

E-mail: michael.materne@dpi.vic.gov.au

Abstract: Lentil has been an important source of protein in many countries where it has been grown for many centuries. In traditional growing regions, lentils are often grown with little or no inputs like fertilizers, pesticides and herbicides on small farms to supply food for the local people. Production and profitability on these farms contrasts with that on farms in more developed countries where lentils have only been grown for a short period and is export orientated

1. INTRODUCTION

Lentil (*Lens culinaris* Medikus ssp *culinaris*) is one of the world's oldest cultivated plants and was domesticated in the Near East from the wild progenitor species *Lens culinaris* ssp. *orientalis* Boiss. over 7,000 years ago. Lentil was domesticated with wheat, barley and other pulses in the "Fertile Crescent" of the Near East and spread in a temporal sequence in all directions from this centre and some of the oldest remains of food plants dated at 7,500–8,500 B.C. are of lentil (Harlan 1971, Cubero 1981, Smartt 1984). It is probable that cultivated lentil achieved its current range in the Old World by about 1,000 BC (Smartt 1984). Lentil was introduced into the Indo-Gangetic plain around 2000 B.C. (Cubero 1981) and now half the world's lentil area is found in South Asia. With the exception of Chile (Barulina 1930), lentil production in the Americas is a relatively recent event. Lentil was first grown in the Palouse region of eastern Washington and northern Idaho of the USA in 1916 and in the prairie of Western Canada in 1969 (Muehlbauer and McPhee 2002). In Australia lentil has only become a crop of significance since 1994.

From a world perspective, lentils are an old crop, consumed for thousands of years in the Indian sub-continent, Middle East, Southern and Eastern Europe and North Africa. Red lentils are preferred in the Indian sub-continent, Middle East and

S.S. Yadav et al. (eds.), Lentil: An Ancient Crop for Modern Times, 173–186.

parts of North Africa such as Egypt, while green lentils are the predominant lentil consumed in the other regions as well as the Middle East. In more recent history, the inhabitants of South America have consumed green lentils, having taken their eating habits from southern Europe. But it has been red lentils which have 'fed the masses' particularly in the Indian-sub continent. Lentils are a staple food in many regions, providing an important source of protein, especially where diets are vegetarian or meat is expensive. On average over 3 million tonnes of lentils are produced worldwide. Most are consumed where they are grown, except in the new world countries of Canada, the United States of America and Australia, and in Turkey, a traditional producer.

Lentil production was initially focused on providing food for local consumption but has now expanded to a crop where significant trade occurs across countries. The current production methods for lentil are diverse and will be explored, with economic considerations, through a comparison between India, representing countries where lentils have traditionally been grown and widely consumed, and Australia, representing highly mechanised countries where production is a recent development and export orientated.

2. A CASE STUDY OF COMMERCIAL CULTIVATION AND PROFITABILITY İN A TRADITIONAL LENTIL PRODUCING COUNTRY – INDIA

2.1. Lentil Production in India

Lentil is an important winter (rabi) pulse crop in India, representing 12% of the area and production of total winter pulses. In the global context, India is the largest producer of lentil, contributing about 27% of the worlds 3.65 million tonnes of lentil production in 2005 (FAOSTAT 2006). The annual output of lentil is now about 0.8–1.1 million tonnes from an area of about 1.4 million hectares, almost double the level of production during the nineteen eighties. Productivity of lentil is about 684 kg/ha compared to an average winter pulse yield of 723 kg/ha. Lentil grows well on the light loamy and alluvial soils of north India and in well-drained light black soils of Madhya Pradesh. The three north Indian states of Uttar Pradesh (51%), Madhya Pradesh (23%) and Bihar (16%) account for about 90% of total lentil production in India with small areas in West Bengal, Rajasthan, Assam, Haryana and Punjab.

The area of lentil in India has increased by about 19% and productivity by 5% between 1995 and 2005 in India. Area expanded in the eastern states of Assam (84%) and West Bengal (19%) and in the central and Western states of Rajasthan (43%) and Madhya Pradesh (29%) rather than in northern India. The largest growths in yield were recorded in Haryana (44%), Rajasthan (27%) and Uttar Pradesh (11%). The net result was an increase in production of about 24% between 1995 and 2005 and India turned from a net importer to an exporter of lentil.

Lentil is an important crop to India's domestic and export economy. It is the only pulse crop with a significant net exportable surplus and demand is increasing for lentil in India. Lentil is consumed mostly in northern and eastern India. Family income and relative price are the key determinants of demand for lentils in India. The estimated long-run elasticity of demand for lentils with respect to income is 0.56 and relative price is −0.87 (Kumar 1998). Although there is a surplus in lentil production, there is an overall supply shortfall of pulses in India. Since the 1980's lentil production has increased to 2.5% per year (FAOSTAT 2006). Projections are that demand and supply will increase in India but the deficit in supply is predicted to increase to 8.4% to 20.6% by 2020 based on a growth in demand of 2.98% per year as proposed by Kumar (1998). Under the current scenario of high domestic prices and supply shortage in international markets, India needs to increase its competitiveness internationally to feed its people into the future.

2.2. Profitability of Lentils in the Farming System

In India lentils are predominantly grown on residual moisture with no irrigation. Variability in yield can be high and is dependent on residual moisture, temperature and disease and pest attack. The total cost of growing lentils is approximately Rs. 8850/ha, with variable costs of Rs. 4710/ha, or about 53% of the total cost in farmers practice (Table 1). Most operations are performed manually, including land preparation and harvest. Generally, farmers do not spray any insecticides and fungicides, and many farmers practice manual weeding. Most of the operations are carried out by family labour with the help of neighboring farmers on an exchange basis. Only harvesting and threshing is done on a contractual basis where 1:11 of the harvested grain is used as payment. Most farmers use their own seed or procure it from neighbours (>95%) as certified seed is not available (Field survey 2005). The average yield obtained in a field study in 2005 is 880 kg/ha under farmers practice. At a selling price of Rs. 16/kg, the gross revenue for lentils is about Rs. 16080/ha and net profit Rs. 7230/ha. The total post harvest losses of lentil during transportation, threshing, winnowing and farm storage have been estimated at 7% of total production.

The uptake of new technologies is limiting lentil productivity and profitability in India. Farmer practice often differs greatly from that proposed from research. Most farmers use less than recommended seeding rates, and fertilizer is either applied in suboptimal doses or not at all. In demonstrations to improve lentil profitability, certified seed was inoculated with rhizobium, herbicides and pesticides were applied at recommended rates, 100 kg DAP and 100 Kg Gypsum fertilizers were applied to the soil, and one irrigation was applied at flowering (Field survey 2005). Harvesting was done manually but threshing was done with a mechanical thresher. Input costs for the recommended practice were higher than current practice but yields were higher. Gross returns were higher and the overall net profit under recommended practice was 58% higher than farmer practice. The cost per hectare was higher under recommended practice but the cost of production per tonne of grain produced was lower than for the recommended practice at Rs. 875/tonne and Rs. 1005/tonne respectively.

Table 1. Cost benefit analysis of lentil cultivation (per ha) in India

Description	Recommended Practice				Farmers Practice			
	Unit	Unit Price (Rs.)	Quantity	Amount (Rs.)	Unit	Unit Price (Rs.)	Quantity	Amount (Rs.)
INPUTS								
Labour	Day			2040	Day			3660
Land preparation	Day	60	3	180	Day	60	3	180
Sowing	Day	60	3	180	Day	60	2	120
Weeding/pesticide/ herbicide application	Day	80	4	320	Day	60	20	1200
Harvesting	Day	60	20	1200	Day	60	20	1200
Threshing	Day	80	2	160	Day	60	16	960
Bullock labour	Pair-day			450	Pair-day			450
Land preparation	Pair-day	150	3	450	Pair-day	150	3	450
Machine cost	Hour			500	Hour			
Threshing	Hour	200	2.5	500				
Seed	Kg	40	30	1200	Kg	20	30	600
Fertilizer				1100				
DAP	Kg	9.8	100	980				
Gypsum	Kg	1.2	100	120				
Herbicide	Number	1	600	600				
Pesticide	Number	1	650	650				
Irrigation	Number	1	700	700				
Total operating costs				7240				4710
Total Fixed Costs				4140				4140
Total cost				11380				8850
OUTPUTS								
Grain yield	Kg	16	1300	20800	Kg	16	880	14080
Fodder yield (byproduct)	Kg	2	1000	2000	Kg	2	1000	2000
RETURNS								
Gross return				22800				16080
Net return				11420				7230
CV of net return (%)				22				30
Cost of production (Rs./tonne)				875				1005
Benefit cost ratio (BCR)				2.00				1.82

Source: Field survey (2005). Note: Rs. 45 = US$1.

Relative profitability is the strongest factor influencing crop choice among farmers. As lentil is a winter crop, the main competing crops are wheat, chickpea, and mustard. Net returns are lower for lentil compared to wheat and mustard, and similar to chickpea (Table 2). However, without subsidies (35% of net returns) the net returns for wheat are lower than for lentil. Therefore, if subsidies for water, electricity and fertilizers were reduced, the comparative profitability of lentil would increase and replace wheat wherever water is not available for irrigation.

Table 2. Relative profitability of competing crops with and without subsidy

Crop	net return with subsidy (Rs)	net return without subsidy (Rs)	% of subsidy
Wheat	10753	6957	35
Chickpea	8536	7955	7
Lentil	8250	7755	6
Mustard	10644	8856	17

Source: Field survey (2005)

Table 3. Yield gap between improved practice and farmers practice in lentil cultivation 2005

Type	Yield (kg)			Net return (Rs./ha)		
Management practice	Improved practice	farmers practice	% increase	Improved practice	farmers practice	% increase
Variety	1224	981	25	10741	7195	49
Weed management	1363	1100	24	13047	11172	17
Fertilizer management	1553	1310	19	12000	9380	28
Rhizobium management	1459	1236	18	14540	11560	26
Irrigation management	1227	1024	20	10332	7892	31
Disease management	1138	780	46	13490	7415	82
Package technology	1456	1037	40	12213	8794	39

Source: AICRP on MULLaRP (2006)

The application of improved practices on farm is the most important way to increase production and profitability of lentil in India. For example the gap between lentil yields on research stations and in on-farm trial yields ranged from 17% to 45% and between on-farm demonstrations and farmer fields ranged from 24% to 69% (AICRP on MULLaRP 2006). Thus the wider adoption of existing technology by farmers has the potential to increase yields by 40% and net returns by 39% (Table 3). Overall, the response to implementing individual management practices ranged from 17% for weed management to 82% for disease management.

Although prices for lentil are lowest at harvest, the relatively small variation in price over time in India indicates that there is little incentive for farmers, wholesalers and retailers to store grain as expenses for storage are higher than the price increase for grain over the storage period. Interestingly, the price index of government support is below minimum prices and therefore is not utilized to support farmers. Average prices have been lowest in the largest lentil producing state Uttar Pradesh and highest in Punjab (about 35% higher).

2.3. Constraints to Lentil Production in India

There is great potential to substantially increase lentil area in India and increase productivity from 684 kg to about 1000 kg/ha if constraints to production are addressed. The use of improved varieties and agronomy will be critical in realizing optimal yields and value adding will help in creating demand and enhancing exports.

Being a dryland crop and cultivated by resource poor farmers, lentil production is limited by a lack of moisture and unfavourable temperatures. Waterlogging and salinity are also locally important. Progress has been made in breeding for tolerance to drought through selection for an appropriate phenology and increased water use efficiency, and winter hardiness through selection for cold tolerance (ICRISAT 2006). The diseases rust, vascular wilt, and ascochyta blight are the key fungal pathogens of lentil in India. Varieties with resistance to rust and ascochyta blight have been released in several states and sources of resistance to vascular wilt are being exploited. Although pea leaf weevil (*Sitona spp.*), the parasitic weed broomrape (*Orobanche spp.*), and to a lesser extent cyst nematode (*Heterodera ciceri*), significantly reduce the yield of lentil, no sources of resistance to these biotic stresses have been identified. Management strategies using seed treatment and foliar fungicides have been developed for the wilt, root rot and rust diseases. Bold seeded lentil varieties with high yield potential have been released (DPL-15, L 4076, LH 84–8, WBL 58) to improve market access.

Water availability is a major constraint to the cultivation of winter crops and as a result, most land is fallowed in winter. Using the fallow land for lentil cultivation would increase production but irrigation infrastructure and water would be required. Lentil is also not a choice crop for irrigation due to its poor response to high inputs of fertilizer and irrigation. Farmers also lack the technical knowledge to successfully grow specialized crops like lentil and thus grow traditional crops like barley and wheat.

Good quality seed of new varieties is critical for increasing lentil production. Certified seed is often not available for farmers and resultant crop failures deter farmers from growing lentil again. Small and scattered demand for seed often makes distribution and marketing costs too high for the seed industry to economically supply certified seed. Seed propagation by villages in collaboration with research stations, private seed companies and government may overcome these barriers.

Most lentil farmers have limited resources and don't have access to the necessary credit to invest in inputs, improved infrastructure or value adding. Farmers also have a lack of grain marketing information that can result in relatively low prices being paid to farmers, especially when export prices are high. The difficulties in marketing small quantities of lentil have also limited the price received by farmers in some areas.

Although India exports a large quantity of lentils ($\sim 120,000$ t pa or 13% of production over the last 10 years), nearly fifty percent of exported Indian lentils are traded to Bangladesh, with trade to Europe and Latin American market being negligible. Improved quality, lower costs of production and export promotion are needed to address limitations in market access.

3. A CASE STUDY OF COMMERCIAL CULTIVATION AND PROFITABILITY IN A RELATIVELY NEW LENTIL GROWING COUNTRY – AUSTRALIA

3.1. Lentil Production in Australia

In Australia lentil is a relatively new crop in terms of introduction with area expanding from less than 1,000 hectares prior to 1994 to a maximum area of 155,000 hectares in 2001. Australia is now a significant exporter of red lentils. Production has been predominantly in the states of Victoria and South Australia in areas with a winter dominant annual rainfall of 350 mm to 450 mm and alkaline, well-drained soils (Materne *et al.* 2002).

The establishment of lentil in Australia was achieved through the development and successful commercialisation of better adapted varieties from ICARDA, Syria, improved agronomic support and skilful and innovative marketing that lead to widespread farmer uptake (Materne and Brouwer 1996). These varieties had improved resistance to ascochyta blight, were more suitable for machine harvest and had an appropriate phenology for southern Australia. These new varieties were up to 68% higher yielding than existing varieties, and lifted average experimental yields to over 2 t/ha in the Wimmera region of Victoria (Brouwer 2002). In Australia, microsperma lentils dominate production as a result of higher and more stable yields and profitability than macrosperma varieties (Materne 2003). However, large markets exist for macrosperma (green) lentils internationally and the potential for these types in Australia will be demonstrated with the imminent release of the well adapted large seeded green lentil variety Boomer.

Difficulties with harvesting lentil were initially a major limitation to farmer uptake but were largely overcome by using improved harvesting technologies, including flexi-fronts, and ensuring seed beds are relatively flat after sowing (Blair 2002). Agronomic promotion of early harvesting, using the right harvester settings to minimise chipped and split grains, and desiccation to control green weeds or uneven ripening were also important. The lack of good weed control in lentils was also a significant barrier to wide adoption of the crop. After determining what chemicals were of value in weed control, registration of chemicals had to be addressed.

Export companies had to overcome marketing barriers as Australia moved from being a net importer to a significant exporter of lentils in a conservative trade with entrenched networks where Canada (green lentils) and Turkey (red lentils) were dominant exporters (Blair 2002). A major effort was needed to quickly change this perception and develop an awareness of Australian lentils. This situation was further exacerbated by a substitution scandal in the early 1990's when Australian vetch had been misrepresented and exported as lentils. Australia is now a significant player in the global trade of lentils and export lentils to the world as a product rather than just a bulk commodity. The Lentil Company identified four key characteristics important to a marketing strategy:

1. **Visual –** colour and size of the lentils and the export packaging
2. **Quality –** purity, cleanliness and freedom from infestation
3. **Palatability –** Cooking time and taste when used in the various country's traditional dishes
4. **Reliability –** ability to execute and perform on contracts and develop long term supply arrangements

Brand identity, consistent high quality and good communications with the end users became important issues in ensuring on going demand for Australian lentils. The production of pulses, including lentils, has also resulted in an expansion in processing companies in the regions where the lentils were grown.

3.2. Profitability of Lentils in the Farming System

The concept of "break" crops for wheat production has existed for many years in Australia, but with the high returns generated from lentils, cereals are now considered as a "break" crop for lentils in some districts. In Australia farm gate prices have fluctuated from between AUS$400 to around AUS$600 per tonne for lentils, depending on demand and variety, with an average price of approximately AUS$450/tonne since 1994. This compares favourably to indicative prices of around AUS$200/t received for peas, wheat and malting barley and in areas where lentils grow well they have been very profitable (Materne *et al.* 2002, Table 4). Subsequently, land

Table 4. Field Crop Gross Margin Guide for South Australia in an average rainfall year (2002)

Crop	Yield t/ha	Value $/tonne	Income $/ha	Variable costs $/ha	Gross margin $/ha
Wheat (APW)	2.4	195	468	171	297
Durum 13%	2.0	250	499	189	310
Malt barley	2.3	200	460	157	303
Feed barley	2.4	150	360	158	202
Milling Oats	2.4	113	270	138	132
Triticale	2.4	170	408	154	254
Lucerne	2	148	504	288	216
Oaten hay	4.1	130	528	362	165
Field peas	1.5	250	375	216	159
Faba beans	1.7	265	451	264	187
Lupins	1.5	194	291	158	132
Lentils	**1.3**	**460**	**598**	**217**	**381**
Vetch	1.6	216	348	176	172
Chickpeas	1.4	389	545	266	279
Canola	1.3	360	468	240	228
Safflower	0.7	313	219	110	109

Source: Crop harvest report 2001/2002 – PIRSA

values and lease payments have increased where lentils can be successfully grown (Long 2002). It has also allowed smaller farmers to lease land with good potential for income and other farming operations to expand as capacity for on farm investment increases (Long 2002, Kearns 2002). As an additional crop option, lentil also provides an opportunity to widen rotations, thereby reducing the incidence of diseases and increasing yield and quality across all crop types (Long 2002). Specific advantages over other pulses include the ability to manage snail contamination and control herbicide resistant ryegrass using weed wipers.

The introduction of lentils as an alternative crop option in Victoria and South Australia has had a massive positive effect on regional economies. The benefits are not only from on-farm production but also extend to additional economic activities, such as transport, processing, and manufacture and sales of equipment and other farming inputs. For example, in Victoria there are ten companies with the capability of cleaning, sizing, bagging, and packing lentils into containers, and increasingly splitting lentils. Regional companies have invested around $25 million in the necessary equipment to transform lentils from a 'farmer dressed' to 'machine dressed' product, creating direct employment for around 80 people (Kearns 2002). Lentils have also played a major part in expansions in silo and machinery manufacture in regional Victoria.

Lentil is estimated to be worth over AUS$100 million pa to Australia and a major pulse crop in Victoria and South Australia. In Victoria lentil is the most significant pulse crop grown, with a gross value of production of about $55 million, with 70% of this being produced in the Wimmera region (Kearns 2002). Furthermore, storage, processing and the transport of lentils adds significantly to their market value. It is estimated that the value of lentil production to total regional output, extra household income and extra consumption in the Wimmera region is 2.5 times the value of lentil production (Kearns 2002). The challenge is now to maintain Australia's competitiveness through improvements in all sectors of the industry, but in particular, in region-based value-adding.

The benefits at the farm level should also not be forgotten. Like other pulses, lentils provide a rotational benefit to subsequent cereal and oilseed crops, with research demonstrating this to be in the order of 0.7 tonnes per hectare, worth approximately $140/ha (Kearns 2002).

3.3. Constraints to Lentil Production in Australia

With the emergence of lentils as a significant industry in Australia problems were not unexpected as the rapid expansion of a new crop is bound to bring out new issues related to varieties, agronomy, disease or quality (Brouwer 2002). However, there is now an improved understanding of the suitability of Australian lentil varieties to the various regions as based on phenological responses and interactions with disease and soil type (Materne 2003, Materne *et al.* 2002). This has greatly assisted advisors and farmers in deciding to introduce lentil into their crop rotations. Effective integrated disease management strategies based on variety resistance, improved agronomic practices (optimum sowing rates and times, improved seed quality) and strategic use of fungicides have also been developed and implemented. It is widely accepted that lentil is sensitive to many abiotic constraints such as low pH, waterlogging, and high levels of boron, salinity and sodicity. Breeding is now focused on improving the tolerance of lentil to these abiotic stresses and in the longer term addressing the constraints of frost, heat and drought to reliable lentil production. Exotic pathogens also pose a significant threat to the lentil production in Australia, but it's being managed through quarantine and preemptive breeding.

For lentils to continue to grow in importance in Australia, the challenge will be to maintain our international competitiveness through continuous improvements in production, quality, value adding and supply chain efficiency (Kearns 2002). Maintaining a high standard of value adding and improving the quality of varieties will be important in maintaining Australia's position in the world market (Kearns 2002). Trading pulses is also more risky than for cereals and standardized international quality measures and specifications are needed to reduce the costs of risk associated with trading lentils. Australia also remains vulnerable to increased exports by major lentil producing countries and economic policies that distort trade and reduce the effective competitiveness of Australian farmers.

4. CONTRASTING COMMERCIAL CULTIVATION AND PROFITABILITY IN LENTIL

4.1. Socioeconomic

In many traditional producing countries lentil is consumed where it is grown or imported to meet demand. As has been shown with India, increasing population size and economic growth will increase demand for lentil in these countries and this drives the need for increased production domestically. A major benefit for increased production is the supply of protein at a price where the nutritional needs of all people are met. In these cases increased competitiveness internationally is needed to feed people domestically. However, in more developed countries such as Canada, the United States of America and Australia, lentil production is export orientated and increased efficiency is essential for farm profitability, viability and continued economic growth. Increasingly this must be achieved with a view to long term environmental sustainability.

Farms in India and other traditional lentil growing countries are small (less than 10 hectares) whereas farms range from 1,000 to many thousands of hectares in more developed countries. Increasing farm size has been fundamental in the increased efficiency and profitability of farms in more developed countries but the necessary reduction in farm numbers and farmers comes with individual and social costs.

Government policies vary widely between countries and may aim for improved self sufficiency and feeding people domestically such as in India, supporting industry to become more profitable through research or support industry directly with subsidies such as with the U.S. Farm Bill. In general government policies are not linked to the level of development in a country but are implemented to meet the needs and priorities of a country.

4.2. Mechanisation

The large-scale production of lentil in countries such as Australia, Canada and USA has been achieved with mechanised harvesting systems, whereas in many traditional lentil producing countries, for example India, lentil is still harvested by hand (Haddad *et al.* 1988, Sarker and Erskine 2002). Nonetheless, hand harvesting is considered a major constraint to lentil production in North Africa and the Middle East and its high cost has caused a large decrease in lentil production in Jordan and Syria (Erskine *et al.* 1991). Varieties suitable for mechanical harvesting, have been released in the Middle east and their use, combined with mechanised harvesting, has increased net returns to growers by an estimated A$200/ha (Sarker and Erskine 2002). A shift towards mechanized farming practices will become increasingly important as economic growth drives up incomes in countries such as India.

Due to their slow growth during winter and short stature, lentil competes poorly with weeds and weed control is a major limitation to growing lentil worldwide (refer the relevant chapter of this book). In many traditional lentil growing countries weeds are removed by hand but this is time consuming and uneconomic in many cases due to the high cost of labour. Tillage reduces the impact of weeds but can delay sowing and increase the risk of erosion and soil degradation (Kukula *et al.* 1983). Production in many developed countries, including North America and Australia, are very dependent on herbicides for weed control. Similarly, disease and pest management in traditional lentil growing areas is less reliant on chemicals and results in lower costs, but also greater losses in production and quality.

4.3. Abiotic and Biotic Constraints to Production

In all countries a defined set of abiotic and/or biotic stresses constrain production. In most cases these are relatively well defined and, although capability may vary, are being addressed. Rather, it is the extension, transfer and ability of farmers to adopt these scientific outputs that is more limited in traditional lentil producing countries. In many developed countries technologies may even be adopted from other areas of the world.

4.4. Adoption of Technologies and Research

In developed countries with highly educated farmers, access to finances and good communication the adoption of research and technology is rapid. Advances are constantly being sought by farmers to increase profitability by at least maintaining their competitive position internationally. However, in poorer countries, farmers are often less educated, do not have the financial capacity to easily adopt new technologies, and communications are more difficult. As shown in the India, the slow adoption or non adoption of technology reduces yields, incomes and ultimately economic growth. In these countries vast gains in profitability can be achieved by implementing existing technologies.

The extension of information and adoption of input based agronomic practices is also limited by the large number of small farms in countries such as India. The larger farms of developed countries are a result of efficiency gains made possible through good access to reliable funds and knowledge for investing on farm. Off farm investment is also a significant component of the farm businesses in many of these countries. In countries such as India, small land holdings, capital and income make access to money and therefore the potential for on farm investment difficult.

In traditional lentil producing countries there are often insufficient mechanisms for the distribution of certified seed to farmers and thus the implementation of superior new varieties. In contrast, more developed countries have a rapid uptake of new varieties, often facilitated by private companies involved in seed production and distribution. In Australia complex systems of marketing and end point royalty collection are now routine. Although limitations to adoption exist, improved production technology and new varieties have significantly enhanced lentil productivity in many traditional producing countries (Sarker and Erskine 2002).

4.5. Competition from other crops

In countries such as India, Turkey and Bangladesh the development of irrigation schemes has caused a shift from growing lentil to the production to high-value crops like vegetables and cotton, or crops such as wheat or rice that are highly responsive to irrigation and fertilizer (Sarker and Erskine 2002). Lentils in all parts of the world face tough competition from cereals, oilseeds and other pulse options. In many cases the higher price of lentil is essential to maintain production because of the lower and less reliable yield of lentil. The low and erratic yield of lentils can be partly attributed to the short history of research in this crop compared to cereals. For example, in most traditional lentil producing countries, most of the lentil area is still occupied by landraces that are vulnerable to a range of biotic and abiotic factors (Sarker and Erskine 2002).

4.6. Marketing

In most countries, market prices are largely dependant on the production of lentil and other protein substitutes like chickpea and peas and consumption internationally. Being mostly a dryland crop, production and therefore price can fluctuate between years. In countries with good communications farmers can make more informed decisions on marketing than in countries where farmers are often isolated and have limited information to support marketing decisions. Improved communications are a key to improving on farm income by maximizing the price farmers receive for grain. However, even with access to information, farmers in more developed countries must still face the uncertainties of international production where lentils are sown and harvested at different times around the world.

5. CONCLUSIONS

Most developed countries are export orientated in terms of lentil production while traditional lentil producing counties such as India need to increase its competitiveness internationally to feed its people. In all countries, governments, researchers, farmers and private industry are aiming to improve competitiveness by reducing the cost of production, increasing total production, the reliability of production and increase prices through improved quality or marketing. It is the ability to develop, transfer and adopt relevant new improved technologies that defines on farm profitability. Small farm sizes, low income, a lack of access to funds, knowledge and infrastructure can prevent on farm investment efficiency gains and increased profitability in traditional lentil producing countries.

REFERENCES

AICRP on MULLaRP (2006) Annual Report (Rabi 2005–06), Indian Institute of Pulses Research, Kanpur-208024

Barulina, H. (1930) Lentils of the USSR and other countries. 40th Supplement to the Bulletin of Applied Botany, Genetics and Plant Breeding, Leningrad: 265–30 (English summary)

Blair, P. (2002) Formative years of the Australian lentil industry – Victoria. In Proceedings of Lentil Focus 2002, pp 19–23 (Ed JB Brouwer) Horsham, Victoria, Australia

Brouwer, J. B. (2002) History of Australian lentil crop improvement. In: Proceedings of Lentil Focus 2002, pp 8–13 (Ed J. B. Brouwer) Horsham, Victoria, Australia

Cubero, J. I. (1981) Origin, taxonomy and domestication. In: Lentils pp 15–38 (Eds C. Webb and G. Hawtin) Commonwealth Agricultural Bureaux, ICARDA

Erskine, W., Diekmann, J., Jegatheeswaran, P., Salkini, A., Saxena, M. C., Ghanaim, A. and Ashkar, F. EL. (1991) Evaluation of lentil harvest systems for different sowing methods and cultivars in Syria. Journal of Agricultural Science 117: 333–338

FAOSTAT Database (2006) Food and Agriculture Organization, Rome. Available at: www.fao.org

Haddad, N. I., Salkini, A. B., Jegatheeswaran, P. and Snowbar, B. A. (1988) Methods of harvesting pulse crops. In: World Crops: Cool season food legumes, pp 431–350 (Ed R. J. Summerfield). Kluwer Academic Publishers

Harlan, J. R. (1971) Agricultural origins: Centers and Noncenters. Science 174: 468–474

ICRISAT(2006) ICRISAT.org

Kumar, Praduman (1998). Food Demand and Supply Projections for India, Agricultural Economics Policy Paper 98–01, Indian Agricultural Research Institute, New Delhi, 1998.

Kukula, S., Haddad, A. and Masri, H. (1983) Weed control in lentils, faba beans, and chickpeas. In: Proceedings of the international workshop on Faba beans, Kabuli chickpeas and lentils in the 1980's. pp 169–177. (Eds M. C. Saxena and S. Varma) ICARDA, 16–20 May 1983, Aleppo, Syria

Kearns, B. (2002) The regional impact of the lentil industry. In Proceedings of Lentil Focus 2002, pp 24–27 (Ed JB Brouwer) Horsham, Victoria, Australia

Long, B. (2002) Formative years of the Australian lentil industry – South Australia. In Proceedings of Lentil Focus 2002, pp 14–18 (Ed JB Brouwer) Horsham, Victoria, Australia

Materne, M. A. (2003) Importance of phenology and other key factors in improving the adaptation of lentil (Lens culinaris Medikus) in Australia. Thesis presented for the degree of Doctor of Philosophy at The University of Western Australia, School of Plant Biology and Centre for Legumes in Mediterranean Agriculture (CLIMA), Faculty of Natural and Agricultural Sciences

Materne, M. A. and Brouwer, J. B. (1996) The new lentil industry in Australia, Factors behind its success. In: Proceedings of First Australian New Crops Conference 1996. (Ed B. C. Imrie, R. A. Bray, I. M. Wood and R. J. Fletcher) (Rural Industries Research and Development Corporation) vol 2 pp 45–52

Materne, M., McMurray, L., Nitschke, S., Regan, K., Heuke, L., Dean, G. and Carpenter, D. (2002) The future of Australian lentil production. In: Proceedings of Lentil Focus 2002, 14–18 (Ed JB Brouwer) Horsham, Victoria, Australia

Muehlbauer, F. J. and McPhee, K. E. (2002) Future of North American lentil production. In: Proceedings of Lentil Focus 2002, pp 8–13 (Ed JB Brouwer) Horsham, Victoria, Australia

Sarker, A. and Erskine, W. (2002) Lentil production in the traditional lentil world. In Proceedings of Lentil Focus 2002, 35–40 (Ed JB Brouwer) Horsham, Victoria, Australia

Smartt, J. (1984) Evolution of grain legumes. I. Mediterranean pulses. Experimental Agriculture 20: 275–296

CHAPTER 12

GENETICS AND CYTOGENETICS OF LENTIL

S. K. MISHRA[1], B. SHARMA[2] AND S. K. SHARMA[3]
[1,3] *National Bureau of Plant Genetic Resources, Pusa Campus, New Delhi 110012, INDIA*
[2] *Former Head, Division of Genetics, Indian Agricultural Research Institute, New Delhi 110012, INDIA*
E-mail: skmishra_gene@rediffmail.com

Abstract: Lentil (*Lens culinaris* Medik.) is an important winter season grain legume grown worldwide in semi-arid regions. Although lentil has been an important crop for centuries, very little attention has been paid in terms of genetic research until recently. A few centers around the world, including India, have started systematic work on genetic and linkage studies in lentil. Inheritance studies involving about three-dozen morphological markers have been completed. A comprehensive linkage map involving molecular markers and a few morphological markers has been developed. In the present paper, an effort has been made to comprehend the current status and recent advances made in terms of genetic, cytogenetic, linkage studies and other related aspects in lentil. The strategies for future research have also been outlined

1. INTRODUCTION

Lentil (*Lens culinaris* Medik.) is one of the most important cool season food legume crops grown worldwide in semi-arid regions. The crop is valued as a high-protein food and for its residues, which are used in animal feeding. The Indian subcontinent is the largest lentil producing region in the world, contributing about 42% of the total world production. Besides high yield potential, its in-built capabilities to withstand moisture stress as well as problematic soil conditions has imparted a very wide adaptability to this crop. In addition, lentil has a distinct export potential superior to most pulses of Indian origin.

Even though lentil has been an important food legume for centuries, very little effort has been made on genetic studies and gene mapping in this crop until recently. In fact genetic markers have not yet been identified in sufficient numbers to create a comprehensive programme on genome mapping. The current linkage maps proposed in lentil consist of small numbers of markers mainly isozymes and DNA markers,

S.S. Yadav et al. (eds.), Lentil: An Ancient Crop for Modern Times, 187–208.

covering a relatively small portion of the lentil genome. An effort, therefore, has been made in the present paper to review the current status of genetic, cytogenetic and other related aspects in lentil.

2. TAXONOMY

Lentil is known by as many as 30 common names in different parts of the world (Kay, 1979). It belongs to the genus *Lens* and tribe Vicieae. Cubero (1981) recognized five species in the genus *Lens (Lens culinaris*, *L. ervoides*, *L. montbretti*, *L. nigricans and L. orientalis.* Later another species *L. odomensis* was also included (Ladizinsky, 1986). Among these, *L. culinaris* Medik. has been regarded as the only cultivated species of lentil. Barulina (1930) sub-divided the cultivated lentil (*L. culinaris*) in to two types, macrosperma and microsperma, primarily based on seed characters. The macrosperma types were characterized by large seeds (6–9 mm diameter), yellow cotyledons, poor pigmentation on flowers and vegetative parts, whereas, the microsperma types were have small seeds (2–4 mm diameter), orange (red)/yellow cotyledons and pigmentation on flowers and vegetative parts. Although, Williams *et al.* (1974) do not support the separation of two types of lentil based on seed size, this concept is still largely accepted by lentil workers. Based on crossability behaviours, the genus *Lens* was classified in two biological species namely, *L. culinaris* and *L nigricans*. Ladizinsky (1979 b) studied seed protein profiles of lentil species from different geographical regions and concluded that *L. culinaris*, *L. orientalis* and *L. nigricans* were closely related whereas *L. ervoides* was different from these. Based on cytogenetic and crossability studies, Ladizinsky *et al.* (1984) recognized two species within the genus *Lens: L. culinaris* and *L. nigricans*. Ladizinsky (1986) while describing the new taxon *L. odemensis*, assigned all the five taxa (Pinkas *et al.* 1985) to the species level. Sharma *et al.* (1995) used RAPD markers to distinguish different lentil taxa representing wild and cultivated lentils. They observed that the ssp. *orientalis* is the closest to the cultivated lentil. On the other hand, *L. ervoides*, was the most divergent wild taxon followed by *L. nigricans*. The genetic similarity between the two latter species was of the same magnitude as between ssp. *orientalis* and cultivated lentil. Sharma *et al.* (1996) confirmed the above findings based on AFLP markers and Ahmad and McNeil (1996) compared a range of techniques for determining the interspecific relationships in lentil.. Van Oss *et al.* (1999) concluded that the genus *Lens* is comprised of seven taxa (*Lens culinaris* ssp. *culinaris*, *L. culinaris* ssp. *orientalis*, *L. odemensis*, *L. ervoides*, *L. nigricans*, *L. tomentosus*, *L. lamottei*). Ferguson *et al.* (1999) reassessed the classification of the genus *Lens* based on evidence relating to crossability behaviour and phylogenetic relationships to identify the morphological markers for taxon delimitation which was also supported by the isozyme and RAPD data. The following classification was proposed based on available information:

L. culinaris Medik
  ssp. *culinaris*
  ssp. *orientalis*

ssp. *tomentosus*
ssp. *odemensis*
Lens ervoides
Lens nigricans
Lens lamottei
On the basis of interspecific hybridization results, Sharma and Chahota (2004) concluded that the wild progenitor of the cultigens, *L. culinaris* ssp. *orientalis*, is a member of primary gene pool and *L. odemensis* of the secondary gene pool. They observed that using embryo rescue techniques, *L. ervoides* can be hybridized with the cultivated lentils and thus can become a member of secondary gene pool. However, *L. nigricans* was incompatible with the cultigen.

3. GEOGRAPHICAL DISTRIBUTION

Although, the geographical distribution of the genus *Lens* is mainly the Mediterranean region, significant variations exist in the distribution of individual species. The species *L. orientalis* has an eastern distribution from Turkey and Israel eastward to Uzbekistan whereas the species *L. nigricans* has the distribution mainly across the northern shores of the Mediterranean from Israel to Spain and into Algeria, Morocco and the Candy Islands. The wild form of the species *L. ervoides* has been reported in Uganda and Ethiopia. The previously recognized species *L. montbretti* is known to have very limited distribution around the headwaters of the Tigris and Euphrates.

4. CENTRES OF DIVERSITY

According to Vishnumittrre (1974), lentil was domesticated in India in the Neolithic Chalcolithic period. Based on phyto-geographic and archeological evidences, lentil belongs to the tribe Vicieae and primary areas of diversity are considered to be South-west Asia and the Mediterranean region. The archaeological evidences also suggest the lentil as one of the primary domesticates in the Near-East arc (Zohary, 1972). There are both wild and cultivated forms in the primary gene pool of lentil. Ladizinsky (1979 a) demonstrated that the species *L. nigricans* belongs to the secondary gene pool. The remaining other species are expected to form the tertiary gene pool. However, there are a few contradictions about the number of species and their inter-relationships (Ladizinsky, 1979 b). However, availability of large diversity is of great help in genetic improvement of lentil.

5. CYTOGENETICS

The cultivated lentil and their wild relatives are self-fertilizing diploids having chromosome number $2n = 2x = 14$. Variable chromosome numbers have been reported in inter-specific hybrids e.g. seven bi-valents in the intra-specific hybrids within the members of *L. culinaris* to five bivalents and one quadrivalent in the F1 hybrids derived from *L. culinaris* and *L. orientalis*. Ladizinsky (1979a) observed variable chromosome associations (univalents to multivalents) in the hybrids derived from *L. culinaris* and *L. nigricans*. Several workers have studied the karyotypes in cultivated lentil, *L. culinaris*, however, the karyotypes reported by different workers were similar. The length of chromosomes ranged from 3.0μ to 9.2 μ. Gupta and Singh (1981) reported two pairs of metacentric, two sub-metacentric and three pairs of acrocentric chromosomes. Ladizinsky (1979a, Ladizinsky *et al.*, 1984) noticed that karyotype of *L. orientalis* was similar to *L. culinaris*. Gupta and Bahi (1983) observed that the karyotype of *L. culinaris* differed from *L. nigricans* by three interchanges.

6. GENETICS OF QUALITATIVE TRAITS

Leaf pigmentation: Pigmented leaf (brown) was reported to be completely dominant over non-pigmented (green) leaf and the gene symbol ***Bl*** has been proposed for pigmented leaf (Kumar, 2002; Mishra 2004).

Chlorina mutant: Vandenberg and Slinkard (1989) observed that a chlorina chlorophyll mutant was determined by a single recessive gene (***chl***) to its normal dominant allele (***Ch***l) which also confirmed later by Viallancourt and Slinkard (1992).

Xantha mutant: Vandenberg and Slinkard (1987) reported that xantha chlorophyll mutant was determined by a single recessive gene (***xan***) to its normal dominant allele (***Xan***) which was subsequently confirmed by Viallancourt and Slinkard (1992).

Foliage colour: Dark green foliage was found to be dominant over light green foliage and the gene symbol ***Dgl*** has been proposed for this trait (Kumar, 2002; Mishra, 2004).

Leaflet size: Inheritance of leaflet size was studied based on analysis of 8272 plants in F_2 generation derived from twenty-four crosses involving parents with broad leaflets and narrow leaflets. The results revealed the incomplete dominance of broad leaflets over narrow leaflets and the gene symbol ***Blf*** has been proposed for broad leaflets (Kumar *et al.*, 2004; Mishra 2004).

Leaf shape: Oval leaf was reported to be monogenic dominant over normal leaf shape and the gene symbol ***Ol*** was proposed for oval leaf shape (Kumar, 2002; Mishra 2004).

Stipule size: Large stipule was found to be incompletely dominant over small stipule and the gene symbol ***Lst*** has been proposed for large stipule (Kumar, 2002).

No. of leaflets: A higher number of leaflets (6–8) was found to be monogenic dominant over a lower number of leaflets (3–4) and the gene symbol ***Hl*** has been proposed for a higher number of leaflets (Kumar, 2002; Mishra, 2004).

Growth habit: The prostate growth habit has been reported to be incompletely dominant over erect growth habit (Ladizinsky, 1979b) and the gene symbol ***Gh*** was proposed for prostrate growth habit. This dominance has since been confirmed by several other workers. Contrary to this, erect growth habit was reported to be completely dominant over prostrate growth habit and the gene symbol ***Ert*** was proposed for erect growth habit (Kumar, 2002; Mishra, 2004).

Epicotyl colour: Ladizinsky (1979c) reported that purple epicotyl was monogenic dominant over green epicotyl. The gene symbol ***Gs*** was proposed for purple epicotyl. However, the same gene symbol has been proposed by other workers (Viallancourt and Slinkard, 1992; Kumar, 2002 and Mishra, 2004) for stem pigmentation. They reported that pigmented stem was dominant over non-pigmented stem.

Tendril formation: Tendrilled plants were found to be monogenic dominant in their genetic control over non-tendrilled plants and the gene symbol ***tnl*** has been proposed for tendrilless leaf (Vandenberg and Slinkard, 1989; Viallancourt and Slinkard, 1992; Kumar 2002; Mishra, 2004).

Plant pubescence: Pubescent nature was found to monogenic dominant over non-pubescent plants and the gene symbol ***Pub*** has been proposed for pubescent nature (Kumar, 2002; Mishra, 2004).

Stem fasciation: Fasciated stem was recorded to be completely dominant over non-fasciated stem and the gene symbol ***Fa*** has been proposed for fasciated stem (Kumar, 2002; Mishra, 2004).

Globe plant type: This is an induced mutant having a compact appearance with globe like phenotype (Plate 1). Analysis of crosses between globe and the normal plant type revealed that the globe type is monogenically dominant over normal plant type. The gene symbol ***Glo*** has been proposed for globe plant type ((Kumar, 2002; Mishra *et al.*, 2002; Mishra, 2004).

Plant height: Analysis of crosses between tall and dwarf plant types indicated that tall plant type behaved as monogenically dominant over dwarf plant type. The gene symbol ***Ph*** has proposed for tallness in lentil (Kumar, 2002; Mishra, 2004).

Flower colour: The flower colour in lentil has been a most variable character in lentil. Overall flower colour depends on the colour of the standard, wings and keel. Ladizinsky (1979c) suggested that the production of coloured (bluish) flower in lentil is governed by a single dominant gene, whereas, the white flower appears under recessive condition. Lal and Srivastava (1975) reported the existence of two genes for controlling the flower colour in lentil. They suggested the gene symbols ***VVPP, vvPP*** and ***VVpp*** for violet, pink and white flowers, respectively. Red flower colour was reported to be monogenically dominant over white flowers and the gene symbol ***P*** was proposed for coloured flowers (Kumar, 2002; Mishra, 2004).

Peduncle length: Long peduncle was reported to be under genetic control of a single dominant gene designated as ***Pdl*** (Kumar, 2002; Mishra, 2004). The short peduncles are produced under homozygous recessive condition.

Pod pubescence: The genetics of pubescence on different plant parts have been studied by different workers. Vandenberg and Slinkard (1989) reported that the production of glabrous pod was determined by a single recessive gene (***glp***) to the pubescent pod (***Glp***). A similar nature of inheritance of pod pubescence was confirmed by Viallancourt and Slinkard (1992). However, Kumar (2002) studied the inheritance of plant pubescence and concluded that a single dominant gene is responsible for production of pubescent-glabrous characters in lentil. He proposed the gene symbol ***Pub***, the recessive counterpart of which creates the glabrous plants.

Pod colour: Monogenic recessive nature of inheritance for green pod colour (***grp***) over red pod (***Grp***) was reported by Vandenberg and Slinkard (1989) which was later confirmed by Viallancourt and Slinkard (1992). Similar findings were noted by Kumar (2002) and Mishra (2004), but they proposed the gene symbol ***Rdp*** to represent red pod colour in lentil.

Pod dehiscence: Pod dehiscence is one of the most important characters contributing to yield losses in lentil. Ladizinsky (1979b) studied the inheritance of pod dehiscence between the crosses *L. culinaris* and *L. orientalis* and reported that pod-indehiscence was controlled by a single recessive gene (***pi***). However, the pod dehiscence is a dominant trait (***Pi***). Such reports were also confirmed by Viallancourt and Slinkard (1992).

Seed coat (testa colour): The seed coat (testa) colour is also an important trait especially for consumption and commercial utilization of lentil. Erskine and Witcombe (1984) classified the ground colour of testa into five groups viz. green, pink, brown, gray and black. This has made the inheritance pattern of seed coat colour more complex. Ladizinsky (1979b) studied the spotting pattern on seed coat and proposed the gene symbol ***Scp*** for seed coat colouration pattern whereas the unspotted seed coat will be produced under recessive homozygous condition (***scpscp***). Vandenberg (1987) and Vandenberg and Slinkard (1990) concluded that the testa colour in lentil is determined by two non-linked dominant genes. They suggested that gray ground colour of testa is determined by the dominant gene, ***Ggc,*** whereas, the tan ground colour is produced by other dominant gene ***Tgc***. The genotypic constitutions for different colours were as follows, ***GgcGgcTgcTgc*** (brown), ***GgcGgctgctgc*** (gray), ***ggcggcTgcTgc*** (tan) and ***ggcggctgctgc*** (green). Vandenberg (1987) reported that black seed coat is determined by one gene (***blsc 1***) in some crosses and by other gene (***blsc 2***) in some other crosses. However, Vallincourt and Slinkard concluded only one dominant gene determining the black seed coat (***Blsc***) in lentil. Emami and Sharma (2000) based on their studies involving the crosses with different testa (black, brown, tan, green) and cotyledon (orange, yellow, dark green) colours reported that although black testa was dominant over non-black testas, its penetrance is not complete, as the expression of testa colour is greatly influenced by the cotyledon colour. They suggested to carry-out detailed genetic analysis using appropriate genotypes for cotyledon and testa colours. The monogenic dominant nature of black testa has been confirmed. The gene symbol ***Blt*** has been proposed for black testa (Sharma *et al.* 2004; Mishra, 2004) although the mode of inheritance proposed remains the same. Monogenic dominant control was reported for mottling on seed coat (***Mot***) over non-mottling (***mot***) in lentil (Kumar, 2002; Mishra, 2004).

Cotyledon colour: Monogenic dominance of orange cotyledon is reported by several workers (Tschermak-Seysenegg, 1928; Wilson *et al.* 1970; Slinkard, 1978; Singh, 1978; Sinha *et al.* 1987, Emami, 1996). However, Emami and Sharma (1996a,b) discovered and confirmed the digenic control of cotyledon colour in lentil. They proposed orange colour is due to the interaction of two dominant genes viz., yellow (***Y***) and brown (***B***). The two genes can produce independently the yellow (***Y-bb***) and brown (***yyB-***) pigments. However, the orange colour will be produced only when both the genes are present in dominant condition (***Y-B-***). Under double recessive conditions (***yybb***), the light green cotyledons are produced, as no pigments are synthesized. Emami and Sharma (2002) discovered a third gene designated as ***Dg*** which leads to the production of dark green cotyledons and this gene behaved as a monogenic recessive to the orange phenotype. This proposal was later confirmed with a voluminous set of data (Sharma *et al.*, 2004; Mishra, 2004). Accordingly, a system of three genes (***Dg***-dark green, ***Y***-yellow, ***B***-brown) has been confirmed to control cotyledon colour in lentil. In the presence of dominant gene ***Dg***, the yellow cotyledon is produced by the gene ***YY***, whereas, the gene ***BB*** produces brown cotyledon. When all the genes are present in dominant condition (***Dg-Y-B-***), the orange (red) cotyledons are produced. However, when the gene for dark green colour is recessive (***dgdg***), irrespective of dominant or recessive genes for yellow/brown/orange cotyledons (***YY or yy; BB or bb; Y-B-***), the dark green cotyledons are developed. The light green cotyledons are produced when the genes for yellow and brown colours are both recessive (***Dg-gyybb***).

Hard seed coat: Ladizinsky (1985) reported that the hard seed coat is a monogenic dominant charatcter and assigned the gene symbol ***Hsc***. This finding has been confirmed by Vaillancourt (1989).

Days to flowering: Sarker *et al.* (1999) reported that the early flowering was determined by a single recessive gene (***sn***). However, the occurrence of early flowering transgressive segregants in F_2 could be attributed to the interaction of the gene ***sn*** and minor genes for earliness.

7. GENETICS OF DISEASE RESISTANCE

Wilt *[Fusarium oxysporumf. Sp. Lentis (Snyder and Hensen)]:* Kamboj *et al.* (1990) reported that the resistance to *Fusarium* wilt in lentil was controlled by two dominant duplicate genes in the variety Pant L 234 whereas it was controlled by two independent dominant genes having complementary effects in the susceptible varieties JL 446 and LP 286. The allelic tests revealed that the five genes were operating in the resistance to *Fusarium* wilt in their materials under study. However, Euzyl *et al.* (1998) reported that the resistance to wilt was conditioned by a single dominant gene after three seasons of testing the $F_{6.9}$, $F_{8.9}$ recombinant inbred lines (RILs) and $F_{2.4}$ progenies in a well-established wilt-sick plot. These populations were developed by crossing between resistant (ILL 5588) and susceptible (L 692-16-$1_{(s)}$) lines. They also proposed the gene symbol ***Fw*** for resistance.

Rust *(Uromyces fabae (Pers.) de Bary):* Rust resistance in lentil is reported to be controlled by a single dominant gene (Sinha and Yadav, 1989). Single, but different dominant genes for rust resistance in different varieties (Pant L 406, Pant L 639, LG 120, UPL 175) was reported by Singh and Singh (1990). Singh and Singh (1992) also reported that resistance is governed by a monogenic dominant gene based on F_2 and F_3 analysis of 21 crosses derived from the crossing of seven resistant and three susceptible parents. Chauhan *et al.* (1996) also observed a single dominant behaviour of rust resistance. Duplicate dominant genes controlling rust resistance was reported by Lal *et al.* (1996). Kumar *et al.* (1997) have reported that resistance to rust in five genotypes (L 178, L 1534, L 2980, L 2991, HPLC 8868) is governed by a single dominant gene whereas in the genotype Precoz, it was found to be controlled by duplicate dominant genes. Kumar *et al.* (2001) studied inheritance of rust resistance in 23 crosses derived from eight resistant and ten susceptible lines in different combinations. They reported that two independent dominant and one recessive genes were imparting resistance to rust in the material under study. For the first time they reported gene symbol as $\boldsymbol{Urf_1}$ (Precoz, L 4603), $\boldsymbol{Urf_2}$ (Pant L 4, L 4147) and $\boldsymbol{urf_3}$ (DPL 21). Chahota *et al.* (2002) studied two crosses (Precoz × L 259; Precoz × Pant L 639) under greenhouse conditions and reported that the resistance was governed by duplicate dominant genes. Mishra *et al.* (2005) have reported the phenomenon of slow rusting in lentil based on evaluation of 305 lentil lines (255 lines from ICARDA, Syria and 50 indigenous lines) at a hot spot location. Mishra (2006) reported monogenic dominant control of rust resistance in lentil. However, the gene for resistance in the variety Precoz was different from that of PL 4 confirming partly the findings of Kumar *et al.* (2001).

Blight** (Ascochyta pisi/A. lentis/A. fabae):** Ahmad *et al.* (1997) reported that the host resistance was controlled by two complementary dominant genes in the wild species namely, *L. ervoides* and *L. odemensis*. Tay and Slinkard (1989) reported that each genotype Laird, ILL 5538 and ILL 5684 had single dominant genes for resistance. Ahmed and Morali (1998) could not observe the Mendelian segregation pattern for virulence in the materials studied.

8. GENETICS OF ABIOTIC STRESSES

Boron: Boron deficiency was identified in highly calcareous soils in India where genetic differences could be demonstrated in chickpea and pigeonpea (Singh *et al.*, 1991). Among nutrient-imbalance issues in cool-season food legumes, boron (B) has received scant attention (Saxena *et al.*, 1994). However, Sakal *et al.* (1988) demonstrated the differential response to boron application in lentil and noticed that the cultivars DL 77–2 and Pant L 406 were the most efficient for boron uptake while the variety L 9–12 being most B inefficient. Similar observations have also been reported from Srivastava *et al.* (2000). Yau (1999) identified the line ILL 5583 as tolerant to boron toxicity. Yau and Erskine (2000) and Hobson *et al.*, (2006) studied the diversity of boron-toxicity tolerance in lentil growth and yield and reported highly significant differences in B-toxicity tolerance between 231 and 310 accessions respectively. On average, accessions from Afghanistan were the most tolerant, followed by those from India, Iraq, Syria, Europe Ethiopia and Nepal.

Iron: Erskine *et al.* (1993) observed significant differences in iron deficiency symptoms in lentil germplasm. Ali *et al.* (1997) studied the inheritance of iron-deficiency based on six crosses involving resistant and sensitive lines. They indicated the resistance was under monogenic dominant control and proposed the gene symbol *Fe* for resistance.

Salt: Ashraf and Waheed (1990) studied the effect of NaCl on germination and emergence of 131 lentil varieties in the greenhouse and reported five varieties producing significantly greater fresh and dry biomass than the others. Detailed study was carried-out by Katerji *et al.* (2001) by taking one sensitive and one tolerant variety from the above study. They concluded that the lentil is most sensitive to salinity and it can be grown on non-saline soils. Maher *et al.,* (2003) investigated over 300 lines for salt tolerance and found considerable variability. However, the main Australian commercial varieties were all low in salt tolerance. Ashraf and Waheed (1998) reported that both additive and non-additive effects were significant for yield components.

9. LINKAGE STUDIES

Genetic maps of agricultural crops are a valuable tool for plant geneticists and breeders. The maps can be used to improve breeding efficiency and tagging of genes by their association with the specific traits, and locating quantitative trait loci (QTL). The first report on linkage in lentil was published by Zamir and Ladizinsky (1984) with two linkage groups. Tadmor *et al.* (1987) determined five linkage groups (2 morphological and 8 isozyme markers) in a cross between *Lens culinaris* and *Lens ervoides*. Havey and Muehlbauer (1989) and Muehlbauer *et al.* (1989) proposed nine linkage groups involving 6 morphological, 8 isozyme loci and 20 RFLP probes using the mapping population derived from an inter-specific cross between *Lens culinaris* and *Lens orientalis*. Weeden *et al.* (1992) developed eleven linkage groups covering 560 cM distance using 64 markers (morphological, isozyme and DNA markers). The comparison between this proposed map with that of *Pisum sativum* revealed eight regions showing linkages among markers were conserved. Tahir *et al.* (1993) compared the data from different studies and proposed 10 tentative linkage groups involving 7 morphological, 25 isozyme, 38 RFLP and 6 other loci. Tahir and Muehlbauer (1994) identified six linkage groups, which included 4 morphological and 17 isozyme loci. Emami (1996) could establish linkage among growth habit, stem colouration and leaf colouration. Euzyl *et al.* (1998) worked extensively on linkage studies of *Lens* involving 177 markers (3 morphological, 89 RAPD, 79 AFLP, and 6 RFLP) using 86 recombinant inbred lines ($F_{6:8}$) obtained from a partially inter-specific cross and proposed seven linkage groups. The map covered 1073 cM of the lentil genome with an average distance of 6.0 cM between adjacent markers. Hoque (2001) constructed four linkage groups in lentil using 12 morphological and one RAPD markers. These markers could cover 217 Kosambi units of lentil genome with an average distance of 8.75 cM Mishra *et al.* (2002) proposed four linkage groups in lentil based on the linkage analysis involving sixteen gene pairs (morphological markers) using F_2 populations of inter-varietal crosses of *Lens culinaris*. They analyzed 496 combinations of different markers. Kumar (2002) constructed four linkage groups in lentil based on morphological markers. Rubeena *et al.* (2003) constructed the first intra-specific linkage map in lentil using 114 markers (100 RAPD, 11 ISSR and 3 RGA) in an F_2 population derived from the cross ILL 5588 x ILL 7537). They proposed nine linkage groups comprising between 6 to 18 markers each with the total coverage of 784.1 cM. The utility of ISSR and RGA markers for mapping in lentil was explored and the primers with an AC repeat motif were found to be more useful. Kahraman *et al.* (2004) proposed nine linkage groups using 175 markers. Average distance between linked markers was 9.1 cM, however, it ranged from 0.3 –21.1 cM. A framework of 130 markers covering 1192 cM distance of lentil genome was used for the analysis of quantitative trait loci (QTLs). Mishra (2004) proposed nine linkage groups in lentil involving 16 morphological markers (including rust resistance) and 11 RAPD primers covering 740.2 cM of the genome. Recently, Mishra (2006) could identify a RAPD marker OPP 15 to be linked with the rust resistance gene (*Urf*) at a map distance of 26.1 cM. Although, the distance is not very close, it can be a beginning point for identification of more closely linked markers for utilization in marker assisted selection.

10. GENETICS OF QUANTITATIVE TRAITS

Genetic divergence: Genetic divergence is the measure of genetic distance among the cultivars or germplasm lines. This divergence may be due to geographic barriers or any other reasons, which may restrict the gene flow, resulting in the formation of distinct groups. In crop improvement programmes, selection of parents for hybridization is a crucial step for harnessing the useful variability for economic purposes. Therefore, breeders and geneticists essentially require information on nature and magnitude of genetic diversity in the materials available at their disposal. The Mahalanobis D^2 technique is a novel and widely used method to work out the genetic divergence. Biswas and Das (1985) estimated genetic diversity over two environments for ten characters in lentil accessions collected from Bangladesh and India and reported that the population from the two countries were divergent from each other. However, the clustering pattern was influenced by the parentage and geographical origin in a few cases while this was not true in others. Balyan and Singh (1986) grouped 48 genotypes of lentil into 12 clusters based on the analysis of nine characters. Based on Nei's average gene diversity, Harvey and Muehlbauer (1989) established that the wild lentils (*L. orientalis* and *L. odemensis*) had greater variability for RFLP and were more close to the cultivated lentil (*L. culinaris*). However, a narrow range of diversity could be obtained within the accessions of *L. ervoides* and *L. nigricans*. Chahota *et al.* (1994) classified 40 genotypes of small seeded lentil (microsperma) into six clusters based on Mahalanobis' D^2 Statistic and Canonical analysis considering 15 traits. The low level of diversity in the cultivated taxon as compared to the wild species was reported in lentil based on allozyme polymorphic survey for 11 loci in 439 accessions (Ferguson and Robertson, 1996). Rathi *et al.* (1998) grouped 21 genotypes into eight clusters based on analysis of six yield and yield contributing traits. They reported that the number of primary branches per plant contributed most towards the total genetic divergence followed by yield per plant. The clustering pattern predicted that the genetic diversity is not necessarily parallel to the geographical diversity. Singh *et al.* (2001) reported eight clusters based on the multivariate analysis of 58 diverse strains of lentil. Jeena and Singh (2001) carried out Hierarchal cluster analysis using 30 genotypes of lentil (28 wild accessions and 2 cultivated) for qualitative (HCA 1), quantitative (HCA 2) and both qualitative and quantitative (HCA 3) traits. The results indicated wide genetic divergence as each analysis yielded four, three and three clusters, respectively. Jeena and Singh (2002) evaluated 61 lentil accessions representing four wild species viz., *L. nigricans* (2), *L. odemensis* (16), *L. ervoides* (24) and *L. orientalis* (19). Based on the analysis of data on 20 quantitative traits, all the accessions could be grouped into four clusters. Interestingly, 58 accessions could be grouped into Cluster 1 while rest of the clusters were mono-genotypic. The study clearly indicated that the genetic diversity was not related to the geographical diversity and species differences. Solanki *et al.* (2002) reported 72 genotypes to be classified into eight and nine clusters under normal and late sown conditions, respectively. Singh *et al.* (2002) reported the genotype x environment interaction on determination of clustering pattern. They carried out D^2 analysis involving 40 genotypes for two consecutive years and grouped them into six and seven clusters, respectively. Rakesh *et al.* (2005) grouped 44 genotypes in to five clusters based on the analysis of data on 15 important characters at two locations. Yadav *et al.* (2005) worked-out genetic divergence based on analysis of data on 50 genotypes under two environments and reported that the genetic diversity was not paralleled to the geographical diversity. Recently, Poonam (2006) grouped 100 lentil genotypes of diverse origin into ten clusters based on analysis of 12 quantitative traits. Out of these 100 accessions, sixty genotypes were also subjected for divergence analysis using ten RAPD primers. Although, these genotypes could be grouped into ten close-knit clusters, there was no parallelism between the two types of the clustering pattern.

Correlation: Adequate knowledge about degree and direction of the association of characters is a pre-requisite for operating an efficient selection programme. Exhaustive studies on interrelationship of characters among themselves and also between yield and yield components have been carried–out. A positive association between seed size and pod size reported by Sharma and Sharma (1978) can be useful in selecting the variability for seed size. Significantly positive genotypic and phenotypic correlations between seed and straw yields have been recorded both in small seeded (*microsperma*) and bold seeded (*macrosperma*) accessions of lentil (Erskine, 1983) indicating the possibility of continued selection for higher seed yield would not adversely affect the straw yield. Sarwar *et al.* (1984) reported positive correlation of seed yield with number of pods per plant, number of primary and secondary branches per plant in the Indigenous as well as in the exotic germplasm. Erskine *et al.* (1985) recorded negative genetic correlation between seed yield and protein content whereas it was positive between cooking time and seed size (Hamdi *et al.* 1991). Although there was positive correlations of seed size with seed impermeability and seed germination. However, the seed impermeability and germination were negatively correlated (Shahi *et al.* 1986). Positive correlations of seed yield per plant with number of primary branches per plant, plant height, number of seeds per plant and 100-seed weight were observed in lentil (Murari *et al.* 1988). Profuse branching and number of pods per plant were positively correlated with seed yield (Zaman *et al.* 1989). A positive and highly significant correlation coefficient was found between seed yield per plant and number of pods per plant (Nigam *et al.* 1990). Hamdi *et al.* (1991) reported positive correlation between seed yield and straw yield. Multiple correlation and regression analysis revealed that the combination of two or three variables such as plant height, number of branches per plant, number of pods per plant was the best method for improving the seed yield. Singh and Singh (1991) observed that plant height was always correlated positively with seed yield per plant in both *microsperma* and *macrosperma* lentils. Seed yield per plant was positively correlated with all the yield components except 100-seed weight in Indigenous lentil germplasm (Pandey *et al.*, 1992). Esmail *et al.* (1994) reported that seed yield was positively and significantly correlated with plant height, number of branches per plant, number of seeds per pod and number of pods per plant, however, it was negatively correlated with flowering duration. Both genotypic and phenotypic positive correlations of seed yield with plant height, number of primary branches per plant, number of pods per plant, protein and methionine contents were observed in 13 parents and their 31 F_1s (Kumar *et al.*, 1995). Seed yield per plant was positively correlated with harvest index (Chauhan and Singh, 2001). Rakesh *et al.* (2005) reported the correlation among yield and yield components (15 characters) in 44 germplasm accessions of lentil. The analysis indicated that that the values of genotypic correlations were slightly higher, in general, than the phenotypic correlations.

Path coefficient: Pods per plant had higher direct effect on seed yield in both Indigenous and exotic germplasm. However, in exotic germplasm, 100 -seed weight showed a high direct effect on seed yield (Sarwar *et al.*, 1984). Luthra *et al.* (1990) reported that biological yield was the main contributor towards seed yield while other characters showed variation in their relative contribution towards seed yield. The path analysis indicated that the number of pods per plant had high positive direct effect on seed yield per plant based on analysis of thirteen hundred germplasm accessions of Indigenous origin (Pandey *et al.*, 1992). However, days to flowering, plant height and number of primary branches per plant had high positive indirect effects via number of pods per plant. Plant height, number of primary branches per plant and number of pods per plant could emerge as direct yield contributors while number of secondary branches per plant, number of pods per plant, and number of seeds per pod influenced the seed yield indirectly via number of primary branches per plant (Kumar *et al.*, 1995).

11. HERITABILITY AND GENETIC ADVANCE

Heritability is an important parameter in the genetic studies of quantitative characters. It is considered as an index of transmissibility of the character(s) from parents to their off-springs. The heritability in a broadsense is the ratio of genetic variance to the total (phenotypic) variance. Thus, the population expressing larger proportion of genetic variability for particular character or group of characters will be more amenable to selection. Although, heritability is an important biometrical estimate, it should be used in conjunction with genetic advance for better understanding and use (Johnson *et al.*, 1955). High heritability coupled with high genetic advance may be the most desirable situation for practical utility. Dixit and Dubey (1985) reported the highest heritability estimate for days to flowering. However, moderate heritability (59.7%) coupled with highest genetic advance in percent of mean (72.9%) was observed for seed yield. Erskine *et al.* (1985) reported highest heritability estimate for average seed weight (98%) followed by cooking time (82%). Lakshmi *et al.* (1986) recorded higher heritability coupled with high genetic advance for germination percentage, hard seed percentage and 100-seed weight. Ali and Johnson (2000) worked-out heritability estimates for winter hardiness under natural and controlled condition. The estimates of narrow sense heritability estimates ranged from 0.32 to 0.71 under field conditions whereas under controlled condition it was maximized at 1.00. Omvir and Gupta (2000) studied heritability in microsperma x macrosperma derived lines. They reported low heritability estimates in poor environments, however, it was higher in the best environment. Rathi *et al.* (2002) reported high heritability estimates along with higher genetic advance for 1000-grain weight. Dayachand (2007) reported higher estimates of broadsense heritability in combination with high genetic advance for days to maturity.

12. COMBINING ABILITY

The success of any breeding programme depends on choice of parents for developing segregating population for selection. Although, *per se* performance has been an important criteria for selection of the parents, good performing parents do not always produce desirable segregants. The information regarding combining ability of the parents for yield and yield related traits has been an important criteria in designing the appropriate breeding methodology. The importance of general combining ability (GCA) variance has been reported for days to flowering (Malhotra *et al.*, 1973, Haddad *et al.*, 1982, Singh and Gupta, 1994); for plant height (Malhotra *et al.*, 1973,Waldia and Chhabra, 1989); for secondary branches per plant (Singh *et al.*, 1975), for primary branches per plant (Singh and Gupta, 1994), for seeds per pod (Gupta and Singh, 1994, Rathi *et al.*, 1994) and for 100-seed weight (Singh and Singh, 1993, Chauhan and Singh, 1993). Preponderance of the non-additive portion of genetic variance for seed yield per plant has been reported by several workers (Singh and Jain, 1971; Chauhan and Singh, 1993). Singh and Singh (2003) carried out combining ability analysis based on an 8x8 diallel. They reported that the parents DPL 62 and K 75 were good general combiners for days to flower, days to maturity, plant height, primary branches per plant, secondary branches per plant, number of pods per plant, 1000-seed weight and grain yield per plant whereas L 830 was good general combiner for earliness.

The specific combining ability (SCA) effect represents the non-additive gene action which is non-fixable in nature. In none of the studies, did a single cross have significant SCA effect for all the characters, SCA effects are found to vary in magnitude with the change of environment. Thus a particular parent may be a good combiner for specific traits. It is also not necessary that the crosses exhibiting positive significant SCA effect involve parents with high × high GCA effects. In most of the cases high SCA crosses result from the high × low, low × high, high × average, average × high GCA combinations. In few rare cases, high × high or low × low GCA combinations may show significant SCA effects. Singh and Singh (1990) reported that both GCA and SCA effects were significant fro all the characters under study across the generations.

13. HETEROSIS

Singh and Jain (1971) reported very low to very high magnitude of heterosis for different characters based the performance of F_1 hybrids developed from 11 parents. They observed that the heterosis for grain yield was mainly due to heterosis in yield contributing characters and genetic diversity among the parents. Goyal *et al.* (1976) recorded highest better parent heterosis for pods per plant (166%) followed by seed yield (146%). Sagar and Chadra (1980) reported highest mid-parent heterosis for yield per plant in nine crosses of lentil. Kamboj (1986) reported heterosis for eight quantitative characters in ten crosses involving five parents. All the hybrids exhibited moderate to high manifestation of standard heterosis except for 100-seed weight for which estimates were negative and high. Erskine *et al.* (1991) worked-out the magnitude of heterosis based on the performance of 50 hybrids over two years under rainfed conditions and concluded that heterosis over better parent was non-significant for grain yield. Interestingly, the magnitude of heterosis was higher in the hybrids involving low yielding parents whereas it was least in the crosses involving high yielding parents. Singh and Singh (1992) reported that heterosis over better parent was maximum for No. of clusters per plant (66.6%) followed by pods per plant (65.3%) and seed yield per plant (15.3%). Kumar *et al.*(1996) recorded high manifestation of heterosis over better parent and standard variety for yield per plant. Dayachand (2007) also reported high magnitude of heterosis over better parent for seed yield per plant.

14. VARIABILITY AND GENETIC RESOURCES

Variability is the hub of any breeding programme. The success of a breeding programme depends upon the nature and magnitude of variability present in the materials. The higher the variability, the greater the chances of attaining better success in selection and vice-versa. Also, among three components of genetic variability namely, additive, dominance and epistatic, the additive component, which is fixable, is of greatest importance in selection breeding. Several studies have been conducted to assess the variability in the germplasm/genetic stocks regarding yield and its components, biotic and abiotic stresses and nutritional parameters (Rajput and Sarwar, 1989; El Attar, 1991; Pandey *et al.*, 1992; Esmail *et al.*, 1994; Kumar *et al.*, 1995; Chauhan and Singh, 1998; Kumar *et al.* 1999; Chakraborty *et al.* 2000, Solanki and Sharma, 2001; Rathi *et al.*, 2002; Hamdi *et al.*, 2003; Mishra *et al.*, 2005; Poonam, 2006 etc.) and a wide range of variability have been reported. Systematic efforts are also being made for augmentation of germplasm through Indigenous collections and acquisition from other countries especially from CGIAR centers. ICARDA, Syria has been the major contributor for enrichment of lentil germplasm. Several potential donors have been identified for different economic traits (Mishra *et al.*, 2005, Sardana, *et al.*, 2005, Singh *et al.*, 2006) in lentil.

15. ROLE OF GENETICS IN VALUE ADDITION

Cotyledon colour has been one of the most important characters for consumer preference. Normally, the lentil variety having orange (red) cotyledon colour is recognized as lentil especially in India. The same is true with yellow cotyledons in a few countries. Occurrence of brown, light green and dark green cotyledons in lentil has rarely been recognized. The recent studies on detailed genetic studies of cotyledon colour have amply demonstrated the trigenic control of cotyledon colour (detailed under the sub-head genetics of qualitative characters). Therefore, genetic manipulation can easily be made for developing varieties with specific cotyledon colour as per consumer demand. The seed size is another parameter which affects the market value of lentil. The genetic studies on inheritance of seed size have direct bearing on varietal development. The protein content and amino acid profile are also important biochemical parameters for deciding the quality of lentil. Reports on heterosis for protein and methionine contents, although old, are of potential value for exploitation in breeding programmes (Kumar *et al.*, 1994). Genetic studies on cooking quality (Erskine *et al.*, 1985) are of special importance for value addition in lentil.

16. RECENT ACHIEVEMENTS

Although, lentil has been an important winter season legume, very little attention was paid in terms of genetic research until recently. However, recognizing its potential to thrive well under adverse agro-edaphic situations, considerable work has been started in a few institutions engaged in lentil research and development with tangible results. Recent studies conducted at the Division of Genetics, Indian Agricultural Research Institute, New Delhi on inheritance and linkage studies involving morphological and molecular markers has paved the way for further research. A large number of multi-marker lines have been developed for their use in future genetic analysis. The joint collaborative efforts of ICARDA, Syria and University of Keil, Germany has led to the development of genetic linkage map using lentil based on micro-satellite, AFLP, RAPD and morphological markers

(ICARDA Annual Report, 2005). Molecular tagging of genes for resistance against biotic and abiotic stresses is a recent initiative.

17. FUTURE OUTLOOK

As discussed above, considerable work on genetic studies has been underway in lentil. There is an immediate need to form a network of researchers working on lentil genome mapping after comprehending the progress made so far. Although several linkage maps have been proposed mostly involving molecular markers, it is essentially required to involve more morphological markers and traits of economic importance for their tagging or establishing tight linkage for practical use in marker assisted selection. The seven linkage groups are still to be identified and defined. This can only be completed if systematic studies are made involving geneticists, cytogeneticists, plant breeders, biotechnologists, plant physiologists and plant protection scientists in a multi-disciplinary mode. There is also a need to create more markers through induced mutagenesis for developing a comprehensive linkage map. The inheritance studies on resistance/tolerance to biotic and abiotic stresses will be useful in designing appropriate breeding methodology based on the regional requirements. Molecular tagging of genes for resistance against biotic and abiotic stresses should receive first priority in order to exploit them in practical breeding programmes with increased selection efficiency and better precision.

REFERENCES

Ahmad, M., Russell, A. C. and McNeil, D. L. 1997. Identification and genetic characterization of different resistance sources to *Ascochyta* blight within the genus *Lens*. *Euphytica*, **97**: 311–315.

Ahmad, M. and McNeil, D. L. 1996. Comparison of crossability, RAPD, SDS-PAGE and morphological markers for revealing genetic relationships within and among Lens species. *Theoretical and Applied Genetics*. **93** (5/6): 788–793.

Ahmed, S. and Morrall, R. A. A. 1998. Inheritance of virulence in *Ascochyta lentis* on lentil. *Lens Newsletter*, **25**(1&2): 67–70.

Ali, A., Riaz- ul-Haque, M. and Bhatti, M. S. 1997. Inheritance of resistance to iron-deficiency chlorosis in lentil. *Lens Newsletter*, **24**(1& 2): 28–29.

Ali, A. and Johnson, D. L. 2000. Heritability estimates for winter hardiness in lentil under natural and controlled conditions. *Plant Breeding*, **119**(3): 283–285.

Annual Report. 2005. International Centre for Agricultural Research in the Dry Areas, Aleppo, Syria. pp.156.

Ashraf, M. and Waheed, A. 1990. Screening of landraces/exotic accessions of lentil (*Lens culinaris*) for salt tolerance at two growth stages. *Plant and Soil*, **128**: 167–176.

Ashraf, M. and Waheed, A. 1998. Genetic basis of salt (NaCl) tolerance in lentil. *Lens Newsletter*, **25**(1&2): 15–22.

Balyan, H. S. and Singh, S. 1986. Genetic divergence in lentil. *Lens Newsletter*, **13** (1): 3–4.

Barulina, E. I. 1930. Lentils of USSR and Other Countries: A Botanico Agronomical Monograph. *Trudy Prikl. Bot. Genet. (Suppl.)*, **40**: 265–304. (cf. S. K. Sharma and R. K. Chahota. 2004. Current status of interspecific hybridization in genus *Lens*. *Journal of Lentil Research*, **1**: 15–18).

Biswas, P. K. and Das, P. K.1985. Genetic divergence in lentil (*Lens culinaris* Medik.). *Annals of Agricultural Research*, **6**: 179–182.

Chahota, R. K., Sharma, S. K. and Lal, C. 1994. Genetic divergence in microsperma lentil. *Legume Research*, **17**(2): 132–134.

Chahota, R. K., Gupta, V. P. and Sharma, S. K. 2002. Inheritance of rust resistance in lentil (*Lens culinaris* Medik.). *Indian Journal of Genetics and Plant Breeding*, **62**(3): 226–227.

Chakraborty, M. and Haque, M. F. 2000. Genetic variability and component analysis in lentil. *Journal of Research, Birsa Agricultural University*, **12**(2): 199–204.

Chauhan, M. P. and Singh, I. S. 1993. Genetics of seed size in lentil (*Lens culinaris* Medik.). Abstr. National Sym. On Plant Breeding Strategies for India in 2000 A.D. and Beyond, 25–27 December, Aurangabad, India, Indian Society of Genetics and Plant Breeding, New Delhi: 220–221.

Chauhan, M. P. and Singh, I. S. 1998. Genetic variability, heritability and expected genetic advance for seed yield and other characters over two years in lentil. *Lens Newsletter*, **25**(1): 3–6.

Chauhan, M. P. and Singh, I. S. 2001. Relationship between seed yield and its component characters in lentil (*Lens culinaris* Medik.). *Legume Research*, **24**(4): 278–280.

Chauhan, M. P., Singh, I. S., Singh, R. S. 1996. Genetics of rust resistance in lentil *Lens culinaris*). *Indian Phytopathology*, **49**(4): 387–388.

Cubero, J. J.1981. Origin, Taxonomy and Domestication. *In*: Lentils (eds. C. Webb and G. Hawtin), CAB, London, U. K.: 15–38.

Daychand. 2007. Genetic Analysis of Yield and Yield Components in Lentil (*Lens culinaris* Medik.). Ph.D. Thesis, to be submitted to CCS University, Meerut, India.

Dixit, P. and Dubey, D. K.1985. Heritability and genetic advance in induced mutants of lentil (*Lens culinaris* Medik.). *Indian Journal of Genetics and Plant Breeding*, **45**(3): 520–524.

El Attar, A. H. 1991. Genetic variability among some exotic lentil germplasm in Egypt. *Bulletin of Faculty of Agriculture*, University of Cairo, **42**(3): 993–1000.

Emami, Mahmoud Khodambbashi. 1996. Genetic Mapping in Lentil (*Lens culinaris* Medik.). Ph. D. Thesis, IARI, New Delhi, India.

Emami, M. Khodambashi and Sharma, B. 1996 a. Digenic control of cotyledon colour in lentil. *Indian Journal of Genetics and Plant Breeding* , **56** (3): 357–361.

Emami, M. Khodambashi and Sharma, B. 1996 b. Confirmation of digenic inheritance of cotyledon colour in lentil (*Lens culinaris* Medik.). *Indian Journal of Genetics and Plant Breeding* , **56** (4): 563–568.

Emami, M. K. and Sharma, B. 2000. Inheritance of black testa colour in lentil (*Lens culinaris* Medik.). *Euphytica*, **115:** 43–47.

Erskine, W. 1983. Relationship between the yield of seed and straw in lentil. *Field Crops Research*, **7**:115–121.

Erskine, W., Williams, P. C. and Nakkoul, H. 1985. Genetic and environmental variation in the seed size, protein, yield and cooking quality of lentils. *Field Crops Research*, **12**: 153–161.

Erskine, W. and Witcombe, J. R. 1984. Lentil Germplasm Catalogue. ICARDA, Aleppo, Syria: pp. 48.

Erskine, W., Saxena, N. P. and Saxena, M. C. 1993. Iron deficiency in lentil: Yield loss and geographical distribution in a germplasm collection. *Plant & Soil*, **151**: 249–254.

Esmail, A. M., Mohamed, A. A., Hamdi, A. and Rabie, E. M. 1994.Genetic variability and heritability for agronomic traits in segregating populations of lentil (*Lens culinaris* Medik.). *Annals of Agricultural Science*, Moshtohor, **32** (3): 1107–1118.

Euzyl, I., Baum, M. and Powell, W., Erskine, W. and Pehu, E. 1998. A genetic linkage map of lentil (*Lens* sp.) based on RAPD and AFLP markers using recombinant inbred lines. *Theoretical and Applied Genetics*, **97**: 83–89.

Ferguson, M. E. and Robertson, L. D. 1996. Genetic diversity and taxonomic relationships within the genus *Lens* as revealed by allozyme polymorphism. *Euphytica*, **91**: 163–172.

Ferguson, M. E., Maxted, N., Slagreen, M. W. and Robertson, L. D. 1999. A reassessment of the taxonomy of *Lens* Mill. (Leguninosae, Papilionodeae, Vicieae). *Bot. J. Linn. Soc.*, 133: 41–49 (cf. S. K. Sharma and R. K. Chahota. 2004. Current status of interspecific hybridization in genus *Lens*. *Journal of Lentil Research*, **1**: 15–18).

Goyal, S. N., Jaimini, S. N. and Tikka, S. B.1976. Heterosis in lentil. *Lens Newsletter*, **3**:1–3.

Gupta, P.K. and Bahi, J. R. 1983. Cytogenetics and origin of some pulse crops *In*: Cytogenetics of Crop Plants (eds. M. S. Swaminathan, P. K. Gupta and U. Sinha). Mac Millan, India: 405–440.

Haddad, N. I., Bogyo, T. P. and Muehlbauer, F. J. 1982. Genetic variance of six agronomic characters in three lentil (*Lens culinaris* Medik.) crosses. *Euphytica*, **31**: 113–120.

Hamdi, A., Erskine, W. and Gates, P. 1991. Relationships among economic characters in lentil. *Euphytica*, **57**:109–116.

Hamdi, A., El-Ghareib, A. A., Shafey, S. A. and Ibrahim, M. A. M. 2003. Genetic variability, heritability and expected genetic advance for earliness and seed yield from selection in lentil. *Ministry of Agriculture Giza, Egypt*, **81**(1): 125–138.

Havey, M. J. and Muehlbauer, F. J. 1989. Linkages between restriction fragment length, isozyme and morphological markers in lentil. *Theoretical and Applied Genetics*, **77**: 395–401.

Hobson, K., Armstrong, R., Nicolas, M., Connor, D. and Materne, M. 2006. Response of lentil (*Lens culinaris*) germplasm to high concentrations of soil boron. *Euphytica*, **151**(3): 371–382.

Hoque, Md. Ekramul. 2001. Inheritance and Gene Mapping Based on Morphological and Molecular Markers in Lentil (*Lens culinaris* Medik.). Ph.D. Thesis, IARI, New Delhi, India.

Jeena, A. S. and Singh, I. S. 2001. Estimation of genetic diversity in lentil germplasm. *Crop Improvement*, **28**(1): 81–88.

Jeena A. S. and Singh, I. S. 2002. Genetic divergence analysis in wild lentils. *Legume Research*, **25**(3): 175–179.

Johanson, H. W., Robinson, H. F. and Comstock, R. E. 1955. Estimation of genetic and environmental variability in soybean. *Agronomy Journal*, **47:** 314–318.

Kahraman, A., Kusmenoglu, I., Aydin, N., Aydogan, A., Erskine, W. and Muehlbauer, F. J. 2004. QTL mapping of winter hardiness genes in lentil. *Crop Science*, **44**: 13–22.

Kamboj, R. K. 1986. Genetics of Resistance to Wilt (*Fusarium oxysporum* f. sp. *lentis*) and Certain Quantitative Traits in Lentil (*Lens culinaris* Medik.). Ph. D. Thesis, G. B. P. U. A. &T., Pantnagar, India.

Kamboj, R. K., Pandey, M. P. and Chaube, H. S. 1990. Inheritance of resistance to *Fusarium* wilt in Indian lentil germplasm. *Euphytica*, **105**: 113–117.

Katerji, N., Hoorn van, J. W., Hamdy, A., Mastrolli, M., Oweis, T. and Erskine, W. 2001. Response of two varieties of lentil to soil salinity. *Agricultural Water Management*, **47**: 179–190.

Kay, D. E. 1979. Food Legumes Trp. Products Institute, Crop and Products Digest No. 3, Her Majesty's Stationary Office, London, pp. 210–223.

Kumar, A., Singh, D. P. and Singh, B. B. 1995. Genetic variability of yield and its components in lentil (*Lens culinaris* Medik.). *Indian Journal of Pulses Research*, **8**(1): 60–66.

Kumar, A., Singh, D. P., Singh, B. B. Kumar, A. 1996. Combining ability analysis in lentil. *Indian Journal of Genetics and Plant Breeding*, **56**(2): 173–176.

Kumar, Rakesh, Kumar, D. and Kumar, S. 1999. Genetic variability in lentil (*Lens culinaris* Medik.). *Annals of Agricultural Research*, **4**(1): 75–77.

Kumar, R., Mishra, S. K. and Sharma, B. 2001. Genetics of rust resistance in lentil (*Lens culinaris* Medik.). *Indian Journal of Genetics and Plant Breeding*, **61**(3): 238–241.

Kumar, Yogesh. 2002. Inheritance and Linkage of Genes for Morphological Traits in Lentil (*Lens culinaris* Medik.). Ph.D. Thesis, CCS University, Meerut, India.

Kumar, Yogesh, Sharma, B., Mishra, S. K., Tyagi, M. C. and Singh, S. P. 2004. Inheritance of leaflet size in lentil (*Lens culinaris* Medik.). *Journal of Lentil Research*, **1**(1): 26–29.

Ladizinsky, G. 1979 a. The origin of lentil and its wild genepool. *Euphytica*, **28:** 179–187.

Ladizinsky, G. 1979 b. The genetics of several morphological markers in lentil. *Journal of Heredity*, **70**: 135–137.

Ladizinsky, G. 1985. The genetics of hard seed coat in the genus *Lens*. *Euphytica*, **34**: 539–543.

Ladizinsky, G. 1986. A new *Lens* from the Middle-East. *Notes from the Royal Botanic Gardens, Edinburgh*, **43**:489–492.

Ladizinsky, G., Braun, D., Goshen, D. and Muehlbauer, F.J. 1984. The biological species of the genus *Lens* L. *Botanical Gazette*, **145**: 253–261. (cf. S. K. Sharma and R. K. Chahota. 2004. Current status of interspecific hybridization in genus *Lens*. *Journal of Lentil Research*, **1**: 15–18).

Lakhani, J. P., Holkar, S. and Misra, R. 1986. Genetics of seedling vigour and hard seed in lentil. *Lens Newsletter*, **13**(2): 10–12.

Lal, S. and Srivastava, R. S. 1975. Inheritance of flower colour in lentils. *Indian Journal of Genetics and Plant Breeding*, **35**: 29–30.

Lal, C., Sharma, S. K. and Chahota, R. K. 1996. Inheritance of rust resistance in lentil. *Indian Journal of Genetics and Plant Breeding* , **56**(3): 350–351.

Luthra, S. K. and Sharma, P. C. 1990. Correlation and path analysis in lentil (*Lens culinaris*). *Lens Newsletter*, **17**(2): 5–8.

Maher, L., Armstrong, R. and Connor, D. 2003. Salt tolerant lentils – a possibility for the future? In Solutions for a better environment: Proceedings of the 11th Australian Agronomy Conference, Geelong, Victoria, Australia, 2–6 February 2003: 0–4

Malhotra, R. S., Balyan, H. S. annd Gupta, P. K. 1978. Crossing techniques in lentil. *Lens Newsletter*, **5**: 7–8.

Mishra, Gyan Prakash. 2006. Inheritance of Rust Resistance and Identification of Molecular Markers for Rust Resistance in Lentil (*Lens culinaris* Medik.). Ph.D. Thesis, IARI, New Delhi, India.

Mishra. S. K. 2004. Final Technical Report on Development of Comprehensive Genetic Linkage Map in Lentil (*Lens culinaris* Medik.). Submitted to NATP(ICAR), Krishi Bhawan, New Delhi, India.

Mishra, S. K., Sharma, B. and Tyagi, M. C. 2002. Final Technical Report on Linkage Studies in Lentil (*Lens culinaris* Medik.). Submitted to ICAR, Krishi Bhawan, New Delhi, India.

Mishra, S. K., Sarker, A., Singh, B. B. and Basandrai, A. 2005. Slow rusting and its potential donors for resistance in lentil (*Lens culinaris* Medik.) *Indian Journal of Genetics and Plant Breeding*, **65**(4): 319–320.

Mishra, S. K., Singh, B. B., Basandrai, A. K., Sarker, A. and Dayachand. 2005. Screening of ICARDA's lentil material for rust resistance in India. *Indian Journal of Plant Genetic Resources*, **18**(1): 49–50.

Murari, K., Pandey, S. L., and Kumar, V. 1988. Simple correlation and multiple regression studies in lentil. *Legume Research*, **11**(2): 101–102.

Muehlbauer, F. J., Weeden, N. F. and Hoffman, D. L. 1989. Inheritance and linkage relationship of morphological and isozyme loci in lentil (*Lens* Miller). *Journal of Heredity*, **80**: 298–303.

Nigam, S. A., Rabie, H. A., Mohamed, M. A. and Mohamed, H. A. 1990. Correlation studies and path analysis in lentil. *Journal of Agricultural Research* (Egypt), **12**(1): 313–331.

Omvir and Gupta, V. P. 2000. Genotype x environment interaction and heritability for quantitative characters in macrosperma x microsperma derived lines of lentil. *Legume Research*, **23**(3): 201–205.

Pandey, A., Singh, D. P. and Singh, B. B. 1992. Evaluation of Indigenous Germplasm for Yield and Yield Components in Lentil (*Lens culinaris* Medik.). N. D. University of Agriculture & Technology, Faizabad, India, Research Bulletin No.1: pp. 45.

Pinkas, R. Zamir, D. and Ladizinsky, G. 1985. Allozyme divergence and evolution in the genus *Lens*. *Plant Systematic Evololution*, **151**: 131. (cf. S. K. Sharma and R. K. Chahota. 2004. Current status of interspecific hybridization in genus *Lens*. *Journal of Lentil Research*, **1**: 15–18).

Poonam. 2006.Chracterization of Lentil Germplasm Through Morphological and Metric Traits and Molecular Markers. Ph.D. Thesis, Bundelkhand University, Jhansi, India.

Rajput, M. and Sarwar, G. 1989. Genetic variability, correlation studies and their implications in selection of high yielding genotypes in lentil. *Lens Newsletter*, **16**(2): 5–8.

Rakesh Kumar, Sharma, S. K., Luthra, O. P. and Sharma, S. 2005. Phenotypic stability of lentil genotypes under different environments. *Annals of Biology*, **21**(2): 155–158.

Rathi, A. S., Sindhu, J. S., Katiyar, R. P. and Katiyar, S. L. 1994. Genetic architecture for grain yield and its components in lentil. Abstr. International Sym. on Pulses Research, 2–4 April, 1994, New Delhi, Indian Society of Pulses Research and Development, Kanpur.

Rathi, A. S., Sindhu, J. S. and Srivastava, S. B. L. 1998. Genetic divergence in lentil. *Indian Journal of Pulses Research*, **11**(2): 137–139.

Rathi, A. S., Sindhu, J. S. and Singh, V. S. 2002. Variability, heritability and genetic advance in lentil. *Legume Research*, **25**(2): 113–116.

Rubeena, R., Ford, P. and Tylor, W. J. 2003. Construction of an intraspecific linkage map of lentil (*Lens culinaris* spp. *culinaris*). *Theoretical Applied Genetics*, **107**: 910–916.

Sagar, P. S. and Chandra, S. 1980. Breeding behaviour and genetic variation for yield in crosses of lentil (*Lens esculenta.*). *Indian Journal of Agricultural Research*, **14**(3): 159–163.

Sakal, R. Singh, A. P. and Singh, R. B. 1988. Differential reaction of lentil varieties to boron application in calcareous soil. *Lens Newsletter*, **15**(1): 27–29.

Sardana, S., Dhillon, B. S.,Singh, Mahendra and Mishra, S. K. 2005. Pulses germplasm: collection, conservation and utilization. *In*: Pulses (eds. G. Singh, H. S. Shekhon and J. S. Kolar), Agrotech Publishing Company, Udaipur, India, pp. 95–144.

Sarker, A., Erskine, W., Sharma, B. and Tyagi, M. C. 1999. Inheritance and linkage relationships of days to flowering and morphological loci in lentil (*Lens culinaris* Medikus subsp. *culinaris*). *The American Genetic Association*, **90**: 270–275.

Sarwar, D. M., Khatoon, F. and Gowda, C. L. L. 1984. Comparative correlation and path analysis in local and exotic germplasm in lentil. *Indian Journal of Genetics and Plant Breeding*, **44**: 201–205.

Saxena, N. P., Saxena, M. C., Ruckenbauer, P., Rana, R. S., El-Fouly, M. M. and Shabana, R. 1994. Screening techniques and sources of tolerance to salinity and mineral nutrient imbalances in cool season food legumes. *In*: Expanding the Production and Use of Cool Season Food Legumes (eds. F. J. Muehlbauer and W. J. Kaiser). Kluwer Academic Publishers, Dordrecht, The Netherlands: 457–471.

Shahi, J. P., Singh, J., Agarwal, I. and Lal, M. S. 1986. Studies on variability for seed size, permeability of seed coat to water and germination in lentil. *Lens Newsletter*, **13**: 14–15.

Sharma, B. and Emami, M. K. 2002. Discovery of new gene causing dark green cotyledons and pathway of pigment synthesis in lentil (*Lens culinaris* Medik.). *Euphytica*, **124**(3): 349–353.

Sharma, B., Tyagi, M. C., Mishra, S. K. and Kumar, Yogesh. 2004. Three-gene control of cotyledon colour in lentil (*Lens culinaris* Medik.). *Journal of Lentil Research*, **1**(1): 1–10.

Sharma, B., Tyagi, M. C., Mishra, S. K. and Kumar, Yogesh. 2004. Confirmation of the inheritance pattern of black testa colour in lentil. *Journal of Lentil Research*, **1**(1): 19–20.

Sharma, S. K. and Sharma, B. 1978. Induced variability for pod and seed size in lentil. *Current Science*, **47**(21): 806–807.

Sharma, S. K., Dawson, I. K. and Waugh, R. 1995. Relationship among cultivated and wild lentils revealed by RAPD analysis. *Theoretical and Applied Genetics*, **91**: 647–654.

Sharma, S. K., Knox, M. R. and Ellis, T. H. N. 1996. AFLP analysis of the diversity and phylogeny of *Lens* and its comparison with RAPD analysis. *Theoretical and Applied Genetics*, **93**: 751–758.

Sharma, S. K. and Chahota, R. K.2004. Current status of interspecific hybridization in genus *Lens*. *Journal of Lentil Research*, **1**: 15–18.

Singh, A. P., Krishna, R., Kumar, R. and Singh, M. P. 2001. Assaying divergence in lentil (*Lens culinaris* Medik.). *Crop Research* (Hisar): **22**(3): 469–473.

Singh, A. P., Sakal, R., Sinha, R. B. and Bhogal, N. S. 1991. Relative response of selected chickpea and pigeonpea cultivars to boron application. *Annals of Agricultural Research*, **12**: 20–25.

Singh, B. B., Mishra, S. K., Sardana, S. and Dixit, G. P. 2006. Lentil and pea. In: Plant Genetic Resources: Foodgrain Crops (eds. B. S. Dhillon, S. Saxena, A. Agarwal and R. K. Tyagi). Narosa Publishing House Pvt. Ltd., New Delhi: 240–254.

Singh, D. P. and Singh, B. B. 1991.Evaluation of exotic germplasm in lentil. *Narendra Deva Journal of Agricultural Research*, **6**(2): 304–306.

Singh, K. B. and Jain, R. P. 1971. Heterosis in lentil (*Lens culinaris* Medik.). *Indian Journal of Agricultural Sciences*, **41**: 678–681.

Singh, I. P. and Singh, J. D. 2003. Combining ability analysis in lentil. *Indian Journal of Pulses Research*, **16**(2): 95–97.

Singh, J. P. and Singh, I. S. 1990. Genetics of rust resistance in lentil (*Lens culinaris*). *Indian Journal of Pulses Research*, **3**(2): 132–135.

Singh, J. P. and Singh, I. S. 1992. Genetics of rust resistance in lentil (*Lens culinaris*). *Indian Journal of Agricultural Sciences*, **62**: 337–338.

Singh, J. P. and Singh, I. S. 1993. Combining ability in lentil. *Indian Journal of Pulses Research*, **6**(1): 25–30.

Singh, T. and Gupta, K. K. 1994. Genetic diversity for yield and related traits in lentil (*Lens culinaris* Medik.). *Plant Archives*, **4**(1): 39–43.

Singh, T. P. Inheritance of cotyledon colour in lentil. 1978. *Indian Journal of Genetics and Plant Breeding*, **38**: 11–12.

Singh, T. P., Singh, K. B. and Malhotra, R. S. 1975. Heterosis and combining ability in lentil. *Indian Journal of Agricultural Sciences*, **45**(6): 259–263.

Singh, Tejbir, Gupta, K. K. and Singh, T. 2004. Genetic diversity for seed yield and related traits in lentil (*Lens culinaris* Medik.). *Plant Archives*, **4**(1): 39–43.

Sinha, R. P., Chaudhary, S. K. and Sharma, R. N. 1987. Inheritance of cotyledon colour in lentil. *Lens Newsletter*, **14**(1&2): 3.

Sinha, R. P. and Yadav, B. P. 1989. Inheritance of resistance to rust in lentil. *Lens Newsletter*, **16**: 41.

Slinkard, A. E. 1978. Inheritance of cotyledon colour in lentil. *Journal of Heredity*, **69**: 139–140.

Solanki, I. S. and Sharma, B. 2001. N-nitroso-N-ethyl Urea induced genetic variability for quantitative characters in lentil (*Lens culinaris* Medik.). *National Journal of Plant Improvement*, **3**: 102–106.

Solanki, I. S., Kumar, K., Malik, B. P.S. and Kumar, R. 2002. Study of genetic divergence under normal and late sown conditions in lentil. *Crop Improvement*, **27**(2): 232–235.

Srivastava, S. P., Bhandari, T. M. S., Yadav, C. R., Joshi, M. and Erskine, W. 2000. Boron deficiency in lentil: Yield loss and geographic distribution in a germplasm collection. *Plant and Soil*, **219**: 147–151.

Tadmor, Y., Zamir, D., and Ladizinsky, G. Genetic mapping of an ancient translocation in the genus *Lens*. *Theoretical and Applied Genetics*, **73**: 883–892.

Tahir, M., Simon, C. J. and Muehlbauer, F. J. 1993. Gene map of lentil: a review. *Lens Newsletter*, **20**(2): 3–10.

Tahir, M. and Muehlbauer, F. J. 1994. Gene mapping in lentil with recombinant inbred lines. *Journal of Heredity*, **85**(4): 306–310.

Tay, J. and Slinkard, A. E. 1989. Transgressive segregation for *Ascochyta* resistance in lentil. *Canadian Journal of Plant Sciences*, **69**: 547.

Tschermak-Seysenegg, E. 1928. Lentil and field bean crosses. *Sityringsber Akad.Wiss. Wein Math. Nat. Ki. I. Abs.,* **137**(3/4): 171–181.(cf. F. J. Muehlbauer, F. J. and A. E. Slinkard, 1987. Breeding methodology. *In:* Lentil (eds. C. Webb and G. Hawtin). CAB-ICARDA, Fernham, England: 69–90.

Vaillancourt, R. 1989. Inheritance and Linkage of Morphological Markers and Isozymes in Lentil. Ph.D. Thesis, University of Saskachewan, Saskatoon, Canada (cf. R. E. Vaillancout and A. E. Slinkard. 1993. Linkage of morphological and isozyme loci in lentil, *Lens culinaris* L. *Canadian Journal of Plant Sciences*, **73**: 917–926).

Vaillancourt, R. and Slinkard, A. E. 1992. Inheritance of new genetic markers in lentil. *Euphytica*, **64**: 227–236.

Van Oss, H., Arnon, Y. and Ladizinsky, G. 1997. Chloroplast DNA variation and evolution in the genus L. Mill. *Theoretical and Applied Genetics*, **94**: 452–457.

Vandenberg, A. 1987. Inheritance and Linkage of Several Quantitative Traits in Lentil. Ph.D. Thesis, University of Saskachewan, Saskatoon, Canada (cf. R. E. Vaillancourt and A. E. Slinkard, 1992, Inheritance of new genetic markers in lentil. *Euphytica*, **64**: 227–236.).

Vandenberg, A. and Slinkard, A. E. 1987. Inheritance of a xantha cholophyll deficiency in lentil. *Journal of Heredity*, **78**: 130.

Vandenberg, A. and Slinkard, A. E. 1989. Inheritance of four new quantitative genes in lentil. *Journal of Heredity*, **80**: 320–322.

Vandenberg, A. and Slinkard, A. E. 1990. Genetics of seed coat colour and pattern in lentil. *Journal of Heredity*, **81**: 484–488.

Vishnumittrre, A. 1974**.** The beginning of agriculture. Paleobotanical evidences in India. *In:* Evolutionary studies in World Crop Diversity and Change in Indian Sub-continent (ed. J. B. Hutchinson). Cambridge, U. K.

Waldia, R. S. and Chhabra, A. K. 1989. Inheritance of some quantitative traits in lentil. *Lens Newsletter*, **16**(1): 6–7.

Williams, T.T., Sanchez, A. M. C. and Jackson, M. T. 1974. Studies on lentils and their variation: the taxonomy of the species. *SABRAO Journal*, **6**: 133–145.

Wilson, V. E., Law, A. G. and Warner, R. L. 1970. Inheritance of cotyledon colour in *Lens culinaris* Medik. *Crop Science*, **10**: 205–207.

Weeden, N. F., Muehlbauer, F. J. and Ladizinsky, G. 1992. Extensive conservation of linkage relationships between pea and lentil genetic maps. *Journal of Heredity*, **83**: 123–129.

Yadav, S. S., Phogot, D. S., Solanki, I. S. and Malik, B. P. S. 2005. Character association and path co-efficient analysis under two environments in lentil. *Indian Journal of Pulses Research*, **18**(2): 147–149.

Yau, S. K. 1999. Boron toxicity in lentil: yield loss and variation between contrasting lines. *Lens Newsletter*, **26**(1&2): 14–17.

Yau, Sui-Kwong and Erskine, W. 2000. Diversity of boron-toxicity tolerance in lentil growth and yield. *Genetic Resources and Crop Evolution*, **47**: 55–61.

Zaman, M. W., Mian, M.A.K. and Rahman, M. M. 1989. Variability and correlation studies in local germplasm of lentil in Bangladesh. *Lens Newsletter*, **16**(1): 17–19.

Zamir, D. and Ladizinsky, G. 1984. Genetics of allozyme variants and linkage groups in lentil. *Euphytica*, **33**: 329–336.

Zohari, D. 1972. The wild progenitor and place of the cultivated lentil. *Lens culinaris*. *Economic Botany*, **26**: 326–332.

CHAPTER 13

MUTATION BREEDING

C. TOKER[1], SHYAM S. YADAV[2] AND I. S. SOLANKI[3]

[1] *Department of Field Crops, Faculty of Agriculture, Akdeniz University, TR-07058 Antalya, Turkey*
[2] *Pulse Laboratory, Division of Genetics, Indian Agricultural Research Institute, New Delhi 110012, India*
[3] *Department of Plant Breeding, Chaudhary Charan Singh Haryana Agricultural University Hisar 125004, Haryana, India*
E-mail: toker@akdeniz.edu.tr

Abstract: When genetic variability is narrowed using traditional breeding methods for a long period, induced mutations are one of the most important approaches for broadening the genetic variation in lentil to circumvent the bottleneck conditions. The aim of this chapter is to review lentil breeding using induced mutations from the beginning of mutation breeding work to the present and to list the outcomes of mutagenesis works on lentils. The number of mutant varieties of all species officially released and recorded in the Food and Agricultural Organization/International Atomic Energy Agency (FAO/IAEA) Mutant Varieties Database is over 2300. From these mutant varieties, more than 265 grain legume cultivars have been developed using induced mutations and have subsequently been released. Gamma rays were the most frequently used technique to alter genes. Many mutant lentils have been mentioned in the available literature while seven mutants have been released for commercial production so far. Mutant lentils have now contributed several million dollars annually to global agriculture. Several specific regional problems in lentil production areas have been coped with using mutant lentil cultivars. Fundamental genetics, physiological and molecular studies will also be come to light using mutant lentils

1. INTRODUCTION

The genus *Lens* Mill. includes seven taxa including the cultivated species *Lens culinaris* Medikus and its wild relatives [*L. orientalis* (Boiss.) Ponert., *L. odemensis* Ladiz., *L. tomentosus* Ladiz., *L. lamottei* Czefr., *L. nigricans* (M. Bieb.) Godr., *L. ervoides* (Brign.) Grande] (van Oss et al. 1997, Ferguson et al. 2000, Ferguson and Erskine 2001, Sarker and Erskine 2006). The cultigen (*Lens culinaris* Medik.) is divided into two subspecies, ssp. *macrosperma* and ssp. *microsperma*, on the basis of seed size by Barulina (see Cubero 1981). This approach has been renewed by Cubero (1981) as race *macrosperma* and race *microsperma*. The breeding methods

S.S. Yadav et al. (eds.), Lentil: An Ancient Crop for Modern Times, 209–224.

in lentil are similar to those utilized in breeding other self-pollinated crops, *i.e.* pure line selection or hybridisation followed by the bulk method, the pedigree method, the single seed descent, or modification of these methods supplemented by mutation breeding and polyploid breeding (Muehlbauer and Slinkard 1981, Muehlbauer et al. 1993, 1996). The use of these traditional breeding methods for a long period may have narrowed available genetic variability. Mutation breeding is one of the most important possible routes to broadening the genetic variation in lentil under bottleneck conditions (Erskine et al. 1998). Other options include use of wild relatives, and molecular genetic approaches which are discussed elsewhere in this book. This chapter summarises the efforts that have gone into creating induced mutations from the pioneering experiments to the release of the most recent lentil cultivar. The chapter also reviews current knowledge on how induced mutation works on lentil.

2. HISTORY OF MUTATION BREEDING

The history of mutation breeding has been well reviewed by van Harten (1998). The earliest description of natural or spontaneous mutants was presented for cereal crops in an ancient book, "*Lulan*". This appeared around 300 BC in China. After this, many aberrant plant forms or variations in plants were discovered from 1590 to C. Darwin's bud variations in 1868. The period between these dates is called the first period of mutations. The second period encompasses the time between the discovery of *X-rays* by W.K. Rontgen in 1895 and application of mutagens from 1897 (the first mutagen treatment) to 1920 (N.I. Vavilov's "law of homologous series of variations" (van Harten 1998). The experiment of L. J. Stadler in the 1920s which used radiation to generate genetic changes in plants (Stadler 1928) initiated "mutation breeding" (Maluszynski et al. 2004). The third period ranges from induction of mutations to the first commercial mutant cultivar, the so called "chlorine-type" in *Nicotiana tabacum.* The fourth period starts with the development of international coordination and some financial assistance by the Food and Agricultural Organization/International Atomic Energy Agency (FAO/IAEA) from 1964 to onwards (van Harten 1998). Plant breeders have therefore been encouraged to use mutation breeding as one of the "peaceful uses of atomic energy". Recently, mutagenesis has received considerable attention for its use in a promising new technique known as "targeted induced local lesions in genomes" (TILLING) (Muehlbauer et al. 2006).

3. NATURE AND TYPES OF MUTATIONS

Mutations are phenotypically classified into two groups (Gaul 1964); (i) macromutations: These are easily detectable in individual plants, phenotypically visible and morphologically distinct and they are qualitatively inherited genetic changes, and occur in major genes or oligogenes, and (ii) micromutations: These result in a small effect that, in general, can be detected only by help of statistical methods and quantitatively inherited genetic changes, and occur in minor genes or polygenes.

Mutations are also divided into groups such as chlorophyll mutations and many other grouped morphological mutations (Table 3).

3.1. Gene Mutations

Mutations are theoretically all changes which occur in DNA sequence and result in changes in the genetic code. A gene mutation or point mutation is the group of all heritable changes which occur within the limits of a single gene. The majority of gene mutations show recessive inheritance but dominant gene mutations occur at a very low frequency (Micke 1999). A frameshift mutation is a deletion and insertion for any number of nucleotides other than three. Paramutations, transmutations and transposable genetic elements or transposons are mutation like effects (van Harten 1998) and widely used in molecular breeding.

3.2. Chromosome Aberrations

Chromosomal mutations span a number of genes (and intergenic regions) and often have a multiplicity of effects. Four types of chromosomal mutations within chromosomes are commonly distinguished: (i) deletions or deficiency, (ii) duplications, (iii) inversions and (iv) translocations. These types of mutations are also called chromosome rearrangements or structural mutations, and are generally not as valuable for plant breeding as gene mutations.

3.3. Chromosome and Genome Alterations

The basic chromosome number of a plant species is indicated by the symbol x. For examples in lentil and tetraploid wheat (*Triticum* sp.), $x = 7$. However, the number of chromosome pairs is seven ($n = 7$) in lentil, while $n = 14$ for tetraploid wheat. The term ploidy refers to the number of sets (each containing x chromosomes) of chromosomes or genomes in a cell, tissue and plant. Possible levels of ploidy include; haploid (n), diploid (2n), triploid (3n) and tetraploid (4n). The subject of ploidy level such as haploidy, polyploidy and allopoliploidy are of considerable importance for fundamental genetics, plant physiology and plant breeding. There may also be additional or B chromosomes that are unstable components of the plant and vary in number (Perfectti and Werren 2001).

3.4. Extra Chromosomal Mutations

Mutations may occur both in chromosomes in the nucleus and outside the nucleus. In a plant cell there are two extranuclear (extra chromosomal) genetic systems, the chloroplast and mitochondrion. Extra-chromosomal mutations bring about leaf variegation, dwarf growth and extrachromosomal genes induce tolerance to herbicides and also cause cytoplasmic male sterility, which is encoded by mitochondrial genome (Lonsdale 1987). The extra chromosomal mutations have considerably importance in practical application of plant breeding.

4. CONSIDERATIONS FOR INDUCED MUTATIONS

The following considerations should be taken into account before starting mutation breeding programmes: (i) Mutations are mostly recessive and they cannot be selected for until the second generation, M_2. Unlike recessive mutations, dominate mutations occur at low frequencies and they can be selected for in the M_1 generation (Micke and Donini 1993). Muehlbauer and Slinkard (1981) reported that mutation breeding is more adaptable for inducing recessive genes than dominant genes. Selection for polygenic traits should be started in individual plant progenies of the M_3 generation after some fixing of multiple homozygotes has commenced (Micke and Donini 1993, van Harten 1998); (ii) Mutations are beneficial with very low frequencies, while the treatments themselves can be detrimental reducing germination, growth rate, vigour and pollen and ovule fertility in the living organisms; (iii) Mutations are randomly induced and they might occur in any gene(s). However, some gene(s) can be more frequently induced to mutate than others; (iv) Mutations can be recurrent. The same gene(s) in a crop plant species may be induced to mutate again and again with different versions potentially having different effects; and (v) Mutations generally have pleiotropic effects due to closely linked gene(s) (Singh 2005).

5. SELECTION OF VARIETY, MUTAGEN AND DOSE

5.1. Which Varieties?

Mutation breeding programmes should be clearly planned and well defined, and large enough to select desirable mutations at the low frequencies likely to be encountered. The variety selected for mutagenesis should in particular be one of the best varieties released recently. At least, two varieties should be used because response to mutagens is different from a variety to another variety. It will be useful in improving specific characters of well adapted and high yielding varieties, which are deficient one or two traits (Anonymous 1977). The varietal group "macrosperma" has been found to be more sensitive to both the mutagen types used than the "microsperma" group (Sharma and Sharma 1986, Reddy and Viswanathan 1993). Sharma and Kharkwal (1982) found that the genotypes in macrosperma group were more responsive to mutagenic treatment and gave a higher frequency of mutated progenies in the M_2. Genotypic response to mutagens has been found to differ within the same group (Sharma and Sharma 1979c).

5.2. Which Mutagens?

The agents that induced mutations are called mutagens and mutagens mainly consist of two different kinds; (i) radiation (physical) and (ii) certain chemical mutagens (Table 1). Mutagens are not only beneficial to create genetic variability in a crop species, but also useful for the effective control of pests during post-harvest storage (Chaudhuri 2002). In addition to the use of induced mutations in plant breeding, there is a tremendous use of induced lentil mutations in fundamental

Table 1. Common mutagens and action mode

1. Physical Mutagens (Radiation)	Action mode	2. Chemical mutagens	Action mode
1.1. Ionising radiation	Breakage of hydrogen bonds and sugar phosphate moiety, cross-linking DNA strands.	2.1. Alkylating agents, i.e. sulphur and nitrogen mustards, ethylene amine (EI), ethylene oxide (EO), ethyl methane sulphonate (EMS), ethyl ethane sulphonate (EES), diethyl sulphate (DES), N-nitroso-N-ethyl urea (NEU) and N-nitroso-N-methyl urea (NMU), N'-methyl-N-nitro-N-nitroso-guanidine (MNNG)	Alkylate phosphate groups, purine and pyrimidine bases. Leads to mis-pairing or loss of bases.
1.1.1. Particulate radiation, i.e. α-rays (Alfa-rays), β-rays (Beta-rays), fast and thermal neutrons		2.2. Acridines, i.e. acriflavine, proflavine, acridine orange, acridine yellow, ethidium bromide	Intercaletes between bases disrupting their alignment and pairing. Results in deletion or addition of bases.
1.1.2. Non-particulate radiation, i.e. X-rays and γ-rays (Gamma-rays)		2.3. Base analogues, i.e. 5-Azocytidine, 5-bromo-deoxyuridine, 2-Aminopurine, Hypoxanthine, Maleic hydrazide, 6-Mercapto purine	Base pair substitution.
1.2. Non-ionising radiation, i.e. UV radiation	Induction of purine or pyrimidine dimers.	2.4. Others, i.e. nitrous acid, hydroxyl amine, sodium azide	Replacement of amino group with a hydroxyl group. Conversion of cytosine to a modified base.

genetics and plant physiology (Sharma and Sharma 1978ab, 1979b, 1981abc, Wilson and Hudson 1978, Miller et al. 1984; Vandenberg and Slinkard 1987, 1989ab, Sinha 1988, 1989ab, Tyagi and Gupta 1991, Sinha and Chowdhury 1991).

Sharma and Kant (1975) treated lentil by gamma rays and N-nitroso N-ethyl urea (NEU) to induce mutations and found that chemical mutagenesis was more successful. Generally chemical mutagens were more efficient than physical mutations for inducing mutations in lentil (Sharma and Kant 1975, Sharma and Sharma 1979ac, 1981, Ravi and Minocha 1987, Sarker and Sharma 1989, Solanki and Sharma 1994, 1999). Among the chemicals, morphological mutation frequency was obtained higher with ethyl methane sulphonate (EMS) than sodium azide

(Gaikwad and Kothekar 2004, Solanki et al. 2004; Solanki 2005, Solanki and Phogat 2005). EMS was observed to be more efficient than sodium azide (Gaikwad and Kothekar 2004) and gamma rays and hydroxylamine (Singh et al. 1989). N-nitroso-N-ethyl urea (NEU) or N-nitroso-N-methyl urea (NMU) was the higher potent than ethyleneimine (EI) and gamma rays based on the frequency of morphological mutation (Sharma and Sharma 1979c, 1981ab, Solanki and Sharma 2000, 2001). Similar findings have been made for other legume species (e.g. soybeans, Carroll et al. 1985).

5.3. Which Doses?

The dose of a chemical mutagen mainly depends on (i) concentration, (ii) duration of treatment, (iii) temperature during treatment (Anonymous 1977). Modifying factors are: (i) pre-soaking, (ii) pH of the solution, (iii) metallic ions, (iv) carrier agents, (v) subsequent washing of seeds (post-washing), (vi) post drying and (vii) storage of treated seeds. To change gene(s) for inducing morphological mutations, EMS doses are between 0.01 and 0.8%. The dose to use in the treatment varies from species to species with very small differences (Siddiqui 1999).

Malik et al. (1998) found that effective dose ranged from 214 to 218 Gy gamma rays for chlorophyll and morphologic mutations, and they found that the 50% lethal dose (LD_{50}) for survival was 250 Gy and radiation sensitivity varied among eight diverse lentil genotypes (Malik et al. 1998). On the other hand, Rajput et al. (1996) found that the lowest chlorophyll mutation frequency occurred at 200 Gy, and the highest occurred at 600 Gy. Paul and Singh (2002) observed that the highest frequency was observed in E 258 at the 150 Gy dose, while the lowest frequency was observed in Pant L 406 at the same dose. The 50% growth reduction (GR_{50}) of primary shoots and useful dose range for mutation breeding in lentil were given as 160–250 Gy and 100–170 Gy for gamma rays, and 9–14 Gy and 50–10 Gy for fast neutrons (N_f), respectively (Anonymous 1977). GR_{50} and a dose close to GR_{50} are considered the optimum dose for lentil by many researchers. Optimum dose produces the maximum frequency of mutations with minimum hazard. An optimum dose can be determined with a preliminary treatment. Overdoses of mutagens will kill too many plants, while under dosing will produce low mutation frequencies. However, lower frequencies may give an advantage of having fewer undesirable background mutations being induced in addition to the mutation being sought. Some factors that influence a mutagen's effects are biological (nuclear volume, chromosome volume and DNA content of variety, and genetic and varietal differences), environmental (oxygen, water status and temperature), and chemical (Anonymous 1977, Sigurbjornsson 1983).

Gamma rays were the most used mutagen to change gene(s) in lentil (Table 2 and 3) due to their easy application. However, as mentioned in the previous section they are not necessarily the most effective. The use of a chemical mutagens requires several procedures such as (i) preparation of seeds, (ii) pre-soaking, (iii) mutagen treatment considering suitable concentration, treatment temperature and time, (iv) post-washing and (v) post-drying; while treatment with physical mutagens includes only two steps which are (i) preparation of seeds and (ii) mutagen treatment. Chemical mutation treatments also require the disposal of left over mutagen which can be highly toxic.

6. PARTS OF LENTIL TO BE TREATED

Although whole plants, seeds, pollen grains, meristems, cells or tissue in culture in crop plants are used (Anonymous 1977, van Harten 1998, Kaul and Nirmala 1999), air dried seeds are the most frequently used part of lentil for mutagenesis (Table 2 and 3). Pollen grains may be used directly such as has been done with pea (*Pisum sativum* L.) (Davies 1984, Saccordo et al. 1993) and maize (*Zea mays* L.) (Neuffer and Chang 1989), and vegetative organs for *in vitro* mutagenesis may be used as well. However, pollen grains are used infrequently because emasculation and pollination are very difficult, time consuming, and pollen survival is short.

Abbo and Ladizinsky (1994) studied genetic aspects of embryo abortion in the genus *Lens*. They found that embryo abortion was not associated with chromosomal aberrations. Irradiation of pollen grains can be beneficial to overcome pre- and post-fertilization problems especially in inter-specific hybridizations. In addition to this, mutations do not induce chimeras when pollens are irradiated (Micke and Donini 1993, Saccardo et al. 1993) whereas in seeds only some of the cell lines may be affected giving a genetically effective cell number of greater than one (Carroll et al. 1988) necessitating delaying selection and requiring a greater number of plants to be screened.

Table 2. Mutant lentils released for commercial production

Mutant variety	Parent variety	Mutagen(s)	Main characters induced	Released	
				Country	Year
S-256 (Ranjan)	B 77	Radiation	Spreading type, high yielding	India	1982
PL 77-2*	BR 25	–	Tolerant to wilt and ascochyta blight	India	1984
Rajendra Masoor 1	–	Gamma rays	Cold tolerant	India	1996
Mutant 17 MM	–	Gamma rays 40 Gy	Seed size	Bulgaria	1999
RH44**	–	EMS	Herbicide (Imidazolinone)-Tolerant	Canada	2006

* Yadav 2005. ** RH44 is one of three mutant varieties (Dr. Vandenberg, pers. comm.).

Table 3. Induced mutant lentils recorded in the available literature

Mutant	Parent variety	Mutagen(s)	Characters induced	Sources
Pod and seed size mutants	L235	Gamma rays 60 Gy	Larger and longer pod and seeds	Sharma and Sharma 1978a
Tendril	L235	Gamma rays 60 Gy or NMU 0.01%	Tendril leaflets	Sharma and Sharma 1978b
Crumpled petal	L235	Gamma rays 100 Gy	Sterile	Sharma and Sharma 1981a
Boat-shaped leaflet and	L258	NMU 0.01%	Boat leaf	Sharma and Sharma 1981b
crinkled leaf	L258	Gamma rays 100 Gy	Crinkle leaf	
Long peduncles	L258	NMU 0.01%	Elongated peduncles	Sharma and Sharma 1981c
Multi-flowers	L235	NMU 0.005%	Sterile	
SKL 2659, HR 73–76, HR 32–35, HR 28–31	–	–	High yield, Earliness	Sharma and Kharkwal 1983a
LM 1, LM 4	–	–	High nodule weight	Rai and Prasad 1983
Compact	T36	Gamma rays and/or NMU	Compact branching	Dixit and Dubey 1986
Dwarf	T36		Dwarf (8–12 cm)	
Staggering	T36		Long branches	
115-1-78, 218–78, 318–78, 514-2-78	–	Gamma rays	High amino acid	Tirdea and Mancas 1986
Dwarf	LL78	Gamma rays 200 Gy	Dwarf (16 cm)	Sinha 1988
Shy mutant	Sehore 74–7	Gamma rays 100 Gy	Dwarf	Sinha 1989a
Male-sterility	LL78	Gamma rays 50–200 Gy	Male-sterile	Sinhac 1989bc
Male sterile lentil	–	Gamma rays 100 Gy	Male sterile	Srivastava and Yadav 2001
A semi-dwarfism	LL78	Gamma rays 50–200 Gy	Plant height = 8.5 cm	Sinha and Chowdhury 1991
Fasciation	L830	Gamma rays 200 Gy + 0.1% EMS	Fasciation	Tyagi and Gupta 1991
M1-30, M1-596	L112	–	Drought tolerant	Salam and Islam 1994
ML-9,	Utfala	Gamma rays 150 Gy	Erect and bush type	Begum et al. 1995

ML-27,	Utfala	Gamma rays 250 Gy	Erect and bush type	
ML-40; ML-42	Utfala	Gamma rays 250 Gy	Erect and synchronous flowering	
Semi-dwarf	Pant L-639	–	Dwarf	Ramesh and Dhananjay 1996
Stunted	Pant L-639		Reduced leaves and pods	
ML-438/8	L-5	–	Lower nitrate reduction	Dutta et al. 1998
Dwarf	P38	–	Dwarf	Tyagi and Ramesh 1998
Bushy dwarf	P38		Bushy dwarf	
AM1-AM40 (40 mutants)	T-36	Gamma rays 50–150 Gy +	Yield criteria	Dubey and Kumar 1999
AML1-AML20 (20 mutants)	K-333	EMS, NMU, DES		
Earliness	P38	Gamma rays	Early maturing	Ramesh and Tyagi 1999
Faciated	P38		Faciation on stem and upper branches	
High yielding	P38		High yielding	
Macrosperma mutants (Six mutants)	HPL 4	–	Earliness, high yield	Sharma and Chahota 1999
AEL 12/30/91	ICARDA-8	Gamma rays 300 Gy	High yielding, earliness	Rajput et al. 2001
AEL 49/20/91	Mansoor-85	Gamma rays 200 Gy		

7. MUTATION BREEDING

7.1. Advantages

Mutation breeding not only creates variability in a crop species, but also shortens the time taken for the development of cultivars via induced mutation compared to those via hybridizations. The average time elapsed from initial mutation treatment to the release of the mutant cultivars was approximately 9 years (Figure 7.1), while this time was more than 9 years for cultivar arising from crossing programmes (Brock, 1977). Moreover mutations induced both qualitative and quantitative characters in a short time altering new alleles of known and previously unknown genes, and modify linkage (Konzak et al. 1977). Further desirable variability could be brought about as new variability in the families *Leguminosae* or *Fabaceae* through induced mutations (Toker and Cagirgan, 2004), whereas variability in hybridization programs is limited to that present in the genotypes/phenotypes of the parents crossed. The existence of mutations is interpreted as supporting N.I. Vavilov's concept of homologous series in heritable variation (Gottschalk 1988). That is, theoretically mutagenesis may create all types of variation that are present in another member of a plant family if the gene exists in the plant treated. It was reported that many mutant lentils were resistant to *uromyces fabae* (Bravo 1983) and out-yielded their parents (Sen 1982, Sharma and Kharkwal 1983b, Salam and Islam 1994, Begum et al. 1995; Ramesh and Tyagi 1999, Sharma and Chahota 1999, Tonev et al. 1999, Mihov et al. 2001). Furthermore mutations are one of the three components of evolution (Sigurbjornsson 1983).

7.2. Disadvantages

The frequency of desirable mutations is very low at about 0.01%. However, mutation frequency will vary for different plant species. Even within a species, cultivars respond differently to mutagen treatment (Sigurbjornsson 1983). Success in mutation breeding depends on methods used handled, effective screening techniques and population grown in M_1 and successive generations. The larger the population in the M_1 is the more success in selection of desirable mutants. Breeders have to screen large populations for desirable mutations. The screening procedures in large populations will require considerable time, labour and other resources. Some mutations have pleiotropic effects due to linked gene(s), other mutations, chromosomal aberrations and deletions. These mutants often have to be backcrossed to parents or adapted varieties. Backcrossing is time consuming work and linkages between genes cannot be broken down easily.

8. MUTANT LENTILS

According to FAO/IAEA Mutant Varieties Database, the number of mutant varieties officially released and recorded is more than 2300 (Jain, 2005). From these mutant varieties, over 265 grain legume cultivars have been released (Bhatia et al. 2001; Maluszynski 2003, Ahloowalia et al. 2004). A variety of common bean (*Phaseolus vulgaris* L.) as a legume, Sanilac, was the first released mutant in Michigan in 1956 (Micke 1988; van Harten, 1998).

Varietal improvement of lentil was initiated as early as 1924 in India (Jeswani, 1988). Similarly, mutation induction work for lentil was probably first initiated in the Indian sub-continent. The first mutant lentil (Table 2) being released there. The Indian Agricultural Research Institute (IARI) has been a pioneer institution for research on induced mutations since 1957, and has released many mutant varieties of legume crops (Ahloowalia et al. 2004; Chopra 2005). The works have been encouraged in order to create useful variation by FAO/IAEA projects (Khan and Shakoor 1977, Ramanujan 1977, Shaikh 1977, Sarma and Kharkwal, 1982, Shaikh et al., 1983, Sarma and Kharkwal, 1983). These mutation breeding efforts created some unique mutations for use in plant breeding programmes. Some mutant

Year	Generation	Application	Progress
1	M_0	• Mutagenic application: Physical or chemical mutagens.	Seeds
1	M_1	• Growing the plants in isolation. • Selection for dominant mutations. • Single plant or bulk harvest.	Chimeric plants
2	M_2	• Growing the plants in single-plant-row or bulk rows • Selection for recessive mutations. • Harvest putative mutants individually. • Single seed descent (SSD) at least two sets. • Bulk harvest the remaining plants.	Segregation for recessive gene(s)
3	M_3	• Growing the plants in traditional sowing density. • Confirmation of the putative mutants. • Continue selection.	Further segregation
4	M_4	• Agronomic evaluation in mini-plots. • Propagation of promising mutants. • Use of mutants in crosses.	Evaluation of genetic stability
5–8	$M_5 - M_8$	• Agronomic evaluation in large plots. • Agronomic evaluation at different locations. • Evaluation of mutants in crosses.	Direct and indirect use of mutants
9	M_9	• Official testing of mutant lines.	Releasing of mutant varieties

Figure 1. Mutation breeding scheme for the improvement of lentil (SSD sets could independently be evaluated for any stress at the target environment)

lentils have been released for commercial production (Table 2). However, in spite of many useful mutant lentils being recorded in the available literature (Table 3) only seven mutant lentil varieties have been released. A comparatively poor figure in comparison to soybean (*Glycine max* L.), groundnut (*Arachis hypogea* L.), pea, common bean, faba bean (*Vicia faba* L.), mung bean [*Vigna radiata* (L.) Wilczek] and chickpea (*Cicer arietinum* L.). On the other hand, special problems in production of lentil will be solved via mutation breeding, i.e. Imidazolinone-Tolerant Lentil Line RH44 (Dr. A. Vandenberg pers. comm.). Herbicide tolerant lentils will act crucial role to overcome weeds in lentil fields in the future. In India, 4 mutant cultivars of blackgram [*Vigna mungo* (L.) Hepper] 8 mutant cultivars of mungbean and 3 of lentil with high yielding capacity have contributed several million dollars annually to the country's agricultural production, *i.e.* for mungbean with an annual value of 64.7 million US$ (Ahloowalia et al. 2004, Chopra 2005).

9. CONCLUSIONS

Genetic variation in available germplasm collections of lentils has been widely used to combat biotic and abiotic stresses. Indigenous lentils are specific ecotypes in the most important lentil production regions of the world and exhibit a marked lack of variability (Erskine et al. 1994). Breeding progress in the cultigen may be limited by this bottleneck which has reduced genetic variability. Although some desirable sources of resistance have been found in the wild species (Erskine et al. 1994, Erskine and Muehlbauer 1995, Tullu et al. 2006; Sarker and Erskine 2006)), there is a difficulty with crosses involving some wild taxa because of post-fertilisation barriers (Muehlbauer et al. 1993). Even if crosses between cultigen and wild relatives are successful, (which is often difficult or even impossible), in addition to the desired gene(s) from wild relative, many undesired gene(s) may be introduced as well. Under this circumstance, a backcrossing programme will be necessary to get rid of undesired gene(s). Therefore, a common and efficient tool to create new and desirable genetic variability in lentil is mutagenesis. Mutant lentils have contributed millions of dollars annually to global agriculture because specific regional production obstacles in lentil have been dealt with using mutant lentil cultivars. Mutations in lentil have also been used to clarify fundamental genetics and physiological processes in lentils.

REFERENCES

Abbo S, Ladizinsky G (1994) Genetic-aspects of hybrid embryo abortion in the genus *Lens* L. Heredity 72: 193–200

Ahloowalia BS, Maluszynski M, Nichterlein K (2004) Global impact of mutation-derived varieties. Euphytica 135: 187–204

Anonymous (1977) Manual on mutation breeding, Second Edition, IAEA, Tech. Rep. Ser. No. 119, Vienna

Begum S, Majid MA, Shaikh MAQ (1995) Selection of promising lentil mutants derived through gamma irradiation. Lens Newsletter 22: 5–8

Bhatia CR, Maluszynski M, Nichterlein K, van Zanten L (2001) Grain Legume cultivars derived from induced mutations, and mutation affecting nodulation. Mutation Breeding Review 13: 1–44

Brock RD (1977) Prospects and perspectives in mutation breeding. In: Muhammed A, Aksel R, von Borstel RC (eds) Genetic Diversity in Plants, Plenum Press, New York, pp 117–132

Carroll, BJ, Gresshoff, PM and Delves, AC. (1988) Inheritance of supernodulation in soybean and estimation of the genetically effective cell number Theoretical and Applied Genetics 76: 54–58

Carroll, BJ, McNeil, DL and Gresshoff, PM (1985) Isolation and properties of soybean [Glycine max (L.) Merr.] mutants that nodulate in the presence of high nitrate concentrations Proceedings of the National Academy of Sciences of the United States of America 82: 4162–4166

Chaudhuri SK (2002) A simple and reliable method to detect gamma irradiated lentil (*Lens culinaris* Medik.) seeds by germination efficiency and seedling growth test. Radiation Physic and Chemistry 64: 131–136

Chopra VL (2005) Mutagenesis: Investigating the process and processing the outcome for crop improvement. Current Science 89: 353–359

Cubero JI (1981). Origin, taxonomy and domestication. In: Webb C, Hawtin G (eds) Lentils, CAB, Slough, pp 15–38

Davies DR (1984) Pollen irradiation and the transfer of maternal genes in *Pisum sativum*. Theoretical and Applied Genetics 67: 245–248

Dixit P, Dubey DK (1986) Three interesting mutant in lentil. Lens Newsletter 13: 5–7

Dubey DK, Kumar S (1999) Mutagenesis in lentil, faba bean and khesari. In: Siddiqui, BA, Khan S (eds) Breeding in crop plants. Mutations & In vitro mutation Breeding, Kalyani Publishers, Ludhiana, pp 35–56

Dutta RK, Mondal MMA, Lahiri BP (1998) Physiological evaluation of advance mutants of lentil in relation to growth, nitrate assimilation and photoharvest. Lens Newsletter 25: 48–51

Erskine W, Muehlbauer FJ (1995) Lentil adaptation to highland winter-sown environments in West Asia and North Africa. In: Keatinge JDH, Kusmenoglu I (eds) Autumn-sowing of lentil in the highlands of West Asia and North Africa, Central Research Institute for Field Crops, Ankara, pp 51–62

Erskine W, Tufail M, Russell A, Tyagi MC, Rahman MM, Saxena MC (1994) Current and future strategies in breeding lentil for resistance to biotic and abiotic stresses. Euphytica 73: 127–135

Erskine W, Chandra S, Chaudhry M, Malik IA, Sarker A, Sharma B, Tufail M, Tyagi MC (1998) A bottleneck in lentil: widening its genetic base in South Asia. Euphytica 101: 207–211

Ferguson ME, Maxted N, van Slageren M, Robertson LD (2000) A re-assessment of the taxonomy of *Lens* Mill. (Leguminosae, Papilionoideae, Vicieae). Journal of the Linnean Society 133: 41–59

Ferguson ME, Erskine W (2001) Lentils (*Lens* L.). In: Maxted N, Bennett SJ (eds) Plant genetic resources of legumes in the Mediterranean Kluwer Academic Publishers, Dordrecht, pp 125–131

Gaikwad NB, Kothekar VS (2004) Mutagenic effectiveness and efficiency of ethyl methane sulphonate and sodium azide in lentil (*Lentil culinaris* Medik.). Indian Journal of Genetics and Plant Breeding 64: 73–74

Gaul H (1964) Mutations in plant breeding. Radiation Botany 4: 155–232

Gottschalk W (1988) Homologous mutation in pea and lentil. Legume Research 11: 32–34

Jain SM (2005) Major mutation-assisted plant breeding programs supported by FAO/IAEA. Plant Cell, Tissue and Organ Culture 82: 113–123

Jeswani LM (1988) Lentil. In: Baldev B, Ramanujam S, Jain HK (eds) Pulse Crops (Grain Legumes), Oxford & IBH publishing Co. Pvt. Ltd., New Delhi, pp 199–214.

Kaul MLH, Nirmala C (1999) Biotecnology: Miracle or mirage. IV *In-vivo* and *in-vitro* mutagenesis. In: Siddiqui, BA, Khan S (eds) Breeding in crop plants. Mutations & In vitro mutation Breeding, Kalyani Publishers, Ludhiana, pp 80–110

Khan MA, Shakoor A (1977) Grain legumes in Pakistan. In: Induced Mutations for the Improvement of Grain Legumes in South East Asia (1975), IAEA-203, International Atomic Energy Agency, Vienna, pp 21–28

Konzak CF, Nilan RA, Kleinhofs A (1977) Artificial mutagenesis as a aid in overcoming genetic vulnerability of crop plants. In: Muhammed A, Aksel R, von Borstel RC (eds) Genetic Diversity in Plants, Plenum Press, New York, pp 163–177

Lonsdale, DM (1987) Cytoplasmic male sterility: a molecular perspective. Plant Physiology and Biochemistry 25: 265–271

Malik IA, Chaudhry MS, Ashraf M, Erskine W (1998) Radio-sensitivity and mutability in lentil (*Lens culinaris* Medik.) as related to seed size. Journal of Genetics & Breeding 52: 9–15

Maluszynski M (2003) Index Issue No. 21–44. Mutation Breeding Newsletter 46: 1–80

Maluszynski M, Szarejko I, Maluszynska J (2004) Mutation techniques. Encyclopedia of Applied Plant Sciences 1–3: 186–201

Micke A (1988) Improvement of grain legumes production using induced mutation. An overview. In: Proceedings of a Workshop on the Improvement of Grain Legume Production Using Induced

Mutations, FAO/IAEA Division, Pullman, Washington, (IAEA, Vienna), pp 1–51

Micke A (1999) Mutations in plant breeding. In: Siddiqui, BA, Khan S (eds) Breeding in crop plants. Mutations & In vitro mutation Breeding, Kalyani Publishers, Ludhiana, pp 1–19

Micke A, Donini B (1993) Induced mutations. In: Hayward MD, Bosemark NO, Romagosa I (eds) Plant breeding: principles and prospects, Chapman & Hall, London, pp 53–62

Mihov M, Mehandjiev A, Stoyanova M (2001) Mutagenesis as a breeding method in lentil. Mutation Breeding Newsletter 45: 35–36

Miller PD, Vaughn KC, Wilson KG (1984) Ethyl methanesulfonate-induced chloroplast mutagenesis in crops. Induction and ultrastructure of mutants. Journal of Heredity 75: 86–92

Muehlbauer FJ (1993) Use of wild species as a source of resistance in cool-season food legumes. In: Singh KB, Saxena MC (eds), Breeding for stress tolerance in cool-season food legumes, ICARDA, A Wiley-Sayce Co-Publication, John Wiley and Sons, Baffins Lane, Chichester, pp 359–372

Muehlbauer FJ, Slinkard AE (1981) Genetics and breeding methodology. In: Webb C, Hawtin GC (eds) Lentils, CAB International, Farnham Royal, Slough, UK, pp 69–90

Muehlbauer FJ, Kaiser WJ, Clement SL, Summerfield RJ (1993) Production and breeding of lentil. Advances in Agronomy 54: 283–332

Muehlbauer FJ, Haddad NI, Slinkard AE, Sarkr B (1996) Lentil. In: Bahl PN, Salimath PM (eds) Genetics, Cytogenetics and breeding of crop plants (vol 1) Pulses and oilseeds, Science Publishers, Inc., Enfield, USA, pp 93–135

Muehlbauer FJ, Cho S, Sarker A, McPhee KE, Coyne CJ, Rajesh PN, Ford R (2006) Application of biotechnology in breeding lentil for resistance to biotic and abiotic stress. Euphytica 147: 149–165

Neuffer MG, Chang MT (1989) Induced mutations in biological and agronomic research. Vortr. Pflanzenzüchtg 16: 165–178

Paul A, Singh DP (2002) Induced chlorophyll mutations in lentil (*Lens culinaris* Medik). Indian Journal of Genetics and Plant Breeding 62: 263–264

Perfectti, F. and Werren, J.H. (2001) The interspecific origin of B chromosomes: experimental evidence. Evolution 55: 1069–1073

Rai R, Prasad V (1983) Salinity tolerance of *Rhizobium* mutants: growth and relative efficiency of symbiotic nitrogen fixation. Soil Biology and Biochemistry 15: 217–219

Rajput MA, Sarwar G, Siddiqui KA (2001) Development of high yielding mutants in lentil. Mutation Breeding Newsletter 45: 35–36

Ramanujan S (1977) Grain legumes in India. In: Induced Mutations for the Improvement of Grain Legumes in South East Asia (1975), IAEA-203, International Atomic Energy Agency, Vienna, pp 29–50

Ramesh B, Dhananjay S (1996) Developmental morphology of induced semidwarf and stunted mutants in lentil. Indian Journal of Genetics & Plant Breeding 56: 335–340

Ramesh B, Tyagi NK (1999) Characteristics and developmental morphology of three agronomically useful mutants in lentil (*Lens culinaris* Medik.). Indian Journal of Agricultural Science 69: 36–39

Reddy VRK, Viswanathan P (1993) Induced mutations in microsperma and macrosperma. Advances in Plant Science 6: 102–115

Saccardo F, Errico A, Crino P, Ocampo, B, Venora G (1993) Mutagenesis and chromosome manipulation for stress tolerance in cool-season food legumes. In: Singh KB, Saxena MC (eds) Breeding for stress tolerance in cool-season food legumes, pp 343–357

Salam MA, Islam MT (1994) Growth, yield and leaf-water attributes of some advanced mutant lentil lines under different soil moisture regimes. Lens Newsletter 21: 32–35

Sarker A, Sharma B (1989) Frequency and spectrum of chlorophyll mutations in lentil (*Lens culinaris* Medik.). Thai Journal of Agricultural Science 22: 107–111

Sarker A, Erskine, W (2006) Recent progress in the ancient lentil. Journal of Agricultural Science 144: 19–29

Sen SN (1982) New lentil mutant variety in West Bengal. Mutation Breeding Newsletter 20: 3–4

Shaikh MAQ (1977) Grain legumes in Bangladesh. In: Induced Mutations for the Improvement of Grain Legumes in South East Asia (1975), IAEA-203, International Atomic Energy Agency, Vienna, pp. 61–70

Shaikh MAQ, Khanum S, Begum S, Ahmed ZU, Majid MA, Zaman KMS (1983) Effects of chemical mutagens on four species of grain legumes. In: Induced Mutations for Improvement of Grain Legumes Production III, IAEA-TECDOC-299, International Atomic Energy Agency, Vienna, pp 77–85

Sharma B, Kant K (1975) Mutation studies in lentils (*Lens culinaris*). Lens Newsletter 2: 17–20

Sharma SK, Sharma B (1978a) Induced variability for pod and seed size in lentil (*Lens culinaris* Medic.). Current Science 47: 806–807

Sharma SK, Sharma B (1978b) Induction of tendril mutations in lentil (*Lens culinaris* Medic.). Current Science 47: 864–866

Sharma SK, Sharma B (1979a) Induced alteration in seed colour of lentil. Indian Journal of Agriculture Science 49: 174–176

Sharma SK, Sharma B (1979b) Leaf mutations induced with NMU and gamma rays in lentil (*Lens culinaris* Medic.). Current Science 48: 916–917

Sharma SK, Sharma B (1979c) Pattern of induced mutability in different genotypes of lentil (*Lens culinaris* Medik.). Zeitscrift fur Pflanzenzuchttung 83: 315–320

Sharma SK, Sharma B (1981a) Note on gamma-ray-induced crumpled mutation in lentil. Indian Journal of Agriculture Science 51: 119–120

Sharma SK, Sharma B (1981b) Note on the leaf variants in lentil. Indian Journal of Agriculture Science 51: 805–807

Sharma SK, Sharma B (1981c) Induced mutations of physiological nature in lentil. Indian Journal of Genetics & Plant Breeding 40: 290–294

Sharma B, Kharkwal MC (1982) Induced mutations in grain legumes. In: Induced Mutations for the Improvement of Grain Legumes Production II, IAEA-TECDOC-260, International Atomic Energy Agency, Vienna, pp 59–64

Sharma B, Kharkwal MC (1983a) Mutation breeding of lentil, cowpea and chickpea. Mutation Breeding Newsletter 21: 5–6

Sharma B, Kharkwal MC (1983b) Mutation studies and mutation breeding in grain legumes. In: Induced Mutations for Improvement of Grain Legumes Production III, IAEA-TECDOC-299, International Atomic Energy Agency, Vienna, pp 65–75

Sharma SK, Sharma B (1986) Mutagen sensitivity and mutability in lentil. Theoretical and Applied Genetics 71: 820–825

Sharma SK, Chahota RK (1999) Agronomic evaluation and dry matter distribution in lentil (*Lens culinaris* Madik.) mutants under different daylengths. Tropical Agriculture 76: 246–249

Siddiqui, BA (1999) Mutagenesis: Tools and techniques- A practical view. In: Siddiqui, BA, Khan S (eds) Breeding in crop plants. Mutations & In vitro mutation Breeding, Kalyani Publishers, Ludhiana, pp 20–34

Sigurbjornsson B (1983) Induced mutations. In: Wood DR (ed) Crop breeding, American Society of Agronomy and Crop Science Society of America, Madison, Wisconsin, pp 153–176

Singh BD (2005) Mutations in crop improvement. In: Singh BD (ed) Plant breeding, principles and methods, Kalyani Publishers, Ludhiana, pp 698–731

Singh D, Singh RM, Singh J (1989) Effect of gamma rays, ethylmethane sulphonate and hydroxylamine on type and frequency of chlorophyll mutations in lentil. Lens 16: 3–5

Sinha RP (1988) Induced dwarf mutant of lentil, RPL-1. Mutation Breeding Newsletter 32: 11

Sinha RP (1989a) Induced shy mutant of lentil (*Lens culinaris* Med.). Current Science 58: 252–253

Sinha RP (1989b) Induced mutant for male-sterility of lentil. Mutation Breeding Newsletter 34: 9–10

Sinha RP, Chowdhury SK (1991) Induced codominant mutations for dwarfism in lentil (*Lens culinaris* Med.). Indian Journal of Genetics & Plant Breeding 51: 370–371

Solanki IS (2005) Isolation of macromutations and mutagenic effectiveness and efficiency in lentil (*Lens culinaris* Medik.). Indian Journal of Genetics and Plant Breeding 65: 264–268

Solanki IS, Sharma B (1994) Mutagenic effectiveness and efficiency of gamma rays, ethylene imine and N-nitroso-N-ethyl urea in macrosperma lentil (*Lens culinaris* Medik.). Indian Journal of Genetics & Plant Breeding 54: 72–76

Solanki IS, Sharma B (1999) Induction and isolation of morphological mutations in different mutagenic damage groups in lentil (*Lens culinaris* Medik.). Indian Journal of Genetics & Plant Breeding 59: 479–485

Solanki IS, Sharma B (2000) Significance and effectiveness of classifying the M_1 material based on mutagenic damage for inducing macro- and micromutations in lentil (*Lens culinaris* Medik.). Journal of Genetics & Breeding 54: 149–155

Solanki IS, Sharma B (2001) Frequency and spectrum of chlorophyll mutations in macrosperma lentil (*Lens culinaris* Medik.). Indian Journal of Genetics and Plant Breeding 61: 283–286

Solanki IS, Phogat DS (2005) Chlorophyll mutation induction and mutagenic effectiveness and efficiency in macrosperma lentil (*Lens culinaris* Medik.). National Journal of Plant Improvement 7: 81–84

Srivastava A, Yadav AK (2001) Gamma ray induced male sterile mutant in lentil. Mutation Breeding Newsletter 45: 22–23

Stadler LJ (1928) Mutations in barley induced by x-rays and radium. Science LXVIII: 186–187

Tirdea G, Mancas D (1986) Study of the aminoacid content of some varieties and mutant lines of lentil (*Lens esculenta* Moench.). Agronomie 28: 67–69

Toker C, Cagirgan MI (2004) Spectrum and Frequency of Induced Mutations in Chickpea. International Chickpea and Pigeonpea Newsletter 11: 8–10

Tonev TK, Mihov MI, Mitova I, Milev G (1999) Dry matter accumulation and chemicals composition in Bulgaria varieties of lentil. I Dry matter accumulation. Bulgarian Journal of Agricultural Science 5: 827–833

Tullu A, Buchwaldt L, Lulsdorf M, Banniza S, Barlow B, Slinkard AE, Sarker A, Tar'an B, Warkentin T, Vandenberg A (2006) Sources of resistance to anthracnose (*Colletotrichum truncatum*) in wild *Lens* species. Genetic Resources and Crop Evolution 53: 111–119

Tyagi BS, Gupta PK (1991) Induced mutations for fasciation in lentil (*Lens culinaris* Med.). Indian Journal of Genetics & Plant Breeding 51: 326–331

Tyagi BS, Ramesh B (1998) Characteristics and developmental morphology of reduced plant height mutants in lentil. Lens Newsletter 25: 6–10

Vandenberg A, Slinkard AE (1987) Inheritance of a xantha chlorophyll deficiency in lentil. Journal of Heredity 78: 130

Vandenberg A, Slinkard AE (1989a) New qualitative genes and linkages in lentil. Canadian Journal of Plant Science 69: 546

Vandenberg A, Slinkard AE (1989b) Inheritance of four new qualitative genes in lentil. Journal of Heredity 80: 320–322

van Harten, AM (1998) Mutation Breeding: Theory and Practical Applications. Cambridge University Press, Cambridge

van Oss H, Aron Y, Ladizinsky G (1997) Chloroplast DNA variation and evolution in the genus *Lens* Mill. Theoretical and Applied Genetics 94: 452–457

Wilson VE, Hudson LW (1978) A lentil mud mutation. Journal of Heredity 69: 357–358

Yadav DS (2005) Lentil. In: Yadav DS (ed) Pulse crops (Production technology), Kalyani Publishers, Ludhiana, pp 271–287

CHAPTER 14

WILD RELATIVES AND BIOTECHNOLOGICAL APPROACHES

PHILIP A. DAVIES[1], MONIKA M. LÜLSDORF[2] AND MAQBOOL AHMAD[1]

[1]*South Australian Research and Development Institute (SARDI), GPO Box 397, Adelaide, SA 5001, Australia*

[2]*Crop Development Centre, University of Saskatchewan, 51 Campus Drive, Saskatoon, SASK S7N 5A8, Canada*

Email: davies.phil@saugov.sa.gov.au

Abstract: Wild species of the genus Lens are an important source of genetic variation for breeding lentil varieties adaptable to new environments and tolerant of biotic and abiotic stresses. The wild species are endemic to a wide range of environments and possess many diverse characteristics. *Lens* species can be divided into three groups, a primary, secondary and tertiary gene pool, according to their inter-crossability. Crosses between members of the different genepools generally fail because the hybrid embryos abort. However, embryo rescue has been used successfully to obtain viable hybrids between groups. It is possible to intercross most of the wild *Lens* species with cultivated lentils using plant growth regulators and/or embryo rescue to allow the growth of hybrid plants. Other biotechnology techniques which may impact on lentil breeding include, micro-propagation using meristamatic explants, callus culture and regeneration, protoplast culture and doubled haploid production. Micropropagation and regeneration from callus culture are relatively well established techniques with further research required for the development of reliable protoplast regeneration and doubled haploid protocols

Abbreviations:

B_5 medium	Culture medium of Gamborg *et al.* (1968)
BAP	6-benzylaminopurine
2,4-D	2,4-dichlorophenoxyacetic acid
GA_3	gibberellic acid
IAA	indole-3-acetic acid
MS medium	Culture medium of Murashige and Skoog (1962)
NAA	α-naphthalene acetic acid
RAPD	Randomly Amplified Polymorphic DNA
TDZ	thidiazuron

S.S. Yadav et al. (eds.), Lentil: An Ancient Crop for Modern Times, 225–240.

1. INTRODUCTION

Lentil is one of the oldest west Asian crops and is still of considerable importance on the Indian subcontinent, in the Middle East, southern Europe, and eastern and northern Africa. On a much smaller scale, it is grown in the New World including Canada, USA and Australia. The total acreage of lentil has grown in the last few years, as has its yield. Production of lentil is estimated at 3.3 million metric tons from an estimated 3.8 million hectares with an average yield of 850 kg/ha (FAOSTAT, 2005). Modern lentil breeding dates back a few decades and is performed at a number of national and international institutions. However, most of the lentil grown by farmers outside the New World is still in the form of land races. These have been selected for adaptation to local conditions and they constitute a valuable source of genetic diversity. Production of widely adapted, high yielding lentil cultivars may cause the extinction of many land races, resulting in an irretrievable loss of genetic diversity.

Another potential source of genetic diversity for the lentil crop is the wild related *Lens* species. The usefulness of these to breeders depends on their genetic relatedness to the cultivated species and the availability of methods for gene transfer. Until recently, only sporadic efforts were made to obtain adequate information on taxonomy, genetics, and evolution of the wild lentil species. This chapter summarises the latest information on taxonomy and genetic variation in the genus *Lens* and includes information on the distribution and ecology of various wild lentil species. It then describes methods for interspecific hybridisation, summarises the interspecific hybrids that have been produced and discusses the potential for improvement of lentils using cell culture technologies.

2. CULTIVATED LENTILS AND WILD RELATIVES

2.1. Taxonomy

The genus *Lens* Miller is a member of the tribe Vicieae, subfamily Papilionacea, family Leguminosae. Beside *Lens*, three other genera are included in the Vicieae: *Vicia* L., *Lathyrus* L., and *Pisum* L. From a morphological point of view, a continuum exists between the genera *Lens* and *Vicia*. However, *Lens* is a much smaller genus, characterised by an annual growth habit, small flowers, calyx deeply divided into subulate, subequal teeth, and a broadly rhomboid compressed legume with one or two orbicular flattened seeds.

The genus *Lens* comprises seven taxa in six species (Ferguson 1998; Ferguson *et al.*, 2000). *Lens orientalis* is the presumed progenitor of *Lens culinaris* and the two species are crossable and produce fully fertile progeny (Muehlbauer *et al.* 2006). According to crossability, phenetic relations and chromosomal diversity, Ladizinsky & Abbo (1993) suggested two biological species in the genus *Lens*: *Lens culinaris* and *Lens nigricans*, with a few subspecies. However, additional information now indicates that some of the proposed subspecies are species in their own right. In 1997, two new species were recognised in genus *Lens*. *Lens tomentosus* was separated from *Lens culinaris* subsp. *orientalis* on the basis of its tomentose, as opposed to puberulent, pods, and a relatively small asymmetrical chromosome which bears a minute satellite (Ladizinsky 1997). *Lens lamottei*, originally described by Czefranove (1971) was found to be the same as a differentiated cytotype identified within *Lens nigricans* by Ladizinsky *et al.* (1983, 1984) and is now recognised as a separate taxon (van Oss *et al.* 1997). Thus as a result of combined evidence of crossability, phenetic relations and morphological markers (Ferguson and Erskine 2001; Ferguson *et al.* 2000), the genus *Lens* consists of the six species listed in Table 1.

From the standpoint of crossability for use in breeding, the *Lens* species can be divided into three groups: *L culinaris* and *L odemensis* make up the primary genepool, *L ervoides* and *L nigricans* belong to the secondary genepool and *L lamottei* and *L tomentosus* belong to the tertiary genepool. (Muehlbauer and McPhee, 2005). Crosses between members of the different genepools generally fail because the hybrid embryos abort. However embryo rescue (see section below) has been used successfully to obtain viable hybrids between groups (Ladizinsky *et al.* 1985). The basic chromosome number of the genus *Lens* is n = 7. All the *Lens* species share more or less the same karyotype, which includes three pairs of metacentric, or submetacentric chromosomes, a pair of metacentric chromosome with a secondary constriction very close to the centromere, and three pairs of acrocentric chromosomes (Ladizinsky & Abbo 1993).

2.2. Geographic Distribution

The main distributional range of the wild lentil species extends from latitude 27 °S to 45 °N and from longitude 70 °E to 15 °W. It includes the Mediterranean basin and extends farther east, up to Tadijikistan. Wild lentils grow almost exclusively in primary habitats where they are not subjected to competition by aggressive colonizer plants. They usually form loose stands in small disjunct populations. The density of plants per site may vary dramatically between years apparently because of climatic conditions.

Table 1. Species of the genus *Lens*

Lens culinaris Medikus
ssp. *culinaris*
ssp. *orientalis* (Boiss.) Ponert
Lens odemensis Ladizinsky
Lens tomentosus Ladizinsky
Lens lamottei Czefranove
Lens ervoides (Brign.) Grande
Lens nigricans (M. Bieb) Godron

Lens ervoides is confined mainly to the Mediterranean region and is relatively common in Israel, Syria, Turkey, the Adriatic coast of Yugoslavia, southern Italy, and more restricted in Spain and Algeria. Isolated populations are also known from Ethiopia and Uganda. *Lens ervoides* usually grows in shady or partially shaded habitats, under a canopy of trees or among shrubs. Ecologically, *Lens ervoides* differs from other wild *Lens* species but may grow close to them when their habitats coincide. *Lens ervoides* has been found adjacent to ssp. *orientalis* in Israel and Turkey and adjacent to *Lens odemensis* in one location in Israel and to *Lens nigricans* in two locations in Yugoslavia.

Lens nigricans is essentially a Mediterranean species, occurring mainly in southern Europe. To the east, it extends to the Crimean Peninsula and Georgia and to the west to La Palma in the Canary Islands, which is also the southern border of this species. This species also occurs sporadically in Algeria, Morocco and on the Italian and French Alps. *Lens nigricans* grows in two different ecological niches: in primary, open or partially shaded and stony habitats, together with other short stature annual legumes, mainly clovers and medics; on limestone, granite and basalt bedrock, from sea level on the Adriatic coast of Yugoslavia up to 1200 m in southern Spain.

The other habitat is typically man-made: abandoned plantations and terraces in Greece, Yugoslavia, France and Spain, in terraced vineyards in the Italian Alps and around ruins in Italy, France and Spain. The populations in these secondary habitats are extremely localised and never extend to adjacent primary habitats which are presumably suitable to *Lens nigricans* (Ladizinsky *et al.* 1985). *Lens odemensis* has only recently been described and was first identified in two locations in Israel, then in Turkey and later identified in herbarium material from other two locations in Turkey and from Chios, the Aegean island. *Lens odemensis* recently was also collected from Syria. *Lens odemensis* grows mostly in open herbaceous habitats together with other annual legumes such as vetch, medic and clover. In Israel and Syria, it occurs on shallow soil and gravel originating from basalt, at altitudes of 700 to 1400 m. In west Turkey, it occurs on calcareous bedrocks, in partially shaded habitats mostly in pine groves from sea level up to 800 m. In southern Turkey, it grows on gravel of basalt and metamorphic rocks.

Lens culinaris ssp. *orientalis* is the wild progenitor of the cultivated lentil. The two are inter-fertile and share the same diagnostic morphological features. Accessions of ssp. *orientalis* occupy the distributional range from Turkey to Tadjikistan and from Iran to the Crimean Peninsula. It is restricted to primary, open or partially shaded habitats on shallow stony soils originating from calcareous, metamorphic and basalt rocks at altitudes ranging from 500 to 1700 m. Mostly it is accompanied by annual vetches, clovers, medics and lathyrus species. Subspecies *orientalis* is also common in the Turkemenian side of the Kopet Dag range, in Uzbekistan and Tadjikistan. The distribution range of ssp. *orientalis* overlaps with those of *Lens odemensis*, *Lens nigricans* and *Lens ervoides*, but they rarely grow side by side. It was found adjacent to *Lens ervoides* in Israel and Turkey but never with *Lens odemensis* or *Lens nigricans* (Ladizinsky 1989).

2.3. Genetic Variation

Several studies have been carried out to evaluate phenetic relations within the genus *Lens*. Conflicting results have emerged which appear to depend on the germplasm and technique used to measure genetic variation (Ahmad and McNeil 1996, Ferguson 1998). Morphologically *Lens lamottei* is closely related to *Lens odemensis* and is practically equally associated with *Lens odemensis* and *Lens culinaris* on the basis of isozyme evidence (Hoffman *et al.* 1986, Ferguson and Robertson 1996); it does however, appear to be the taxon most distantly related to all other taxa according to RAPD marker analysis (Ferguson 1998). Evaluation of species relations of *Lens tomentosus* by biochemical or molecular techniques have not been reported. High genetic diversity has been reported within *Lens nigricans*, *Lens odemensis* and ssp. *orientalis* relative to cultivated lentil. *Lens lamottei* and *Lens ervoides* are the only species reported as having a similar or more restricted genetic base than cultivated lentil (Ferguson 1998).

The geographical distribution of genetic variation as revealed by molecular techniques has been mapped in four wild *Lens* taxa. Centres of diversity as well as areas of low genetic diversity have been identified (Ferguson 1998). For *Lens culinaris* ssp. *orientalis*, two centres of diversity exist, one in south-eastern Turkey and north-western Syria, the other in southern Syria and northern Jordan. *Lens culinaris* ssp. *orientalis* accessions from Iran, central Asia and northern Turkey are genetically all very similar and correspond to the common cytotype identified by Ladizinsky *et al.* (1984). The centre of diversity of *Lens odemensis* overlaps with the southern centre of diversity of *Lens culinaris* ssp. *orientalis* in southern Syria and northern Jordan. A region of high genetic diversity exists for *Lens ervoides* along the eastern Mediterranean coast, but the populations from the coastal region of the former Yugoslavia have a particularly narrow genetic base.

A clear centre of diversity exists for *Lens nigricans* in south-west Turkey with areas of low diversity along the coast of former Yugoslavia, France and Spain. Centres of diversity for *Lens* are also characterised by high population density.

2.4. Morphological Features

Morphological traits are the most useful criterion for species identification. The main differential characteristics of *Lens ervoides*, *L nigricans*, *L odemensis, L culinaris* ssp. *orientalis* and *Lens culinaris* ssp. *culinaris* are listed in Table 2 (Ahmad *et al.* 1997a).

2.5. Domestication

Barulina (1930) was the first to suggest that small seeded cultivated lentil originated from *Lens orientalis* and also suggested that the centre of origin of the cultivated lentil was in the mountainous regions of the Hindo-Kush and Himalayas. Lentil

Table 2. Morphological differences between *Lens* species

Morphological trait	*Lens ervoides*	*Lens nigricans*	*Lens odemensis*	*Lens culinaris* ssp. *orientalis*	*Lens culinaris* ssp. *culinaris*
Plant habit	single or branched column, ascending-erect	single or branched column, decumbent-ascending	single or branched column, decumbent-ascending	single or branched column, decumbent-ascending or erect	single or branched column, ascending-erect or erect
Leaves	4–6 leaflets per leaf, rachis 5–15 mm, ending in leaflet in lower leaves and in tendril in upper leaves	6–8 leaflets per leaf, rachis 8–25 mm, ending in tendril in the upper leaves	6–8 leaflets per leaf, rachis 8–20 mm, ending in tendril in the upper leaves	6–8 leaflets per leaf, rachis 5–25 mm, ending in tendril in the upper leaves	10–16 leaflets per leaf, rachis 20–50 mm, ending in tendril in the upper leaves
Stipules	linear to semi-hastate, 1–3 mm	semi-hastate, 3–5 mm, strongly dentate at the base, perpendicular with parallel position to the stem	semi-hastate, 2–4 mm, slightly dentate at the base, usually horizontal to the stem	lanceolate entire, 2–3 mm	lanceolate entire, 2–4 mm, sometimes with a slight appendage at the base
Peduncle	1–2 flowered, rarely aristate, 22–40 mm	1–3 flowered, aristate, 20–40 mm	1–2 flowered, aristate, 15–35 mm	1–3 flowered, aristate, 12–30 mm	1–3 flowered, aristate, 32–35 mm
Calyx	2–3 mm, teeth shorter than corolla	5–8 mm, teeth as long or longer than corolla	4–6 mm, teeth as long as corolla	4–6 mm, teeth little shorter than corolla	4–7 mm, teeth longer or shorter than corolla
Pod	pubescent or glabrous, rhomboid 8–11 × 3.5–5 mm, 1–2 seeds	glabrous, rhomboid 9–12 × 4–6 mm, 1–2 seeds	glabrous, rhomboid 7–11 × 4–6 mm, 1–2 seeds	glabrous, rarely pubescent, rhomboid 7–11 × 4–6 mm, 1–2 seeds	glabrous, rarely pubescent, rhomboid 1–20 × 4–12 mm, 1–2 seeds
Seed	gray-brown, diameter 2–3 mm	mottled black-brown, diameter 2.5–3.5 mm	mottled gray-brown, diameter 2.5–3.5 mm	mottled gray-brown, diameter 2.5–3.5 mm	variety of colours mottled or plain, diameter 3–9 mm

was utilised by man during the early stages of the Neolithic Revolution. Remains of lentil seeds in archaeological digs suggest that it was one of the first plants to be exploited by man (Zohary and Hopf 1988). The oldest seed remains come from the Middle East, hence the prevailing idea that lentil domestication occurred here, together with that of other pulses and cereals.

The wild progenitor ssp. *orientalis* is at least as common in central Asia as in the Middle East, if not more. All the analysed material from that region are of the common crossability group and share the standard chromosome arrangements, which could be taken as support for Barulina's view that central Asia was the centre for lentil domestication but Zohary (1972) and Williams *et al.* (1974) argued that cultivated lentil had its origin in the Near East arc where it was cultivated with other vegetables as early as the 7th millennium B.C. This evidence for the Near East origin comes from archaeological remains.

2.6. Potential as Genetic Resources

Wild relatives are an integral part of the gene pool of crop plants. They may possess genetic diversity, such as resistance to various diseases and better tolerance to environmental stresses, which is lacking in cultivated crops. Rational exploitation of the wild gene pool depends on the genetic affinities between the crop plant and its wild relative, and on the availability of methods for gene transfer.

The genetic potential of the wild lentil gene pool has not yet been thoroughly estimated. Sources of resistance to the major foliar disease of lentil, rust, the most important soil-borne disease of lentil, vascular wilt and ascochyta blight have been identified in the wild gene pool (Ahmad *et al.* 1997b). Resistance to vascular wilt and ascochyta blight have also been found in *Lens culinaris* ssp. *orientalis* (Bayaa *et al.* 1994, 1995). Greater resistance to cold tolerance has been found in *Lens culinaris* ssp. *orientalis* than in the cultivated lentil (Hamdi *et al.* 1996). The wild gene pool, particularly *Lens odemensis* and *Lens ervoides* also show greater resistance to drought in terms of low relative reduction in yield with drought stress (Hamdi and Erskine 1996).

3. INTERSPECIFIC HYBRIDIZATION

3.1. Crossability Potential

Crossability is defined by Ladizinski (1992) as the potential for intercrossing individuals belonging to different taxa and for producing embryos or seeds that can give rise to an F_1 plant. The potential to cross within the genus *Lens* is hampered by crossability barriers within the species as well as between species (Ladizinsky & Abbo 1993, Ladizinsky 1997, Ferguson *et al.* 2000). *L culinaris* ssp.*orientalis* is considered to be the wild progenitor of the cultigen and most accessions are readily cross-able within the species (Ladizinsky *et al.* 1984). However, exceptions were reported by Ladizinsky & Abbo (1993) and van Oss *et al.* (1997) who identified *L culinaris* ssp.*orientalis* accession S76 from Syria as compatible with only two accessions (S74 and S138) but incompatible with all other accessions due to hybrid embryo abortion.

Table 3 presents an overview of all the different crosses attempted in the genus *Lens*, the success of hybrid production, fertility of hybrids and factors critical for success. Partially fertile hybrids can be obtained from many crosses within the genus (see Table 3) with varying degrees of hybrid fertility. In many cases, application of GA_3 (gibberellic acid) or rescue techniques due to embryo abortion was required. Only four crosses have not resulted in hybrids to date, *L culinaris ssp orientalis* × *L ervoides or L nigricans* (Ladizinsky *et al.* 1984), *L culinaris ssp tomentosus* × *L lamottei* (van Oss *et al.* 1997), and *L culinaris ssp odemensis* × *L ervoides* (Ladizinsky *et al.* 1984). In all of these crosses, either GA_3 was not applied or embryo rescue techniques were not attempted at the time.

3.2. Crossability Barriers

Ladizinsky (1992) explained that success in lentil crosses depends on the interaction between the parental genomes in the hybrid zygote, embryo or endosperm and between the hybrid tissue and the surrounding maternal tissue. The crossability between lentil and its wild relatives is hampered by pre- and post-fertilization barriers. Problems arise with chromosome pairing in many crosses, for example between *L culinaris* × *L tomentosis* (Ladizinsky 1997). Another common problem is that hybrid embryos cease to grow about 7–14 days after pollination due to endosperm degeneration and thus need rescuing in order to obtain viable hybrids. Hence, *L culinaris* × *L ervoides* or *L culinaris* × *L nigricans* crosses need embryo rescue techniques in order to develop mature hybrid plants (e.g. Abbo and Ladizinsky 1991, Cohen *et al.* 1984). In some *L culinaris* × *L culinaris* ssp *orientalis* crosses, the hybrid embryo ceased growing but the endosperm shows no sign of disintegration (Ladizinsky 1992). In contrast, Abbo and Ladizinsky (1991) observed that the endosperm was found to be either abnormal or lacking in *L culinaris* × *L culinaris* ssp *orientalis* crosses. Hybrids showed varying degrees of fertility usually due to chromosome translocations and subsequent problems with chromosome pairing at meiosis (Ladizinsky *et al.* 1984). These problems can occur in the F_1 and also persist into later generations causing partial or complete sterility. For example, in crosses of *L culinaris* cv. Eston × *L ervoides* L01-827A, 150 F_1 seeds were obtained but only 85 (57%) could be advanced to F_2 (Fiala 2006). Fertility is often very low with little viable pollen produced in anthers and varies depending on the accession in *L culinaris* × *L culinaris* ssp *orientalis* crosses from 2–69% (Ladizinsky *et al.* 1984. Ladizinsky *et al.* (1984). Albino seedlings can occur in the F_1 generation and thus also prevent hybridization success (Ladizinsky & Abbo 1993).

Table 3. Intra- and inter-specific crosses in the genus lentil[1]

Parent 1	Parent 2	Hybrid Status	Critical factors	References
culinaris	*orientalis*	Mostly fertile	Karyotype Embryo rescue GA_3 Environment	Ladizinsky *et al.* 1984 Cohen *et al.* 1984 Abbo and Ladizinsky 1991 Ladizinsky & Abbo 1993 Ahmad *et al.* 1995 van Oss *et al.* 1997 Fratini and Ruiz 2004
culinaris	*odemensis*[2]	Partially fertile	GA_3 Embryo rescue	Goshen *et al.* 1982 Ladizinsky *et al.* 1984 Fratini and Ruiz 2006
culinaris	*tomentosus*[3]	Partially fertile	Karyotype Embryo rescue	Ladizinsky & Abbo 1993
culinaris	*ervoides*	Partially fertile	Embryo rescue GA_3	Cohen *et al.* 1984 Ladizinsky *et al.* 1985 Ahmad *et al.* 1995 Fiala 2006 Fratini and Ruiz 2006
culinaris	*lamottei*	Partially fertile	Embryo rescue	Fiala 2006
culinaris	*nigricans*	Partially fertile	Embryo rescue GA_3	Cohen *et al.* 1984 Ladizinsky *et al.* 1985 Ahmad *et al.* 1995 Fratini and Ruiz 2006
orientalis	*odemensis*	Partially fertile		Ladizinsky *et al.* 1984 Goshen *et al.* 1982
orientalis	*tomentosus*[4]	Sterile	Karyotype Embryo rescue Temp. > 28 C	Ladizinsky & Abbo 1993 Ladizinsky 1997 van Oss *et al.* 1997
orientalis	*ervoides*	None obtained		Ladizinsky *et al.* 1984
orientalis	*nigricans*	None obtained		Ladizinsky *et al.* 1984
tomentosus	*lamottei*	None obtained	Not attempted	van Oss *et al.* 1997
odemensis	*ervoides*	None obtained		Ladizinsky *et al.* 1984
ervoides	*nigricans*	Partially fertile		Ladizinsky *et al.* 1984 Ladizinsky & Abbo 1993
ervoides	*lamottei*	Partially fertile		Ladizinsky *et al.* 1984
nigricans	*lamottei*	Sterile		Ladizinsky *et al.* 1984 van Oss *et al.* 1997

[1] Accessions are listed as Parent 1 or 2 regardless of the direction of the crosses

[2] Initially described as *L nigricans* with horizontal stipule type but later designated as *L culinaris* ssp *odemensis* (Ladizinsky *et al.* 1984, Ferguson *et al.* 2000)

[3] Initially described as *L culinaris* ssp *orientalis* with tomentose pods and later designated as *L culinaris* ssp *tomentosus*

[4] Initially described as *L culinaris* ssp *orientalis* No. 133 and later designated as *L culinaris* ssp *tomentosus* (Ladizinsky & Abbo 1993)

3.3. Hybrid Embryo Rescue Protocols

Ahmad *et al.* (1995) reported obtaining viable hybrids between *L culinaris* × *L ervoides, L culinaris* × *nigricans, L culinaris* × *L culinaris* ssp *odemensis,* and *L culinaris* ssp. *culinaris* × *L culinaris* ssp *orientalis* by applying 50 – 400 ppm GA_3 to the developing pods four and ten days after pollination. Hybrid embryos from interspecific lentil crosses often abort 7–14 days after pollination due to hybrid endosperm breakdown or chromosome abnormalities, resulting in shriveled, non-viable seeds. Cohen *et al.* (1984) were the first to report that hybrid embryos could be rescued by culturing the ovules on an agar solidified MS medium supplemented with 100 g l^{-1} sucrose, 0.2 mg l^{-1} IAA (indole-3-acetic acid), 0.5 mg l^{-1} GA_3, and 0.5 mg l^{-1} zeatin. Seven to 10 days later, embryos were excised and transferred to MS medium with 30 g l^{-1} sucrose and 0.3 mg l^{-1} zeatin and sub-cultured on the same medium 2 weeks later (Ladizinsky *et al.* 1985). Fratini and Ruiz (2006) developed a protocol in which hybrid ovules were rescued 18 days after pollination using a medium consisting of MS salts, 1 μM IAA, 0.8 μM kinetin, 1% sucrose and 0.8% agar. Two weeks later, embryos were excised and cultured on the saume medium for another 2 weeks followed by transfer to culture tubes until plantlet development. They obtained 6 hybrids between *L culinaris* × *L culinaris* ssp *orientalis*, 2 *L culinaris* × *L nigricans* hybrids, and 1 *L culinaris* × *L ervoides* plant. The authors compared different techniques for obtaining lentil interspecific hybrids including crossing without embryo rescue, crossing without rescue but applying GA_3 to the developing pod (Ahmad *et al.* 1995) and embryo rescue using an improved embryo rescue medium (Cohen *et al.* 1984). Even though the number of hybrids obtained with their improved medium was low, all other methods failed to produce mature hybrids except for one *L culimaris* × *L culinaris* ssp *odemensis* hybrid obtained with the rescue medium of Cohen *et al.* (1984). Fiala (2006) also obtained *L culinaris* × *L ervoides* hybrids using the Cohen *et al.* (1984) protocol. In addition, one viable *L culinaris* ssp *culinaris* × *L lamottei* hybrid was also produced in this study. However, the hybrid plantlet could not be rooted directly and was subsequently rooted via micrografting (Gulati *et al.* 2001). Improving the embryo rescue protocol to obtain an improved crossing efficiency seems to be a critical step in overcoming hybrid embryo abortion in the genus *Lens*.

3.4. Shoot Regeneration

Due to the difficulties in obtaining interspecific lentil hybrids, a technique is often required to quickly multiply shoots prior to attempting root induction. *In vitro* propagation from apical meristems of lentil was first reported by Bajaj (1979) who found that shoot regeneration occurred on MS medium (Murashige and Skoog, 1962) supplemented with 2 mg l^{-1} IAA plus 0.5 mg l^{-1} kinetin. Shoot regeneration from meristematic explants using BAP (6-benzylaminopurine) was later reported by Polanco *et al.* (1988) where seedling shoot tips, first nodes and immature seeds cultured on MS medium containing BAP at 2.25 or 0.225 mg l^{-1}

with NAA (α-naphthalene acetic acid) at 0.186 or 0.0186 mg l^{-1} and with or without GA_3(1 mg l^{-1}), produced multiple shoots.

Malik and Saxena (1990) observed that TDZ (thidiazuron) was more effective than kinetin or zeatin for the induction of multiple shoots in axenic seedling cultures with greatest numbers of shoots and greatest percentage regeneration occurring between 10 to 30 μM TDZ. Further optimisation of the shoot regeneration protocol was made by Ahmad *et al.* 1997b) who found that optimal shoot regeneration was obtained using MS medium lacking sucrose and containing 2.89 μM GA_3 plus 1.11 μM BAP.

Ye *et al.* (2002) confirmed that BAP and TDZ induce multiple shoot formation in axenic seed cultures and also found that MS salts produced more shoots and larger shoots than Gamborg's B_5 medium (Gamborg *et al.* 1968) and that an additional 750 mg l^{-1} $CaCl_2$ was necessary to minimise shoot tip necrosis. Fratini and Ruiz (2002) reported higher numbers of shoots using TDZ but subsequent rooting was inhibited after the use of this growth regulator. Hence, the authors recommended zeatin for shoot induction probably due to the lower carry-over effect of the natural cytokinin.

3.5. Root Regeneration and Grafting

The induction of root growth in *in vitro* lentil cultures, which is critical for obtaining whole plants after embryo rescue, has proven more difficult than with many other plant species. In the first report of lentil tissue culture (Bajaj and Dhanju, 1979) no details were provided about root growth. Later reports on embryo rescue (Cohen *et al.*, 1984; Ladizinsky *et al.* 1985) described root growth from rescued embryos on MS medium containing 0.2 mg l^{-1} IAA, 0.2 mg l^{1} IAA and 30 g.l^{-1} sucrose. In the first report of plant regeneration from lentil callus tissue (Williams and McHughen, 1986) roots were not obtained *in vitro* but shoots produced *in vitro* were successfully rooted on sand in a mist chamber.

Polanco *et al.* (1988) found that roots could be regenerated from shoots on media containing either 2 mg l^{-1} IAA or 0.186 mg l^{-1} (1 μM) NAA. Root induction varied between 0 and 86% depending on genotype and explant. Ahmad *et al.* (1997b) found that MS medium with 5.37 μM NAA produced optimal rooting across a range of *Lens* species and their F_1 interspecific hybrids. Polanco and Ruiz (1997) reported that BAP had a strong inhibitory effect on root growth and that 2 mg l^{-1} (11.42 μM) IAA induced roots on 4.6 – 39.3% of shoots cultured.

Fratini and Ruiz (2002) found that both TDZ and BAP inhibited root initiation when used during the shoot induction phase. They recommended reducing the time that shoots are exposed to these growth regulators to a minimum. In contrast, Ye *et al.* (2002) observed no inhibition of rooting after shoot induction with BAP. In this study, shoot tips from the cultivated and wild lentil developed roots on medium with 1.5 mg/l NAA. However, differences between rooting capacities of different species was observed with *L ervoides* shoots rooting at 83% whereas *L nigricans* shoots only rooted at 52%. In an earlier study, Ye *et al.* (2000) obtained 70% and

74% rooting of *L culinaris* × *L ervoides* and *L culinaris* × *L culinaris* ssp*orientalis* hybrids, respectively.

Fratini and Ruiz (2003) determined that by placing the apical end of nodal stem segments in the culture medium ("inverted") rather than the basal end, the frequency of root induction was significantly increased. The highest rooting percentage was 95% which was obtained with inverted stem segments placed on MS medium containing 3% sucrose, 5 μM IAA and 1 μM kinetin. This compares with 11% rooting when stem segments were placed in the normal orientation on the same medium.

A later study by Newell *et al.* (2006) clearly demonstrated that aeration, rather than shoot orientation, is the critical factor resulting in increased rooting frequency. They showed that up to 100% of lentil shoot cuttings could produce roots if the proximal cut end was well aerated.

Gulati *et al.* (2001) developed a micrografting method in which lentil shoots were inserted into decapitated seedling root stock, lining up the exposed vascular tissues. The advantages of this technique are that a short time (less than two weeks) is required for rooting and that any growth regulator can be used during shoot induction, giving success rates of 84 to 96%. This technique was used to root hybrids from crosses involving *L culinaris* × *L lamottei* with 53% efficiency (Fiala 2006).

3.6. Callus Culture and Somatic Embryogenesis

Callus culture and subsequent regeneration by somatic embryogenesis or organogenesis is necessary for genetic transformation, enhanced recombination between genomes of interspecific hybrids and for *in vitro* selection at the cellular level. The initial report of lentil tissue culture by Bajaj and Dhanju (1979) also described callus production from excised meristems on MS medium supplemented with 1 to 2 mg l^{-1} 2,4-D (2,4-dichlorophenoxy acetic acid) but no regeneration from callus was observed. The first report of regeneration from callus was by Williams and McHughen (1986) who found that callus could be produced on MS medium with most combinations of 2,4-D, kinetin and GA_3 at concentrations of 0.1, 1 and 10 mg l^{-1}. Shoots from callus tissue were only found to regenerate on media containing 10 mg l^{-1} kinetin and either 1 or 0.1 mg l^{-1} GA_3.

Callus derived from immature embryo tissue was reported by Saxena and King (1987) using medium with between 1 to 10 mg l^{-1} 2,4-D. This callus was observed to be embryogenic. Polanco *et al.* (1988) observed callusing of shoot tip, first node and leaf explants on MS medium supplemented with either 2,4-D, BAP, NAA or IAA with 2,4-D giving the greatest callusing response. Rozwadowski *et al.* (1990) successfully produced callus colonies from epicotyl protoplasts using complex KM8P medium supplemented with a combination of five growth regulators (2.2 μM 2, 4 − D + 2.7 μM NAA + 2.2 μM BAP + 2.3 μM kinetin + 1.4 μM GA_3) or three growth regulators (5.4 μM NAA + 2.2 μM 2, 4 − D + 2.2 μM BAP). However none of these callus colonies regenerated shoots or embryos.

The first report of lentil somatic embryogenesis was by Saxena and King (1987). Immature embryo explants were cultured on either B5A (B5 + 500 mg l^{-1} ammonium nitrate) or MS medium supplemented with 1 – 10 mg l^{-1} 2,4-D. Callus growth was best at 2,4-D concentrations between 1 – 5 mg l^{-1} and the B5A medium produced more organised callus than MS. Callus initiated on medium containing 1 mg l^{-1} 2,4-D and subcultured to medium without 2,4-D but supplemented with 1 mg l^{-1} BAP and 0.25 mg l^{-1} IAA produced club shaped embryoids. The embryoids were transferred to B5A medium without growth regulators and with the addition of 70 mg l^{-1} glutamine to promote embryo development. Embryos that developed well-defined shoot and root axes were able to germinate on a modified B_5 medium (B_5A) free of growth regulators.

3.7. Protoplast Culture and Hybridisation

Somatic hybridization using protoplast fusion has the potential to overcome pre- and post-zygotic barriers to interspecific hybridisation (Davey *et al.* 2005). It is possible to regenerate plants from a number of legume species including *Pisum* (Ochatt *et al.*, 2000), *Trifolium* (Gresshoff, 1980), *Lotus* (Ahuja *et al.* 1983) and *Melilotus* (Luo and Jia, 1998) and asymmetric protoplast fusion has been used for *Medicago* improvement (Tian and Rose, 1999; Yuko *et al.* 2006). However there are no reports of successful growth or regeneration of protoplasts of *Lens* species. Rozwadowski *et al.* (1990) cultured protoplasts from lentil epicotyl tissue and around 6% of protoplasts developed into cell colonies. However there are no reports of successful plant regeneration from lentil protoplasts.

4. HAPLOIDS AND DOUBLED HAPLOIDS

Doubled haploids are an important breeding tool in many crop species including wheat, barley, rice, maize and canola. The implementation of doubled haploids increases selection efficiency and allows new varieties to be bred up to five years faster than with conventional breeding methods alone. Haploids may be produced either from immature pollen cells, immature egg cells or following asymmetric chromosome elimination after interspecific hybridisation. A recent review of the literature on doubled haploid production in the Fabaceae (Croser *et al.*, 2006) indicates that none of these approaches have been successful to date for producing lentil haploid plants, but the early stages of isolated microspore division have been observed.

5. SUMMARY

The wild relative species of cultivated lentils are a significant source of genetic variation available for improvement of the relatively narrow genetic base of this crop. The wild species are endemic to a wide range of environments and possess many diverse characteristics including disease resistances and abiotic stress tolerances which may benefit cultivated lentils. It is possible to intercross most of the wild *Lens* species with cultivated lentils using plant growth regulators and/or embryo rescue to allow the growth of hybrid plants. There is enormous potential to exploit these hybrids for the improvement of cultivated lentil germplasm. Other biotechnology techniques including doubled haploid production and regeneration from protoplast culture are much less developed but there has been significant groundwork done to expect that these technologies, particularly doubled haploids, may be of benefit to lentil improvement programs within the next decade.

REFERENCES

Abbo S and Ladizinsky G (1991) Anatomical aspects of hybrid embryo abortion in the genus Lens L. Bot Gaz 152 (3): 316–320.

Ahmad M, and DL McNeil DL (1996) Comparison of crossability, RAPD, SDS-PAGE and morphological markers for revealing genetic relationships within and among *Lens* species. Theoretical and Applied Genetics 93: 788–793.

Ahmad M, Fautrier AG, McNeil DL, Burritt DJ and Hill GD. (1995) Attempts to overcome postfertilization barrier in interspecific crosses of the genus *Lens*. Plant Breeding 114:558–560.

Ahmad M, McNeil DL and Sedcole JR (1997a) Phylogenetic relationships in Lens species and their interspecific hybrids as measured by morphological characters. Euphytica 94(1):101–111.

Ahmad M, Fautrier AG, McNeil DL, Hill GD, Burritt DJ (1997b) *In vitro* propagation of *Lens* species and their F_1 interspecific hybrids. Plant Cell, Tissue & Organ Cult 47: 169–176.

Ahuja PS, Hadiuzzaman S, Davey MR and EC Cocking EC (1983) Prolific plant regeneration from protoplast derived tissues of *Lotus corniculatus* L. (birdsfoot trefoil). Plant Cell Reports 2:101–104.

Bajaj YPS and Dhanju MS(1979) Rescue of plants from apical meristem tips of some legumes. Current Science 48:906–907.

Barulina, H. (1930) Lentil of the USSR and other countries, Suppl. 40th Bull. Appl. Bot. Genet. Plant Breed. Leningrad.

Bayaa B, Erskine W and Hamdi A (1994) Response of wild lentil to Ascochyta fabae f.sp. lentis from Syria. Genetic Resources and Crop Evolution 41: 61–65.

Bayaa B, Erskine W and Hamdi A (1995) Evaluation of a wild lentil collection for resistance to vascular wilt. Genetic Resources and Crop Evolution 42: 231–235.

Cohen D, Ladizinsky G, Ziv M and Muehlbauer FJ (1984) Rescue of interspecific lens hybrids by means of embryo culture. Plant Cell, Tissue & Organ Cult 3: 343–347.

Croser JS, Lulsdorf M, Davies PA, Clarke H, Bayliss K, Mallikarjuna N, Siddique K (2006) Towards doubled haploid production on the fafaceae: progress and constraints. Critical Reviews in Plant Science 25:139–157.

Czefranove, Z. (1971) Novosti Systematischeski Vyssich Rastenii 8: 184–191.

Davey MR, Anthony P, Power JB and Lowe KC (2005) Plant protoplasts: status and biotechnological perspectives. Biotechnology Advances 23:131–171.

FAO (2005) "FAOSTAT database" Food and Agriculture Organisation of the United Nations. http://faostat.fao.org/faostat/collections?ver-sion = ext&hasbulk = 0&subset = agriculture.

Ferguson, M. (1998) PhD Thesis: *Studies of genetic variation within the genus lens*. School of Biological Sciences, University of Birmingham, Birmingham, UK.

Ferguson ME, and Robertson LD (1996) Genetic diversity and taxonomic relationships within the genus *Lens* as revealed by allozyme polymorphism Euphytica 91: 163–172.

Ferguson ME, Maxted N, van Slageren M and Robertson LD (2000) A re-assessment of the taxonomy of *Lens* Mill. (Leguminosae, Papilionoideae, Vicieae). Bot J Linnean Soc 133: 41–59.

Ferguson ME and Erskine W (2001) Lentils (*Lens* L.). In: Maxted N and Bennett SJ (eds.), Plant Genetic Resources of Legumes in the Mediterranean, pp. 125–131. Kluwer Academic Publishers, Dordrecht, The Netherlands.

Fiala JV (2006) Transferring resistance to *Colletotrichum truncatum* from wild lentil species to cultivated lentil species (*Lens culinaris* subsp *culinaris*). MSc thesis, University of Saskatchewan, Saskatoon, Canada, 131 pp.

Fratini R and Ruiz ML (2002) Comparative study of different cytokinins in the induction of morphogenesis in lentil (*Lens culinaris* Medik). In vitro Cell Dev Biol – Plant 38: 46–51.

Fratini R and Ruiz ML (2003) A rooting procedure for lentil (*Lens culinaris* Medik) and other hypogeous legumes (Pea, chickpea and *Lathyrus*) based on plant polarity. *Plant Cell Reports* 21:726–732.

Fratini R and Ruiz ML (2004) Intra-specific and inter-sub-specific crossing in lentil (*Lens culinaris* Medik.) Can. J. Plant Sci 84: 981–986.

Fratini R and Ruiz ML (2006) Interspecific hybridization in the genus *Lens* applying *in vitro* embryo rescue. Euphytica 150: 271–280.

Gamborg OL, Miller RA, Ojima K (1968) Nutrient requirements of suspension cultures of soybean root cells. Experimental Cell Research 50:151–158.

Goshen D, Ladizinsky G, Muehlbauer FJ (1982) Restoration of meiotic regularity and fertility among derivatives of *Lens culinaris* × *L nigricans* hybrids. Euphytica 31: 795–799.

Gulati A, Schryer P, McHughen A (2001) Regeneration and micrografting of lentil shoots. In vitro Cell Dev Biol – Plant 37: 798–802.

Gresshoff PM (1980) *In vitro* culture of white clover: callus, suspension, protoplast culture and plant regeneration. *Bot Gaz* 141:157–164.

Hamdi A and Erskine W (1996) Reaction of wild species of the genus Lens to drought. Euphytica 91:173–179.

Hamdi A, Küsmenoglu I and Erskine W (1996) Sources of winter hardiness in wild lentil. Genetic Resources and Crop Evolution. 43: 63–67.

Hoffman D, Soltis D, Muehlbauer F and Ladizinsky G (1986) Isozyme Polymorphism in *Lens* (Leguminosae). Systematic Botany 11: 392–402.

Ladizinsky G. (1989) Ecological and genetic considerations in collecting and using wild relatives. In: Brown A.H.D,. Marshall D.R., Frankel O.H., and Williams L.T., (Eds.), The Use of Plant Genetic Resources, pp. 297. Cambridge.

Ladizinsky G (1992) Crossability relations. Monograph on Theoretical and Applied Genetics pp. 15–31.

Ladizinsky G (1993) Wild Lentils. Critical Rev Plant Sci. 12(3): 169–184

Ladizinsky G (1997) A new species of Lens from south-east Turkey. Botanical Journal of the Linnean Society 123: 257–260.

Ladizinsky G, Braun D and Muehlbauer FJ (1983) Evidence for domestication of Lens nigricans (M. Bieb.) Godron in southern Europe. Botanical Journal of Linnean Society 87: 169–176.

Ladizinsky G, Braun D, Goshen D and Muehlbauer FJ (1984) The biological species of the genus *Lens* L. [*Lens nigricans*]. Bot Gazette 145 (2): 253–261.

Ladizinsky G, Cohen D and Muehlbauer FJ (1985) Hybridization in the genus *Lens* by means of embryo culture. Theor Appl Genet 70: 97–101.

Ladizinsky G and Abbo S (1993) Cryptic speciation in *Lens culinaris*. Genet Res & Crop Evol 40: 1–5

Luo JP and Jia JF (1998) Plant regeneration from callus protoplasts of the forage legume *Astragalus adsurgens* Pall. Plant Cell Reports 17:313–317.

Malik KA and Saxena PK (1990) Thidiazuron induces high-frequency shoot regeneration in intact seedlings of pea (*Pisum sativum*), chickpea (*Cicer arietinum*) and lentil (*Lens culinaris*). Aust J Plant Physiol 19:731–740.

Muehlbauer FJ and McPhee KE (2005) In: Genetic resources, chromosome engineering and crop improvement, Vol. 1, Grain Legumes, Ch. 8, Lentil (*lens culinaris* Medik), pp 219–230. Eds. RJ Singh, and PP Jauhar. Taylor & Francis, Boca Raton.

Muehlbauer FJ, Cho S, Sarker A, McPhee KE, Coyne CJ, Rajesh PN and Ford R (2006). Application of biotechnology in breeding lentil for resistance to biotic and abiotic stress. Euphytica 147 (1–2): 149–165.

Murashige T and Skoog F (1962) A revised medium for rapid growth and bioassays with tobacco tissue cultures. Physiol. Plant. 15:473–497.

Newell C, Growns D and McComb J (2006) Aeration is more important than shoot orientation when rooting lentil (*Lens culinaris* Medik) cv. 'Digger' microcuttings *in vitro*. In vitro Cell. Dev. Biol. – Plant 42:197–200.

Ochatt SJ, Mousset-Déclas C, and Rancillac M (2000) Fertile pea plants regenerate from protoplasts when calluses have not undergone endoreduplication. Plant Science 156:177–183.

Oss H van, Aron Y and Ladizinsky G (1997) Chloroplast DNA variation and evolution in the genus *Lens* Mill. Theor Appl Genet 94: 452–457.

Polanco MC and Ruiz ML (1997) Effect of benzylaminopurine on *in vitro* and *in vivo* root development in lentil, *Lens culinaris* Medik. Plant Cell Reports 17:22–26.

Polanco MC, Pelaez MI and Ruiz ML (1988) Factors affecting callus and shoot formation from *in vitro* cultures of *Lens culinaris* Medik. Plant Cell Tissue Organ Culture 15:175–182.

Rozwadowski KL, Saxena PK and King J (1990) Isolation and culture of *Lens culinaris* Medik. Plant Cell Tissue Organ Culture 15:175–182.

Saxena PK and King J (1987) Morphogenesis in lentil: plant regeneration from callus cultures of *Lens culinaris* Medik. via somatic embryogenesis. Plant Science 52:223–227.

Tian D and Rose RJ (1999) Asymmetric somatic hybridization between the annual legumes *Medicago truncatula* and *Medicago scutellata*. Plant Cell Reports 18:989–996.

Williams .T, Sanchez AMC and Jackson MT (1974) Studies on lentils and their variation, I. The taxonomy of the species. SABRAO Journal 6, 133–145.

Williams DJ and McHughen A (1986) Plant regeneration of the legume *Lens culinaris* Medik (lentil) *in vitro*. Plant Cell Tissue Organ Culture 7:149–153.

Ye G, McNeil DL, Conner AJ and Hill GD (2002) Multiple shoot formation in lentil (*Lens culinaris*) seeds. New Zealand J Crop & Hort Sci 30: 1–8.

Ye G, McNeil DL, Conner AJ and Hill GD (2000) Improved protocol for the multiplication of lentil hybrids without genetic change by culturing single node explants. SABRAO J Breeding & Genet 32 (1): 13–21.

Yuko M, Mitsuru K, Takamizo T, Kanbe M, Inami S and Hattori K (2006) Interspecific hybrids between *Medicago sativa* L and annual *Medicago* containing alfalfa weevil resistance. Plant Cell Tissue Organ Culture 84:79–88.

Zohary D (1972) The wild progenitor and the place of origin of the cultivated lentil *Lens culinaris*. Economic Botany 26, 326–332.

Zohary D and Hopf M. (1988) Domestication of Plants in the Old World. Clarendon Press, London.

CHAPTER 15

BREEDING METHODS AND ACHIEVEMENTS

MICHAEL MATERNE[1] AND DAVID L. MCNEIL[2]

[1]*Grains Innovation Park, The Department of Primary Industries, Private Bag 260, Horsham, Victoria 3401, Australia*

[2]*School of Agricultural Science, University of Tasmania, Private Bag 54, Hobart, Tasmania 7001, Australia*

E-mail: michael.materne@dpi.vic.gov.au

Abstract: Lentil breeding has a relatively short history, however, since the inception of the ICARDA breeding program in 1977 substantial gains have been made in overcoming regional bottlenecks in germplasm diversity. This program has since been supplemented by breeding programs in both developing (eg India) and developed countries (Australia and Canada). These programs have had substantial success in improving tolerance to both biotic (disease) and abiotic stress as well as improving regional adaptation. The Australian breeding program is detailed indicating differences and similarities with other programs. In recent years the need to concurrently develop agronomic approaches and breeding has lead to a greater collaboration among breeders and agronomists

1. INTRODUCTION

Lentil breeding has a relatively short history compared to many of the major crops such as cereals. In many traditional lentil producing countries most of the lentil area is still occupied by landraces that are vulnerable to a range of biotic and abiotic factors (Sarker and Erskine 2002). However, regional breeding programs have now been established around the world and cultivars are being released that offer great advantages to farmers. Unfortunately, in many cases the distribution and adoption of new cultivars is still limiting the potential benefits of breeding in developing countries.

2. FORMATIVE YEARS FOR LENTIL BREEDING

The first major initiative to improve lentil began with the establishment of the International Center for Agricultural Research (ICARDA) in 1977. Based at Aleppo, Syria, it is a centre supported by the Consultative Group in International Agricultural

S.S. Yadav et al. (eds.), Lentil: An Ancient Crop for Modern Times, 241–253.

Research (CGIAR), an international group of donor agencies, scientists, and administrators from developed and developing countries. Among its activities, ICARDA has the responsibility for lentil improvement internationally (Germplasm Program Annual Report for 1999.

ICARDA has played an essential role in the collection, characterization of landraces, the development of germplasm for countries around the world and cultivars for direct release. In order to understand the structure of the world lentil collection, landraces held at ICARDA were characterised into four major regional groups identified through analysis of variability in quantitative and qualitative morphological traits (Erskine *et al.* 1989). These were the Levantine group (Egypt, Jordan, Lebanon and Syria), the northern group (Greece, Iran, Turkey, USSR, Chile), the Indian subcontinent group and the Ethiopian group. Erskine *et al.* (1994b) later found that the dissemination of lentil around the world has resulted in the selection of different regionally specific balances between photoperiod and temperature for the control of flowering (Erskine *et al.* 1994b). For example, cultivated lentil spread from West Asia to the Indo-Gangetic plain around 2,000 BC (Erskine and Saxena 1993). Lentil landraces originating from West Asia flower much later in Pakistan and India than the local landraces, and their reproductive development begins when conditions are increasingly hot and dry in that environment (Erskine and Saxena 1993) (Table 1).

Although half the world area of lentil is found in South Asia, it is based on a narrow genetic base of exclusively pilosae lentil types with a reduced sensitivity to photoperiod and increased sensitivity to temperature than landraces from West Asia (Erskine *et al.* 1998). This "daylength bottle neck" limited the flow of lentil germplasm into the Indo-Gangetic plain and has been implicated in creating the low

Table 1. Lentil varieties emanated from ICARDA supplied materials and released by national programs (Sarker and Erskine 2002)

Region	Country	No. of varieties	Reason for release
Asia	Bangladesh, India, Nepal, Pakistan, China, Afghanistan, Iran, Iraq, Syria, Lebanon, Jordan, Yemen, Turkey	35	High yield; wilt, rust and ascochyta blight resistance; good standing ability, high biomass, early maturity, winter-hardiness
Africa	Ethiopia, Egypt, Morocco, Libya, Tunisia, Algeria, Lesotho, Sudan	24	High yield; wilt and rust resistance; early maturity, tolerance to excess moisture
The Americas	Argentina, Canada, Ecuador, USA	6	High yield; rust and ascochyta blight resistance; good standing ability
Oceania	Australia, New Zealand	9	High yield, ascochyta blight resistance; good standing ability
Europe	Portugal	2	High yield
Caucasus	Georgia	1	High yield

yield potential and disease susceptibility of lentil in that region. As in other regions, ICARDA was instrumental in expanding the gene pool for breeding in South Asia by intercrossing diverse lines and supplying inbred and segregating populations to South Asia for evaluation, selection and release.

ICARDA provides inbred lines, segregating populations and elite nurseries to lentil researchers around the world. As a result, ICARDA's lentil program has been very successful in changing the productivity of the crop across the world, with 77 of its lines having been released as local varieties in 29 countries by 2002 (Sarker and Erskine 2002). Lines were released in traditional lentil growing areas but were also the basis for new industries in developed counties such as Australia (Materne and Brouwer 1996). In the formative years of the ICARDA breeding program, Dr William Erskine, lentil breeder, conducted an international breeding program targeting all major lentil production regions and worked closely with scientists in these countries to identify constraints to production and in exploiting ICARDA germplasm. ICARDA scientists have also facilitated collaborative research programs with breeders in developed counties to the benefit of lentil improvement internationally.

3. BREEDING SUCCESSES

One of the major achievements of ICARDA's collaborative research is the breaking of an ancient 'bottleneck' as presented by the narrow genetic base of lentil in South Asia (Erskine *et al.* 1998). The genetic base of lentil in this region has been broadened through introgression of genes from ICARDA germplasm (Sarker and Erskine 2002). Early, high yielding and disease resistant varieties, such as Barimasur-2 and Barimasur-4 have been released in Bangladesh (Sarker *et al.* 1999a, 1999b), high yielding varieties with resistance to fungal diseases released in Pakistan (Tufail *et al.* 1995) and extra-early and extra-bold lines have been developed in India to fit in different cropping systems (Chauhan and Singh 1995, Sarker and Erskine 2002). The medium-maturity cultivar Shekher (ILL 4404) is now being grown in the mid-hills region of Nepal, a new area for lentils (Sarker and Erskine 2002).

3.1. Phenology

Lentils are grown in a broad range of climates and within each region there are variations in climate, soils, diseases and pests that impact on the performance of a genotype. The success of ICARDA in supplying landraces and breeding lines for countries around the world has been discussed. A major focus of initial evaluation was the identification of cultivars with regionally specific flowering responses to provide the basis for adaptation to the climatic variables of an environment. In traditional lentil growing regions the optimal time to flowering response was represented in the indigenous landraces for the growing system in use there. However, but in new production areas the optimal flowering response was unknown and needed to be demonstrated by looking at diverse responses over several years in evaluation trials where grain yield was measured. Where major changes in crop agronomy was also introduced (eg tolerance of diseases) there was also a need to reestablish the optimal flowering physiology under the altered circumstances. Key varieties in new production areas included Crimson and Eston in USA, Laird in Canada, and Digger and Northfield in Australia. Alternatively, a photothermal model and climatic data were used to select genotypes suited to winter sowing in the highlands of central and eastern Anatolia, Turkey (Keatinge *et al.* 1995, 1996).

3.2. Tolerance to Abiotic Stresses

In Syria and Australia, and most likely all other lentil growing areas, selection for yield under variable rainfed conditions has increased water use efficiency in lentil through an increased response to moisture availability (Murinda and Saxena 1983, Erskine and Saxena 1993, Materne 2003). However, breeding has increasingly focused on addressing abiotic and biotic constraints, particularly disease. In the USA and Turkey (Central Anatolia), large yield increases have been achieved by sowing lentil in winter rather than spring using genotypes tolerant to cold temperatures during winter (Saker *et al.* 1988, Erskine *et al.* 1981, Kusmenoglu and Aydin 1995, Hamdi 1996, Muehlbauer and McPhee 2002).

Although generally adapted to alkaline soils, lentil growth can be affected by hostile subsoil factors such as high pH, toxic levels of boron and salinity and sodicity (Yau 1999, Yau and Erskine 2000, Saxena *et al.* 1993, Hobson *et al.* 2006). Although variation in tolerance to these factors has been identified, breeding to target these stresses is to our knowledge relatively limited. Breeding lines with improved tolerance to boron, derived from ILL2024 have been developed in Australia and based on controlled environment experiments could improve yields by up to 91% in the target regions (Hobson *et al.* 2006). Similarly, lines with improved tolerance to NaCl have been developed and are soon to be released in Australia. These lines look to have great potential as they are the highest yielding entries in advanced yield trials (Materne *et al.* 2005). In contrast to Australia, boron deficiency has been identified as a limitation to lentil production on soils in Nepal and selection for tolerance occurs in this country (Srivastava *et al.* 1999, 2000).

3.3. Resistance to Disease

Of the diseases that occur and have proven destructive in lentils, breeding has had the greatest impact on delivering cultivars with improved resistance to Fusarium or vascular wilt, caused by *Fusarium oxysporum f.sp. lentis* Vasd. and Srin., rust, caused by *Uromyces vicia-fabae* (Pers.) Schroter, ascochyta blight caused by *Ascochyta lentis* Vassilievsky, Anthracnose caused by *Colletotrichum truncatum* (Schwein.) Andrus and Moore and stemphylium blight caused by *Stemphylium botrysum* Wallr. and botrytis grey mould, caused by *Botrytis cinerea* and *Botrytis fabae*.

Cultivars with resistance to Fusarium wilt have been developed by ICARDA and released in Middle Eastern countries such as Syria. ILL5588 (PI592998, Talia 2) has been the major source for resistance, registered as lentil germplasm resistant to vascular wilt (Fusarium wilt) and released as a cultivar (Erskine *et al.* 1996).

The rust resistant varieties Bakria (ILL4605), Bichette (ILL5562) and Hamira (ILL6238) have been released in Morocco and yields of up to 2.8 t/ha have been achieved (Sarker and Erskine 2002). In Ethiopia lentil varieties like Adaa and Alemaya have been released that have a high level of resistance to rust and to the wilt root rot complex (Sarker and Erskine 2002). Rust resistance breeding and/or evaluation is also important in India and South America.

Breeding for resistance to ascochyta blight has been a major success for lentil breeders internationally. ILL5588 (PI592998, Talia 2) has been registered as lentil germplasm resistant to vascular wilt (Fusarium wilt) and to ascochyta blight (Erskine *et al.* 1996) and as Northfield, an ascochyta blight resistant cultivar, in Australia (Ali 1995). Other cultivars released with improved resistance to ascochyta blight include Rajah (ILL6343) in New Zealand (Russell 1994a, b), Manserha 89 in Pakistan (Erskine and Saxena 1993, Erskine *et al.* 1994a), Pant L4 (Singh *et al.* 1994) and Masoor 93 (Tufail *et al.* 1995) in India and CDC Milestone, CDC Glamis, CDC Grandora, CDC Sovereign, CDC Vantage and CDC Robin were the first released in Canada (Vandenberg *et al.* 2001, 2002a, b, c, d, e) and Nipper in Australia.

In Canada, cultivars such as Robin have been released that have moderate resistance to anthracnose derived from Indianhead (Vandenberg *et al.* 2002e). The combination of resistance to anthracnose and ascochyta blight makes this cultivar a significant advance in breeding for Canada.

Breeding for resistance to Stemphylium blight has been a major success for collaborative breeding programs between ICARDA and the Bangladesh government. Bari-Masur varieties with resistance to Stemphylium blight are making a major impact in Bangladesh (Ashutosh Sarker pers. Comm.).

The first lentils with resistance to *Botrytis cinerea* were identified in the USSR (Khare 1981). The first cultivar with improved resistance to *Botrytis cinerea* was Masoor 93, which was released in Pakistan (Tufail *et al.* 1995). Genotypes with resistance to *Botrytis cinerea* and *Botrytis fabae* were identified in Australia (Materne *et al.* 2002a, Materne *et al.* 2006). Using Indianhead as a resistant source, the resistant cultivar Nipper was developed and commercialised in Australia in 2004 to provide farmers with a low risk cultivar in areas where botrytis grey mould can cause total crop loss in susceptible cultivars (Bretag 2000).

3.4. Harvestability

The large-scale production of lentil in countries such as Australia, Canada and USA has been achieved with mechanised harvesting systems, whereas in many traditional lentil producing countries lentil is still harvested by hand (Haddad *et al.* 1988, Sarker and Erskine 2002). Nonetheless, hand harvesting is considered a major constraint to lentil production in North Africa and the Middle East and its high cost has caused a large decrease in lentil production in Jordan and Syria (Erskine *et al.* 1991). Joint ICARDA and national breeding programs have produced new varieties suitable for mechanical harvesting, such as Idlib 1 and Idlib 2 in Syria, Rachyya in Lebanon, IPA 98 in Iraq and Sayran 96 in Turkey. Varieties suitable for mechanical harvesting, have been released in the Middle east and their use, combined with mechanised harvesting, has increased net returns to growers by an estimated A$200/ha (Sarker and Erskine 2002). Varieties with improved characteristics were a key reason for the development of a lentil industry in Australia (Materne and Brouwer 1996).

3.5. Weed Management

Due to their slow growth during winter and short stature, lentil competes poorly with weeds and weed control is a major limitation to growing lentil worldwide. Lentil production in many countries is dependent on herbicides for weed control. Cultivars with good early vigour, as well as improved tolerance to current herbicides that can cause crop damage (Materne *et al.* 2002b, Muehlbauer and Slinkard 1983) and herbicides that are used less frequently in lentils such as ALS-inhibitors (Holm *et al.* 2007) offer potential to improve weed control in lentil. Crop topping and weed wick wiping techniques have also been developed to control difficult weeds in Australia (Preston 2002) but will be dependent on having varieties with earlier uniform maturity and uniform height to maximize success.

4. BREEDING METHODOLOGY

Breeding methodology at ICARDA is based on a bulk population method with single plant selection at F4. F3 derived segregating populations and inbred lines are then distributed internationally. This low cost method has enabled ICARDA to develop populations for a diverse range of environments. This method has also been adopted in other countries around the world. In Australia this method is utilized along with single seed selection at F2 or F3 and summer increase prior to evaluation. In Canada single plants are selected as early as the F1 stage in complex crosses or at F2 in simple crosses (Albert Vandenberg pers. comm.). In most cases evaluation is conducted at diverse locations in the target growing region of a country with screening for major abiotic and biotic constraints (Figure 1).

Screening agronomic traits such as height and maturity and physical seed characteristics are usually a routine part of breeding programs during early generation multiplication and evaluation phases. Mass selection for physical seed characteristics is done by hand picking, using sieves or mechanization using small scale equipment such as gravity tables or colour sorters. Screening for biotic and abiotic stresses has been highly successful in lentil and is expanding. Screening may be a routine part of early generation multiplication and evaluation such as for ascochyta blight in Australia, where disease causes natural selection for resistant plants in the field and mass selection for seed resistance is done using a colour sorter to eliminate

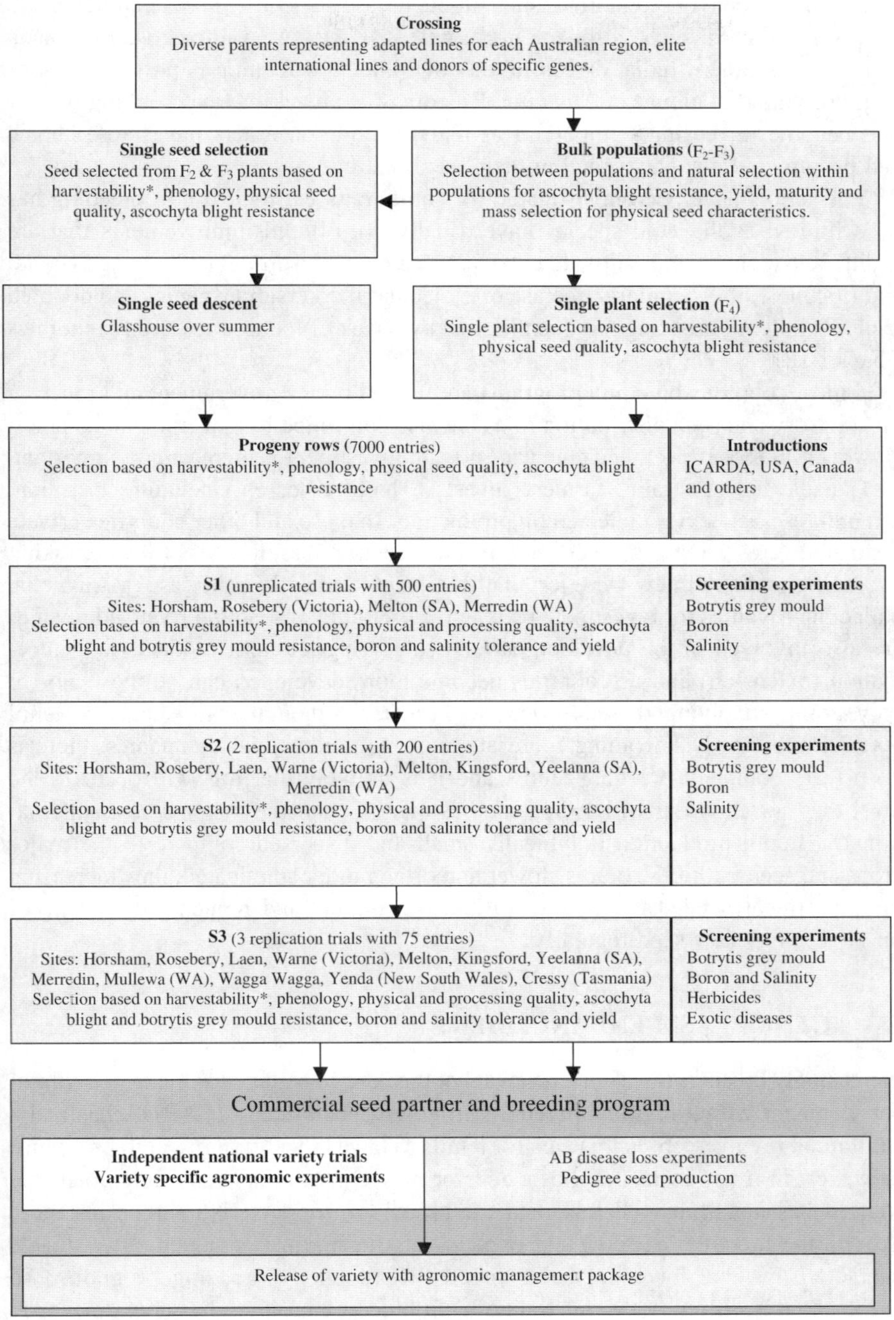

Figure 1. Representation of the lentil breeding program in Australia (high yield, tall, lodging resistant, non shattering, low pod drop, uniform maturity, disease & abiotic stress resistant)

seeds with ascochyta blight blemishes. In other cases, specific disease nurseries are grown in the field, such as for Fusarium wilt at ICARDA, Anthracnose in Canada and rust in northern India, or controlled environment screening is pursued such as with botrytis grey mould in Australia. Breeding for processing and cooking quality will become increasingly important as markets and consumers have more choice and become more sophisticated in their specifications.

Wild lentils have been investigated for some traits but their use in breeding has been limited as the wild species have usually not offered improvements that are significantly above the cultivated species, for example drought tolerance (Hamdi and Erskine 1996). Similarly, the use of molecular markers, transgenics and doubled haploids in lentil breeding has been limited by a lack of focused research (genomics Chapter 18).

In most countries breeding programs are funded by the government and much of the advisory and seed distribution roles are also controlled by government agencies. However, in more developed countries private investment is increasingly important. In Canada and Australia, farmers invest in lentil research, including breeding, through research levies collected on production. In these and other countries private companies are increasing investment in the variety release process by undertaking the multiplication of new varieties, distribution of seed and, in the case of Australia, collecting royalties for investment back into agricultural research. Private companies are also involved in the development and sale of agricultural products and information (extension) and as countries become more developed can be the major or only group supplying these services to farmers. Although components of lentil research are slowly becoming more attractive to commercial companies, there is no private company breeding lentils and it is unlikely that this will occur to the level that currently exits for crops such as oilseeds and the major cereals internationally. Lentil production is typically small and dispersed compare to the major crops and seed multiplication is slower and often more complicated, thus increasing the investment needed to get new varieties to market and reducing the desire of farmers to buy new seed regularly.

5. GERMPLASM ENHANCEMENT

Increasingly breeders have become aware of the need to better coordinate germplasm enhancement within regional breeding programs. Historically ICARDA has had a substantial involvement in this area for lentils as have other lentil research programs. However, in many instances the search for new characteristics and new tolerance sources has not been well linked to the breeding programs. Numerous instances exist in the literature of screening experiments for a range of potentially valuable characteristics that have not been directly taken up by breeding programs. To overcome this difficulty the Australian lentil breeding program has now developed a more formalised approach to incorporate germplasm enhancement using traditional and biotechnological approaches which is outlined in Figure 2. With this approach it is hoped to more rapidly locate and most importantly, incorporate, new

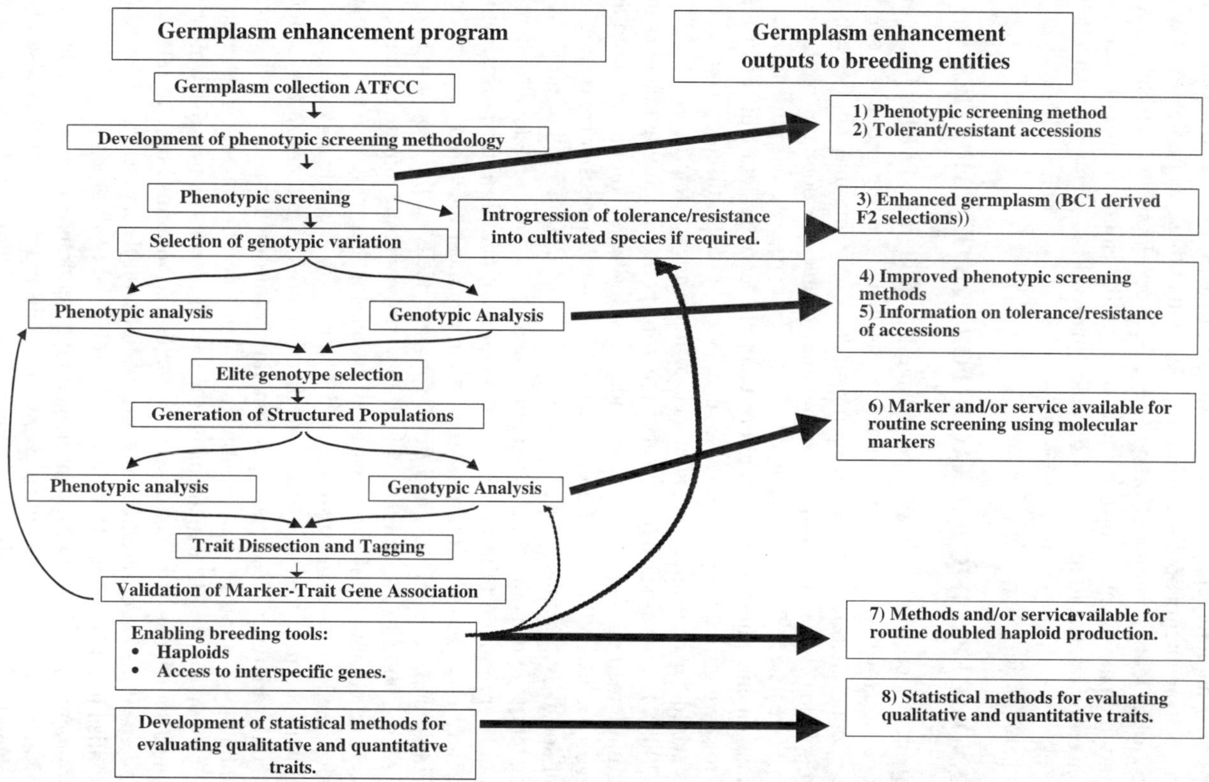

Figure 2. Representation of the lentil germplasm enhancement program and its links to the breeding program in Australia

characteristics into final cultivars. It is also intended that through this mechanism the breeding program is well positioned to collaborate with and provide adapted backgrounds for new biotechnological advances that may occur (eg Genetic Modification, doubled haploids etc.). It is also expected that by linking to breeding and germplasm enhancement approaches in other pulses that these technologies may be transferred to the lentil breeding efforts. For example developments in tissue culture of peas may indicate avenues for this method in lentils, isolation of sequence data from other crops related to a specific tolerance may aid in locating similar alleles in lentils.

6. GENOME SPECIFIC PACKAGES FOR LENTIL VARIETIES

Enhancements in breeding and agronomy have led to increased yields for lentils internationally. However, agronomic practices used in breeding programs can be quite specific and therefore the genetic gains made in breeding may not be realized in farmer's fields if the agronomy used does not enable the potential of the new variety to be realised. This will occur wherever genotype by environment, specifically management, interactions occur. For example lentil varieties differ in their response to sowing time for yield, disease severity and quality, disease management strategies based on resistance and tolerance to herbicides (Materne *et al.* 2002b, Materne 2003). Lentils are also likely vary in their response to farming practices that are being implemented such as a shift to wider row spacings, stubble retention and zero tillage systems, rolling of paddocks post sowing to enable easier harvesting and the use of specially formulated fertilizers (McNeil *et al.* 2006). Further progress in increasing lentil yields is thus likely to require increased concurrent development of genetics and agronomy. In this concurrent development paradigm for pulses, which are frequently used as rotation crops, the agronomy must take into account both the individual crop requirements and the agronomic needs of the cropping system (McNeil *et al.* 2006). In Australia the need for variety specific agronomy has been recognised and all new lentil varieties are evaluated for their response to the major agronomic practices used in Australia, and a variety management package made available with the new varieties to ensure that farmers can maximize the benefits of new lentil varieties. These packages contain the information to optimise grain yield and quality benefits of new varieties in a range of environments. The research concentrates on agronomic management aspects of new varieties for which we have limited knowledge. To maximise efficiencies and minimise time to release of the management package, research runs in parallel with the last stage of breeding, commercialisation of varieties (McNeil *et al.* 2006).

Development of variety specific packages involves interaction of breeders and agronomists late in the breeding cycle. However, substantial benefits may also occur from much earlier collaborations during germplasm development phases of breeding and two major opportunities exist. Firstly when new agronomy systems become plausible, genotypes must be identified which are optimally matched to the new agronomic practice, for example stubble retention, zero tillage systems with wide row spacings and sowing with satellite guidance and automatic steering in Australia (McNeil *et al.* 2006). Secondly, when new genes become available (for example, altered boron membrane transporters) that may provide possible improvements in traits of interest (either through GM, non-GM biotechnology or traditional selection approaches) their specific interactions with the environment, environmental requirements for desired trait expression and opportunities for agronomic application need elucidation (McNeil *et al.* 2006).

It is clear that to benefit from 'genome specific agronomy' approaches requires the concurrent input of agronomy and breeding research in collaboration with extension and farming systems experts to ensure maximal development for 'real world' situations.

REFERENCES

Ali, S. M. (1995) Register of Australian grain legume cultivars. Lens culinaris (lentil) cv. Northfield. Australian Journal of Experimental Agriculture 35: 1181–1182

Bretag, T. W. (2000) Botrytis grey mould in lentils 2000. On the Pulse – A farmer guide to sowing pulses (Ed. J Brand), Agriculture Victoria, VIDA Horsham

Chauhan, M. P. and I. S. Singh. 1995. Introgression of genes for bold seed size from macrosperma type into Indian microsperma lentils. LENS Newsletter, 22: 3–4

Erskine, W. and Saxena, M. C. (1993) Problems and prospects of stress resistance breeding in lentil. In: Breeding for stress tolerance in cool-season food legumes. pp 51–62. (Eds K. B. Singh and M. C. Saxena) John Wiley and Sons

Erskine, W., Adham, Y. and Holly, L. (1989) Geographic distribution of variation in quantitative traits in a world lentil collection. Euphytica 43: 97–103

Erskine, W., Myveci, K., and Izgin, N. (1981) Screening a world lentil collection for cold tolerance. LENS Newsletter 13: 19–27

Erskine, W., Tufail, M., Russell, A., Tyagi, M. C., Rahman, M. M. and Saxena, M. C. (1994a) Current and future strategies in breeding lentil for resistance to biotic and abiotic stresses. Euphytica 73: 127–135

Erskine, W., Hussain, A., Tahir, M., Bahksh, A., Ellis, R. H., Summerfield, R. J., and Roberts, E. H. (1994b) Field evaluation of a model of photothermal flowering responses in a world lentil collection. Theoretical and Applied Genetics 88: 423–428

Erskine, W., Diekmann, J., Jegatheeswaran, P., Salkini, A., Saxena, M. C., Ghanaim, A. and Ashkar, F. EL. (1991) Evaluation of lentil harvest systems for different sowing methods and cultivars in Syria. Journal of Agricultural Science 117: 333–338

Erskine, W., S. Chandra, M. Chaudhury, I. A. Malik, A. Sarker, B. Sharma, M. Tufail and M. C. Tyagi. 1998. A bottleneck in lentil: widening the genetic base in South Asia. Euphytica, 101: 207–211.

Germplasm Program Annual Report for 1999 (1999) About ICARDA and CGIAR. Germplasm Program Annual Report for 1999. International Center for Agricultural research in the Dry Areas, Allepo, Syria.

Haddad, N. I., Salkini, A. B., Jegatheeswaran, P. and Snowbar, B. A. (1988) Methods of harvesting pulse crops. In: World Crops: Cool season food legumes, pp 431–350 (Ed R. J. Summerfield). Kluwer Academic Publishers

Hamdi, A. and Erskine, W. (1996) Reaction of wild species of the genus Lens to drought. Euphytica. 1996. 91: 2, 173–179

Hamdi, A., Kusmenoglu, I. and Erskine, W. (1996) Sources of winter hardiness in wild lentil. Genetic Resources and Crop Evolution. 1996. 43: 1, 63–67

Hobson, K., Armstrong, R., Nicolas, M.,Connor, D. and Materne, M. (2006) Response of lentil (*Lens culinaris*) germplasm to high concentrations of soil boron. Euphytica 151, 371–382

Holm, F. A., Vandenberg, A. and Slinkard, A. E. (2007) Lentil plants having increased resistance to Imidazolinone herbicides. IP Australia wbsite

Keatinge, J. D. H., Aiming, Qi, Kusmenoglu, I., Ellis, R. H., Summerfield, R. J., Erskine, W. and Beniwal, S. P. S. (1995) Defining critical weather events in the phenology of lentil for winter sowing in the west Asian highlands. Agriculture and Forest Meteorology 74: 251–263

Keatinge, J. D. H., Aiming, Qi, Kusmenoglu, I., Ellis, R. H., Summerfield, R. J., Erskine, W. and Beniwal, S. P. S. (1996) Using genotypic variation in flowering responses to temperature and photoperiod to select lentil for the west Asian highlands. Agriculture and Forest Meteorology 78: 53–65

Kusmenoglu, I. and Aydin, N. (1995) The current status of lentil germplasm exploitation for adaptation to winter sowing in the Anatolian highlands. In Autumn-sowing of lentil in the highlands of West Asia and North Africa, pp 63–71 (Eds J. D. H. Keatinge and I. Kusmenoglu) Ankara: CRIFC

Materne, M. A. (2003) Importance of phenology and other key factors in improving the adaptation of lentil (Lens culinaris Medikus) in Australia. Thesis presented for the degree of Doctor of Philosophy at The University of Western Australia, School of Plant Biology and Centre for Legumes in Mediterranean Agriculture (CLIMA), Faculty of Natural and Agricultural Sciences

Materne, M. A. and Brouwer, J. B. (1996) The new lentil industry in Australia, Factors behind its success. In: Proceedings of First Australian New Crops Conference 1996. (Ed B. C. Imrie, R.A. Bray, I.M. Wood and R. J. Fletcher) (Rural Industries Research and Development Corporation) vol 2 pp 45–52

Materne, M., Bretag, T., McMurray, L., Nitschke, S. and Lindbeck, K. (2002a) Genetic variability in lentil for resistance to botrytis grey mould (Botrytis ssp.) In: Plant breeding for the 11th millennium (Ed J. A. McComb) Proceedings of the 12th Australasian Plant Breeding Conference, Perth, W. A., 15–20th September, 2002. pp 809–811

Materne, M., McMurray, L., Nitschke, S., Regan, K., Heuke, L., Dean, G. and Carpenter, D. (2002b) The future of Australian lentil production. In: Proceedings of Lentil Focus 2002, 14–18 (Ed JB Brouwer) Horsham, Victoria, Australia

Materne, M., Regan, K., McMurray, L., Nitschke, S., Dean, G., Heuke, L., and Matthews, P. (2006) Breeding for NaCl tolerance and improved adaptation in lentil. In: 'Breeding for success: Diversity in action' (Ed C.F. Mercer) Proceedings of 13th Australiasian Plant Breeding Conference, Christchurch, new Zealand 18–21 April 2006. pp 1198–1203

McNeil, L., Brand, J., Materne, M. and Jones, B (2006) Genome specific packages for pulses in Australia. In: "Economic and environmental value of grain legumes" (Ed A Schneider) Grain Legumes no 45. pp 25–26.

Muehlbauer, F. J. and McPhee, K. E. (2002) Future of North American lentil production. In: Proceedings of Lentil Focus 2002, pp 8–13 (Ed JB Brouwer) Horsham, Victoria, Australia

Muehlbauer, F. J. and Slinkard, A. E. (1983) Lentil improvement in the Americas. In: Proceedings of the international workshop on Faba beans, Kabuli chickpeas and lentils in the 1980's. pp 351–366. (Eds M. C. Saxena and S. Varma) ICARDA, 16–20 May 1983, Aleppo, Syria.

Murinda, M. V. and Saxena, M. C. (1983) Agronomy of faba beans, lentils, and chickpeas. In: Proceedings of the international workshop on Faba beans, Kabuli chickpeas and lentils in the 1980's. pp 229–244. (Eds M. C. Saxena and S. Varma) ICARDA, 16–20 May 1983, Aleppo, Syria

Preston, C. (2002) Managing an eternal pest – weeds. In: Proceedings of Lentil Focus 2002, pp 69–73 (Ed J. B. Brouwer) Horsham, Victoria, Australia

Russell, A. C. (1994a) Three new pulse cultivars for New Zealand's arable industry. In: Proceedings of the Annual Conference of the Agronomy Society of New Zealand 24: 125–128

Russell, A. C. (1994b) 'Rajah' lentil (Lens culinaris Medik.) New Zealand Journal of Crop and Horticultural Science 22: 469–470

Sakar, D., Durutan, N. and Meyveci, K. (1988) Factors which limit the productivity of cool season food legumes in Turkey. In: World Crops: Cool season food legumes, pp 137–145 (Ed R. J. Summerfield). Kluwer Academic Publishers

Sarker, A. and Erskine, W. (2002) Lentil production in the traditional lentil world. In Proceedings of Lentil Focus 2002, 35–40 (Ed JB Brouwer) Horsham, Victoria, Australia

Sarker, A. W. Erskine, M.S. Hassan, and N. Debnath (1999a) Registration of "Barimasur-2" Lentil. Crop Sci., 39: 875

Sarker, A., W. Erskine, M.S. Hassan, M.A. Afzal and A.N.M.M Murshed (1999b) Registration of "Barimasur-4" Lentil. Crop Sci., 39:876

Saxena, N. P., Johansen, C., Saxena, M. C. and Silim, S. N. (1993) The challenge of developing biotic and abiotic stress resistance in cool-season food legumes. In: Breeding for stress tolerance in cool-season food legumes. pp 245–270. (Eds K. B. Singh and M. C. Saxena) John Wiley and Sons

Singh, I. S., Singh, J. P., Singh, A. K. and Chauhan, M. P. (1994) Pant Lentil 4: a high yielding, rust, wilt and blight resistant variety for the North Western Plains of India. LENS Newsletter 21: 8–9

Srivastava, S. P., Joshi, M., Johansen, C. and Rego, T. J. (1999) Boron deficiency of lentil in Nepal. LENS Newsletter 26: 22–24

Srivastava, S. P., Bhandari, T. M. S., Yadav, C. R., Joshi, M. and Erskine, W (2000) Boron deficiency in lentil: yield loss and geographic distribution in a germplasm collection. Plant and Soil 219: 147–151

Tufail, M., Ahmad, M. and Ali, A. (1995) Masoor-93: an ideal combination of characters for Punjab province, Pakistan. LENS Newsletter 22: 50–52

Vandenberg, A., Kiehn, F. A., Vera, C., Gaudiel, R., Buchwaldt, L., Kirkland, K. J., Morrall, R. A. A., Wahab, J. and Slinkard, A. E. (2001) CDC Milestone lentil. Canadian Journal of Plant Science 81: 113–114

Vandenberg, A., Kiehn, F. A., Vera, C., Gaudiel, R., Buchwaldt, L., Dueck, S., Morrall, R. A. A., Wahab, J. and Slinkard, A. E. (2002a) CDC Glamis lentil. Canadian Journal of Plant Science 82: 103–104

Vandenberg, A., Kiehn, F. A., Vera, C., Gaudiel, R., Buchwaldt, L., Dueck, S., Morrall, R. A. A., Wahab, J. and Slinkard, A. E. (2002b) CDC Grandora lentil. Canadian Journal of Plant Science 82: 105–106

Vandenberg, A., Kiehn, F. A., Vera, C., Gaudiel, R., Buchwaldt, L., Dueck, S., Morrall, R. A. A., Wahab, J. and Slinkard, A. E. (2002c) CDC Sovereign lentil. Canadian Journal of Plant Science 82: 107–108

Vandenberg, A., Kiehn, F. A., Vera, C., Gaudiel, R., Buchwaldt, L., Dueck, S., Morrall, R. A. A., Wahab, J. and Slinkard, A. E. (2002d) CDC Vantage lentil. Canadian Journal of Plant Science 82: 109–110

Vandenberg, A., Kiehn, F. A., Vera, C., Gaudiel, R., Buchwaldt, L., Dueck, S., Morrall, R. A. A., Wahab, J. and Slinkard, A. E. (2002e) CDC Robin lentil. Canadian Journal of Plant Science 82: 111–112

Yau, S. K. (1999) Boron toxicity in lentil: yield loss and variation between contrasting lines. LENS Newsletter 26: 14–17

Yau, S. K. and Erskine, W. (2000) Diversity of boron-toxicity tolerance in lentil growth and yield. Genetic Resources and Crop Evolution 47: 55–61

CHAPTER 16

VARIETAL ADAPTATION, PARTICIPATORY BREEDING AND PLANT TYPE

I. S. SOLANKI[1], SHYAM S. YADAV[2], AND P. N. BAHL[3]

[1]*Department of Plant Breeding, Chaudhary Charan Singh Haryana Agricultural University, Hisar 125004, India*

[2]*Pulse Laboratory, Division of Genetics, Indian Agricultural Research Institute, New Delhi 110012, India*

[3]*A-9, Nirman Vihar, New Delhi 110092, India*

E-mail: solanki255@rediffmail.com

Abstract: The need for adaptation to environments is modified by a need to yield well across a range of seasons and changing microenvironments that can lead to large genotype environment interactions. These interactions may be linked to specific physiological or other traits of the plant which are under genetic control and may be understood. Consequently, different breeding schemes (e.g., farmer participation or research station directed) may be needed in different situations. Similarly under different agro-ecological situations different types of plants may need to be selected (e.g., well watered vs. rainfed). A range of possible factors that affect the ideal adaptation and approach are discussed in this chapter as a means to better understand the process of lentil adaptation that has taken place and continues to take place around the world

1. INTRODUCTION

Adaptation is the process by which plants/varieties/organisms become more suited to survive and function in a given environment. The history of adaptation of lentil in India is well reflected in naturally existing plant types. The luxuriant growth, profuse production and simultaneous shedding of flowers/immature pods and indeterminate and long duration of growth, low harvest index and photosensitivity are factors that reflect a domestication process aimed at maximum survival rather than maximum production. The term adaptation refers to the relationship between the plant and its environment, and thus, can be used in two ways, i.e., to describe both a process and condition. The process is one of the modifications to suit new environmental conditions, and the condition is the result of that process. In the latter case, the

S.S. Yadav et al. (eds.), Lentil: An Ancient Crop for Modern Times, 255–274.

process may be unknown or ignored and the adaptation describes the present performance of a population in one or more environments. The prevalent practice of cultivating one or more pulses together mixed with cereals and millets speaks of adaptation to subsistence farming (Lal 2001).

A common breeding objective in lentil is to develop cultivars with high and stable yield over a range of production environments. Inevitably breeders must sacrifice performance in some environments in order to attain an acceptable level of overall performance, that is, broad adaptation involves specific sacrifices, and limits are placed on performance in particular environments because of the range of environments, which the breeder must consider. Cultivar stability is yield variability over years at a site and cultivar adaptability is yield variability across sites averaged over years. Lentil breeders must be concerned with both stability and adaptability when selecting among breeding lines. Stability and adaptability should be closely related if genotype × environment interactions are caused by unpredictable environmental variables such as rainfall, rather than by predictable ones, such as soil type differences that vary across sites but not years.

In choosing among possible cultivars, a farmer would be interested mainly in their relative stability at the farm site, or conversely in the relative amount of risk associated with the use of each for a given yield level at that site. Where genetic differences in performance are relatable to factors associated with particular locations, they can be exploited by development of regional breeding or selection programmes if adequate resources exist. However, current resources in India (most Asian countries) do not permit such a proliferation of programmes for the food legumes, in general, and lentils, in particular. Therefore, a strategy to develop broadly adapted cultivars is most appropriate. Given that lentil is generally grown under conditions in which moisture stress is virtually assured in most years, it is likely that cultivars may require broad stability even if they possess narrow adaptation. In certain regions, special problems are encountered like, rust and wilt in north west plain zone (NWPZ) and north east plain zone (NEPZ) of India, wilt in CZ and stem and root rot in rice fallows of northern India. For such regions, regional breeding for narrow adaptations can be more productive. Development of pest and disease resistant varieties is a major objective of all the lentil improvement programmes. An emphasis on protective breeding is not surprising : the problems are economically significant, the objectives are easily defined, and selection can be conducted in discrete programmes. Genetically resistant varieties have many advantages over other forms of control.

2. IMPORTANCE OF ADAPTATION

The phenotype of an individual is the resultant effect of its genotype and the environment in which it develops. Furthermore, the effects of genotype and environment may not be independent. This interplay of genetic and non-genetic effects on the phenotypic expression is called genotype-environment (GE) interaction. Some genotypes perform well in a wide range of environments, while others need specific environments to express their genetic potential. A combination of stability in earliness and high-yielding ability is a very important attribute of any genotype (e.g. L 9–12, LH84–8 and Precoz) to be released for general cultivation as a variety.

Farmers are concerned with the production of particular varieties over time and space. As a result, the central objective of lentil scientists is the improvement of the productivity of lentil in specific environments and improving its adaptation to a range of production environments. This may involve genetic and/or environmental modifications designed to alleviate limits of productivity or adaptation. Genetic improvement involves a modification of the genotype to produce a more appropriate phenotypic expression in particular environments. Agricultural production inevitably involves modification of the environment to relieve a particular stress or limit to productivity. This may either negate or alter the need for genetic modification or change the probability of its success in the short or longer term. In practice, improvements in productivity and adaptation depend upon the manipulation of both the genetics and the environment of the plant, and concurrent manipulation is likely to result in optimum performance. Thus, genetic manipulation is only one aspect of crop improvement, and may not be the most appropriate means of resolution of the primary limit to productivity or adaptation.

Plant breeders have observed that adaptation of lentil cultivars is highly area-specific. Since lentils had been domesticated and then widely disseminated throughout the Mediterranean region, Asia and Europe by the Bronze Age, and now, almost throughout the world, it seems probable that tremendous genetic variation exist within the genus for adaptation to environment (Summerfield 1981). Lentil (*Lens culinaris* Medikus subsp. *culinaris*), an annual diploid ($2n=2x=14$) plant is a grain legume well adapted to cool conditions. It is normally sown after autumn rains in the Mediterranean region, or after the monsoon rains, i.e. onset of winter, in India and Pakistan, and grows throughout the winter (*rabi*) season. The crop is reported to be intolerant to extreme heat and cold, which could be why it is confined to higher elevations in tropical countries such as Ethiopia and Mexico, or is grown during the spring at higher elevations in temperate countries such as Iran and Turkey. The crop tolerates drought better than waterlogged soils. For severely drought prone areas selection for early flowering is required (Silim *et al.* 1993).

Improvement through breeding for larger crop yields always includes a conscious or unconscious attempt to produce varieties or populations that are able to make more productive use of the environments for which they are intended. A breeder can expect the greatest benefit only if resistance attributes for *Ascochyta* blight, fusarium wilt, pests, drought, frost, lodging, etc., are incorporated into genotypes which are physiologically well adapted to particular environments. However, the ways in which the climate of a region determine what species will thrive, and the effects of seasonal and annual vagaries of weather on economic yields, are poorly understood in most crops (Monteith 1977), and in lentils hardly at all. Lentil land races must result from centuries of natural and artificial selection in environments, which embody diverse agro-climates and edaphic conditions at various latitudes and altitudes.

3. VARIETAL ADAPTATION FOR DIFFERENT ENVIRONMENTS

Lentil plants have a broad range of characteristics that may be influenced by genetics and environment and a range of these interactions and how they may affect productivity are discussed below. More detailed description of many factors are included in other chapters. The detection of the genetic difference component of the interaction is improved when plants are grown in environments, which maximize the difference in response between genotypes, i.e. in environments outside the range of conditions to which particular genotypes or populations are adapted. For the identification of lentil genotypes which are well adapted not only to seasonal changes in the aerial environment but also to different seraphic conditions (e.g. hot, dry or saline soils), it is also important to consider the symbiotic association with *Rhizobium*. A nodulated legume can obtain at least part of its nitrogen requirements from symbiotic fixation (Chapter 8). Unfortunately, the symbiotic relationship has frequently been ignored in studies of interactions between genotypes and environment in lentils.

3.1. Effects of Climatic Factors

Lentil genotypes are sensitive to geographical and seasonal differences in photoperiod and temperature. In India, sowing in October produces the largest yields but vegetative growth is terminated progressively sooner, and yields are smaller as sowing is delayed (Saxena and Singh 1977). Nevertheless, a large proportion of Indian crops are not sown until after the harvest of paddy in December. Similar benefits of sowing earlier than farmers would normally do, have been recorded in Syria (Hawtin *et al.* 1979). Crops sown early in India and Pakistan on conserved moisture in soil, may exploit chance showers of rain but, unless resistant genotypes or appropriate chemicals are used, fungal diseases such as *Fusarium oxysporum* (wilt) can negate all other advantages. Under irrigated conditions in north west plain zone of India, timely sown crop (second fortnight of November) results in the best expression of seed yield and its component characters (Solanki and Singh 2000).

The major traits of adaptation for lentil producing large yields in low rainfall Mediterranean-type environment (short season) are early flowering, and pod and seed set before the onset of terminal drought. Early phenology together with rapid ground cover and dry matter production allows greater water use in the post flowering period (Siddique *et al.* 2001). The development of earlier flowering cultivars with greater dry matter production together with improved agronomic packages will increase and stabilize lentil yields in low rainfall environments (Siddique *et al.* 1998). The yield under drought conditions is strongly correlated with early vigour and early pod set which enables the plants to escape drought (Leport *et al.* 2003). Drought tolerance is closely related to the distribution of root systems in the soil. Stem length, root length and lateral root number were highly correlated, both amongst themselves and with yield. The line, ILL 6002, exhibited significantly superior root and shoot traits and yield, and therefore, is a valuable germplasm for breeding drought tolerant cultivars (Sarker *et al.* 2005).

Some lentil varieties produce higher yields in environments, which are conducive to high yields, as expected, whilst others yield better-than-average in poor environments but fail to exploit better conditions (Malhotra *et al.* 1971). In general, varieties which mature relatively early, which are likely to be small-seeded types in India, may be the least stable, but the most desirable in different locations. Sowing date studies reported for lentils tell us little about those characters which contribute to high and stable yields, or otherwise, in different photo-thermal regimes.

Lentil plants are, usually less than 70 cm tall but, if flowering is delayed and crop durations are extended by cool ($<10\,^{\circ}C$) temperatures, their indeterminate habit can result in excessively tall plants. Branching pattern depends on genotype and on population density. Sparse branching may be associated with bold seeds and deep roots. Variation in both the number and size of a particular plant organ can be analysed in terms of two variables which may or may not be independent: the rate and the duration of growth (Monteith 1977). When the size or number of organs is fixed genetically, a change in growth rate associated with warmer or cooler temperatures may be offset by a proportional change in duration, so that the net effect may be small. However, if the rate of growth is limited by some non-genetic factor(s), such as the supply of C or N, a change in growth rate in association with change in temperature may not be compensated for by differences in growth duration.

Lentil seedlings, owing to hypogeal germination, are less likely to be killed by freezing, wind or insect damage, or grazing, than if germination were epigeal. If the young shoots are damaged, new buds can be initiated from the nodes below ground. Rates of germination, emergence (hypocotyl elongation) and early seedling growth depend markedly on temperature. Lentil seeds can germinate throughout a wide range of temperature, whether in the light or in darkness, or in constant or diurnally-fluctuating regimes. The optimum temperature range varies with genotype and the age and size of seeds. Smaller seeds, which have a greater surface area: volume ratio than larger seeds, can germinate more rapidly than larger ones at cool temperatures (15 to 25 °C). Young plants can withstand a severe frost, but may be killed if this is prolonged, or repeated (Slinkard 1979), or accompanied by desiccating winds. Cold nights can lessen water uptake, hence genotypes adapted to cold conditions may also tolerate the physiological drought (Steponkus 1978). One lentil genotype (WH 2040), introduced into USA from Greece, has been reported to withstand intense cold ($-23\,^{\circ}C$) at the seedling stage, without the protection of snow cover, and to establish adequate crop stands thereafter (Wilson and Hudson 1978).

In the countries surrounding the Mediterranean sea, the period of vegetative growth in the lentil crop coincides with progressively lengthening days and warmer temperatures. In comparison, crops in the Indian sub-continent will experience at this stage of development shortening days and cool, or even cold (0 to 2 °C) air temperatures (Sinha 1977). Crop growth rates are very low during early vegetative growth, especially when air temperatures are cool (Saxena 1979), and although some genotypes grow more rapidly than others, this species produces only meagre yields per unit of crop time.

Seed set in controlled environments is improved when relative humidity is maintained close to 50% in 16h photoperiods (Muehlbauer 1979) and in a temperature regime of 27° day to 21 °C night. The first (lowermost) fruits to mature usually contain heavier seeds than those which ripen later, although seed size depends also on genotype, the proportion of reproductive load in single-seeded pods, soil fertility and maturation environment. Notwithstanding these additional factors, an acropetal decline in seed size could indicate that pod-filling is limited by the availability of assimilates, N or other nutrients. If yield were limited by sink size, rather than the supply of nutrients, it may be reasonable to expect that later-formed fruits would be equally as well filled as those, which matured earlier (Sheldrake and Saxena 1979).

Sinha (1977) reported that lentils produce seed yields as large as 3.5 t/ha. Variations between genotypes and/or environments have been described by statistical procedures, such as correlation, rather than seeking to explain how environmental variations in time affect the physiological and morphological processes, and hence growth, development and yield (Bunting 1975). Lentil seed yields have been found to be positively associated with branching, the number of flowers and pods per node or per plant and the number of seeds per pod and sometimes, plant height, days to flowering and 100-seed weight (Table 1). For plant height, days to flowering and the number of seeds per pod both types of associations, i.e. positive and negative with final yield have been reported. Some workers have suggested that harvest index may be a useful selection criterion for large yields.

Donald and Hamblin (1976) suggested that it might be useful in plant breeding practice to distinguish between 'isolation environments', 'competition environments' and 'crop environments' and to recognise that plants with different characters will produce large economic yields in each situation. Selecting early generations in an 'isolation environment' (well spaced plants) may be the most appropriate for harvest index. It seems prudent that, in seeking adaptation to environment in lentils, evaluate the proposals recently advocated for cereals, not only with respect to the data which are collected as a matter of routine in yield trials but also in more specific studies which, hitherto, have evaluated HI in 'competition' and/or 'crop environments' (Singh 1977, Solanki and Singh 2000).

The majority of cultivars released so far in lentil are selections from germplasm and not from hybridization programmes. Most of these cultivars are well adapted to only restricted regions, others are more broadly adaptable and some can exploit less favourable environments (Kumar *et al.* 2005, Solanki 2001). Large genotype × environment interactions have been recorded for most characters which contribute to variations in seed yield (Kumar *et al.* 2005, Solanki 2001, Yadav *et al.* 2002), and we can not hope to identify with confidence the main effects and interactions of climatic factors such as photoperiod and temperature on the more responsive components of yield, morphological or reproductive, until these relationships have been studied more carefully. Winter cold has a smaller effect on yield than rainfall, with no consistent overall effect, but with differences over regions (Erskine and El-Ashkar 1993).

Table 1. Associations of different traits with seed yield in lentil (compiled from Huyghe (1998))

Trait	Workers reporting	
	Positive association	Negative association
Days to flowering	–	Joshi *et al.* 2005, Muehlbauer 1974, Singh and Dixit 1970, Wilson 1977
Plant height (cm)	Kumar *et al.* 2002, Vir and Gupta 2002, Solanki *et al.* 2002	Muehlbauer 1974, Singh and Dixit 1970, Wilson 1977
Branches/plant	Kumar *et al.* 2002, Om Vir and Gupta 2002, Solanki 1999, 2006, Solanki *et al.* 2002, Yadav *et al.* 2005	–
No. of flowers/plant	Singh and Singh 1976, Wilson 1977	–
Clusters (Nodes)/ plant	Om Vir and Gupta 2002, Solanki 1999, 2006	–
Pods/cluster (node)	Kumar *et al.* 2002, Om Vir and Gupta 2002, Singh and Singh 1976, Wilson 1977	–
Pods/plant	Joshi *et al.* 2005, Kumar *et al.* 2002, Om Vir and Gupta 2002, Singh and Singh 1976, Solanki 1999, 2006, Solanki *et al.* 2002, Wilson 1977, Yadav *et al.* 2005	–
Seeds/pod	Joshi *et al.* 2005, Kumar *et al.* 2002, Om Vir and Gupta 2002, Singh and Singh 1976, Wilson 1977	Muehlbauer 1974, Singh and Dixit 1970, Wilson 1977
100-seed weight (g)	Om Vir and Gupta 2002, Solanki *et al.* 1992	–
Biological yield/plant (g)	Kumar *et al.* 2002, Om Vir and Gupta 2002, Solanki 1999, 2006, Yadav *et al.* 2005	–
Harvest index (%)	Kumar *et al.* 2002, Solanki 2006, Yadav *et al.* 2005	–

3.2. Effects of Water and Soil Factors

Whether the lentil crop responds favourably to irrigation or not depends largely on local edaphic conditions and rainfall patterns. Irrigation at the branching stage and again at pod formation increased yields in one season, but had no effect in subsequent season when there was well distributed rainfall (Sinha 1977). Irrigation may not be necessary in systems where lentils are grown after the harvest of paddy (Singh and Virmani 1974), and provision for adequate surface drainage may be advantageous in Mediterranean regions where rainfall can be expected during crop growth (Oweis *et al.* 2004). Appropriate plant populations in dry land farming systems should be maintained. While dense populations may intercept radiant energy more effectively than sparse stands, they may also deplete the soil profile of water more rapidly, and so eventually produce smaller economic yields.

Of course, this may not occur with short duration cultivars, which, by maturing rapidly, may avoid substantial stress (Silim *et al.* 1993). Yields are less responsive to row spacing than to sowing density. The narrowest row spacing (0.2 m), generally gives the greatest seed yields, which decrease linearly with increased row spacing (Silim *et al.* 1990).

Pulse crops are often grown on marginal lands under conditions of agronomic neglect (Jain 2005). Lentils are still grown in edaphic conditions which are not very different from those in their native habitats, i.e. dry, coarse-textured, stony hill side sites. Thus selection pressures have continued to be for adaptation to drought, poor fertility and competition with pests, pathogens and weeds. Lentil crops respond to better management, fertilizer and different nutrient applications. Siddique and Loss (1999) reported that there is no effect of sowing depth on crop phenology, nodulation or dry matter production in lentil over locations and cropping seasons. Sowing at depth (4–6 cm) may improve crop establishment where moisture from summer and autumn rainfall is stored in subsoil below 5 cm, by reducing damage from herbicides applied immediately before or after sowing, and by improving the survival of *Rhizobium* inoculated on the seed due to more favourable soil conditions at depth (Siddique and Loss 1999). Brandt (1999), while evaluating the management practices for black lentil green manure for the semi-arid Canadian prairies observed that if green manuring is practiced, early incorporation with lentil leaf strips is the most promising management system. However, even with improved water management practices, green manuring did not demonstrate a consistent advantage over summer fallow, which may be required to offset the added economic costs required to enact this practice.

Lentil productivity is influenced positively by the rate and method of phosphate placement (band placement) in different environments (Tawaha and Turk 2002). The effect of organic materials on soil physical environment and yield sustainability in rainfed rice-lentil sequence found the sole application of FYM as the best followed closely by the conjunctive use of FYM and inorganic fertilizer (Sarkar *et al.* 1995). Neither crop rotation nor tillage practice have a measurable impact on lentil diseases, but epidemics of *Ascochyta lentis* and *Botrytis cinerea* were most severe in treatments with the densest plant stands. The trends of tillage, rotation and environment over years demonstrated that regardless of tillage or crop rotation practices, the annual environment was the most important factor limiting the severity of disease and the prevalence of causal agents in the complex (Bailey *et al.* 2000). Chickpea, lentil and dry pea yielded 76%, 77% and 90%, respectively, of their fallow field yields when grown on stubble indicating that the pulse crops have excellent potential for intensifying cropping systems in the dry semi-arid prairie by replacing summer fallow in crop rotations. The tillage system had no consistent effect on plant densities, which were generally adequate (Miller *et al.* 2001).

The function of plant breeding is to produce varieties that exploit fully the potential of the environment. The breeder's objective should be to produce a variety with a large yield potential, not one with a large yield. To increase lentil yields from rainfed farmlands, involves objectives which include responsiveness to improved agronomic practices; adaptation to the edaphic and aerial environment and to water expectation and storage, nutritional and culinary qualities to meet consumer requirements and preferences. The opportunity to breed a new variety that will be well adapted throughout a wide geographical area occurs rarely, and when it does, it must be followed by diversification to provide varieties which can exploit local potential. Examples of very widely adapted varieties of lentil in several geographical areas with highly stable yields are available; L9-12, a *microsperma* variety, and Precoz, a *macrosperma* variety that possess earliness and high yield potential, is grown in a number of countries around the globe.

4. MULTILOCATION TESTING AND VARIETAL IDENTIFICATION

Multilocation testing is the evaluation of a genotype (introduction or developed through breeding) at a number of locations for its yield stability and adaptability. Multilocation testing of genotypes may be undertaken for one or more years depending upon the objective(s) of the breeding programme. The genotypes which have been observed superior in different station trials for seed yield, pest-disease resistance, lodging resistance, quality characters, etc. and only those with acceptable seed quality are evaluated further in the multilocation trials first in the state and then at the national/international level. The performance of the genotypes tested in multilocation trials in the state decides which genotypes are to be entered in the advanced varietal trials (AVTs). In these trials conducted for at least three years, the genotypes found superior to the existing cultivars (checks) in seed yield, or at par in seed yield but superior in some other trait, are identified for further testing at the farmers' fields. The differential performance of varieties in regional yield trials permits evaluation of the genotype × environment interactions.

With the establishment of the International Research Centres, like ICARDA, ICRISAT, IITA, AVRDC, etc., international yield trials are now conducted with different pulse crops (ICARDA for lentils). Through the international nurseries, it is possible to identify varieties with adaptation to a broad range of environmental conditions, as well as to identify those with adaptation to specific environments. The nurseries also serve to disseminate superior germplasm to breeders in a wide range of locations around the world.

After the identification of a variety 'on-farm trials', also known as 'field verification trials' or 'demonstration trials', are conducted at the farmers' fields. They are unreplicated, large-plot tests, conducted with the identified variety and the local land race or cultivar. These trials are conducted at a large number of locations. Selecting suitable locations for the trials covers a wide range of environments. The objectives of the trials are to grow the newly identified varieties under farmers' conditions to assess the yield and to obtain the farmers' views. New varieties are generally tested for two or three years. If superior in yield and acceptable to the farmers, a new variety will be considered for release. The cultivar is considered for release for commercial cultivation in an area, region, or country, if it is superior to the existing cultivar in yield, or if it has a good yield and is superior in quality or some other special characteristic such as height and lodging resistance, which have implications for mechanical harvesting. At the time of release, the cultivar is named as for instance, Sapna (LH84-8), and the results are published in popular journals (Brouwer 1995, Solanki *et al.* 1992).

4.1. Agroclimatic Zones for Lentil Cultivation in India

For pulse cultivation, India is sub-divided into five zones, i.e. north west plain zone (NWPZ), north east plain zone (NEPZ), north hill zone (NHZ), central zone (CZ) and southern zone (SZ). Lentil is grown in all, except the SZ where it is cultivated sporadically. Among different states, it is mainly cultivated in UP, MP, Bihar and West Bengal, which together contribute more than 80% area and production of this crop. Traditionally, the northern belt has been the area of small-seeded lentils. Several high-yielding varieties resistant to rust have been released for northern hills and plains. During the eighties, concerted efforts were made to develop the bold-seeded varieties and consequently a large number of bold-seeded varieties suitable for northern plains were developed and released, e.g. LH84-8, L4076, K75, DPL15, LH82-6, DPL62, etc. In central parts of India, bold seeded types are preferred. High yielding varieties with tolerance to wilt have been developed for this region e.g. JL3, IPL81, L4076, K75, etc. This crop has a great promise in rice fallows of the northern plains (Singh 2006).

5. FARMERS PARTICIPATION IN VARIETAL DEVELOPMENT

In lentil, to date no new variety has been developed/released through extensive farmer participation. However, for a crop like lentil that is widely grown in developing countries this approach offers many advantages in developing and disbursing better adapted varieties. Farmer participatory approaches for the identification or breeding of improved crop cultivars can be usefully categorized into participatory varietal selection (PVS) and participatory plant breeding (PPB). PVS is a more rapid and cost effective way of identifying farmer-preferred cultivars if a suitable choice of cultivars exists. If this is impossible, then the more resource-consuming PPB is required. PPB can use, as parents, cultivars that were identified in successful PVS programmes. Compared with conventional plant breeding, PPB is more likely to produce farmer acceptable products, particularly for marginal environments (Witcombe 1996).

In most developing countries, only a few farmers in marginal areas have adopted improved cultivars, often because they have not been exposed to acceptable alternatives to their landraces. Alternative approaches for identifying cultivars that are acceptable to resource-poor farmers have been suggested and tried by a number of workers. Chambers (Chambers 1989) reviewed the few examples of providing the farmers with 'a basket of choices' of varied genetic material none of which used lentil(Maurya *et al.* 1988, Sperling *et al.* 1993, Weltzein *et al.* 1996, Joshi and Witcombe 1996). All of these are examples of participatory varietal selection, since farmers were evaluating near-finished or finished products. In contrast, participatory plant breeding is the selection by farmers of genotypes from segregating generations. There are few examples of PPB in the literature (Sthapit *et al.* 1996). The methods of PVS and PPB have been outlined and their impact on biodiversity has been reviewed by Witcombe *et al.* (1996).

5.1. Participatory Varietal Selection (PVS)

Most cultivars grown by farmers in developing countries like India are old and only a few of the released cultivars are grown widely. One of the main reasons for low cultivar replacement rates is that farmers have inadequate exposure to new cultivars. If adoption rates are to be improved, farmers need to try a wide range of novel cultivars in their fields through their involvement in PVS programmes. The cultivars should be selected not only from the target region but also from other regions or countries. A successful participatory varietal selection programme (PVS) has four phases:

(a) Identification of farmers' requirements

Farmers' requirements have to be identified first so that they can be given more appropriate genetic material to test. This can be done by using several methods, either separately or in combination. The methods include participatory rural appraisal, the examination of farmers' crops around harvest time, and the pre-selection, by farmers, of varieties from trials of many entries, grown either on a research station or on a farm. In areas where there is a diversity of landraces in farmers' fields and where resources allow, the local germpalsm can be collected and grown in a trial, on station or on farm, with recommended cultivars as the control.

(b) Search for suitable released material and advanced lines

A search is made for cultivars that most closely meet the important identified characteristics, particularly those relating to maturity, plant height, agro-ecological niche, and grain quality. Cultivars are selected from new and old releases at any level (national, zonal), and from pre-release material at an advanced stage of testing.

(c) Experimentation on farmers' fields

Various testing and evaluation systems can be employed (e.g. Joshi 1996, Deshmukh 2005) and they vary greatly in the extent of farmer participation (Table 2). One of the simplest methods of PVS used was in rice, where farmers were offered small quantities of seed from various varieties (released, pre-released and advanced lines) to grow under their own conditions without intervention from researchers.

Table 2. Methods of varietal selection with varying degrees of farmer participation [after Witcombe *et al.* (1996)]

Sr. No.	Methods in increasing order of farmer participation	Evaluation includes
1.	Researcher-managed and evaluated on-station trials; farmers may visit station to identify farmer-acceptable material	Yield data; possibly farmer evaluation
2.	Researcher-managed on-farm trials, replicated design; farmers may be involved in evaluation	Yield data; possibly farmer evaluation
3.	Farmer-managed, replicated design, on-farm trials, with scientists' supervision; several entries per farmer	Yield data; farmers' perceptions
4.	Farmer-managed unreplicated design, on-farm trials; one cultivar per farmer; replication across farmers	Yield data; farmers' perceptions
5.	Trials as in 4	Farmers' perceptions only
6.	Farmer-managed trials; no formal design either within a farm or across farmers	Informal, anecdotal

(d) Wider dissemination of farmer-preferred cultivars

PVS is usually conducted with farmers situated in small geographical area. However, to be of value results must be extended to a larger area. This is easier with existing released varieties. For participatory approaches to be more cost-effective, data on farmer perceptions and demand for seed need to be considered by varietal release committees, rather than the almost total reliance presently placed on yield data from scientist managed yield trials.

5.2. Participatory Plant Breeding (PPB)

PPB, in which farmers select from segregating material, is a logical extension of the PVS. However, PPB is more resource-consuming. PPB needs to be used when the possibilities of PVS have been exhausted, or when the search process fails to identify any suitable cultivars for testing. PPB can exploit the results of PVS by using identified cultivars as parents of crosses. PPB methods are poorly documented, and there are only a few examples in the literature and none for lentils (Worede and Mekbib 1993, Sthapit *et al.* 1996).

For predominantly self-pollinated crops such as lentils, there are many PPB methods that have different degrees of farmers' participation (Table 3). In all methods, plant breeders are facilitators of the research, have an essential role in disseminating the results and managing varietal release and linking the PPB program to other breeding programs.

6. ADAPTATION OF PLANT TYPE

Plant type is the morphological appearance of a plant and is an important characteristic for the identification and characterization of cultivars (Singh and Singh 1994). The ideal plant type is that combination of structure and developmental traits of

Table 3. Methods of plant breeding in predominantly self-pollinating crop with varying degrees of farmer participation [after Witcombe *et al.* (1996)]

Sr. No.	Methods in increasing order of farmer participation	Site specificity
1.	All generations grown by plant breeders on station; farmers involved at pre-release stage or even after release	Wide adaptation targeted; early generations may all be in single location followed by multi locational testing
2.	Early generation (F_2 or F_3) in farmers' fields; all other generations and procedures with plant breeder on station (Thakur 1995)	Single location testing site for F_2 or F_3
3.	Best advanced lines at F_7 or F_8 given to farmers for testing; closest method to participatory varietal selection since farmers given nearly finished product (Galt 1989)	Easy to test best advanced lines across locations
4.	From F_3 or F_4 onwards farmers and plant breeders collaborate to select and identity the best material on farm (and also on station); farmers select; plant breeders facilitate the process; release proposal prepared by plant breeders (Sthapit *et al.* 1996)	Possible to run selection procedures on early generations in more than on location
5.	Breeder gives F_3 and F_4 material to farmers: all selection left to farmers; at F_7 to F_8 or later, breeders monitor diversity in farmers' fields and identify best material to enter in conventional trials (Salazar 1992)	Extremely easy to run selection schemes in many locations
6.	Trained expert farmers make crosses and do all selection with or without assistance from breeders; breeders can place best material in conventional trials	Specific to farmers' requirements

the genotype which are best adapted to a particular environment and suit to the cultivator's need better than other genotypes of that particular crop (Lal 2001). Traits that determine the structure of the plant are mainly stem/plant height and branch numbers and sizes. All of the traits are highly influenced by the environment and contain subcomponents (e.g. internode length) that vary as the plant grows. Hence assigning an absolute value for a specific cultivar may be of limited value. Branching patterns differ from erect compact, having a narrow branch angle, to prostrate or spreading types with several intermediate types. The lentil plant may have few or many primary branches, which arise directly from the main stem, and often many secondary branches, which arise from the primary branches. The production of branches is highly affected by plant population and at high population levels, branching is greatly suppressed. Therefore, the combination of plant height, branching and environment may result into a number of plant types like, tall, dwarf, bushy, erect, semi-erect, semi-spreading, spreading, etc. All these plant types are cultivar specific, but genotype × environment interaction plays an important role. If a cultivar is erect growing, it may behave as semi-erect if space planted.

Farmers, with large acreages in lentil cultivation, desire mechanization of cultural operations. While machines developed for cereals have been modified and used, they have not been entirely satisfactory. One reason for lack of satisfactory mechanization is poor plant type. Tall and erect types are the ideal plant type in lentil for mechanized harvesting. However, spreading/ semi-spreading plant types are very useful in rainfed areas to avoid evaporational losses from the soil, and thus, are helpful in conserving precious soil moisture.

Donald and Hamblin (1976) introduced the concept of plant ideotypes . The most significant point in the ideotype concept is the performance of a plant in isolation and in population. In many species, particularly cereals or determinate crops where the yield of the main shoot is a major contributing factor, the relationship between the individual plant yield and yield of a population, remains reasonably valid. However, this does not necessarily happen in indeterminate species such as lentil. Lentil tends to become more vegetative if the unstressed crop is irrigated at the flowering stage and ambient temperatures are not high as in north India. Also the partitioning of assimilates and particularly of nitrogen, is rather poor. All of these factors should be considered when suggesting an ideotype for lentil.

Traditionally, lentil has been grown mostly as a rainfed crop all over the world. Therefore, the centres for selection have naturally, and rightly so, attempted to select for rainfed conditions. For such a situation, a small plant population density of spreading/semi-spreading plant type is desirable to enable plants to reach maturity before the soil moisture is depleted, with a provision that the plant has the capacity to enlarge itself if more water becomes available. Consequently, the character of enlargement in growth in response to water availability at the time of the reproductive phase could prove detrimental to seed yield. Secondly, in such plant types, most of the yield is derived from secondary and tertiary branches. When such plant types are grown in a high population density, they fail to produce their reproductive branches and hence yield remains static. This means that selection for poor water regimes and for assured water availability would require different approaches. Possibly, for poor water availability, a low plant population with a capacity to respond if available water increased due to occasional rain would be a correct objective. However, selection for assured water availability would require a different approach.

Selection for adequate water availability will have to be done under conditions where soil moisture does not become limiting at a critical stage of crop growth. In most cases while lentil plants are being selected for assured moisture supply, the crop will have to be given irrigation at flowering. Those plants, which do not resume vigorous vegetative growth, will be the right plants for irrigated conditions.

There appears to be not much information available on the relationship between plant type (morphological features) and nitrogen harvest index in lentil, although such information is necessary for economic yield. However, there are features, which may have an important bearing on this aspect. First, the sequential senescence and shedding of leaves without necessarily contributing nitrogen to developing pods and seeds, is a limitation. Second, when the pods develop, they mostly derive assimilates from the leaves in the axils of which they are borne. Consequently, the leaves of non-flowering nodes do not substantially contribute their nitrogen to pods and seeds. Possibly, a slow senescence of the whole plant during the development of pods might be an indicator of uniform mobilization from different parts of the plant.

It is usually recognized that the harvest index (HI) in lentil is low. A relatively erect plant type with shorter internodes might be a suitable ideotype, particularly for irrigated agro-ecosystems under good management conditions. Breeding for erectness combined with tall growth by hybridization between conventional spreading and tall types with sturdy stem and resistance to lodging has given good results. Solanki *et al.* (1992) constructed a model plant type with the help of correlation and path coefficient analysis and the one emerging from a comparison of plant morphology of high and low yielding lines as well as the seed yield pattern of lines with high and low levels of expression of seed yield components, and all three agree to a large extent (Table 4). The architecture of this model plant was tall height with a higher number of secondary branches and pods per plant and high seed weight. Similarly, Om Vir *et al.* (2002) postulated a plant ideotype bearing higher biomass, harvest index, 100-seed weight, pods per plant, fertile nodes per plant, pods per node, seeds per pod and plant height which could enhance the seed yield of lentil. The *macrosperma* cultivar, Matilda, well adapted to the Wimmera

Table 4. Identification of model plant characteristics based on three parameters [after Solanki *et al.* (1992)]

Character	Comparison of highest and lowest yielding lines (an average of 20 lines in each case)			Pattern of seed yield (g per plant) in lines with high and low levels of expression of the quantitative traits			Correlation coefficients with yield per plant
	Highest	Lowest	Difference	Highest	Lowest	Difference	
Yield per plant (g)	2.993	1.200	1.793**	–	–	–	–
Leaf area (cm^2)	2.21	1.89	0.32	2.126	1.756	0.370	0.192
Plant height (cm)	36.40	33.00	3.40*	2.374	1.528	0.846**	0.387**
Number of primary branches	4.80	4.88	-0.08	2.169	2.239	−0.070	−0.078
Number of secondary branches	10.42	7.36	2.66*	2.420	1.596	0.824*	0.404**
Clusters per plant	34.6	25.48	9.12	2.083	1.616	0.467	0.160
Pods per plant	54.92	34.28	20.64*	2.448	1.830	0.618	0.350*
Seeds per pod	1.64	1.72	−0.08	2.049	1.759	0.290	−0.070
100-seed weight (g)	2.80	1.94	0.86**	2.438	1.661	0.777*	0.339*

*and**: Significant at 5 and 1 per cent level of probability, respectively.

region of Victoria, Australia has semi erect growth habit and shorter plant type, especially when rains occur late in the growing season (Brouwer 1995).

Huyghe (1998) reviewed genetics and genetic modifications of plant architecture in grain legumes. Crop architecture may be modified to improve adaptation of crops to different environments and to increase seed yield and its stability. The main peculiarities of grain legume architecture are: (1) the indeterminate growth habit, which may lead to a prolonged growth cycle with consequences on maturation; and (2) strong within plant competition between reproductive and vegetative growth. Flowering date is of major importance for adaptation of a crop to environmental conditions. The branching pattern may be directly affected independently of other architectural modifications. Leaf size and structure contribute to the leaf area index of the crop and may influence the light interception efficiency. The determinate growth habit modifies the duration of the growth cycle and the assimilate partitioning, while the dwarfism may improve the adaptation to a range of environments through a reduction of the risk of lodging. The pod walls may contribute to pod photosynthesis but they account for a large proportion of the pod weight at harvest. This reduces the crop harvest index.

Many lentil breeders consider what might be the ideal plant type for their target area, and often construct a crop ideotype. Such a thought process can focus attention on critical problems. For instance, in areas where greater plant height and erectness are needed for mechanized harvest, particularly in irrigated agro-ecosystems under good management, those traits would receive greater attention in the choice of parents for hybridization and subsequent selection.

However, in moisture-limited (rainfed) agro-ecosystems, breeding and selection might emphasize rapid emergence in relatively dry seedbeds; a spreading growth habit to rapidly cover the soil surface, reduce evaporation from the soil surface and compete with weeds; along with early maturity to escape hot, dry conditions that usually prevail late in the growing season. Ideotypes for irrigated agro-ecosystems under good management might be late maturing to take advantage of the long growing season made possible by the supplemental water.

It would be difficult to visualize a single plant type able to maximize seed yield over the diverse agro-ecological zones (rainfed and irrigated) in which lentil is grown. Nevertheless, for specific regions and breeding programmes, the formulation of ideotypes may serve a useful purpose. Most schemes of ideotype focus on tall plant habit and increased branch and pod number with more seeds per pod. A physiologically possible approach in lentil might be to develop compact plant types capable of greater biological yields per unit area, then focus on improving their harvest index. Greater compactness of lentil plants would allow establishment of larger plant populations and make it possible to intercept a larger portion of solar radiation, compete with weeds, use more of the available moisture, and increase seed yield per unit area.

The biological yield might be improved by developing tall, erect, compact plant types to permit increased plant populations and therefore, increase biological yield per unit area. Other possibilities might be increased branch number and longer duration of flowering and seed-filling. A more favourable partitioning of the biological yield might be possible through increased pod number, seeds per pod, or seed size.

Based on the above discussion it is possible to formulate, different plant types of lentil for diverse agro-ecosystems (Lal 2001).

6.1. Plant Type for Dryland/Rainfed Environments

Deep root system with high root volume; semi-spreading to spreading growth habit; good water retention capacity and restricted transpiration; pubescent foliage; profuse branching; reduced biomass; high harvest index; short growth duration.

6.2. Plant Type for Optimum Moisture Irrigated Conditions

Erect to semi-erect growth habit with short internodes; compact plant type with restricted branching; high biomass production capacity; relatively longer growth duration.

6.3. Plant Type for Sole Cropping

Spreading to semi-spreading growth habit for rainfed areas and erect with shorter internodes for irrigated and assured rainfall areas; high biomass production capacity.

6.4. Plant type for Inter-Cropping

Erect and compact growth habit with shorter internodes; good low light photosynthesis.

6.5. Plant type for Multiple/Relay Cropping

Quick germination; early vigour; early flowering with longer reproductive phase; early and synchronous maturity; quick senescence; high responsiveness to inputs; resistance to diseases and insect pests (Table 5).

Table 5. Suitable varieties of lentil for various inter-cropping systems

Intercropping	Suitable varieties
a. Mustard + lentil	Bihar [Ranjan, PL406, PL639, PL209, BR25] Madhya Pradesh [K75, JLS1, PL406, PL639]
b. Barley + lentil	Madhya Pradesh [K75, JLS1, PL406, PL639]
c. Linseed + lentil	Madhya Pradesh [K75, JLS1, PL406, PL639]

REFERENCES

Bailey, K.L., Gossen, B.D., Derksen, D.A. and Watson, P.R. 2000. Impact of agronomic practices and environment on diseases of wheat and lentil in south-eastern Saskatchewan. *Canadian Journal of Plant Science*, 80(4): 917–927.

Brandt, S.A. 1999. Management practices for black lentil green manure for the semi-arid Canadian prairies. *Canadian Journal of Plant Science*, 79(1): 11–17.

Brouwer, J.B. 1995. *Lens culinaris* (lentil) cv. Matilda. *Australian Journal of Experimental Agriculture,* 35(1): 117.

Bunting, A.H. 1975. Time, phenology and yield of crops. *Weather*, 30: 312–325.

Chambers, R. 1989. Institutions and practical change. Reversals, institutions and change. *In*: Farmer First (Eds. R. Chambers, A. Parcy and I.A. Thrupp). London: Intermediate Technology Publications: 181–195.

Deshmukh, R.B. 2005. Advances in major pulse crops research - Success stories. Souvenir: 4th International Food Legumes Research Conference, Oct. 18–22, New Delhi, India: 25–26.

Donald, C.M. and Hamblin, J. 1976. The biological yield and harvest index of cereals as agronomic and plant breeding criteria. *Advances in Agronomy*, 28: 361–405.

Erskine, W. and El-Ashkar, F. 1993. Rainfall and temperature effects on lentil (*Lens culinaris*) seed yield in Mediterranean environments. *Journal of Agricultural Science,* 121(3): 347–354.

Galt, D.L. 1989. Joining FSR to commodity programme breeding efforts earlier: increasing plant breeding efficiency in Nepal. Agricultural Administration (Research and Extension) Network: Network, Paper 8. London: Overseas Development Institute.

Hawtin, G.C., Singh, K.B. and Saxena, M.C. 1979. Some recent advances in the understanding and improvement of *Cicer* and *Lens*. *In*: Summerfield, R.J. and Bunting, A.H. (Eds.). Advances in Legume Science, London; HMSO.

Huyghe, C. 1998. Genetics and genetic modifications of plant architecture in grain legumes: a review. *Agronomie*, 18 (5–6): 383–411.

Jain, H.K. 2005. Evaluation of humans, grain legumes and cultures. 4th International Food Legumes Research Conference. Oct. 18–22, New Delhi, India: 3–6.

Joshi, A. and Witcombe, J.R. 1996. Farmer participatory crop improvement. II. Participatory varietal selection, a case study in India. *Experimental Agriculture*, 32: 461–477.

Joshi, K.D., Rana, R.B., Subedi, M., Kadayat, K.B. and Sthapit, B.R. 1996. Addressing diversity through farmer participatory variety testing and dissemination approach: A case study of *Chaite* rice in the western hills of Nepal. *In*: Using Diversity, Enhancing and Maintaining Genetic Resources on Farm. Proceedings of a workshop held on June 19–21, New Delhi, India. (Eds. L. Sperling and M.L. Loevinsohn), IDRC, New Delhi, India: 158–178.

Joshi, N., Singh, S. and Singh, I. 2005. Variability and association studies in lentil. *Indian Journal of Pulses Research,* 18(2): 144–146.

Kumar, R., Sharma, S.K., Malik, B.P.S., Dahiya, A. and Sharma, A. 2002. Correlation studies in lentil (*Lens culinaris* Medik.). *Annals of Biology*, 18(2): 121–123.

Kumar, R., Sharma, S.K., Luthra, O.P. and Sharma, S. 2005. Phenotypic stability of lentil genotypes under different environments. *Annals of Biology,* 21(2): 155–158.

Lal, S. 2001. Plant type concept in pulses. Souvenir: National Symposium on Pulses for Sustainable Agriculture and Nutritional Security. April 17–19, New Delhi, India: 56–62.

Leport, L., Turner, N.C., French, R.J., Thomson, B.D. and Siddique, K.H.M. 2003. Physiological response of cool-season grain legumes to drought in the low rainfall mediterranean environment of South-Western Australia. Management of agricultural-drought: agronomic and genetic options: 163–172.

Malhotra, R.S., Singh, K.B., Bhullar, G.S. and Sethi, S.C. 1971. Phenotypic stability in lentil. *Indian Journal of Genetics and Plant Breeding,* 31: 21–25.

Maurya, D.M., Bottrall, A. and Farrington, J. 1988. Improved livelihoods, genetic diversity and farmers' participation: a strategy for rice-breeding in rainfed areas of India. *Experimental Agriculture*, 24: 311–320.

Miller, P.R., McDonald, C.L., Derksen, D.A. and Waddington, J. 2001. The adaptation of seven broadleaf crops to the dry semiarid prairie. *Canadian Journal of Plant Science,* 81(1): 29–43.

Monteith, J.L. 1977. Climate. *In*: Alvin, P. de T. (Ed.) Ecophysiology of tropical crops; Academic Press, New York, pp. 1–27.

Muchlbauer, F.J. 1974. Seed yield components in lentil. *Crop Science,* 14: 403–406.

Muchlbauer, F.J. 1979. Selection procedures. International Seminar on Lentils. Aleppo, ICARDA. (mimeo).

Om Vir and Gupta, V.P. 2002. Analysis of relationships of yield factors in *macrosperma* × *microsperma* derivatives of lentil. *Legume Research* 25(1): 15–20.

Oweis, T., Ahmed, H. and Mustafa, P. 2004. Lentil production under supplemental irrigation in a Mediterranean environment. *Agricultural Water Management,* 68 (3): 251–265.

Salazar, R. 1992. MASIPAG: alternative community rice breeding in the Philippines. *Appropriate Technology*, 18: 20–21.

Sarkar, S., Singh, S.R., Dasgupta, M.K. (ed.), Ghosh, D.C. (ed.), Das-Gupta, D. (ed), Majumdar, D.K. (ed.), Chattopadhyay, G.N. (ed.), Ganguli, P.K. (ed.), Munsi, P.S. (ed.) and Bhattacharya, D. 1995. role of organic materials on soil physical environment and yield sustainability in rainfed rice-lentil sequence in eastern Uttar Pradesh. Proceedings of National Symposium on Sustainable Agriculture in Sub-humid Zone. March 3–5, India: 58–63.

Sarker, A., Erskine, W. and Singh, M. 2005. Variation in shoot and root characteristics and their association with drought tolerance in lentil landraces. *Genetic Resources and Crop Evolution,* 52(1): 87–95.

Saxena, M.C. 1979. Lentil plant ideotype. International Seminar on Lentils. Aleppo, ICARDA. (mimeo).

Saxena, M.C. and Singh, H.P. 1977. Research on winter pulses. Experimental Station Technical Bulletin. No. 101: 23–42, Pantnagar, India.

Sheldrake, A.R. and Saxena, N.P. 1979. Comparison of earlier- and later-formed pods of chickpea (*Cicer arietinum* L.). *Annals of Botany,* 43: 467–473.

Siddique, K.H.M. and Loss, S.P. 1999. Studies on sowing depth for chickpea (*Cicer arietinum* L.), faba bean (*Vicia faba* L.) and lentil (*Lens culinaris* Medik.) in a Mediterranean-type environment of south western Australia. *Journal of Agronomy and Crop Science,* 182(2): 105–112.

Siddique, K.H.M., Loss, S.P., Pritchard, D.L., Regan, K.L., Tennant, D., Jettner, R.L. and Wilkinson, D. 1998. Adaptation of lentil (*Lens culinaris* Medik.) to Mediterranean-type environments: effect of time of sowing on growth, yield and water use. *Australian Journal of Agricultural Research*, 49(4): 613–626.

Siddique, K.H.M., Regan, K.L., Tennant, D. and Thomson, B.D. 2001. Water use and water use efficiency of cool season grain legumes in low rainfall Mediterranean-type environments. *European Journal of Agronomy,* 15(4): 267–280.

Silim, S.N., Saxena, M.C. and Erskine, W. 1990. Seedling density and row spacing for lentil in rainfed Mediterranean environments. *Agronomy Journal,* 82 (5): 927–930.

Silim, S.N., Saxena, M.C. and Erskine, W. 1993. Adaptation of lentil to the Mediterranean environment. I. Factors affecting yield under drought conditions. *Experimental Agriculture*, 29(1): 9–19.

Singh, T.P. 1977. Harvest index in lentil (*Lens culinaris* Medik.). *Euphytica*, 26: 833–839.

Singh, B.B. 2006. Project Coordinator's Report. All India Coordinated Research Project on MULLaRP. *Annual Group Meet* (*Rabi*, 2005–06), Sept. 12–14, Kolkata, India: 1.

Singh, B. and Dixit, R.K. 1970. Genetic variability in some quantitative characters in lentil (*Lens esculenta* Moench). *Madras Agricultural Journal*, 57: 227–230.

Singh, V. and Singh, P. 1976. Path analysis for yield components in lentil. *Lens,* 3: 6–7.

Singh, J.P. and Singh, I.S. 1994. Evaluation of lentil germplasm for plant type, initial flowering and disease resistance. *Lens Newsleter*, 21 (1): 5–7.

Singh, K.B. and Virmani, S.S. 1974. Pulses. *Punjab Agricultural University, Ludhiana*, India.

Sinha, S.K. 1977. Food legumes: distribution, adaptability and biology of yield. *FAO Plant Production and Protection Paper No. 3*. pp. 124.

Slinkard, A.E. 1979. Tolerance of lentils to environmental stress. International Seminar on Lentils. Aleppo, ICARDA. (mimeo).

Solanki, I.S. 1999. Association analysis for grain yield and other quantitative traits in lentil. *Indian Journal of Pulses Research,* 12(2): 243–246.

Solanki, I.S. 2001. Stability of seed yield and its component characters in lentil (*Lens culinaris*). *Indian Journal of Agricultural Sciences,* 71 (6): 414–416.

Solanki, I.S. 2006. Comparison of correlations and path coefficients under different environments in lentil (*Lens culinaris* Medik.). *Crop Improvement,* 33(1): 70–73.

Solanki, I.S. and Singh, V.P. 2000. Genetic variability for yield and its components as influenced by planting dates in lentil. *Indian Journal of Pulses Research,* 13 (1): 52–53.

Solanki, I.S., Singh, V.P. and Waldia, R.S. 1992. Model plant type in lentil (*Lens culinaris* Medik.). *Legume Research,* 15(1): 1–6.

Solanki, I.S., Waldia, R.S., Singh, V.P., Malik, B.P.S. and Kakkar, P.S. 1992. Sapna (LH84-8): a lentil cultivar for the North West Plain Zone of India. *Lens Newletter,* 19(2): 12–13.

Sperling, L., Loevinsohn, M.E. and Ntabomvra, B. 1993. Rethinking the farmers role in plant breeding: local bean experts and on-station selectin in Rwanda. *Experimental Agriculture,* 29: 509–519.

Steponkus, P.L. 1978. Cold hardiness and freezing injury of agronomic crops. *Advances in Agronomy,* 30: 51–98.

Sthapit, B.R., Joshi, K.D. and Witcombe, J.R. 1996. Farmer participatory crop improvement. III. Participatory plant breeding: a case study of rice in Nepal. *Experimental Agriculture,* 32: 487–504.

Summerfield, R.J. 1981. Adaptation to Environments. *In*: Webb, C., Hawtin, G. (Eds.) Lentils. Commonwealth Agricultural Bureaux London: 91–110.

Tawaha, A.M. and Turk, M.A. 2002. Lentil (*Lens culinaris* Medic.) productivity as influenced by rate and method of phosphate placement in a Mediterranean environment. *Acta Agronomica Hungarica,* 50 (2): 197–201.

Thakur, R. 1995. Prioritization and development of breeding strategies for rainfed lowlands: a critical appraisal. *In*: Proceedings of the IRRI Conference. Fragile lives in Fragile Eco-systems. Los Banos, Philippines, IRRI: 817–824.

Weltzien, R.E., Whitaker, M.L. and Dhamotharan, M. 1996. Diagnostic methods for breeding pearl millet with farmers in Rajasthan. *In*: Enhancing and Maintaining Genetic Resources on Farm. Proceedings of a Workshop, June 19–21, 1995, New Delhi, India: 127–139. (Eds. Sperling, L. and Loevinsohn, M.L.). International Development Research Centre, New Delhi, India.

Wilson, V.E. 1977. Components of yield and seed characteristics in lentil. *Horticultural Science,* 12: 555–556.

Wilson, V.E. and Hudson, L.W. 1978. Registration of WH-2040 lentil germplasm. *Crop Science,* 18: 1097.

Witcombe, J.R., Joshi, A., Joshi, K.D. and Sthapit, B.R. 1996. Farmer participatory crop improvement. I. Varietal selection and breeding methods and their impact on biodiversity. *Experimental Agriculture,* 32: 445–460.

Worede, M. and Mekbib, H. 1993. Linking genetic resource conservation to farmers in Ethiopia. *In*: Cultivating knowledge, Genetic Diversity, Farmer Experimentation and Crop Research. (Eds. de Boef, W., Amanor, K. and Wellard, K.). Intermediate Technology Publications, London: 78–84.

Yadav, S.S., Phogat, D.S., Solanki, I.S. and Malik, B.P.S. 2002. Impact of different environments on genetic variation in lentil. *Indian Journal of Pulses Research,* 15 (2): 181–182.

Yadav, S.S., Phogat, D.S., Solanki, I.S. and Tomer, Y.S. 2005. Character association and path coefficient analysis under two environments in lentil. *Indian Journal of Pulses Research,* 18 (2): 147–149.

CHAPTER 17

LENSOMICS: ADVANCES IN GENOMICS AND MOLECULAR TECHNIQUES FOR LENTIL BREEDING AND MANAGEMENT

REBECCA FORD[1], BARKAT MUSTAFA[1], PRAHBPREET INDER[1], RUBEENA SHAIKH[2], MICHAEL MATERNE[3], AND PAUL TAYLOR[1]

[1] *BioMarka, Faculty of Land and Food Resources, The University of Melbourne, Victoria 3010, Australia*

[2] *Department of Crop and Soil Sciences, Washington State University, 291 Johnson Hall, P.O. Box 646420, Pullman, WA 99164, USA*

[3] *Grains Innovation Park, The Department of Primary Industries, Private Bag 260, Horsham, Victoria 3401, Australia*

Email: rebeccaf@unimelb.edu.au

Abstract: Lentil is a self-pollinating diploid (2n = 14 chromosomes) annual cool season grain legume produced as a high protein food source throughout the world. Several lentil genome maps are available and recent progress towards a consensus map has been made by employing robust locus markers that are derived from the model legume *Medicago truncatula* and other legume genomes. Such markers are co-dominant and will likely be useful across a broad lentil genetic background for marker-assisted trait selection. Candidate trait-associated genes are under investigation, particularly for disease resistance, and these are soon likely to become available for validation against pathogen populations and in differing environments using transgenic approaches. For this, reliable transformation systems have been developed. However, further effort is required to develop a robust and high-throughput full regeneration system for transformant lentil plants. The near future of Lensomics will include further candidate gene characterisation through transcriptome and reverse genetic techniques. These studies will be conducted to uncover genes responsive to biotic and abiotic stimuli as well as those governing desirable seed quality traits, such as size, shape and colour. Furthermore, proteomic and metabolomic approaches will be employed to derive information on the functional mechanisms involved

S.S. Yadav et al. (eds.), Lentil: An Ancient Crop for Modern Times, 275–289.

1. INTRODUCTION

Since lentil is historically grown in areas of the world where funding for genetic research is scarce there is paucity in the development and implementation of molecular techniques into lentil breeding in comparison to cereal and other crop species. Indeed, many of the sought after traits are simply inherited and maybe selected through visual phenotyping more cost effectively than through molecular analysis due to the initial complexities and cost of implementing molecular markers in a breeding program.

Molecular tools have rather recently been employed by several research teams for assistance with breeding by understanding the genetic basis of many traits and for selection against major production constraints such as susceptibility to important foliar fungal disease. Consequently, much of this chapter will focus on the development of molecular markers and the identification of gene sequences associated with resistance to fungal foliar pathogens, as well as the development of advanced technologies such as genetic transformation and transcript profiling. These are techniques that are still somewhat in their infancy in lentil, when compared to the less genetically orphaned crop species. However, marker technologies, phenotyping capabilities and the development of mapping populations have progressed to a stage where rapid and extensive uptake of molecular genotyping should occur in lentil breeding within the next 5 to 10 years.

The implementation of markers and genetically transformed materials into lentil breeding programs must be cost efficient and only employed for the accurate and fast selection/introduction of otherwise difficult to select/absent traits. Markers associated with, or transgenic plants carrying functional genes that code for the genetic mechanisms governing abiotic stresses such as drought, frost, cold, boron, salinity, herbicide tolerance as well as biotic constraints such as ascochya blight, botrytis grey mould, anthracnose, rust, fusarium wilt, stemphylium blight, helicoverpa and bruchids would greatly benefit the global lentil economy. Subsequent to the implementation of the first 'high value' markers, it will become economically attractive to use a larger number of markers that cover a wide range of traits.

2. GENOTYPING AND MAPPING

2.1. Map Progress

In order to identify regions of the genome associated with traits of interest and to subsequently select for those regions and potentially identify the candidate genes responsible, a detailed genome linkage map is sought. The initial *Lens* genetic linkage maps were constructed using morphological and isozyme markers (Zamir and Ladizinsky 1984; Tadmor et al. 1987). The first map comprising DNA-based markers was produced by Havey and Meuhlbauer (1989). Subsequent maps created with either intersubspecific or interspecific crosses were those of Weeden et al. (1992), Tahir et al. (1993), Tahir and Muehlbauer (1994) and Vaillancourt and Slinkard (1993). With the advent of PCR based markers, the number of mapped markers across the *Lens* genome increased dramatically. The first extensive map comprised 177 random amplified polymorphic DNA (RAPD), amplified fragment length polymorphism (AFLP), restriction fragment length polymorphism (RFLP) and morphological markers was constructed using a RIL population from a cross between a cultivated *L. culinaris* ssp. *culinaris* cultivar and a *L. culinaris* ssp. *orientalis* accession (Eujayl et al., 1998). The major reason for using distantly related parents was due to the limited polymorphism detected within the cultivated gene pool (Ford et al. 1997).

The first intraspecific lentil map was reported by Rubeena et al. (2003) and comprised 114 RAPD, inter simple sequence repeat (ISSR) and resistance gene analogue (RGA) markers. A subsequent intraspecific linkage map was constructed using a F_2 population from a cross between ILL7537 (ascochyta blight-resistant) and ILL6002 (ascochyta blight-susceptible). The map comprised 72 markers (38 RAPD, 30 AFLP, 3 ISSR and one morphological), and spanned a total length of 412.5 cM at a LOD score of 4.0 and a maximum recombination fraction (θ) of 0.25 (Rubeena et al. 2006). Two quantitative trait loci governing resistance to the fungal pathogen *Ascochyta lentis* were identified on linkage groups I and II, respectively, at which dominant and partial dominant gene action was observed. These QTL may represent the effects of the two major dominant genes previously reported to be responsible for *A. lentis* resistance in ILL7537 (Nguyen et al. 2001), however, the underlying candidate resistance genes are yet to be isolated from these QTL regions.

PCR based markers that are inherited in a co-dominant fashion, as well as markers that originate from known gene sequence, have enabled the very recent development of transferable and function-associated lentil genome maps. Such maps are not only applicable across multiple genetic backgrounds (genotypes) but also enable the direct association of specific genome regions with predicted gene function. Short sequence repeat (SSR; microsatellite) markers are particularly useful because they are unilocus and multi-allelic, being produced from amplification of the repeat region between flanking primers. SSR markers may be sourced from within known gene sequences, making them useful for future function association.

SSR markers have previously been used for genotyping in soybean (Rongwen et al. 1995), field pea (Ford et al. 2002) and chickpea (Winter et al. 1999). Several suites of SSR markers have been developed from the genomic sequence of the ubiquitous accession ILL5588 (Australian cultivar Northfield; Zavodna et al. 2000; Hamwieh et al 2005; P. Inder, unpublished). For the construction of the SSR marker sequence libraries, genomic DNA was bonded onto a nylon membrane and hybridized with radiolabeled oligonucleotide repeats; namely $(GT)^{10}$, $(GA)^{10}$, $(GC)^{10}$, $(GAA)^{8}$, $(TA)^{10}$ and $(TAA)^{5}$. Recently, several of these markers were successfully transferred across the genetic backgrounds of elite Australian cultivars (Table 1). This database will play an important role in seed typing for quality assurance in the domestic breeding program, cultivar integrity in commercial production and commercial export certification.

Table 1. A rudimentary genotype database constructed for 10 elite Australian lentil accessions using SSR amplification profiles and fluorescent capillary electrophoresis

	SSR 107		SSR 204		SSR 48		SSR 80	
	Allele 1	Allele 2	Allele 1	Allele 2	Allele 1	Allele 2	Allele 1	Allele 2
Northfield	169	169	194	194	174	174	163	163
Indianhead	131	131	202	202	193	193	151	151
Digger	131	131	202	202	191	191	163	163
Boomer	133	133	194	194	191	191	147	147
ILL7537	131	131	196	197	191	199	165	165
ILL6788	127	127	194	194	174	174	143	143
Palouse	188	188	200	200	199	199	143	143
ILL2024	133	133	196	196	203	203	143	143
Nipper	169	169	194	194	174	174	163	163
Nugget	131	131	196	196	191	191	163	163

The first *Lens* sp. map to include a SSR marker was that of Durán et al. (2004). Recently, Hamwieh et al (2005) added 39 SSR and 50 AFLP markers to the map constructed by Eujayl et al. (1998) to produce a comprehensive *Lens* map comprising 283 genetic markers and covering 715 cM. Subsequently, the first lentil map, that contained 18 SSR markers as well as 79 intron-targeted amplified polymorphic (ITAP) gene-based markers, was constructed using a F_5 RIL population developed from a cross between ILL5722 (Australian cultivar Digger) and ILL5588. The map comprised seven linkage groups that varied from 80.2 cM to 274.6 cM in length and spanned a total of 928.4 cM (Phan et al. 2006a).

2.2. Toward a Lentil Consensus Linkage Map

A lentil consensus map will comprise a set of robust and transferable markers from which genetic distance can be measured and compared across different genetic backgrounds. This will enable the tracking of gene recombination events for the building of superior genotypes. In order to construct a consensus map, previously constructed genome maps may be anchored with a common set of genetic markers that span the representative linkage/chromosome groups. Also, orthologous markers that are transferable between related legume species will enable rapid generation and anchoring of maps in species such as lentil where there is little pre-existing genomic information. So far, seven morphological markers have been mapped in lentil of which only four (cotyledon colour *Yc*, anthocyanin pigmentation in stem *Gs*, seed coat pattern *Scp* and pod dehiscence-indehiscence *Pi*) have been placed on multiple maps. Recently, Rubeena et al. (2006) anchored seven linkage groups with those of a previously published map (Rubeena et al. 2003). For this, 22 RAPD and two ISSR markers were transferred among populations. Of more use in map anchoring, due to their stability and reproducibility, will be the newer SSR and ITAPS markers. Phan et al. (2006a) compared ESTs from phylogenetically distant

M. truncatula, *Lupinus albus*, and *Glycine max* species to produce 500 intron-targeted amplified polymorphic (ITAP) markers. They also used 126 *M. truncatula* cross-species markers to generate comparative genetic maps of lentil (*Lens culinaris* Medik.) and white lupin (*Lupinus albus* Linn.) (Phan et al 2006b). Subsequently, Phan et al (2006b) used 18 common SSR markers to join the new map with another pre-existing comprehensive lentil map (Hamwieh et al 2005). Comparative mapping was also conducted that enabled the visualisation of a macrosyntenic relationship between lentil and the model genome *M. truncatula* (Phan et al 2006b; Figure 1). The composite lentil map will serve as a foundation for the future use of genomic and genetic information in lentil genetic analysis and breeding for traits such as pod indehiscence, flower colour, seed coat pattern, seed shape, fusarium wilt, ascochyta blight, botrytis grey mould and virus resistance and flowering responses.

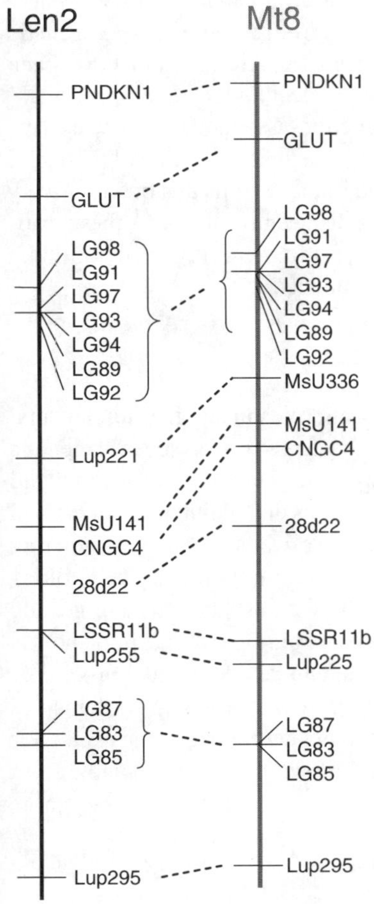

Figure 1. Evidence of simple homologeous and conserved macrosynteny between *Medicago tuncatula* (linkage group 8) and lentil (linkage group 2). Reproduced from Phan et al (2006b)

3. MARKER-ASSISTED BREEDING

Eujay et al. (1998) first identified markers suitable for the selection of a simply inherited disease resistance trait loci for fusarium wilt resistance (*Fw*). Subsequently, Ford et al (1999) identified RAPD markers that were close and flanking the major dominant locus for ascochyta blight resistance in the ILL5588 accession (*Ral*1/*AbR*$_1$). Chowdhury et al (2001) also developed RAPD markers that flanked the recessive ascochyta blight resistance locus in the cultivar Indianhead (*ral*2). More recently, markers have been identified that also flank the codominant ascochyta blight resistance loci in ILL7537 (Rubeena et al. 2006) and Tullu et al (2003) identified markers linked to the anthracnose resistance locus in accession PI320937 (*LCt*-2). The most recent report of markers developed to select for disease resistance were those reported by Hamwieh et al (2005), who identified close and flanking markers for the *Fw* locus in ILL5588.

Research has also focused on the conversion of arbitrary markers to sequence-specific markers that are reproducible and robustly transferable among genetic backgrounds (Nguyen et al. 2001). Although several SCAR markers have been developed and validated among genotypes, they do not select for the genes specifically governing the traits of interest. Rather, the newer function-associated molecular maps that are being developed (Phan et al. 2006a, b), will enable direct selection of the actual candidate genes. Together with knowledge of the genomic regions that quantitatively account for genetic portions of particular phenotypes (Rubeena et al. 2006), these maps will enable accurate selection of multiple gene traits, for future trait/gene pyramiding (Tar'an et al. 2003) and adaptation to various environments.

4. GENETIC ENGINEERING

Applications of genetic engineering can play important roles in solving fundamental challenges faced in classical breeding. Through the targeting of specific genetic pathways or expression of known functional genes, transgenic approaches may aid in increased yield by improving agronomic traits, such as enhancing pest, stress and herbicide resistance. Improvements could also be made in the quality of the crop, including its food and feed characteristics. Thus, genetic engineering technology provides an important adjunct to classical breeding. In the post genomic era, genetic engineering is a key suite of tools used to answer basic biological question such as gene function and their roles in various physiological and developmental processes. Reliable, efficient and reproducible regeneration and transformation systems are prerequisites for assessing the effect of altering genomes with novel genes and their associated functions and also in exploiting the vast knowledge know available from the model crop genomes.

4.1. Cell and Tissue Culture

The grain legumes have been less amenable to manipulation in tissue culture (McClean and Grafton 1989), and generally are more recalcitrant to *in vitro* regeneration and transformation (DeKathen and Jacobsen 1990) than may other crop species. However, routine protocols for obtaining stable transformants are now available for the major grain legumes such as the common bean (*Phaseolus vulgaris*), soybean (*Glycine max*), pea (*Pisum sativum*), peanut (*Arachis hypogea*), and alfalfa (*Medicago sativa*) as well as the model legume *Medicago truncatula* (Puonti-Kaerlas et al. 1990; Christou 1992; Russell et al. 1993, Trick et al. 1997). For lentil, limited research has been conducted in developing useful and stable transformation and regeneration protocols.

Organogenesis and somatic embryogenesis are the two common methods used for regeneration of complete plants in tissue culture. Somatic embryogenesis involves the production of a bipolar structure with root and shoot axis and a closed vascular system. This involves the induction of embryogenic callus and development of these cells into embryos by manipulating culture conditions including media and growth regulators. Somatic embryos originate from single cells and thus are an excellent target for transformation systems, and have been successfully used in the genetic transformation of the legume relative, soybean (Trick et al. 1997). In lentil to date there is only one report of successfully achievement of stimulated somatic embryogenesis (Saxana and King 1987). In general, the efforts to achieve somatic embryogenesis have failed due to the embryo not proceeding beyond the characterised globular and heart shapes.

Organogenesis describes the processes by which cells and tissues are de-differentiated, leading to the production of shoot or root primordium whose vascular systems are often connected to the parent tissues. The stages involved in complete plantlet regeneration via *de novo* organogenesis include shoot bud formation, shoot development and rooting of the shoots. Bajaj and Dhanju (1979) first reported direct shoot organogenesis in lentil from apical meristems using media containing kinetin (Kin). Shoot bud formation was achieved relatively easily in lentil and this initial report of shoot organogenesis in lentil was followed by many others in which a variety of explants such as apical meristem (Bajaj and Dhanju 1979), stem nodes (Polanco et al. 1988, Sing and Raghuvanshi 1989, Ahmed et al. 1997), cotyledonary node (Warkentin and McHughen 1993, Sarker et al. 2003b), epicotyls (Williams and McHughen 1986), decapitated embryo, embryo axis and immature seeds (Polanco and Ruiz 2001) were used. However, shoot regeneration from leaf tissue has not yet been reported. Explants derived from mature seeds have subsequently been preferred as the explant of choice mainly because of their year-round availability. So far, Murashige and Skoog (MS) salts medium has been the most commonly reported medium for lentil regeneration. Several cytokinins such as Kin, 6-benzylaminopurine (BAP), N-phenyl-N'-1, 2, 3-thiadiazol-5-yl-urea (TDZ) and auxins such as 2,4-dichlorophenoxyacetic acid (2,4-D), α-Naphthalene acetic acid (NAA) , indol-3-butyric acid (IBA) and indol-3-acetic acid (IAA) have been used for direct or indirect organogenesis. Also, gibberellic acid (GA3) has been used to alter shoot length *in vitro* (Sarker et al. 2004).

An efficient and reproducible rooting protocol is necessary to obtain viable plants from *in vitro* regenerated shoots. Often species, genotype and/or explant dependent, the success may also be greatly influenced by the media and phytohormones used. Induction of functional roots on *in vitro* formed plantlets has become the most difficult stage in developing a complete and robust lentil regeneration system (Fratini and Ruiz 2002), and to date no reproducible rooting protocol has been reported. In lentil, as for other pulse crops, increasing concentrations of cytokinins in the culture media, mainly BA or TDZ, has resulted in a higher number of shoots regenerated. However, shoot length and subsequent rooting was greatly reduced (Mohamed et al. 1992; Gulati and Jaiwal, 1994; Prakash et al. 1994; Sanago et al., 1996; Polisetty et al. 1997; Subhadra et al. 1998, Sarker et al. 2003b). Using high concentrations of cytokinin to induce shoot bud formation resulted in stunted elongation of shoots that lacked shoot apical meristems and vascular connections. This led to subsequent difficulty in regenerating roots from such shoots. The inhibitory effect of BA on rooting has been well documented by Polanco and Ruiz (1997) via *in vivo* and *in vitro* studies of lentil seedlings. The strong inhibitory effect of this cytokinin on root growth, was demonstrated by a drastic *in vivo* reduction of the mitotic index of the root meristem. Furthermore, Malik and Saxena (1992), observed a progressive decrease in root development with an increase in TDZ concentration. Only stunted primary roots developed on media with TDZ of 5mM or higher, and prolonged exposure for three to five weeks, resulted in callus production from the primary root. Fratini and Ruiz (2002) subsequently found that TDZ and BA inhibited root formation by inducing root swelling and stunting, and at higher concentrations caused callusing of the main root. A continued culture on TDZ induced shoots without a shoot apical meristem, resulting in fused shoots and no plant regeneration. However, through limiting the culture period on TDZ and transferring the regenerated shoots to a growth-regulator-free medium prior to rooting, whole plants were eventually obtained (Fratini and Ruiz 2002).

Sarker et al (2003b) observed 30% root induction on shoots regenerated in the presence of BAP and Kin when shoots were cultured on MS medium supplemented with 25 mg/l IBA. However, roots did not develop from the base of the regenerated shoots but at a level slightly higher than the cut ends. Furthermore, it was observed that once inside the auxin-rich medium, the tip of the roots callused, blocking further growth of the roots and histological study by freeze microtomy showed that the roots did not form vascular bundle connection with the shoot. Fratini and Ruiz (2002) subsequently reported a 95% rooting efficiency by culturing nodal segments of lentil with an axillary bud in an inverted orientation in media with 5 μM IAA and 1 μM Kin, and concluded that the improvement in rooting success was due to polarity. However they were only able to regenerate about two shoots per explants. Most recently Newell et al. (2006) reported a 100% rooting rate from lentil nodal microcuttings by placing them in an inverted fashion in media composed of sphagnum peat, coarse river sand and perlite at a 0.5:2:2 ratio, and concluded that the improved rooting efficiency was due more to aeration than polarity. Studies are required to test the applicability of this rooting procedure for a transformation compatible regeneration system where individual transformed cells first need to be induced to form shoot primordia requiring the use of cytokinin followed by lengthy exposure to selection media to kill non transformed cells.

4.2. Transformation

Genetic transformation of lentil tissues has been reported using several different gene transfer methods. In particular, several foreign genes have been introduced into lentil protoplast by electroporation, lipofection or PEG treatment (Maccarrone et al 1992a, b, 1993, 1995a, b). Chowria et al. (1995; 1996) reported *in planta* electroporation-mediated transformation of nodal meristems and 20% of the branches that grew from the nodal meristems were chimeric. However, the segregation ratios in the putatively transgenic R2 populations were strongly biased against transgene presence or expression. Also, lentil protoplast were electroporated with an aim to reduce lipoxygenase activity by antisense RNA mediated gene silencing (Maccarrone et al 1995b), however, no plants were regenerated from transformed protoplast-derived callus.

The susceptibility of lentil to *Agrobacterium* was first demonstrated by Warkentin and McHughen (1992) through the production of tumors on lentil stems and shoots apices *in vivo* and *in vitro*. Warkentin and McHughen (1992; 1993) later evaluated a number of explants (shoot apices, epicotyl, root, cotyledons, and cotyledonary nodes) and observed transient GUS expression at all wound sites except the cotyledonary nodes and the axils of the cotyledonary petioles. Sarker et al (2003a) also confirmed that cotyledonary nodes were not suitable for *Agrobacterium*-mediated transformation as multiple shoot regeneration occurred from pre-existing meristems in the explant. *Agrobacterium*-mediated transformation efficiency in lentil varies with genotype and strain used, with the EHA101, EHA105 and GV2260 strains being used most commonly. Mahmoudian et al (2002) reported that vacuum infiltration by *Agrobacterium* improved transformation efficiency while Hoque et al (2003) reported that sonication and vacuum infiltration improved transformation efficiency, as measured by expression of the GUS gene. Sarker et al (2003a) reported obtaining transformed lentil shoots from decapitated embryo explants using *Agrobacterium* strain LBA4404, harboring the pBI121 plasmid. Southern blot analysis was later used to confirm integration of the transgene in the lentil genome (B. Mustafa, unpublished). Selection of transformed tissue was done using 50–200 mg/l kanamycin and the transformation efficiency was between 1.5 and 1.9% (Sarker et al. 2003). Using *Agrobacterium*, Barton et al (1998) produced T1 seed of lentils transformed with a 35S-bar-GUS construct and confirmed the stable transfer of the bar gene in lentil plants grown in glasshouse and screen house experiments.

The first report of gene transfer in lentil using particle bombardment was by Öktem et al (1999) who used cotyledonary node explants. Almost 50% of the bombarded explants showed transient GUS expression at 24 hours after bombardment. Chimeric stable expression was observed in regenerated shoots without selection pressure. Following, Gulati and McHughen (2000) reported the bombardment of lentil cotyledonary nodes with the pCAMBIA1201 plasmid carrying the GUS and hpt genes. GUS and PCR assays detected putative transformants however no transformed shoots were recovered. Gulati et al (2002) reported fertile transgenic lentil plants after bombarding lentil cotyledonary nodes with a plasmid containing a mutant acetolactate synthase gene (ALS) which confers resistance to sulfonylurea herbicides. Putative transgenic shoots were regenerated on MS media with 4.4 μM BAP, 5:2 μM GA3 and chlorsulfuron. The regenerants were micrografted, successfully transferred to soil and the T0 and selfed progeny plants were screened using metsulfuron herbicide leaflet painting. PCR and Southern hybridisation were used to confirm the survivor T1 transformants.

5. FUTURE PROSPECTS IN LENTIL GENOMICS

The extensive research that has identified genes for use in the transformation of cereals, cotton, soyabeans, canola and other crop species will be of value for lentil production. Examples currently used in agricultural production systems and of benefit for lentil include genes that confer resistance to insects (eg Bt Cotton), herbicide tolerance and virus resistance. Genes currently being evaluated that may have future impact for lentil include those that potentially confer drought or frost tolerance, or non specific disease resistance. Currently, the largest barriers to the use of transgenic lentils are consumers views on transgenics, IP restrictions and the large costs of meeting regulatory requirements for their release.

More tools are becoming available in order to further understand the functional genetic components governing traits of interest and hence aid in selection of the optimal genome fragments in advanced breeding programs. Several of these tools are based on the principle of reverse genetics in which a gene sequence or its expression is altered to study the effect on the phenotype in a particular environment and to compare this to the wild type. Methods such as transposon mutagenesis (Tisser et al. 1999), target induced local lesions in genomes (TILLING) analysis (Henikoff et al. 2003) and post-transcriptional gene silencing (PTGS) through RNA interference (Voinnet 2002) are possible avenues for future lentil functional genomics studies.

Mutagenic lentil populations have been developed for the purposes of studying gene 'knock-out' effects using gamma rays and chemical treatments such as ethyl methane sulfonate (EMS). The TILLING procedure employs an EMS-generated mutant library within which point mutations are sought to provide differentials in enzyme cleavage points. In legumes, TILLING has been applied in the model crop genomes of *Lotus japonicus* (3697 mutant plants; Perry et al. 2003) and *Medicago truncatula* (2000 mutant plants; VandenBosch and Stacey 2003). Furthermore, a population was recently developed in lentil at the Department of Primary Industries, Horsham, Australia (M. Materne, unpublished).

Transcript profiling used to identify genes associated with traits of interest has been applied to the pulse genomes (Muehlbauer et al. 2006). In particular, the cDNA-AFLP technique was used to identify candidate genes for resistance to

Ascochyta rabiei causing ascochyta blight in chickpea (Cho et al. 2005). The flavanone-3-hydroxylase (*F3H*) gene was qualitatively differentially expressed resistant and susceptible plants. Alternatively, the microarray technique has recently been used to identify genes associated with resistance to ascochyta blight in lentil. For this, a cDNA microarray, named the *Pulse Chip* was developed which comprised 565 expressed sequence tags (EST) from a chickpea cDNA library enriched for reaction to *A. rabiei*, 156 ESTs from a *Lathyrus* cDNA library enriched for reaction to *A. pinodes* and 41 lentil ESTs and RGAs from the GenBank database (Coram and Pang 2005). The pulse chip was employed to study expression profiles in the resistant ILL7537 and susceptible ILL6002 lentil genotypes at 6, 24, 48, 72 and 96 hours after inoculation with *Ascochyta lentils*. Key differential genes included; a proline-rich protein (LS0156) for cell wall strengthening, a super oxidase dismutase enzyme (U116) for antioxidant defence, a salicylic acid binding protein (U174), a Snakin-2 antimicrobial protein (U278) and a Bet VI type pathogenesis-related protein (LS481) (Mustafa et al. 2006). Validation of functionality will follow through QRT-PCR and PTGS analyses. This will likely be achieved through the already developed transgenic and *in vitro* shoot regeneration methods.

6. CONCLUSIONS

Lentil is a genetic orphan compared to many larger crop species. However, rapid advances in the development and use of molecular tools in the breeding of lentil is expected in the short to medium term. A consensus map is now available in lentil that can form the basis of a more saturated genetic map for use in mapping genes conferring morphological characteristics, tolerance to abiotic stresses, resistance to pests and diseases and improved quality. Lentil will benefit greatly from genomic research in other species and by its close relationships with the model species, for which much genomic information and tools are available. The lack of haploid technologies in lentil has necessitated the slower and more costly development of mapping populations using single seed descent. However, a large number of populations have been developed in lentil that can be used to map genes for many of the worlds economically important traits. The population used by Phan et al (2006b) alone could be used to develop markers for seed characteristics, resistance to ascochyta blight, fusarium wilt, botrytis grey mould and virus resistance, flowering responses and adaptation (Materne 2003). Currently the political, social and regulatory environment is limiting the development of transgenic cultivars more than the capability of the scientists.

REFERENCES

Ahmad M, Fautrer AG, McNeil V, Hill GD, Burritt DJ (1997) *In vitro* propagation of *Lens* species and their F1 interspecific hybrids. Plant Cell Tissue and Organ Culture 47:169–176

Bajaj YPS, Dhanju M S (1979) Regeneration of plants from apical meristem tips of some legumes. Current Science 48:906–907

Barton J, Smith PMC, Fletcher N, Walker R, Leece E and Chappel S (1998) Methods for transformation and regeneration of other Pulses. In Cooperative Research Centre for Legumes in Mediterranean Agriculture Annual Report 1997–98. p 53

Cho S, Chen W, Muehlbauer FJ (2005) Constitutive expression of the flavone 3-hydroxylase gene related to pathotype-specific ascochyta blight resistance in *Cicer arietinum* L. Physiological and Molecular Plant Pathology 67:100–107

Chowdhury MA, Andrahennadi CP, Slinkard AE, Vandenberg A (2001) RAPD and SCAR markers for resistance to ascochyta blight in lentil. Euphytica 118:331–337

Chowrira G, Akella V, Lurquin PF (1995) Electroporation-mediated gene transfer into intact nodal meristems *in planta*: Generating transgenic plants without *in vitro* tissue culture. Molecular Biotechnology 3:17–23

Chowrira G, Akella V, Fuerst PE, Lurquin PF (1996) Transgenic grain legumes obtained by *in planta* electroporation-mediated gene transfer. Molecular Biotechnology 5:85–96

Christou P (1992) Genetic engineering and *in vitro* culture of crop legumes. Technomic Publishing, Pennsylvania, USA. pp 307

Coram TE, Pang ECK (2005) Expression profiling of chickpea genes differentially regulated during a resistance response to *Ascochyta rabiei*. Plant Biotechnology Journal 4:647

DeKathen A, Jacobsen HJ (1990) *Agrobacterium tumefacians*-mediated transformation of *Pisum sativum* L. using binary and cointegrate vectors. Plant Cell Reports 9:276–279

Durán Y, Fratini R, García P, Pérez de la Vega M (2004) An intersubspecific genetic map of *Lens*. Theoretical and Applied Genetics 108:1265–1273

Eujayl I., Baum M, Powell W, Erskine W, Pehu E (1998a) A genetic linkage map of lentil (*Lens* sp.) based on RAPD and AFLP markers using recombinant inbred lines. Theoretical and Applied Genetics 97,83–89.

Ford R, Pang ECK, Taylor PWJ (1997) Diversity analysis and species identification in *Lens* using PCR generated markers. Euphytica 96:247–255

Ford R, Pang ECK, Taylor PWJ (1999) Genetics of resistance to ascochyta blight (*Ascochyta lentis*) of lentil and identification of closely linked molecular markers. Theoretical and Applied Genetics 98:93–98

Ford R, Le Roux K, Itman C, Brouwer JB, Taylor PWJ (2002) Genome-specific sequence tagged microsatellite site (STMS) markers for diversity analysis and genotyping in *Pisum* species. Euphytica 124:397–405

Fratini R, Ruiz ML (2002) Comparative study of different cytokinins in the induction of morphogenesis in lentil (*Lens culinaris* Medik.). *In Vitro* Cellular and Developmental Biology – Plant 38:46–51

Gulati A, Jaiwal PK (1994) Plant regeneration from cotyledonary node explants of mungbean (*Vigna radiata* (L.) Wilczek). Plant Cell Reports 13:523–527

Gulati A., McHughen A. (2000) Regeneration and particle bombardment-mediated genetic transformation of lentil (*Lens culinaris* Medik).In: Proceedings of 6th International Congress of Plant Molecular Biology, Quebec, Canada, June18th–24th, 2000

Gulati A, Schryer P, McHughen A (2001) Regeneration and micrografting of lentil shoots. *In Vitro* Cellular and Developmental Biology – Plant 37:798–802

Gulati A, Schryer P, McHughen A (2002) Production of fertile transgenic lentil (*Lens culinaris* Medik) plants using particle bombardment. *In Vitro* Cellular and Developmental Biology – Plant 38:316–324

Hamwieh A, Udupa SM Choumane W, Sarker A, Dreyer F, Jung C, Baum M (2005) A genetic linkage map of *Lens* sp. based on microsatellite and AFLP markers and the localization of fusarium vascular wilt resistance. Theoretical and Applied Genetics 110:669–677

Hoque MI., Hassan F, Sarker RH, Kisecker H, Jacobsen H-J (2003) Lentil improvement through biotechnology. In: *In Vitro* Culture, Transformation and Molecular Markers for Crop Improvement Eds. Islam AS. Science Publishers, Inc. Enfield, USA. pp 175–192

Havey M H, Meuhlbauer F J (1989) Linkages between restriction fragment length, isozyme and morphological markers in lentil. Theoretical and Applied Genetics 77: 839–843

Henikoff S, Till BJ, Comai L (2004) Tilling. Traditional mutagenesis meets functional genomics. Plant Physiology 135:630–6

Maccarrone M, Dini L; Di Marzio L, Di Giulio A, Rossi A; Finazzi Agrò, A (1992a) Interaction of DNA with cationic liposomes: ability of transfecting lentil protoplasts. Biochemical and Biophysiological Research Communications 186:1417–1422

Maccarrone M, Veldink GA, Vliegenthart JFG (1992b) Inhibition of lipoxygenase activity in lentil protoplasts by monoclonal antibodies introduced into the cells via electroporation. European Journal of Biochemistry 205:995–1001

Maccarrone M, Dini LA, Rossi A, Finazzi Agrò, A (1993) Gene transfer to lentil protoplasts by lipofection and electroporation. Journal of Lipid Research 3:707–716

Maccarrone M, Veldink GA, Finazzi Agrò A, Vilegenthart JFG (1995a) Lentil root protoplasts: a transient expression system suitable for coelectroporation of monoclonal antibodies and plasmid molecules. Biochemical and biophysiological Acta 1243:136–142

Maccarrone M, Hilbers MP, Veldink GA, Vliegenthart JFG, Finazzi-Agrò, A. (1995b) Inhibition of lipoxygenase in lentil protoplasts by expression of antisense RNA. Biochemical and biophysiological Acta 1259:1–3

Mahmoudian M, Yücel M, H A Öktem (2002) Transformation of lentil (*Lens culinaris* M.) cotyledonary nodes by vacuum infiltration of *Agrobacterium* tumefaciens Plant Molecular Biology Reporter 20:251–257

Malik KA, Saxena PK (1992) Thidiazuron induces high frequency shoot regeneration in intact seedlings of pea (*Pisum sativum*), chickpea (*Cicer arietinum*) and lentil (*Lens culinaris*). Australian Journal of Plant Physiology 19:731–740

Materne M A (2003) Importance of phenology and other key factors in improving the adaptation of lentil (*Lens culinaris* Medikus) in Australia. PhD thesis, The University of Western Australia

McClean P, Grafton KF(1989) Regeneration of dry bean (*Phaseolus vulgaris* L.) via organogenesis. Plant Science 60:117–122

Mohamed MF, Read PE, Coyne DP (1992) Plant regeneration from *in vitro* culture of embryonic axis explants in common and tepary beans. Journal of the American Society of Horticultural Science 117:332–336

Muehlbauer FJ, Cho S, Sarker A, McPhee KE, Coyne CJ, Rajesh PN, Ford R (2006) Application of biotechnology in breeding lentil for resistance to biotic and abiotic stress. Euphytica 147:149–165

Mustafa B M, Coram T E, Pang ECK, Taylor PWJ, Ford R (2006) Unraveling *Ascochyta lentis* resistance in lentil. Ascochyta 2006 conference. 2nd–5th July, France. http://www.grainlegumes.com/default.asp?id_biblio = 350

Newell C, Growns A, McComb DJ (2006) Aeration is more important than shoot orientation when rooting lentil (*Lens culinaris* Medik.) Cv. 'Digger' microcuttings *in vitro*. *In Vitro* Cellular and Developmental Biology – Plant 42:197–200

Nguyen T, Brouwer JB, Taylor PWJ, Ford R (2001) A novel source of resistance in lentil (*Lens culinaris* ssp. *culinaris*) to ascochyta blight caused by *Ascochyta lentis*. Australasian Plant Pathology 30:211–215

Öktem HA, Mahmoudian M, Eyidooan F, Yücel M (1999) GUS gene delivery and expression in lentil cotyledonary nodes using particle bombardment. Lens Newsletter 26:3–6

Perry JA, Wang TL, Welham TJ, Gardner S, Pike JM, Yoshida S, Parniske M (2003) A TILLING reverse genetics tool and a web-accessible collection of mutants of the legume *Lotus japonicus*. Plant Physiology 131:866–871

Phan HTT, Elwood SR, Hane JK, Ford R, Materne M, Oliver RP (2006a) Extensive macrosynteny between *Medicago truncatula* and *Lens culinaris* ssp. *culinaris*. Theoretical and Applied Genetics. DOI 10.1007/s001222-006-0455-3

Phan HTT, Ellwood SR, Ford R, Thomas S, Oliver R (2006b) Differences in syntenic complexity between *Medicago truncatula* with *Lens culinaris* and *Lupinus albus*. Functional Plant Biology 33:775–782

Polanco MC, Pelaez MI, Ruiz ML (1988) Factors affecting callus and shoot formation from *in vitro* cultures of *Lens culinaris* Medik. Plant Cell Tissue and Organ Culture 15:175–182

Polanco MC, Ruiz ML (1997) Effect of benzylaminopurine on *in vitro* and *in vivo* root development in lentil (*Lens culinaris* Medik.). Plant Cell Reports 17:22–26

Polanco MC, Ruiz ML (2001) Factors that affect plant regeneration from *in vitro* culture of immature seeds in four lentil (*Lens culinaris* Medik.) cultivars. Plant Cell Tissue and Organ Culture 66:133–139

Polisetty R, Paul V, Deveshwar JJ, Khetarpal S, Suresh K, Chandra R (1997) Multiple shoot induction by benzyladenine and complete plant regeneration from seed explants of chickpea (*Cicer arietinum* L.). Plant Cell Reports 16:565–571

Puonti-Kaerlas J, Eriksson T, Engstrom P (1990) Production of transgenic pea (*Pisum sativum* L.) plants by *Agrobacterium tumefaciens*-mediated gene transfer. Theoretical and Applied Genetics 84:443–450

Rongwen J, Akkaya MS, Bhagwat AA, Lavi U, Cregan PB (1995) The use of microsatellite DNA markers for soybean genotype identification. Theoretical and Applied Genetics 90:43–48

Rubeena, Taylor PWJ, Ades PK, Ford R (2006) QTL mapping of ascochyta blight (*Ascochyta lentis*) resistance in lentil (*Lens culinaris*). Plant Breeding 125:506–512

Rubeena., Taylor PWJ, Ford R (2003) Molecular mapping the lentil (*Lens culinaris ssp. culinaris*) genome. Theoretical and Applied Genetics 107:910–916

Russell DR; Wallace K, Bathe J, Martinell B, McCabe D (1993) Stable transformation of *Phaseolus vulgaris* via electric-discharge mediated particle acceleration. Plant Cell Reports 12:165–169

Sanago MHM, Shattuck VI, Strommer J (1996) Rapid plant regeneration of pea using thidiazuron. Plant Cell Tissue and Organ Culture 45:165–168

Sarker RH, Biswas A, Mustafa BM, Mahbub S, Hoque MI (2003) *Agrobacterium*-mediated transformation of lentil (*Lens culinaris* Medik.) Plant Tissue Culture 13:1–12

Sarker RH, Mustafa BM, Biswas A, Mahbub S, Nahar M, Hashem R, Hoque MI (2003) *In vitro* regeneration in lentil (*Lens culinaris* Medik.) Plant Tissue Culture 13:155–163

Saxena PK, King J (1987) Morphogenesis in lentil plant regeneration from callus Science 52:223–227

Singh RK, Raghuvanshi SS (1989) Plantlet regeneration from nodal segment and shoot tip derived explants of lentil. *Lens* Newsletter 16:33–35

Subhadra, Vahishat RK, Chowdhury JB, Singh M, Sareen PK (1998) Multiple shoots from cotyledonary node explants of non-nodulating genotype (ICC435M) of chickpea, *Cicer arietinum* L. Indian Journal of Experimental Biology 36:1276–1279

Tadmor Y, Zamir D, Ladizinsky G (1987) Genetic mapping of an ancient translocation in the genus *Lens*. Theoretical and Applied Genetics 73:883–892

Tahir M, Muehlbauer FJ (1994) Gene mapping in lentil with recombinant inbred lines. Journal of Heredity 85:306–310

Tahir M, Simon CJ, Muehlbauer FJ (1993) Gene map of lentil: A review. *Lens* Newsletter 20:3–10

Tar'an B, Buchwaldt L, Tullu A, Banniza S, Warkentin TD, Vandenberg A (2003) Using molecular markers to pyramid genes for resistance to ascochyta blight and anthracnose in lentil (*Lens culinaris* Medik). Euphytica 134:223–230

Tissier AF, Marillonnet S, Klimyuk V, Patel K, Torres MA, Murphy G, Jones JDG (1999) Multiple independent defective suppressor-mutator transposon insertions in *Arabidopsis*: A tool for functional genomics. The Plant Cell 11:1841–1852

Trick NU, Dinkins RD, Santarem ER, Samoylo RDV, Meurer CA, Walker DR, Parrot WA, Finer JJ, Collins GB (1997) Recent advances in soybean transformation. Plant Tissue Culture and Biotechnology 3: 9–26

Tullu A, Buchwaldt T, Warkentin T, Taran B, Vandenberg A (2003) Genetics of resistance to anthracnose and identification of AFLP and RAPD markers linked to the resistance gene in PI320937 germplasm of lentil (*Lens culinaris* Medik). Theoretical and Applied Genetics 106:428–434

Vaillancourt RE, Slinkard AE (1993) Linkage of morphological and isozyme loci in lentil, *Lens culinaris* L. Canadian Journal of Plant Science 73:917–926

VandenBosch KA, Stacey G (2003) Summaries of legume genomics projects from around the globe. Community resources for crops and models Plant Physiology 131:840–865

Voinnet O (2002) RNA silencing: Small RNAs as ubiquitous regulators of gene expression. Current Opinion in Plant Biology 5:444

Warkentin TD, McHughen A (1992) *Agrobacterium tumefaciens*-mediated beta-glucuronidase (GUS) gene expression in lentil (*Lens culinaris* Medik.) tissues. Plant Cell Reports 11:274–278

Warkentin TD, McHughen A (1993) Regeneration from lentil cotyledonary nodes and potential of this explant for transformation by *Agrobacterium tumefaciens*. *Lens* Newsletter 20:26–28

Weeden NF, Muehlbauer FJ, Ladizinsky G (1992) Extensive conservation of linkage relationship between pea and lentil genetic maps. Journal of Heredity 83:123–129

Williams DJ, McHughen A (1986) Plant regeneration of the legume *Lens culinaris* Medik (lentil) *in vitro*. Plant Cell Tissue and Organ Culture 7:149

Winter P, Pfaff T, Udupa SM, Huettel B, Sharma PC, Sahi S, Arreguin-Espinoza R, Weigand F, Muehlbauer FJ, Kahl G (1999) Characterization and mapping of sequence-tagged microsatellite sites in the chickpea (*Cicer arietinum* L.) genome. Molecular and General Genetics 262:90–101

Zamir D, Ladizinsky G (1984) Genetics of allozyme variants and linkage groups in lentil. Euphytica 33:329–336

Závodná M, Kraic J, Paglia G, Gregova E, Morgante M (2000) Differentiation between closely related lentil (*Lens culinaris* Medik.) cultivars using DNA markers. Seed Science and Technology 28:217–219

CHAPTER 18

LENTIL DISEASES

PAUL TAYLOR[1], KURT LINDBECK[2], WEIDONG CHEN[3], AND REBECCA FORD[4]

[1]*Center for Plant Health and BioMarka, Faculty of Land and Food Resources, The University of Melbourne, Victoria 3010, Australia*
[2]*Grains Innovation Park, The Department of Primary Industries, Private Bag 260, Horsham, Victoria 3401, Australia*
[3]*USDA-ARS, Grain Legume Genetics and Physiology Research Unit, 303 Johnson Hall, Washington State University, Pullman, WA 99164-6402, USA*
[4]*BioMarka, Faculty of Land and Food Resources, The University of Melbourne, Victoria 3010, Australia*
E-mail: paulwjt@unimelb.edu.au

Abstract: Fungal diseases of lentils are the most important biological constraint to productivity. *Ascochyta lentis* (ascochyta blight) and *Fusarium oxysporum* f. sp. *lentis* (fusarium wilt) are the major fungal pathogens that can cause severe losses in most lentil growing regions of the world. Fungal diseases such as botrytis grey mould (*Botrytis fabae* and *B. cinerea*), rust (*Uromyces viciae-fabae*), stemphylium blight (*Stemphylium botryosum*), and anthracnose (*Colletotrichum truncatum*) are also important in some growing seasons in particular countries when environmental conditions are conducive for infection. Lentil plants can also be infected by a range of viruses but generally the affect on yield is not as great as that caused by fungal pathogens. Lentil yellows disease caused by bean leaf roll virus (BLRV), beet western yellows virus (BWYV), or subterranean clover red leaf virus (SCRLV) is widespread throughout the world. Other important virus diseases of lentil include bean yellow mosaic (BYMV), pea seed borne mosaic (PSbMV), cucumber mosaic (CMV), alfalfa mosaic (AMV) and broad bean stain (BBSV). Integrated disease management practices including use of resistant cultivars, modified cultural practices and use of fungicides or insecticides can reduce the impact of these diseases on lentil production

1. INTRODUCTION

Lentil plants are affected by a wide range of pathogens with fungal diseases being the most important. These decrease productivity through infection and damage to leaves, stems, roots and pods, and reduce marketability by discoloring seed. Most

S.S. Yadav et al. (eds.), Lentil: An Ancient Crop for Modern Times, 291–313.

major economically important diseases are found in all lentil growing regions of the world eg., ascochyta blight, whereas, diseases like fusarium wilt and anthracnose have not been detected in some major lentil producing countries such as Australia. As well, certain virulent pathotypes of a pathogen have restricted geographical range (eg. Ct0 pathotype of anthracnose in Canada). Therefore, it is important for industry and quarantine personnel to be aware of the range of pathogens that can infect lentil, and be able to identify the cause(s) of a disease outbreak, or be able to detect exotic pathogens on imported seed or plant parts. Early correct identification and detection of lentil pathogens will prevent incursion of exotic diseases into areas where the particular disease does not exist.

The major fungal diseases of lentil are described with descriptions of the causal organism, symptoms produced on the plant, epidemiology, and disease management and control. Viruses are then described followed by a listing of minor lentil diseases.

2. ASCOCHYTA BLIGHT

Ascochyta blight, caused by *Ascochyta lentis* Bond. and Vassil, is one of the most important biotic constraints to lentil production (Figure 1a). Able to attack all above ground plant parts at any growth stage under favorable conditions, the disease causes reduction in yield and seed quality. The disease is prevalent throughout the world and has been reported to cause yield losses of up to 70%, 30–50% and 50% in Canada, USA and Australia respectively (Gossen and Morrall, 1983; Kaiser, 1992; Brouwer *et al.*, 1995).

There are two stages within the *A. lentis* life-cycle, the asexual or anamorph stage and the sexual or teleomorph stage. The asexual stage is characterized by the production of pycnidia in the lesions on infected plants, which are 175–300 μm in diameter with a minute round osteole (Bondartzeva-Monteverde and Vassilievsky; 1940 cited by Agrawal and Prasad, 1997). The pycnidia release conidia which are cylindrical, straight or rarely curved, round at the ends with a median septum. The teleomorph (*Didymella lentis*) was observed for the first time on over wintered lentil straw in 1992 in Idaho, USA (Kaiser and Hellier, 1993), confirming the heterothallic nature of *A. lentis*, with two distinct mating types. Compatible mating types (MAT1-1 and MAT1-2) are required for the development of fertile pseudothecia and viable ascospores. Kaiser *et al.* (1997) differentiated the teleomorphs of *A. fabae* and *A. lentis* on the basis of a pathogenicity test and morphology. They also distinguished ascospores of *A. fabae* from *A. lentis* using molecular markers. Kaiser (1997) identified both mating types in isolates from Australia, Canada, Italy, Morocco, New Zealand, Pakistan, Spain, Syria, Turkey and USA. Consequently, *D. lentis* was proposed as a new species that was distinct from *D. fabae*.

2.1. Symptoms

The symptoms of the disease include lesions on leaves, petioles, stems and pods (Figure 1a). The irregularly shaped lesions on leaves, petioles and stem are tan

Figure 1. Diseases of lentil. a) Ascochyta blight of pods – *Ascochyta lentis*; b) rust on leaves – *Uromyces viciae-fabae*; c) Stemphylium blight of leaves – *Stemphylium botryosum*; d) Botrytis grey mould of pods – *Botrytis fabae*; e) White mold – *Sclerotinia sclerotiorum*

and darker brown on pods and seeds. Black pycnidia are visible in the centre of mature/older lesions. In severe infection, lesions can girdle the stem, leading to breakage and subsequent death of all tissues above the lesion. Heavily infected seeds are shriveled and discoloured with whitish mycelium and pycnidia (Kaiser and Hannan, 1986).

2.2. Epidemiology

Cool, wet weather is conducive to *A. lentis* infection, disease development and spread. The disease affects all the aerial parts of the plant and is seed borne. Conidia may be dispersed under rain splash up to 15 cm (Pedersen *et al.*, 1994) while wetness periods of 24–48 hrs and temperatures of 10–15°C are optimum for infection (Pedersen and Morrall, 1994). The pathogen may also be dispersed by wind blown infected leaflets (Pedersen *et al.*, 1994) and through infected seed (Kaiser and Hannan, 1986). Kaiser and Hannan (1986) found *A. lentis* infection in seeds of 46 accessions from 30 countries. They also found the pathogen on seed from countries where the disease was previously unrecorded. The pathogen was found to survive in infected pods and seeds for over three years when stored under optimum conditions. However, the fungus lost viability after 21 weeks on seeds and 29 weeks on pods when buried at a soil depth of 16 cm. Kaiser *et al.* (1989) stored infected lentil seed at 20, 5, −18 and −160 to −196°C and proposed that the pathogen would survive in infected lentil seed as long or longer than the seed remained viable.

The initial infection process was studied by Roundhill *et al.* (1995). Detached leaves inoculated with a spore suspension of *A. lentis* conidia germinated within six hours of inoculation, and germ tubes and appressoria developed after 10 hours. A penetration peg then pierced the cuticle, often near the junction of two epidermal cells and cytoplasm aggregated adjacent to the infection hypha. Within 40 hours, the plasmalemma was disrupted and by 52 hours, the cytoplasm and nucleus broke down and the cell became largely occupied by the fungus with only remnants of the protoplasm present. Once the epidermis was fully colonized, the pathogen invaded the mesophyll with macroscopic symptoms being evident by day nine.

2.3. Management and Control

The most economical and sustainable strategies to control ascochyta blight are through resistance breeding along with cultural practices. Losses due to ascochyta blight can be minimised by crop rotation, early sowing to escape moist weather at harvest, the use of disease free seed and burning of diseased debris from the previous crop (Nene *et al.*, 1988). A three-year break between lentil crops reduced the amount of inoculum in the soil. Sun drying of lentil seed was found to be useful in controlling seed borne inoculum (Beniwal *et al.*, 1989). Hot water and dry heat treatment at 55 °C for 25 minutes and 70 °C for 24 hrs respectively inhibited fungal growth in the seed however, seed germination declined drastically with the hot water treatment (Ahmed and Beniwal, 1991).

A large number of fungicides have been evaluated for control of seed-borne infection with benomyl, carbendazim, carbathiin, ipodion and thiobendazole reported to be effective in varying degrees (Morrall, 1988; Bretag, 1989). Metalaxyl and thiram have also been found to reduce fungus growth but thiram did not control the disease effectively in the field (Bretag, 1989). In particular, Kaiser and Hannan (1987) reported greater seedling emergence from infected lentil seed and increased yield after treatment with thiobendazole and benomyl but thiobendazole showed a phytotoxic effect at 3 g or more a.i./ kg of seed. The effects of foliar application of fungicides were studied by Beauchamp *et al.* (1986) who reported that captafol, chlorothalonil, folpet and metiram completely inhibited conidia germination at 32 μg/ml or less. Seed yield increased and seed infection was reduced using single applications of these fungicides at early bloom to early pod set.

Lentil breeding programs have developed resistant cultivars however, knowledge of pathogenic diversity is important when choosing appropriate isolates to screen for resistance. Many studies have shown pathogenic diversity among isolates by assaying a set of host-specific differential genotypes or cultivars. Ahmed *et al.* (1996) studied the virulence of 84 *A. lentis* isolates from Canada and 16 isolates from other countries and found that isolates collected in 1978 were less virulent than those collected in 1992. This increase in virulence over time may have been due to genetic recombination and/or host genotype–directed selection for specific virulence in the pathogen population. This highlights the need to evaluate the host reaction to the disease with highly virulent isolates in order to identify the most robust resistance sources. Alternatively, screening of germplasm in the field where breeding lines are exposed to the local population of *A. lentis* can account for variability within the pathogen population (Tivoli et al. 2006).

Nasir and Bretag (1997a) divided a collection of Australian *A. lentis* isolates into six pathotypes based on quantitative differences in pathogenicity. However, there is concern as to whether true pathotype differences exist or if the differences observed in disease severity are a measure of the natural distribution of aggressiveness within a population, ranging from low to high (Taylor and Ford 2007). Banniza and Vandenberg (2006) reported that the host reaction of 16 lentil genotypes to 65 isolates of *A. lentis* collected in Canada resulted in a continuum of severity of infection. These results indicated natural variation of aggressiveness in the population without any distinct pathotypes. Taylor and Ford (2007) defined a pathotype as a subclass or group of a pathogen distinguished from others of the same species by its pathogenicity on a specific host (genotype) ie a qualitative difference in disease severity. Whereas, aggressiveness reflects the natural variation in pathogenicity or level of disease (measured quantitatively) within the pathogen population. However, since resistance to *A. lentis* has been found to be controlled by specific resistance genes (Ford et al., 1999; Nguyen et al., 2001), there is the likelihood that pathotypes of *A. lentis* have evolved that have qualitative differences on lentil genotypes.

3. FUSARIUM WILT

Fusarium wilt of lentil is an important disease reported in every continent where lentil is grown except Australia (Beniwal *et al.*, 1993; Tosi and Cappelli, 2001). The disease may cause complete crop failure under favorable conditions for disease development, and can be the major limiting factor for lentil cultivation in certain areas (Chaudhary and Amarjit, 2002). The common name lentil wilt has been used to describe many general wilting and dying symptoms. Hence a number of pathogens have been reportedly associated with lentil wilt (Khare, 1981) possibly because of the difficulty in species identification and confusion in the fusarium taxonomy. Strictly speaking, the causal organism of vascular wilt of lentil is *Fusarium oxysporum* Schlecht. Emend. Snyder & Hansen f. sp. *lentis* Vasudeva and Srinivasan. Although its sexual state has not been found, it is generally believed to belong to the Hypocreales of Ascomycetes.

In culture, the mycelium of the pathogen is hyaline, septate and much branched. Growth patterns on media vary from fluffy to appressed and vary in color from no color to pink. *F. oxysporum* f. sp. *lentis* produces three kinds of spores: microconidia; multi-septate macroconidia, which have a distinct foot cell and a pointed apical cell; and chlamydospores (Khare 1980). Microconidia are ovoid or kidney-shaped, hyaline and usually one celled. Macroconidia are long with pointed apical cell and notched basal cell, and two to seven celled. Chlamydospores are oval or spherical, one-celled, and thick walled, formed singly in macroconidia or apical or intercalary in the hyphae.

3.1. Symptoms

Fusarium wilt usually occurs near or at reproductive stages (flowering to pod-filling) of crop growth. Symptoms include wilting of top leaves that resemble water deficiency, stunting of plants, shrinking and curling of leaves from the lower part of the plants that progressively move up the stems of the infected plant. Plants finally become completely yellow and die. Root symptoms include reduced growth with marked brown discoloration, tap root tips that are damaged and proliferation of secondary roots above the area of tap root injury. Discoloration of vascular tissue in the lower stem may not always be visible. However, in India, the disease has also been reported to occur at the seedling stage. General symptoms at the seedling stage include seed rot and sudden drooping more like wilting and damping off (Khare, 1980).

Field diagnosis should be done in connection with field cropping history. Recent lentil production especially with a history of fusarium wilt will indicate potential wilt problems. Suspect stunted and wilted plants should be carefully removed from the soil so that the roots can be checked for reduced growth without external fungal growth. External fungal growth indicates the presence of other diseases such as collar rot. Lower stems should be split to check for vascular discoloration. Although vascular discoloration is not always symptomatic of fusarium wilt the presence of discoloration would confirm the disease. Culturing of infected plant tissue in the laboratory should be done with caution because of the possible presence of other saprophytic *Fusarium* spp. that appear similar to *F. oxysporum* f. sp. *lentis*. A pathogenicity test on lentil is necessary to confirm *F. oxysporum* f. sp. *lentis*.

3.2. Epidemiology

Like many other formae specilaes of *F. oxysporum*, the pathogen has a very limited host range as it only infects lentil in nature. In inoculation studies, *F. oxysporum* f. sp. *lentis* was unable to infect cowpea, french-bean, bengal gram, lathyrus, mungbean, uribean, pea, soybean or red gram (Khare, 1980). The disease is favored by warm and dry conditions (Bayaa and Erskine, 1998) with an optimal temperature of 22–25 °C.

F. oxysporum f. sp. *lentis* is a soilborne pathogen, although seed infestation and infection is common. The chlamydospores can survive in soil either in dormant form or saprophytically for several years without a suitable host. A survey of soil samples from Sangod Tehsil of Kota, Rajasthan, India, found that *F. oxysporum* f. sp. *lentis* was the most prevalent lentil pathogen (Chaudhary and Amarjit, 2002). Synergistic interaction between *F. oxysporum* f. sp. *lentis* and root knot nematode *Meloidogyne javanica* was observed in lentil cultivars resistant or susceptible to fusarium wilt (De *et al.*, 2001). Presence of the nematode significantly increased wilt incidence, caused significant reduction in shoot length, root length and nodulation in both susceptible and resistant cultivars (De *et al.*, 2001).

3.3. Control Methods

The most economical means to control fusarium wilt of lentil is through the use of resistant cultivars (Bayaa et al., 1997; Stoilova and Chavdarov, 2006). Resistant or moderately resistant lentil cultivars (DPL 58, DPL 61 and DPL 62) significantly reduced wilt incidence and severity of root rot, and increase grain yield (Chaudhary and Amarjit, 2002). Studies in genetics of resistance to fusarium wilt will eventually help to produce more resistant lentil cultivars (Eujayl et al., 1998). Selecting cultivars that mature early and adjusting the planting date if possible can reduce disease incidence by escaping a portion of lentil growth from weather conditions favorable to the disease. The most suitable planting dates vary according to the different production regions. Use of clean seed for sowing and/or the use of fungicidal seed treatments can eliminate or reduce contaminating inoculum sources. Since the pathogen has a very restricted host range, a three to five year rotation will help reduce inoculum level in the field.

Although *in vitro* and greenhouse tests showed fungicides were effective against fusarium wilt, field applications were not always practical because of the cost and technical difficulty of incorporating chemicals into soil during the growing season. Seed treatment with benomyl fungicide reduced the incidence of fusarium wilt. Biological control has been a focus of recent research. El-Hassan and Gowen (2006) tested three formulations to enhance efficacy of the biocontrol agent *Bacillus subtilis*, and found that formulations with either talc or glucose significantly decreased disease severity and showed enhanced plant growth promoting activity by increasing root length. Efficacy and practicality of biocontrol in the field remains to be worked out.

4. BOTRYTIS GREY MOULD

Botrytis grey mould (BGM) of lentil, caused by the fungal pathogens *Botrytis fabae* (Sard) (teleomorph: *Botryotinia fabae*) and *B. cinerea* (Pers.: Fr.) (*Helotiales, Sclerotiniaceae*) (teleomorph: *Botryotinia fuckeliana*), is a serious, but sporadic disease. Knights (1987) first reported the disease in Australia on lentil in the wet year of 1983 and since then the disease (Figure 1d) has caused considerable damage to commercial lentil crops grown throughout Victoria and South Australia (Lindbeck et al. 2003). In Canada, the disease was first reported in 1970 (Morrall et al. 1972) with serious epidemics of botrytis stem and pod rot occurring from 1992 to 1994 (Morrall 1997). A series of cool, wet summers in those years provided ideal conditions for botrytis epidemics to occur. Elsewhere around the world the disease has been recorded on lentil in the USA in 1964 (Wilson and Brandsberg 1965) and in New Zealand in 1987 (Cromey et al. 1987). Botrytis grey mould has also been reported as being a serious problem throughout the sub-continent including Bangladesh (Gowda and Kaul 1982), Nepal (Karki 1993) and Pakistan (Bashir and Malik 1988, Iqbal et al. 1992). Brouwer et al. (2000) found only *Botrytis cinerea* to be a problem in lentil production in Pakistan but not the rest of the Indian subcontinent, despite *B. fabae* being a common disease in the region. On the South American continent, the disease has been reported as a production constraint in Colombia (Bascur 1993). *B. cinerea* has also been isolated off infected and dying plants in Chile (France et al. 1988) and from lentil with symptoms in the field in northern Egypt (Hamdi and Hassanein 1996). In Europe, the production of lentil is claimed to be limited by low profitability and its susceptibility to *Botrytis* in wet climates (Carrouee et al. 2000).

Colonies of *B. cinerea* grow quickly, reaching 6.0 cm diameter and more in 10 days at 20°C on oatmeal agar, at first hyaline but later becoming grey to greyish brown (Domsch et al. 1980). Conidiophores arise irregularly and often in patches, without a basal swelling, frequently 2 mm or more long, mostly 16–30 μm thick, branched, often with a stipe and a rather open head of branches, smooth, clear, brown below, paler near the apex, with the ends of the branches often quite colourless. The conidia are ellipsoidal or obovoid, often with a slightly protuberant hilum; colourless to pale brown, smooth, 6–18 × 4–11 μm (mostly 8–14 × 6–9 μm) (Ellis and Waller 1974a). Sclerotia are black and usually smaller and thinner than those of *Sclerotinia sclerotiorum*. *Botrytis cinerea* is distributed worldwide, but occurs mainly in humid temperate and subtropical regions (Domsch et al. 1980).

Ellis and Waller (1974b) provided a description of *B. fabae*. Conidiophores are not normally found on leaves under field conditions but develop and produce conidia in a humid chamber. In culture on bean leaf agar conidium production is encouraged by the presence of relatively high concentrations of inorganic salts such as sodium nitrate. Sclerotia are formed abundantly in culture, discrete or sometimes confluent, mostly 1–1.7 mm in diameter, rarely up to 3 mm. Conidia are always much larger than those of *Botryotinia fuckeliana*, 14–29 × 11–20 μm (mostly 16–25 × 13–16 μm).

4.1. Symptoms

All aboveground plant parts of lentil can be affected by botrytis grey mould. Depending on the location of the crop, symptoms may initially appear either on flowers and pods (Figure 1d), or lower in the crop canopy. The most damaging symptoms become apparent after the crop has reached canopy closure and a humid microclimate is produced under the crop canopy. The disease first appears on the lower foliage as discrete lesions on leaves which are initially dark green, but turn greyish-brown, then cream as they age, that enlarge and coalesce to infect whole leaflets. Severely infected leaves senesce and fall to the ground. These can often act as a secondary source of inoculum by lodging in leaf and stem axils and initiating stem infections. If the canopy remains humid for extended periods infection can spread to the lower stems which quickly become girdled and covered with a furry layer of conidiophores, eventually causing stem death and whole plant infection.

Death of plants can often occur before the onset of flowering and pod fill. Infection will continue to spread resulting in patches of dead plants within crops (Bayaa and Erskine 1998). When the weather turns dry and the infected plants are disturbed, clouds of spores are released into the air. Flowers can show symptoms of infection with typically grey mouldy growth present on petals, causing flower death (Bayaa and Erskine 1998). Pods which become infected will be covered in the grey mouldy growth, rot, and turn brown when dried out. Seeds within these pods will fail to fill properly (Davidson et al. 2004). Infected seeds are discoloured and shrivelled (Bayaa and Erskine 1998). When infected seeds are sown seedling blight can occur. Seedling blight is characterised by the prolific grey mycelial growth of the pathogen on the hypocotyl at the soil line (Morrall 1997). This stage of the disease also has the potential to spread along seedling rows as the pathogen spreads from plant to plant (Morrall 1997), reducing seedling populations.

4.2. Epidemiology

There are several main sources of inoculum of botrytis grey mould, these include; seed-borne inoculum, sclerotia, mycelium in old infected trash, and alternate host plants. In Australia, *B. cinerea* and *B. fabae* have been frequently isolated off lentil seed (T. Bretag and K. Lindbeck unpublished data). Under Canadian conditions *B. cinerea* has been found to be highly seed-borne and can affect seed viability, seedling emergence and crop establishment (Morrall 1997). *Botrytis cinerea* has also been isolated from lentil seed in India (Rajendra et al. 1987), Spain (Diaz and Tello 1994) and the USA (Kaiser 1992). Sclerotia are considered the main survival structure for both *B. cinerea* and *B. fabae*. Sclerotia produced by *B. fabae* were considered an important source of inoculum for chocolate spot of faba beans (Harrison 1979), but only on the soil surface where the bodies are exposed to sunlight and produce conidia. In addition, sclerotia have the ability to produce conidia over an extended period of time. Under laboratory conditions sclerotia of *B. cinerea* were found to continue to sporulate for approximately 12 weeks after the first crop of conidia (Nair and Nadtotchei 1987). Resting mycelium in old host plant debris may survive and produce conidia under humid conditions for extended periods (Bayaa and Erskine 1998). Both *B. cinerea* and *B. fabae* can survive in lentil trash on the soil surface for at least 12 months under Australian conditions (K. Lindbeck, unpublished data).

The development of botrytis grey mould epidemics is largely determined by the prevailing environmental conditions during periods of inoculum production and dispersal in the presence of the host. The coinciding of all these events can result in the development of an epidemic very quickly when compared to most other diseases (Jarvis 1980b). It is generally assumed that for *B. cinerea*, inoculum is always present in the field and that production, liberation and dispersal of inoculum is an ongoing process (Jarvis 1980b), for *B. fabae* this principle will not always apply given its restricted host range. Environmental conditions and canopy density have also been shown to be primary factors that influence the development of botrytis grey mould epidemics in lentil crops (Kaiser 1992, Morrall 1997, Bailey et al. 2000). A dense crop canopy, especially following canopy closure, and humid conditions following rain favour the sporulation and dispersal of *B. cinerea* on decaying lentil tissue (Kaiser 1992), and its appearance is often characteristic of a lentil crop with rank growth (Morrall 1997). There have been numerous studies on identifying the optimum temperatures and relative humidities for disease development by *B. cinerea* and *B. fabae* on other host crops, namely chickpea and faba bean. Temperatures ranging from 15–25 °C and RH > 95% have been found to be optimal for initiation and development of disease (Harrison 1980, Wilson 1937, Tripathi and Rathi 1992, Rewal and Grewal 1989) particularly at flowering and after canopy closure (Lindbeck et al. 2002).

Botrytis cinerea has a broad crop host range collectively, including faba bean, chickpea, field pea, lupin and pasture legumes such as lucerne and clover. Other host species include a wide range of ornamental and horticultural crops. This provides the pathogens with a wide geographic distribution and alternate host mechanism. *B. cinerea* is known to have over 200 host plant species including many ornamental, horticultural, field crop and weed species (Jarvis 1980a). The wide host range of *B. cinerea* is likely to make the role of alternate hosts an important part in survival from one season to the next (Davidson et al. 2004). Unlike *B. cinerea*, *B. fabae* is known to have a more restricted host range. Yu (1945) found only four plant species were able to become infected, (ie, produce lesions) after inoculating 28 species of leguminous plants with conidia of *B. fabae*, these included *V. faba* L., *Pisum sativum* L., *P. sativum* var. *arvense* Poir, and *Vicia sativa* L. Other recorded hosts of *B. fabae* include *Phaseolus vulgaris* (Ellis and Waller 1974b).

4.3. Control Methods

Practices that have been effective in crop canopy management can be used to avoid the creation of a microclimate which encourages disease epidemics (Bretag and Materne 1998a). Practices that delay or avoid the formation of a dense canopy include the adjustment of sowing dates and rates, use of wider row spacing to increase air flow, weed control and optimum fertiliser use, particularly avoiding high nitrogen levels (Bayaa and Erskine, 1998; Lindbeck et al. 2002). A program of stubble reduction may also be undertaken by grazing, burning or burying, to reduce the carryover of infected stubble into the following season. In addition, potential alternate host plants can be controlled to reduce the early build up of disease inoculum (Lindbeck et al. 2002). Lentils should also not be grown adjacent or into a lentil, faba bean, chickpea, vetch or lathyrus stubble (Lindbeck et al. 2002).

Farmers can reduce the risk of seedling blight and disease carry-over by retaining seed only from disease free crops for sowing the following year, and using seed with less than 5% infection. Seed treatments with fungicides such as benomyl, carboxin, chlorothalonil or thiabendazole can reduce seed-borne inoculum levels (Bayaa and Erskine 1998, Bretag and Materne 1998b, Lindbeck et al. 2002, Morrall, 1997). Foliar fungicides are recommended in Australia for control of botrytis grey mould in lentil crops (Lindbeck et al. 2002, 2003); however, Bayaa and Erskine (1998) stated that fungicide control for grey mould in lentil was uneconomic. Carbendazim, chlorothalonil, mancozeb and procymidone are the products widely used in Australia (Lindbeck et al. 2002). Iqbal et al. (1992) evaluated 14 fungicides and found that benomyl, thiabendazole and tridemorph were the most effective against *B. cinerea*.

Resistance to botrytis grey mould is poorly understood, but requires a better understanding to enable different sources of resistance to be utilised and subsequent pyramiding of resistance genes (Tivoli et al. 2006). Resistant lentil germplasm has been identified in Australia (Bretag and Materne 1999, Lindbeck et al. 2003), Canada (Kuchuran et al. 2003), Nepal (Karki, 1993) and Pakistan (Erskine et al. 1994, Tufail et al. 1993). Testing in Australia has found variability within Australian lentil germplasm for resistance to botrytis grey mould (Bretag and Materne 1999) and a breeding program to improve the resistance to the disease is currently underway (Lindbeck et al. 2003). The lentil variety 'Nipper' was released in 2006 from the Australian lentil breeding program with resistance to both botrytis grey mould and ascochyta blight.

5. LENTIL RUST

Rust, caused by *Uromyces viciae-fabae* (Pers.) Schroet, (*Uredinales, Pucciniaceae*) is regarded as the most important foliar disease of lentil (Figure 1b) (Erskine et al. 1994). Complete crop failures can occur due to this disease (Beniwal et al. 1993). Rust of lentil is widespread globally, but is considered to be a production problem in Algeria, Bangladesh, Canada, Ethiopia, India, Italy, Morocco, Pakistan, Nepal, Syria and Turkey (Erskine et al. 1994). The disease also occurs widely in South America including Argentina, Brazil, Chile, Colombia, Ecuador, and Peru (Bascur 1993).

Rust is an autoecious fungus, completing its life cycle on lentil. The aecia of *U. viciae-fabae* are amphigenous or hyphyllous, usually in groups surrounding the pycnia or sometimes scattered, cupulate, 0.3–0.4 mm diam. The aeciospores are spheroidal, 18–26 μm diam.; wall hyaline, verrucose, 1 μm thick. Uredia are amphigenous and on the petioles and stems, scattered, cinnamon, 0.5–1 mm diam. Uredospores are ellipsoidal or obovoidal 22–28 × 19–22 μm; wall luteous to sienna, very finely echinulate, 1–2.5 μm thick; pores 3–4, equatorial or occasionally scattered on *Lathyrus*. Telia are like the uredia but black and larger: 1–2 mm diam. Teliospores are ellipsoidal, obovoidal or cylindrical, rounded or subacute above, 25–40 × 18–26 μm; wall chestnut, smooth, 1–2 μm thick at the sides, 5–12 μm thick above; pedicels sienna to luteous, up to 100 μm long. (Laundon and Waterson 1965).

5.1. Symptoms

Rust starts with the formation of yellowish-white pycnidia and aecial cups on the lower surface of leaflets and on pods, singly or in small groups in a circular form (Agrawal et al. 1993). Later, brown uredial pustules emerge on either surface of leaflets, stem and pods (Figure 1b). Pustules are oval to circular and up to 1 mm in diameter. They may coalesce to form larger pustules (Bayaa and Erskine 1998). The telia, which are formed late in the season, are dark brown to black, elongated and present mainly on branches and stems. In severe infections leaves are shed and plants dry prematurely (Bakr 1993), the affected plant dries without forming any seeds in pods or with small shriveled seeds. The plant has a dark brown to blackish appearance, visible in affected patches of the paddock or in the whole paddock if totally infected (Beniwal et al. 1993).

5.2. Epidemiology

The disease generally starts from low-lying patches in the paddock and radiates towards the border (Bayaa and Erskine 1998). Lentil seed may be contaminated with pieces of rust-infected leaf, stem and pericarps, which can act as primary inoculum for the recurrence of the disease in most years (Khare 1981, Agrawal et al. 1993). Rust may also perpetuate on weed hosts from where it may infect lentil crops by windborne teliospores. High humidity, cloudy or drizzly weather with temperatures 20 to 22 °C favour disease development (Agrawal et al. 1993). The disease generally occurs during the flowering /early podding stage. Aeciospores germinate at 17–22 °C and infect other plants forming either secondary aecia at temperatures of 17 – 22 °C or uredia at 25 °C. Uredosori develop later in the season and are rapidly followed by telia (Beniwal et al. 1993). After harvest, aecia and uredia present on lentil trash die out, but teliospores tolerate high temperatures and allow the fungus to survive the summer. At lower temperatures, uredospores could be an important means of survival (Bayaa and Erskine 1998). Uredomycelium is highly resistant to heat and sunlight and is probably important for continued development and survival of rust in hot, dry conditions. The predominant form of survival will vary with the environment and location (Bayaa and Erskine 1998). Teliospores germinate at 17–22 °C without a resting period and cause new outbreaks of the disease each season.

There are 70 recorded hosts of *U. viciae-fabae* including lentil, chickpea, field pea, *Lathyrus* spp and *Vicia* spp. (Parry and Freeman 2001). Degrees of host specialisation and pathogenic variability do exist within populations of *U. viciae-fabae* worldwide. Much research has been performed regarding race identification within *U. viciae-fabae* over many years with conflicting outcomes regarding the suggestion of forma speciales within the species.

5.3. Control Methods

Cultural control methods currently recommended for control of *U. viciae-fabae* include: control of volunteer plants over summer; isolation of new season crops from old host crop stubbles (MacLeod 1999) and destruction of old lentil stubbles (Prasada and Verma 1948). Early studies on the control of lentil rust in India found seed treated with Agrosan (phenylmercury acetate) to control seed-borne inoculum (Prasada and Verma 1948). Singh (1985) found Vigil (diclobutrazole), applied as a seed dressing prevented the appearance of *U. viciae-fabae* up to 70 days following inoculation with uredospores; bayleton (triadimefon) prevented disease appearance up to 40 days post inoculation and the untreated control was severely infected with rust 35 days after inoculation. Experiments investigating the use of foliar fungicides for rust control by Agarwal et al. (1976) found Hexaferb (Ferric dimethyldithiocarbamate) and Dithane M-45 to give the best control of *U. viciae-fabae* in experimental plots at Jabalpur, India. In addition, Dithane M-45 also increased plot yield by 82% and grain weight by 24% when compared to the untreated control. The use of host plant resistance is the best means of rust control (Bayaa and Erskine 1998). Genetic differences among genotypes and sources of resistance have been reported worldwide, with several rust resistant lines available. Resistance to rust is reported to be controlled by a single dominant gene (Sinha and Yadav 1989). Studies in factors influencing the mechanism of resistance to rust in lentil (Reddy and Khare 1984) reported that resistant cultivars contained more leaf surface wax, P, K, S, Zn, Fe, Cu levels of phenols than susceptible cultivars which had higher levels of amino acids, N, Mn, and sugars. Structurally there were no significant differences found between resistant and susceptible cultivars.

6. STEMPHYLIUM BLIGHT

Stemphylium blight of lentil is caused by the pathogen, *Stemphylium botryosum* Wallr (*Pleosporales, Pleosporaceae*) (teleomorph: *Pleospora herbarum* (Fr) Rab.). The disease has been reported on lentil from Bangladesh (Bakr 1993), Canada (Morrall 2003), Egypt, Syria (Bayaa and Erskine 1998) and the USA (Wilson and Brandsberg 1965). The disease has the potential to cause yield losses of up to 62% under conducive conditions (Figure 1c) (Bakr 1993).

Conidiophores of *S. botryosum* have 1–7 septate, 20–72 × 4–6 μm, pale brown to brown, with a swollen apical sporogenous cell 7–11 μm diam., and slightly roughened toward the apex. They possess a single apical pore 5–8 μm diam. Conidia are oblong, olive to brown, ovoid to subdoliiform, occasionally constricted at 1–3 transverse septa and at the 1–3 longitudinal septa if complete, 19.5 × 28.5 μm with a single basal pore 8 μm diam. and a roughened outer wall. Ascostromata are scattered, immersed to erumpent in the tissue of the host, 100–500 μm in diam. Asci are 90–250 × 20–50 μm containing eight ascospores, cylindrical to slightly club shaped. Ascospores are light to yellow brown, ellipsoid to club shaped with 7 septate, slightly constricted at the three primary transverse septa, muriform and 26–50 × 10–20 μm (Booth and Pirozynski 1967).

6.1. Symptoms

Symptoms of stemphylium blight start with the appearance of small pin-headed light brown to tan coloured spots on leaflets. Under ideal conditions the small spots enlarge rapidly, covering the entire leaflet surface within a 2–3 day period. The infected tissue appears light cream in colour, often with angular patterns of lighter and darker areas that spread across, or long, the entire leaflet (Morrall 2003; Figure 1c). The affected foliage and stems gradually turn dull yellow, giving a blighted appearance to the crop (Bakr 1993). The infected leaves can be abscised rapidly, leaving only the terminal leaflets on the stems. The stems bend down, dry and gradually turn ashy white, but pods remain green. White mycelial growth can sometimes be seen on the infected stems (Bakr 1993).

6.2. Epidemiology

Important sources of *S. botryosum* inoculum include infected crop debris and infected seed. Infected crop debris can be a source of primary inoculum in the form of air-borne ascospores or as resting mycelium, based on the studies of the pathogen on other host crops such as alfalfa (Gilchrist 1990). *Stemphylium botryosum* is known to be carried on seed (Booth and Pirozynski 1967) and *Stemphylium* spp. has been isolated off lentil seed in Australia (Nasir and Bretag 1997), but the significance of seed-borne *S. botryosum* inoculum on disease initiation in lentil is not clearly understood (Mwakutuya 2006). Bakr (1993) has reported from Bangladesh that the pathogen commences infection when the ambient night temperature remains above 8°C, and the mean day temperature exceeds 22 °C. The RH inside the crop canopy must also reach 94%. In India, Singh and Singh (1993) found that an average mean temperature of 18 °C ± 2 °C and RH of 85−90% in the morning was favourable for disease development and spread. Most recently, in Canada Mwakutuya (2006) found that symptom development of *S. botryosum* was optimised after 48 h of leaf wetness at temperatures above 25 °C. The host range of *S. botryosum* is wide and includes a large number of ornamental, horticultural and field crop species. These include lentil (Bakr 1993), lupin (Tate 1970), tomato (Bashi and Rotem 1975) spinach (Koike et al. 2001), alfalfa, clover (Smith 1940), lettuce (Tate 1970), apple, onion and gladiolus (Booth and Pirozynski 1967).

6.3. Control Methods

There is little published information available regarding cultural control methods for *S. botryosum* in lentil. Being a stubble-borne disease strategies such as destruction of old crop residues, and crop rotation would assist in decreasing potential inoculum sources. In Bangladesh, delayed sowing was found to significantly decrease the incidence of stemphylium blight in lentil, but later sowing resulted in reduced crop yields and heavy infection by *U. viciae-fabae* (Bakr 1993). Foliar fungicides have been found to be effective in the management of stemphylium blight. In Bangladesh the application of Roval 50 WP was found to effectively control the disease when applied three times at weekly intervals starting from the initiation of the disease (Bakr 1993). In other horticultural crops, such as asparagus and garlic, *Stemphylium* spp. has been successfully controlled using chlorothalonil (Meyer et al. 2000), tebuconazole and procymidone (Basallote-Ureba et al. 1998). Sources of host plant resistance have been identified in screening nurseries in Bangladesh. The resistant varieties 'Barimasur 3' and 'Barimasur 4' were released with resistance to *S. botryosum* (Sarker et al. 1999a, b). Studies by Chowdhury et al. (1997) found lentil cultivars with resistance to *S. botryosum* had a higher number of epidermal hairs, thicker cuticle, thicker epidermal cell layer and thicker cortical layers. In addition, resistant lines were also found to have fewer stomata than susceptible cultivars.

7. ANTHRACNOSE

Anthracnose is an important disease of lentil, caused by *Colletotrichum truncatum* (Schwein.) Andrus and Moore and has been reported from Bangladesh, Canada, Ethiopia, Morocco, Syria and Canada (Anderson et al. 2000; Kaiser et al. 1998). Although *C. truncatum* has not been recorded as a pathogen on lentil in Australia, isolates have been recorded from soybean (*Glycine max*), *Xanthium occidentale* and peanut (*Arachis hypogaea*) growing in northern Australia. These non-host isolates were shown to be genetically different to the lentil infecting isolates from Canada using molecular markers and morphological descriptors (Ford et al. 2004). However, under optimal inoculation conditions the soybean isolate can infect lentil and faba bean leaves and stems. Further gene sequencing and pathogenicity testing of the Australian isolates of *C. truncatum* on lentil is required to validate the taxonomy of this group of isolates. In the mean time, the Australian lentil industry remains under threat from this exotic disease, which causes severe yield and seed quality loss under epidemic conditions.

In culture, the mycelium of *C. truncatum* growing on PDA at 25 °C is dark brown to black in colour with setae being rare. The conidia are ellipsoidal, hyaline and aseptate with one rounded and one pointed end and are $16.0–20.0 \times 3.0–5.0\,\mu m$ in size. Setae are generally acicular, most swollen at the base, tapered to the apex and comprised one or two septa. Mycelial appressoria and appressoria are produced directly from the germ tube, are generally brown, clavate and occasionally irregular. No teleomorph has been found in the wild (Kaiser et al. 1998) however, the teleomorph was recently induced under laboratory conditions and named *Glomerella truncate* (Armstrong-Cho and Banniza, 2006).

7.1. Symptoms

Irregularly shaped, light brown necrotic lesions start to develop on lower stems and gradually increase in number and size until they coalesce and give the stems a blackish brown appearance. Lesions on leaves are circular with few acervuli in the middle of each lesion and premature leaf drop begins at early flowering. Conidia form in acervuli on infected plants, and secondary spread of conidia to neighboring plants occurs by rain splash. The fungus penetrates the vascular tissue, which results in plant wilting, and large brown patches of dying plants become evident in the field after flowering.

7.2. Epidemiology

The disease is favored by high humidity and temperatures of 25–30°C, and is seed borne but has not been shown to be transmitted from seed to seedling. The pathogen is capable of surviving for up to four years as microsclerotia in crop residue and may become active again when in contact with fresh host tissue, spreading as conidia with rain splash and on plant debris through wind dispersal between crops (Buchwaldt et al. 1996).

A study on the infection process of leaves inoculated with a spore suspension, by Chongo et al. (2002) found that in the initial infection phase, conidia germinated within 3–6 hours after infection (hai) and formed appressoria at 6–12 hai. By 24 hai infection hyphae infected epidermal cells inter- and intra-cellularly. Differential host cell reaction was observed by the resistant cultivar 2 to 3 days after infection. Hyphal spread was slower and phenolic compounds accumulated more quickly in the resistant line, resulting in fewer, smaller lesions than in the susceptible cultivar.

7.3. Control Methods

The current disease management practices are based primarily on application of foliar fungicides such as chlorothalonil or benomyl (Chongo et al. 2002). However, seed treatment with fungicides such as benomyl or thiabendazole provides complete control of the seed-borne fungus. Breeding for resistance has suffered from a lack of highly resistant germplasm to include in breeding programs. In Canada, Buchwaldt et al. (2004) found only 16 out of 1,771 accessions of lentil to contain resistance to anthracnose after field and glasshouse screening. As well two pathotypes of *C. truncatum* were identified with the Ct0 pathotype isolated more frequently from commercial seed samples than the Ctl pathotype, although both pathotypes were isolated with similar frequency from plants in commercial fields planted to susceptible cultivars. Pathotype Ct0, to which no resistance has yet been identified, presents a high risk to lentil production in Canada and potentially worldwide.

PCR-based diagnostics tests have been developed to detect the pathogen in plant tissues (Ford et al. 2004) and are a valuable and reliable alternative to conventional seed health testing methods. These tests can be applied directly to suspect infected tissues taken from the field, to identify the pathogen much faster and potentially more accurately than traditional culturing techniques.

8. VIRUSES

Lentil plants can be infected by a range of viruses but generally the affect on yield is not as great as that caused by fungal pathogens (Beniwal et al. 1993). Viruses tend to be transmitted by aphids and/or seed infection thus controlling the insect vector, planting disease-free seed and use of resistant cultivars will aid in controlling these diseases. The following viruses of lentil are found in the major lentil growing regions of the world.

Lentil yellows disease is caused by several related luteoviruses such as bean leaf roll virus (BLRV), beet western yellows virus (BWYV), or subterranean clover red leaf virus (SCRLV). BLRV was first reported in Australia in 1999 (Freeman, unpublished) and BWYV was recently reported in Iran (Makkouk et al. 2001). The causal viruses are transmitted in a persistent manner by aphids, but not by seed. Epidemic spread of this disease is always associated with high aphid vector populations. The initial symptoms on leaves of virus infected lentil plants show interveinal chlorosis, which intensifies with time until the whole leaf becomes yellow. Other symptoms include leaf rolling, reduction in leaf size and significant reduction in pod setting.

Bean yellow mosaic is caused by the bean yellow mosaic virus (BYMV) which belongs to the potyvirus group. The virus has a wide host range and is transmitted through sap and by aphids in the nonpersistent manner. Transmission of the virus in lentil seed has not been reported. Leaf symptoms of infected plants include mild mosaic followed by leaf narrowing. The new growth of leaves from leaf axils are narrow, elongated and light green. Early infections adversely affect plant growth and yield (Beniwal et al. 1993).

Pea seed borne mosaic is caused by the pea seedborne mosaic virus (PSbMV) which belongs to the potyvirus group. The virus is seed borne in lentil, but also infects faba beans and field peas (Makkouk et al. 1993; Latham and Jones 2001) and is transmitted in a nonpersistant manner by aphids. The disease is characterized by a mild mosaic and malformation of leaves. The affected plants show little stunting and twisting of stems. Seeds from the affected plants are smaller than normal and deformed (Beniwal et al. 1993).

Cucumber mosaic virus (CMV) belongs to the cucumovirus group. CMV has a very wide host range and can be transmitted through sap, by aphid species in a nonpersistent manner and infected seed (Fletcher et al. 1999). Leaf symptoms include vein clearing followed by mild systemic mottle. Infected plants show stunting as the disease advances.

Alfalfa mosaic virus (AMV) is an alfamovirus and has a wide host range. The virus is spread by aphid species as well as being seed transmitted (Latham and Jones 2001). In areas where large aphid populations occur, crop losses can be high due to reduced plant growth and seed yield. Symptoms can be variable depending on the stage of growth at infection; environmental conditions and the host however, in lentil plants the leaves become twisted, deformed and stunted leading to necrotic tip growth.

Broad bean stain virus (BBSV) is a comovirus known to occur in Europe, North Africa and West Asia however, this virus has yet to be detected in Australia. BBSV naturally infects faba bean, dry pea and vetch and is transmitted through sap, seed and by weevils. The disease is characterized by a mild mottling on the leaves, which is not easily recognizable because of the small size of the lentil leaf. The affected plants are reduced in growth, which is easily recognized, especially when they are compared with healthy plants. Seeds from infected plants occasionally show dark staining on the seed coat (Beniwal et al. 1993).

9. MINOR DISEASES

Other minor diseases of lentil include White mould caused by *Sclerotinia sclerotiorum* (Lib.) de Bary that occurs from early flowering to pod setting where conditions are wet and cool (Figure 1e). The pathogen produces sclerotia that can survive in the soil until cool moist soil conditions exist that induce the sclerotia to germinate and produce apothecia that release ascospores to reinfect lentil crops. Other root and basal stem fungal diseases include collar rot caused by *Sclerotium rolfsii* Sacc., *Rhizoctonia solani* Kuhn; and dry root rot caused by *Rhizoctonia bataticola* (Taub.) Butler, in which the perfect stage of the fungus is *Macrophomina phaseolina* (Tassi) Goid. Also Pythium seedling and root caused by *Pythium aphanidermatum* (Edson) Fitzp. and *P. ultimum* Trow; and, black root rot caused by *Fusarium solani* (Mart.) Appel & Wr.

Minor foliar fungal diseases include Powdery mildew caused by *Erysiphe polygoni* DC (conidial stage, *Oidium* sp.) that has been recorded in Cyprus, Ethiopia, India, Siberia, Sudan, Syria, Tanzania and the former USSR; Downy mildew caused by *Peronospora lentis* Gäumannn from Egypt, France, India and Syria; Alternaria blight caused by *Alternaria alternata* (Fr.) Keissler that has limited range in Egypt, Ethiopia and India. The following fungi have been recorded as pathogens on lentil: *Cercospora lensi*,*Cladosporium herbarum* (Pers.) Link, *Cylindrosporium* sp., *Helminthosporium* sp., *Phoma medicaginis* (Malbr. & Roum.), *Septoria* sp., *Stemphylium hotryosum* Walr.

REFERENCES

Agarwal SC, Prasad KVV (1997) Diseases of lentil. Science Publishers Inc., USA.

Agarwal SC, Khare, MN, Agarwal PS (1976) Control of lentil rust by use of fungicides. Indian Phytopathology 29:90–91

Agarwal SC, Singh K, Lal SS (1993) Plant Protection of Lentil in India. In: Lentil in South Asia (pp 147–165). Erskine W and Saxena MC (eds), International Centre for Agricultural Research in the Dry Areas, Aleppo, Syria.

Ahmed S, Beniwal SPS (1991) Ascochyta blight of lentil and its control in Ethiopia. Tropical Pest Management 37: 3668–373.

Ahmed S, Morrall RAA, Sheard JW (1996) Virulence of *Ascochyta fabae* f.sp. *lentis* on lentil. Canadian Journal of Plant Pathology 18: 354–361.

Anderson KL, Buchwaldt L, Chongo G, Gossen BD, Morrall RAA, Pearse PG (2000) Diseases of lentil in Saskatchewan 1999. Canadian Plant Disease Survey 80: 96–98.

Armstrong-Cho CL, Banniza S (2006) *Glomerella truncata* sp. nov., the teleomorph of *Colletotrichum truncatum*. Mycological Research 110: 951–956.

Bailey KL, Gossen BD, Derksen DA, Watson PR (2000) Impact of agronomic practices and environment on diseases of wheat and lentil in southeastern Saskatchewan. Canadian Journal of Plant Science 80(4): 917–927.

Bakr MA (1993) Plant Protection of Lentil in Bangladesh. In: Lentil in South Asia (pp 177–186). Erskine W and Saxena MC (eds), International Centre for Agricultural Research in the Dry Areas, Aleppo, Syria.

Banniza S, Vandenberg A (2006) Investigations into the population structure of *Ascochyta lentis* in western Canada. Proceedings of the 1st International Ascochyta workshop on grain legumes. Le Tronchet, France.

Basallote-Ureba M, Prados-Ligero AM, Melero-Vara JM (1998) Effectiveness of tebuconazole and procymidone in the control of *Stemphylium* leaf spots in garlic. Crop Protection 17(6): 491–495.

Bascur GB (1993) Lentil and Faba Bean in Latin America: Their Importance, Limiting Factors and Research. ICRDA, Aleppo, Syria.

Bashi E, Rotem (1975) Sporulation of *Stemphylium botryosum* f. sp. *lycopersici* in tomatoes and of *Alternaria porri* f. sp. *solani* in potatoes under alternating wet-dry regimes. Phytopathology 65: 532–535.

Bashir M, Malik BA (1988) Diseases of major pulse crops in Pakistan: a review. Tropical Pest Management 34: 309–314.

Bayaa B, Erskine W (1998) Diseases of lentils. In: The Pathology of Food and Pasture Legumes (pp. 423–471) Allen DJ and Lenné JM (eds.) CAB International and ICRISAT, Wallingford, UK.

Bayaa B, Erskine W, Singh M (1997) Screening lentil for resistance to Fuarium wilt: methodology and sources of resistance. Euphytica 98:69–74.

Beauchamp CJ, Morrall RAA, Slinkard AE (1986) The potential for control of ascochyta blight of lentil with foliar-applied fungicides. Canadian Journal of Plant Pathology 8: 254–259.

Beniwal SPS, Seid A, Tadesse N (1989) Effect of sun drying of lentil seeds on the control of seedborne *Ascochyta lentis*. LENS Newsletter 16: 26–28.

Beniwal SPS, Bayaa B, Weigand S, Makkouk K, Saxena MC (1993) Field Guide to Lentil Diseases and insect Pests. International Centre for Agricultural Research in the Dry Areas, Aleppo, Syria, 106pp.

Bondartzeva-Monteverde VN, Vassilievsky NI (1940) A contribution to the biology and morphology of some species of Ascochyta on Leguminosae. *Acta Institute of Botany Academic Science USSR* 1938, Ser II, pp. 345–376 (cited by Agrawal and Prasad, 1997).

Booth C, Pirozynski KA (1967) *Pleospora herbarum*. C.M.I. Descriptions of Pathogenic Fungi and Bacteria, No. 150.

Bretag TW (1989) Evaluation of fungicides for the control of ascochyta blight in lentils. Tests of agrochemicals and cultivars. Annals of Applied Biology (Suppl.): 44–45.

Bretag TW, Materne MA (1998a) Lentil: Effect of seeding rate on foliar disease and grain yield. In: 'Summary of Field Research in the Wimmera 1998'. (Ed. J Hyett) pp. 35–36. (Department of Natural Resources and Environment: Horsham, Victoria).

Bretag TW, Materne MA (1998b) Lentil: Evaluation of 15 lines for resistance to botrytis grey mould. In: 'Summay of Field Research in the Wimmera 1998'. (Ed. J Hyett) pp. 37–38. (Department of Natural Resources and Environment: Horsham, Victoria).

Bretag TW, Materne MA (1999) Evaluation of 36 lentil lines for resistance to ascochyta blight and botrytis grey mould. In: 'Summary of Field Research in the Wimmera 1999'. (Ed. J Hyett) pp. 38–40. (Department of Natural Resources and Environment: Horsham, Victoria).

Brouwer JB, Bretag TW, Materne MA (1995) Coordinated improvement program for Australian lentils. Proceeding of 2nd European Conference on Grain Legumes, Copenhagen, Denmark. p. 25.

Brouwer JB, Sharma B, Malik BA, Hill GD (2000) Region 6: Asia-Pacific: Meeting the challenge. In: Linking research and marketing opportunities for pulses in the 21st century. Proceedings of the third international food legumes research conference. Current Plant science and biotechnology in agriculture, Vol 34, pp 115–129 (Ed R. Knight) Kluwer Academic Publishers.

Buchwaldt L, Morrall RAA, Chongo G, Bernier CC (1996) Windborne dispersal of *Colletotrichum truncatum* and survival in infested lentil debris. Phytopathology 86: 1193–1198.

Buchwaldt L, Anderson KL, Morrall RAA, Gossen BD, Bernier CC (2004) Identification of lentil germplasm resistant to *Colletotrichum truncatum* and characterization of two pathogen races. Phytopathology 94: 236–243.

Carrouee B, Gent GP, Summerfield RJ (2000) Production and uses of grain legumes in the European Union. In: Linking research and marketing opportunities for pulses in the 21st century. Proceedings of the third international food legumes research conference. Current Plant science and biotechnology in agriculture, Vol 34, pp 79–97 (Ed R. Knight) Kluwer Academic Publishers.

Chaudhary RG, Amarjit K (2002) Wilt disease as a cause of shift from lentil cultivation in Sangod Tehsil of Kota, Rajasthan. Indian Journal of Pulses Research 15: 193–194.

Chongo G, Gossen BD, Bernier CC (2002) Infection by *Colletotrichum truncatum* in resistant and susceptible lentil genotypes. Canadian Journal of Plant Pathology 24: 81–85.

Chowdhury AM, Ahmed A, Zaman M (1997) Studies on the defence structural factors of some susceptible and resistant varieties of lentil plants. Journal of Mycopathological Research 35: 35–39.

Cromey M G, Mulholland RI, Russell AC, Jermyn WA (1987) *Ascochyta fabae* f. sp. *lentis* on lentil in New Zealand. New Zealand Journal of Experimental Agriculture 15: 235–238.

Davidson JA, Pande S, Bretag TW, Lindbeck KD, Krishna-Kishore G (2004) Biology and control of*Botrytis* spp. in legume crops. In: Botrytis: Biology, Pathology and Control, (pp 295–318) edited by Y. Elad, B. Williamson, P. Tudzynski and N. Delen. Dordrecht, The Netherlands: Kluwer

De RK, Ali SS, Dwivedi, RP (2001) Effect of interaction between *Fusarium oxysporum* f. sp. *lentis* and *Meloidogyne javanica* on lentil. Indian Journal of Pulses Research 14: 71–73.

Diaz D, Tello JC (1994) A fungal inventory of lentil seeds (*Lens culinaris* Medik.) harvested in Castilla-La Mancha. Boletin de Sanidad Vegetal, Plagas 20(4): 857–870 (abstr).

Domsch KH, Gams W, Anderson TH (1980) Compendium of Soil Fungi. Academic Press, London

El-Hassan SA, Gowen SR (2006) Formulation and delivery of the bacterial antagonist *Bacillus subtilis* for management of lentil vascular wilt caused by *Fusarium oxysporum* f. sp. *lentis*. Journal of Phytopathology 154: 148–155.

Ellis MB, Waller JM (1974a) *Sclerotinia fuckelina,* C.M.I. Descriptions of Pathogenic Fungi and Bacteria, No. 431

Ellis MB, Waller JM (1974b) *Botrytis fabae*. C.M.I. Descriptions of Pathogenic Fungi and Bacteria, No. 432.

Erskine W, Tufail M, Russell A, Tyagi MC, Rahman MM, Saxena MC (1994) Current and future strategies in breeding lentil for resistance to biotic and abiotic stresses. Euphytica 73: 127–135.

Eujayl I, Erskine W, Bayaa B, Baum M, Pehu, E (1998) Fusarium vascular wilt in lentil: Inheritance and identification of DNA markers for resistance. Plant Breeding 117: 497–499.

Fletcher JD, Russell AC, Butler RC (1999) Seed-borne cucumber mosaic virus in New Zealand lentil crops: yield effects and disease incidence. New Zealand Journal of Crop and Horticultural Science 27: 197–204.

Ford R, Pang ECK, Taylor PWJ (1999) Genetics of resistance to ascochyta blight (*Ascochyta lentis*) of lentil and identification of closely linked molecular markers. Theoretical and Applied Genetics 98: 93–98.

Ford R, Banniza S, Photitia W, Taylor PWJ (2004) Morphological and molecular discrimination of *Colletotrichum truncatum* causing anthracnose on lentil in Canada Australasian Plant Pathology 33: 559–569.

France I A, Sepulveda P, Tay UJ (1988) Identification of *Botrytis cinerea* Pers. on lentil (*Lens culinaris* Med.). Agricultura Tecnica Santiago 48(2): 158–160 (abstr).

Gilchrist DG (1990) Stemphylium Leaf Spot. In: Compendium of Alfalfa Diseases (2nd ed) (pp 17–20), D Stuteville and D Erwin (eds), APS Press.

Gossen BD, Morrall RAA (1983) Effect of ascochyta blight on seed yield and quality of lentils. Canadian Journal of Plant Pathology 5: 168–173.

Gowda CLL, Kaul AK (1982) Lentil. In: Pulses in Bangladesh (pp 51–106). Bangladesh Agricultural Research Institute, Joydebpur, Dhaka and Food and Agriculture Organisation of the United Nations.

Hamdi A, Hassanein AM (1996) Survey of Fungal Diseases in North Egypt. LENS Newsletter 23(1/2): 52–56.

Harrison JG (1979) Overwintering of *Botrytis fabae*. Transactions of the British Mycological Society 72(3): 389–394.

Harrison JG (1980) Effects of environmental factors on growth of lesions on field bean leaves infected by *Botrytis fabae*. Annals of Applied Biology 95: 53–61.

Iqbal SM, Hussain S, Malik BA (1992) *In vitro* evaluation of fungicides against *Botrytis cinerea* of lentil. Lens Newsletter 19: 49–51.

Jarvis WR (1980a) Taxonomy. In: The Biology of Botrytis. J. R. Coley-Smith, K. Verhoeff and W. R. Jarvis, Academic Press: 1–17.

Jarvis WR (1980b) Epidemiology. In: The Biology of Botrytis. J. R. Coley-Smith, K. Verhoeff and W. R. Jarvis (eds), Academic Press: 219–250

Kaiser WJ (1992) Fungi associated with the seeds of commercial lentils from the U.S. Pacific Northwest. Plant Disease 76(6): 605–610.

Kaiser WJ (1997) Intra- and international spread of ascochyta pathogens of chickpea, faba bean and lentil. Canadian Journal of Plant Pathology 19: 215–224.

Kaiser WJ, Hannan RM (1986) Incidence of seedborne Ascochyta lentis in lentil germplasm. Phytopathology 76: 355–360.

Kaiser WJ, Hannan RM (1987) Seed treatment fungicides for control of seedborne *Ascochyta lentis* on lentil. Plant Disease 71: 58–62.

Kaiser WJ, Hellier BC (1993) *Didymella* sp. The teleomorph of *Ascochyta lentis* on lentil straw. *Phytopathology* 83: 692. (*Abstr.*).

Kaiser WJ, Wang BC, Rogers JD (1997) *Ascochyta fabae* and *A. lentis*: Host specificity, teleomorphs (*Didymella*), Hybrid analysis, and taxonomic status. *Plant Disease* 81: 809–816.

Kaiser WJ, Mihov M, Muehlbauer FJ, Hannan RM (1998) First report of anthracnose of lentil incited by *Colletotrichum truncatum* in Bulgaria. Plant Disease. 82: 128.

Kaiser WJ, Stanwood PC, Hannan RM (1989) Survival and pathogenicity of Ascochyta fabae f. sp. lentis in lentil seeds after storage for four years at 20 to −196 °C. Plant Disease 73: 762–764.

Karki PB (1993) Plant Protection of Lentil in Nepal. In: Erskine W and Saxena MC (eds.) Lentil in South Asia. (pp. 187–193) ICARDA, Aleppo, Syria.

Khare MN (1981) Diseases of Lentils, *In*: Lentils (Eds. C Webb, G Hawtin), Farnham Royal, U.K., pp 163–172.

Khare M N (1980) Wilt of Lentil. Jawaharlal Nehru Krishi Vishwa Vidyalaya, Jabalpur, M.P India 155 pp.

Knights EJ (1987) Lentil: A potential winter grain legume crop for temperate Australia. Journal of the Australian Institute of Agricultural Science 53: 271–280.

Koike ST, Henderson DM, Butler EE (2001) Leaf spot disease of spinach in California caused by *Stemphylium botryosum*. Plant Disease 85: 126–130.

Kuchuran M, Banniza S, Vandenberg B (2003) Evaluation of lentil varieties for resistance to Botrytis gray mould. In: Proceedings of Pulse Field Days, 6–7 January 2003, Saskatoon, Canada (p 89).

Latham LJ, Jones RAC (2001) Incidence of virus infection in experimental plots, commercial crops, and seed stocks of cool season crop legumes. Australian Journal of Agricultural Research 52: 397–413.

Laundon GF, Waterson JM (1965) *Uromyces viciae-fabae*. CMI Descriptions of Pathogenic Fungi and Bacteria No. 60 Commonwealth Mycological Institute, Kew, UK.

Lindbeck KD, Bretag TW, Materne MA (2003) Breeding for resistance to *Botrytis fabae* in Australian lentils. In: Proceedings of 8th International Congress of Plant Pathology, Christchurch, New Zealand, p 291.

Lindbeck KD Materne MA, Davidson JA, McMurray L, Panagiotopoulos K (2002) Lentil Disease Management Strategy for Southern Region GRDC. Pulse Australia and state departments extension article.

MacLeod W (1999) Faba bean: Rust disease. Agriculture Western Australia Farmnote 114/96.

Makkouk KM, Kumari SG, Bos L (1993) Pea seed-borne mosaic virus: occurrence in faba bean (*Vicia faba*) and lentil (*Lens culinaris*) in West Asia and North Africa, and further information on host range, transmission characteristics, and purification. European Journal of Plant Pathology 99: 115–124.

Makkouk KM, Fazlali Y, Kumari YSG, Farzadfar S (2002) First record of Beet western yellows virus, Chickpea chlorotic dwarf virus, Faba bean necrotic yellows virus and Soybean dwarf virus infecting chickpea and lentil crops in Iran. Plant Pathology 51: 387–387.

Meyer MP, Hausbeck MK, Podolsky R (2000) Optimal fungicide management of purple spot of asparagus and impact on yield. Plant Disease 84: 525–530.

Morrall RAA (1997) Evolution of lentil diseases over 25 years in western Canada. Canadian Journal of Plant Pathology 19(2): 197–207.

Morrall RAA (1998) Using thiobendazole to control seed-borne *Ascochyta* in lentil. Canadian journal of Plant Pathology 10: 370 (Abstr).

Morrall RAA (2003) Diseases of Lentil. In: Diseases of Field Crops in Canada (3rd ed) (pp 191–200) (KL Bailey, BD Gossen, RK Gugel and RAA Morrall eds),The Canadian Phytopathological Society.

Morrall RAA, McKenzie DL, Duczek LJ, Verma PR (1972) A qualitative survey of diseases of some specialty crops in Saskatchewan in 1970 and 1971: Sunflower, safflower, buckwheat, lentil, mustards and field pea. Canadian Plant Disease Survey 52(4): 143–148.

Mwakutuya E (2006) Epidemiology of Stemphylium blight on lentil (*Lens culinaris*) in Saskatchewan. MSc Thesis, Department of Plant Sciences, University of Saskatchewan, Saskatoon.

Nair, NG Nadtotchei A (1987) Sclerotia of Botrytis as a Source of Primary Inoculum for Bunch Rot of Grapes in New South Wales. Journal of Phytopathology 119: 42–51.

Nasir M, Bretag TW (1997a) Pathogenic variability in Australian isolates of *Ascochyta lentis*. Australasian Plant Pathology 26: 217–220.

Nasir M, Bretag TW (1997b) Prevalence of *Ascochyta fabae* f.sp. *lentis* on lentil seed from Victoria, Australia. Australasian Plant Pathology 26: 117–120.

Nene YL, Hanounik SB, Qureshi SH, Sen B (1988) Fungal and bacterial foliar diseases of pea, lentil, faba bean and chickpea. In: Summerfield RJ (ed) World crops: cool season food legumes. Kluwer Academic publishers, the Netherlands pp 577–589.

Nguyen TT, Taylor PWJ, Brouwer JB, Pang ECK, Ford R (2001) A novel source of resistance in lentil (*Lens culinaris* ssp. *culinaris*) to ascochyta blight caused by *Ascochyta lentis*. Australasian Plant Pathology 30: 211–215.

Parry R, Freeman A (2001) *Uromyces viciae-faba*. In: Pathogens of the Temperate Pulse *Genera Cicer, Lathyrus, Lens, Lupinus, Pisum*, and *Vicia* Volume 1 : Pathogen Pest Data Sheets. The State of Victoria, Department of Natural Resources and Environment, p 441.

Pedersen EA, Morrall RAA (1994) Effect of cultivar, leaf wetness duration, temperature and growth stage on infection and development of ascochyta blight of lentil. Phytopathology 84: 1024–1030.

Pedersen EA, Morrall RAA, McCartney HA, Fitt BDL (1994) Dispersal of conidia of *Ascochyta fabae* f sp. *lentis* from infected lentil plants by simulated wind and rain. Plant Pathology 43: 50–55.

Prasad R, Verma UN (1948) Studies on lentil rust, *Uromyces fabae* (Pers.) de Bary in India. Indian Phtyopathology 1:142–146.

Rajendra P, Chaudhary CB, Prasad R (1987) Seed-borne mycoflora of lentil. Lens Newsletter 14(1–2): 20–22.

Reddy RR, Khare MN (1984) Further studies on factors influencing the mechanism of resistance to lentil (*Lens culinaris* M.) to rust (*Uromyces fabae* (Pers.) de Bary). LENS Newsletter 11:29–32.

Rewal N, Grewal JS (1989) Effect of temperature, light and relative humidity on conidial germination of three strains of *Botrytis cinerea* infecting chickpea. Indian Phytopathology 42: 79–83.

Roundhill SJ, Fineran BA, Cole ALJ, Ingerfeld M (1995) Structural aspects of ascochyta blight of lentil. Canadian Journal of Botany 73: 485–497.

Sarker A, Kumar J, Rahman MM, Hassan MS, Zaman W, Afzal MA, Murshed ANMM (1999a) Registration of 'Barimasur-3' lentil. Crop Science 39:5, 1536.

Sarker A, Erskine W, Hassan MS, Afzal MA and Murshed ANMM (1999b) Registration of 'Barimasur-4' lentil. Crop Science 39:5, 876.

Singh K (1985) Effect of seed treatment on lentil rust (*Uromyces fabae)* development. LENS Newsletter 12:26–27.

Singh JN, Singh AD (1993) Effect of environment on the development and spread of *Stemphylium* blight on lentil. Indian Phytopathology 46: 252–253

Sinha RP, Yadav BP (1989) Inheritance of resistance to rust in lentil. LENS Newsletter 16:41.

Smith OF (1940) Stemphylium leaf spot of red clover and alfalfa. Journal of Agricultural Research 61: 831–846.

Stoilova T, Chavdarov P (2006) Evaluation of lentil germplasm for disease resistance to Fusarium wilt (*Fusarium oxysporum* f. sp. *lentis*). Journal of Central European Agriculture 7:121–126.

Tate KG (1970) A foliage disease of blue lupin caused by *Stemphylium botryosum* Wallr. New Zealand Journal of Agricultural Research 13: 710–716.

Taylor PWJ, Ford R (2007) Biology of ascochyta blight of cool season food and feed legumes. European Journal of Plant Pathology (In press).

Tivoli B, Baranger A, Avila CM, Banniza S, Barbetti M, Chen W, Davidson J, Lindeck K, Kharrat M, Rubiales D, Sadiki M, Sillero JS, Sweetingham M, Muehlbauer FJ (2006) Screening techniques and sources of resistance to foliar diseases caused by major necrotrophic fungi in grain legumes. Euphytica 147: 223–253.

Tosi L, Cappelli C (2001) First report of *Fusarium oxysporum* f. sp. *lentis* of lentil in Italy. Plant Disease 85: 562.

Tripathi, HS, Rathi YPS (1992) Epidemiology of botrytis gray mold of chickpea. Botrytis Gray Mold of Chickpea: Summary Proceedings of the BARI/ICRISAT Working Group Meeting to Discuss Collaborative Research on Botrytis Gray Mold of Chickpea, Joydebpur, Bangladesh, ICRISAT, Patancheru, Andhra Pradesh, India.

Wilson AR (1937) the chocolate spot disease of beans (*Vicia faba* L.) caused by *Botrytis cinerea* Pers. Annals of Applied Biology 24: 258–288.

Wilson VE, Brandsberg J (1965) Fungi isolated from diseased lentil seedlings in 1963–64. Plant Disease Reporter 49(8): 660–662.

Yu T F (1945) The red-spot disease of broad beans (Vicia faba L.) caused by *Botrytis fabae* Sardina in China. Phytopathology 35(12): 945–954.

CHAPTER 19

ABIOTIC STRESSES

MICHAEL MATERNE[1], DAVID McNEIL[2], KRISTY HOBSON[1], AND REBECCA FORD[3]

[1]*Grains Innovation Park, The Department of Primary Industries, Private Bag 260, Horsham, Victoria 3401, Australia*

[2]*School of Agricultural Science, University of Tasmania, Private Bag 54, Hobart, Tasmania 7001, Australia*

[3]*BioMarka, Faculty of Land and Food Resources, The University of Melbourne, Victoria 3010, Australia*

E-mail: michael.materne@dpi.vic.gov.au

Abstract: There has been a large focus on biotic stresses in lentil as these cause obvious and serious reductions in yield and quality. However, increasingly abiotic stresses are being identified as major factors involved in the low and unreliable yield of lentils in many countries. Within each growing region, variations in climate, soils, and interactions between climate and soil affect lentil productivity and quality directly, or indirectly though their influence on foliar and soil borne diseases, pests and rhizobia. Furthermore, the impact of a specific stress can be influence by the relative tolerance of a cultivar and/or effect of particular cultural control methods

1. INTRODUCTION

The distribution and quantity of rainfall and the temperature regime are the main determinants of where and in what season lentil is grown around the world. In West Asia and North Africa (WANA) and Australia the climate is characterised by cold, wet winters, rapidly rising temperatures in spring and hot, dry summers (Erskine and Saxena 1993, Erskine *et al.* 1994a, Materne 2003). In winter in these regions low radiation levels and temperatures restrict vegetative growth but increase water use efficiency (Silim *et al.* 1993, Hamdi and Erskine 1996). Lentil is winter sown at altitudes below approximately 850 m, typically in areas that receive an annual rainfall of 300 to 450 mm (Erskine and Saxena 1993, Silim *et al.* 1993, Erskine *et al.* 1994a, Materne *et al.* 2002). However, lentil is spring sown in colder, high altitude areas of West Asia, such as Central Turkey, and in USA, Europe, Canada, Chile and Argentina, where it is grown on stored moisture supplemented by

S.S. Yadav et al. (eds.), Lentil: An Ancient Crop for Modern Times, 315–329.

rainfall during the spring and summer growing seasons. In the subtropical regions of Pakistan, India, Nepal and Bangladesh, lentil is grown as a winter ('rabi') crop on residual soil moisture (Ali *et al.* 1993, Muehlbauer *et al.* 1995).

Within each growing region, variations in climate, soils, and interactions between climate and soil affect lentil productivity and quality directly, or indirectly though their influence on foliar and soil borne diseases, pests and rhizobia. Furthermore, the impact of a specific stress can be influence by the relative tolerance of a cultivar and/or effect of particular cultural control methods. For example, late sowing to avoid extremely cold temperatures in the USA and Turkey (Muehlbauer and McPhee, 2002). The influence of abiotic stresses can occur over a broad region such as occurs with drought, or can be highly heterogeneous within a small area such as occurs with soil toxicities in south eastern Australia (Nuttal *et al.* 2003a). To maximize lentil production and quality, the major abiotic stresses must be identified in a particular region and managed through cultural practices to avoid the stress and/or the release of tolerant cultivars.

2. DROUGHT

A lack of water is the major limitation to lentil production worldwide and in it's most severe form drought can result in crops with no economic grain yield. Drought is a regular and severe constraint to crop yields in many areas of the world where lentils are grown (McWilliam 1986). Autumn or winter sown crops in Mediterranean environments are likely to experience intermittent drought during their vegetative growth period and terminal drought during their reproductive period when temperatures are increasing and rainfall is decreasing. Spring sown crops in Mediterranean environments, and winter sown crops in the semi arid tropics, experience progressively increasing drought stress during the growing season. In South Asia, drought may also occur during plant establishment if sowing is delayed and plant roots cannot reach subsoil moisture (Bhattarai *et al.* 1988, Rahman and Mallick 1988).

In order to understand the structure of the world lentil collection, landraces held at the International Center for Agricultural Research in the Dry Areas (ICARDA) were characterised into four major regional groups identified through analysis of variability in quantitative and qualitative morphological traits (Erskine *et al.* 1989). These were the Levantine group (Egypt, Jordan, Lebanon and Syria), the northern group (Greece, Iran, Turkey, USSR, Chile), the Indian subcontinent group and the Ethiopian group. Erskine *et al.* (1994b) later found that the dissemination of lentil around the world has resulted in the selection of different regionally specific balances between photoperiod and temperature for the control of flowering. These regionally specific flowering responses provide the basis for adaptation to the climatic variables of an environment for which abiotic and biotic constraints can be evaluated and addressed. Shrestha *et al.* (2006) found differences in the response of genotypes to water deficits imposed during flowering and podding that also indicate differences between lentil groups in growth and water use. Water deficits reduced seed yield by up to 60% in crossbreds selected for South Asia and the South Asian cultivar, Simal, but seed yield increased with water deficit at flowering in the West Asian genotype, Cassab.

Compared to other pulses, lentil is relatively tolerant of drought and is grown in drier areas of the WANA region than faba bean and chickpea. These lentil growing regions may have as little as 250 mm of annual rainfall (Muehlbauer *et al.* 1985, Saxena 1983, Silim *et al.* 1993, Erskine *et al.* 1994a, Hamdi and Erskine 1996). However, lentil's adaptation to drought is typically through avoidance, with forced senescence and crop maturity induced by high temperatures, severe drought stress or a combination of both (Erskine and Saxena 1993, Erskine *et al.* 1994a).

In the Mediterranean and temperate world regions, lentil yields are very dependent on available soil moisture during the growing season. For example, the total growing season rainfall accounted for 41 to 55% of the total variation in seed yield in lentil in WANA over many sites and seasons (Erskine and Saxena 1993) and 56% of the total variation in seed yield among 11 diverse lentil genotypes grown at two sites in south eastern Australia over four years (Materne 2003). At one site in Syria, total growing season rainfall accounted for up to 80% of the variation in seed yield over several seasons (Erskine *et al.* 1994a). This would also be true of subtropical areas and in cold, high altitude and/or latitude areas, where levels of stored moisture and snow melt and rainfall during the spring and summer growing seasons, respectively, have a major impact on yield.

In Syria and Australia, and most likely all other lentil growing areas, selection for yield under variable rainfed conditions has increased water use efficiency in lentil through an increased response to moisture availability (Murinda and Saxena 1983, Erskine and Saxena 1993, Materne 2003). However, to breed lentil cultivars that are tolerant to drought, it is necessary to identify genetic variability for traits that are associated with low water availability tolerance. Various mechanisms, including high early vigour and early flowering and maturity have been proposed for escaping terminal drought. For example, in a dry year in Syria, the highest yielding lentil genotypes produced a large amount of biomass, flowered early and had a brief, rapid seed-filling phase (Silim *et al.* 1993). Time to flowering accounted for 49% of the variation in seed yield, indicating that avoidance through early flowering, such as with ILL6035, was the key to minimising the effects of drought stress (Silim *et al.* 1993). Early flowering is important for drought escape but it cannot be utilised if genotypes have a relatively low mean yield over many seasons and sites, due to an inability to respond to increasing available soil moisture (Turner *et al.* 2001, Materne 2003). In Syria the medium flowering genotypes ILL4400 and ILL4401 had a high seed yield in most seasons compared to other landraces studied (Murinda and Saxena 1983) but they were intolerant to drought in a separate study (Silim *et al.* 1993). Similarly, early flowering and maturing genotypes were high yielding at low rainfall, low yielding environments in Australia but the yields of these genotypes was low compared to the best medium rainfall genotypes over eight sites and five years (Materne 2003). Inherently late, temperature responsive and photoperiod insensitive genotypes for flowering were the most broadly adapted genotypes in Australia and were proposed for improving drought avoidance as time to flowering is optimal in most years but potentially earlier in drought years when temperatures are often higher (Materne 2003).

Wild lentils, particularly *L. culinaris* ssp. *orientalis*, are often found in habitats characterised by low average rainfall (Erskine and Saxena 1993, Erskine *et al.* 1994a) but they produce markedly less biomass and seed yield than cultivated lentils under dry conditions (Hamdi and Erskine 1996). Hamdi and Erskine (1996) found that the species *L. odemensis* Ladizinsky had the least reduction in seed yield when drought and non drought treatments were compared. However, ILL1861 (cultivated lentil) was the highest yielding genotype under drought conditions.

Lentil, together with chickpea, had a higher degree of osmotic adjustment compared to other grain legumes studied by Leport *et al.* (1998), especially with water deficits during flowering (Shrestha *et al.* 2006). This indicated that avoidance may not be the only mechanism for drought tolerance in lentil and chickpea. Deeper rooting has been advocated as a way of increasing legume productivity under moisture limiting conditions (Buddenhagen and Richards 1988, Turner *et al.* 2001). Genotypic differences in the length and spread of roots have been reported for lentil but not in association with water usage (Saxena *et al.* 1993, Saxena and Hawtin 1981). Short duration cultivars of chickpea and field pea have faster early root growth rates, but this is not sustained during seed fill, leading to early root senescence and the loss of effective root length (Saxena *et al.* 1993). This is also likely to be true of lentil and, as previously shown, early flowering is important for drought avoidance but this may limit the potential for greater root length as a drought tolerance mechanism. In Australia, where subsoils have high levels of boron and salt, cultivars with tolerance to these factors may be better able to utilise subsoil moisture and assist in alleviating drought stress (Materne *et al.* 2002).

Screening for drought tolerance and associated mechanisms may be successfully achieved through limiting water supply but as with early flowering as an escape mechanism, care must be taken in ensuring that selection is not for major regional adaptational traits. Hence, screening multiple genotypes under controlled environment conditions may be an initial useful step, without interference from other potentially adaptive physiological responses. Late sowing was not successful as a screening method in Syria as it altered the vegetative and phenological development of the crop through changes in temperature and photoperiod. (Erskine and Saxena 1993, Erskine *et al.* 1994a). In Australia, late sowing improved the relative performance of late flowering genotypes that perform poorly in drought years (Materne 2003).

Within a defined growing season, the early sowing of lentil often produces the highest seed yield in Mediterranean type environments such as in Italy, Syria, Egypt, Libya and Western Australia, Ethiopia, sub tropical environments in India and Bangladesh, and in North America and South America (Muehlbauer *et al.* 1995, Materne 2003). As with many crops, early sowing has been advocated in many areas to avoid rising temperatures and drought during the reproductive period and maximize yields. However, early sowing can expose crops to increased weed competition, diseases and abiotic stresses. For example in Australia, early sowing increased the prevalence of the diseases *Ascochyta lentis* and *Botrytis spp* and lodging (Materne 2003, Knights 1987). Early sowing increased the incidence of rust (*Uromyces fabae*) (Singh and Dhingra 1980, Mittal 1997), fusarium wilt (*Fusarium oxysporum* f.sp. *lentis*) (Kannaiyan and Nene 1975, Mittal 1997), *Botrytis cinerea* (Knights 1987), root rot (*Rhizoctonia solani* and *Macrophomina phaseolina*), downy mildew (*Peronospora lentis*), ascochyta blight (*Ascochyta lentis* Vassilievsky) (Knight *et al.* 1989, Mittal 1997) and collar rot (*Scerotia sclerotiorum*) (Agrawal *et al.* 1976, Mittal 1997). Early sowing also increased nodule damage by the insect *Sitona crinitus* (Weigand *et al.* 1992), and increased the number and dry weight of weeds (Mishra *et al.* 1996) and infestations of the parasitic weed *Orobanche* spp. in Syria (Silim *et al.* 1991, Hezewijk *et al.* 1987). Excessive vegetative growth and severe lodging due to high rainfall reduced yield of early sown lentil in Ethiopia (Bejiga 1991). In the USA, Turkey (Central Anatolia), Canada, Chile and Argentina, lentil is sown as early as possible in spring provided the soil is not too wet (Muehlbauer *et al.* 1995). In some of these areas, large yield increases can be achieved by sowing lentil in winter rather than spring if problems associated with a lack of winter hardiness, increased incidence of diseases and weed control issues can be addressed (Kusmenoglu and Aydin 1995, Erskine *et al.* 1996, Muehlbauer and McPhee 2002). Thus, addressing constraints to early sowing is another effective method of improving tolerance to terminal drought.

It has also been postulated (see chapter 8) that drought stress may act through reduction in fixation capability either through poorer rhizobium survival in droughted soils or more adverse effects of drought on lentil fixation than total productivity. The proportion of N assimilated from the atmosphere (%Ndfa) may decline with increasing increasing drought stress. This may mean that the symbiosis is preferentially sensitive to drought stress.

3. WATERLOGGING

In comparison with cereals, lentil, field pea and chickpea are intolerant to waterlogging at germination and have severely depressed vegetative growth, especially of roots (Crawford 1977 and Thomson unpublished data in Jayasundara *et al.* 1998). Excess water during winter can reduce lentil yields in Mediterranean environments and likewise if it occurs in late spring and summer in subtropical environments. In Nepal sowing of lentil is delayed to avoid waterlogging and *Lathyrus sativus* is preferred for early sowing. However, in most areas waterlogging can only be avoided by not growing the crop in prone areas or physically altering the environment using drainage systems or raised beds. The sensitivity of lentil (and potentially also lentil nitrogen fixation) to waterlogging and anaerobic conditions accounts for the poor response of the crop to irrigation, but responsive genotypes have been identified with large root parenchyma (Erskine and Saxena 1993, Erskine *et al.* 1994a). Among a range of germplasm, ACC-36115, ACC-215711, ACC-36140, ACC-215348, NEL-944, FLIP-89-63L and FLIP84-78 were most tolerant to waterlogging 30 days after sowing and early maturing lines were more sensitive to waterlogging than late maturing lines (Bejiga and Anbessa 1995). ILL3490 was tolerant to waterlogging in glasshouse studies in Australia (J. Clements pers. comm) but lines with greater field tolerance have been observed in Nepal (D. McNeil pers. comm.).

4. TEMPERATURE

Temperature has a major influence on the growth and development of plants, pathogens and symbiotic organisms such as rhizobia. Temperature has been shown to influence the evolution and adaptation of lentil spread worldwide. For example, Erskine (1996) found that among 171 lentil genotypes from Syria and Turkey, large-seeded (macrosperma) accessions had a longer reproductive period than small-seeded (microsperma) accessions in Syria. They concluded that larger seeded accessions were higher yielding in cooler seasons due to a longer seed filling period, but were lower yielding at higher temperatures. The clustering of traits that define the phenological adaptation of lentil to an ecological environment indicated that local environments have been important in the evolution of the species (Erskine *et al.* 1989). The dissemination of lentil into new environments has thus caused selection for different regionally specific balances between photoperiod and temperature for the control of flowering (Erskine *et al.* 1994b). For example, cultivated lentil spread from West Asia to the Indo-Gangetic plain around 2,000 BC (Erskine and Saxena 1993). Lentil landraces originating from West Asia flower much later in Pakistan and India than the local landraces, and their reproductive development begins when conditions are increasingly hot and dry in that environment (Erskine and Saxena 1993). Similarly, European landraces are too late to flower and mature in West Asia and Australia (Materne 2003). Thus changes in phenology will change the risk of exposure to frost, drought and high and low temperatures during sensitive growth stages. The timing of flowering is a particularly important event as it determines the duration of the vegetative phase (sowing to flowering), which establishes the potential of the crop, and at the same time determines the climatic conditions that the crop will be exposed to during reproductive growth (Lawn *et al.* 1995).

Temperature has a major influence on the general adaptation of lentils but extremes in temperature during the growth and reproduction of lentils can result in a more specific and dramatic effect. High temperatures during spring are major constraints to lentil production in West Asia, North Africa and Australia and cold winters limit production to spring sowing in higher altitude or latitude regions (Erskine and Saxena 1993).

4.1. Low Temperature

In the USA and Turkey (Central Anatolia), large yield increases have been achieved by sowing lentil in winter rather than spring (Saker *et al.* 1988, Erskine *et al.* 1981, Kusmenoglu and Aydin 1995, Erskine *et al.* 1996, Muehlbauer and McPhee 2002).

However, problems associated with a lack of winter hardiness have been encountered and the genetic mechanisms governing tolerance are under investigation (Kahraman *et al.* 2004).

Genotypes that can survive temperatures below freezing during vegetative growth have been selected in field environments in different regions of the world where they survived temperatures as low as −26.8 degrees Celsius (°C)(Erskine *et al.* 1981, Hambdi *et al.* 1996, Stoilova 2000, Chen *et al.* 2006) and in controlled environment experiments using a cold treatment of −15°C (Ali *et al.* 1999). In the study of Hambdi *et al.* (1996) a total of 245 accessions of wild lentil, 10 of cultivated lentil and three accessions of *Vicia montbretii* (syn. *L. montbretii*) were evaluated for winter hardiness in Syria and Turkey where absolute minimum temperatures were −16°C and −18.9°C respectively. Accessions of *L. culinaris* subsp.*orientalis* exhibited the highest level of winter hardiness, on average; whereas, accessions of *L. nigricans* subsp.*ervoides* were the most susceptible. Correlations revealed that winter hardiness was concentrated among accessions originating from high elevation areas. A cold tolerance nursery (LICTN) was established by ICARDA in 1987/88 and the selected tolerant genotypes were distributed internationally for evaluation (Malhotra and Saxena 1993). Efforts to screen for cold tolerance and develop molecular markers have been hindered by the variability in the screening environment caused by large differences between sites and seasons and within sites (Erskine and Saxena 1993, Muehlbauer pers comm). However, Erskine *et al.* (1999) identified one gene that controlled tolerance to radiation frost injury in lentil using recombinant inbred lines. A random amplified polymorphic DNA (RAPD) marker, OPS-16750, was linked to the locus for radiation-frost tolerance (Frt) trait at 9.1 centimorgans (cM). In the USA, winter hardy types have been selected and evaluated with a range of agronomic practices to initiate winter sowing (Chen *et al.* 2006).

In some Mediterranean environments frost during the reproductive period can cause major economic losses by killing flowers, pods and seeds with associated reductions in seed yield and quality. Currently, the only mechanism for limiting frost damage is using later flowering genotypes or delaying sowing to avoid frosts. However in Australia, the benefits of these strategies can be small or non existent in years when frosts occur but they result in lower yields in all other years (Materne 2003).

4.2. High Temperatures

Hot or dry weather during flowering and pod fill present severe constraints to the productivity of lentil crops in many regions of the world, including the Mediterranean (Erskine 1985). Summerfield *et al.* (1989) reported that under controlled conditions, progressively warmer temperatures post flowering restricted vegetative growth, accelerated progress towards reproductive maturity and reduced seed yield. Rhizobia are also susceptible to higher temperatures, particularly when conditions are moist (Malhotra and Saxena 1993). Earlier flowering may be a mechanism for avoiding high temperatures. However, earlier flowering may expose crops to a greater risk of frost damage in some environments, particularly if flowering is early and the crop produces inadequate biomass to sustain large seed yields (Turner *et al.* 2001, Materne 2003).

5. NUTRIENT TOXICITIES

Although generally adapted to alkaline soils, lentil growth can be affected by hostile subsoil factors such as high pH, toxic levels of boron and salinity and sodicity (Nuttall *et al.* 2001, Nuttall *et al.* 2003a, Yau 1999, Yau and Erskine 2000, Saxena *et al.* 1993). Lentil is traditionally grown on neutral to alkaline soils. For example, much of India's lentil is grown in regions with moderate to highly alkaline soils (pH 7.5–9.0) although lentils are also grown on the slightly acidic soils (pH 5.5–6.5) of the Andean foothills (Knights 1987). The crop thrives in warm sandy soils but may produce excessive vegetative growth at the expense of seed yield when grown on rich, moister soils (Smartt 1984, Muehlbauer *et al.* 1995). In Australia, lentil is best adapted to alkaline grey cracking clay and red brown earth soils in medium rainfall (350–450 mm/year) areas (Materne 2002b). However, seed yields have been compromised where soil pH was below 6.0 ($CaCl_2$) and clay content was less than 15% (Siddique *et al.* 1999). Pulse crops are generally considered to be more sensitive to subsoil constraints than cereal crops (Jayasundara *et al.* 1998). Alleviating such toxicity problems through soil modification is not an economic or practical solution and hence, if lentil is to be grown in these regions, the breeding of more tolerant lentil cultivars is considered the best approach to maximise yields.

5.1. Boron Toxicity

Boron (B) toxicity is increasingly being recognised as a problem in the arid areas of West Asia and Australia where lentil is widely grown (Yau 1999, Yau and Erskine 2000, Hobson *et al.* 2006). In the alkaline-soil cropping regions of southern Australia, high concentrations of soil B have been identified as a limitation to crop growth and grain yield (Ralph 1991). The highest concentrations of B in this region have been found at depths between 40 and 100 cm (Cartwright *et al.* 1984; Nuttall *et al.* 2003b, Hobson *et al.* 2004), although concentrations at shallower depths (10 to 20 cm) may also affect lentil due to their high sensitivity. Levels as low as 4 ppm have produced visual toxicity symptoms on lentil 26 days after sowing (Chauhan and Asthana 1981). In this same experiment, it took 47 and 37 days to see visual symptoms of boron toxicity on barley and oats respectively. The amelioration of B toxicity through soil modification is not an economic or practical solution. Hence the breeding of more tolerant cultivars is considered the best approach to minimise yield losses (Rathjen *et al.* 1999).

High field spatial variability in the distribution of soil B (Ryan *et al.* 1998; Nuttall *et al.* 2003a) makes field screening risky and difficult to interpret. In contrast controlled-environment screening provides an efficient means of identifying tolerant germplasm from large numbers of accessions, prior to confirming results in the field. In glasshouse studies ILL5883 was most tolerant to B as based on seedling symptoms and seed yield (Yau 1999, Yau and Erskine 2000). Accessions from Afghanistan were the most tolerant, followed by those from India, Iraq, Syria, Europe, Ethiopia, and Nepal (Yau and Erskine 2000). Hobson *et al.* (2003) identified lines with better tolerance to B than ILL5883 among landraces from Ethiopia (ILL2024), Afghanistan (ILL213A, ILL1818, ILL1763, ILL1796) and the Middle East (ILL5845), whilst accessions from Europe had the least tolerance. These origins of tolerance are consistent with wheat (Moody *et al.* 1988), winter barley (Yau 2002) and field pea (Bagheri *et al.* 1994). Current Australian lentil cultivars were intolerant to concentrations of soil B that occur in the Mallee region of Australia as reflected in severe effects on both above and below ground biomass accumulation and grain yield (Hobson *et al.* 2006). B-tolerance identified at the seedling stage persisted through to maturity and resulted in higher seed yield. ILL2024 and ILL213A had the least symptoms in the seedling stage and the greatest seed yield in soluble B concentration of 18.2 mg/kg (Hobson *et al.* 2006). While the two tolerant accessions were generally characterised by an ability to partially exclude B from shoot tissues, ILL2024 was more tolerant in terms of leaf toxicity symptoms but ILL213A was more tolerant in terms of growth maintenance at high leaf-B concentrations (Hobson, unpublished data).

When grown in a reconstituted core resembling 'natural' high soil B distribution, ILL2024 had no significant reduction in yield where high subsoil-B occurred (18.2 mg/kg) from 30 or 10 cm in the profile under controlled moisture. In comparison, the current Australian lentil cultivar Cassab, had significantly reduced yields of 32 and 91% when high subsoil-B occurred at 30 and 10 cm respectively (Hobson *et al.* 2004).

The development and release of B tolerant cultivars offers great potential for improving seed yield in large areas of southern Australia and thus for the expansion of the crop. Investigations are currently being conducted in Australia to determine the inheritance and benefits of boron tolerance.

Boron deficiency has been identified as a limitation to lentil production on soils in Nepal (Srivastava *et al.* 1999). Differences in the response of lentil genotypes to applied boron have been reported in the field in India (Sakal *et al.* 1988). Variation in tolerance to boron deficiency has been identified in lentil, with germplasm from South Asia exhibiting the least symptoms and those from the Middle East exhibiting the most severe symptoms (Srivastava *et al.* 2000). Gahoonia *et al.* (2005) found that lentil lines exist with different root morphologies that may be better able to scavenge micronutrients such as B from the soil and give 10–20% yield increases. There is increasing evidence that the same genetic mechanisms are likely to control tolerance to both boron deficiency and toxicity, predominantly boron exclusion (Yau and Erskine 2000, Dannel *et al.* 2002). This dual control may restrict the adaptability of lentils from South Asian countries such as Bangladesh to Australian soils. The two lentil accessions from Nepal that were tested by Hobson *et al.* (2003) were both found to be intolerant of high soil boron.

5.2. Salinity

Salinity occurs mainly in arid and semi-arid regions, where evaporation considerably exceeds precipitation, such as in West and Central Asia and in Australia, and in coastal areas because of the ingress of seawater (Saxena *et al.* 1993). The use of irrigation has also led to salinization of productive lands in the Indo-Gangetic Plain of South Asia, WANA, western USA and Australia (Saxena *et al.* 1993). Secondary salinity is of increasing importance in the dryland cropping areas of Australia, including south eastern and Western Australia where subsoil salinity is a result of cropping on soil laid down over an ancient sea bed (McWilliam 1986). Legumes are relatively sensitive to salt and lentil is comparatively more sensitive than field pea and faba bean and similar to chickpea (Saxena *et al.* 1993). Salinity is not generally considered to be a problem for lentil production worldwide (Erskine *et al.* 1994a), although the major lentil growing areas of the world are regions with a high frequency of saline or sodic soils (Saxena *et al.* 1993). Response to salinity is affected by many other environmental factors such as soil water status, relative humidity, temperature and nutrition (Saxena *et al.* 1993, Lachaal *et al.* 2002).

Variation in tolerance to $MgSO_4$, NaCl, Na_2SO_4 and $MgCl_2$ was identified in the USDA World Lentil Collection (Jana and Slinkard 1979) but the level of tolerance was considered insufficient for breeding to continue (Muehlbauer and Slinkard 1983). The NaCl tolerant accessions DL443 and Pant L406 (Rai *et al.* 1985), ILL5845, ILL6451, ILL6788, ILL6793 and ILL6796 (Ashraf and Waheed 1990) and LG128 (ILL3534) (Maher *et al.* 2003) have been identified. A positive correlation was observed between degrees of salt tolerance at different stages of growth in the glasshouse (Ashraf and Waheed 1993a) and tolerance was based on the typical halophytic mechanism of salt inclusion (Ashraf and Waheed 1993, Shah and Muhammad 1976). The inheritance of salt tolerance has been investigated but the results were inconclusive, although recessive genes were implicated (Ashraf and Waheed 1998). In the glasshouse, tolerant accessions were unaffected by NaCl concentrations commonly found in the southern Mallee of Victoria but the growth of Australian cultivars was severely reduced (Maher *et al.* 2003). Therefore the development and release of tolerant cultivars offers great potential for improving seed yield in these areas. However, cultivars that are tolerant to both NaCl and B will be required in areas such as the southern Mallee of Victoria where both toxicities occur. High yielding breeding lines have been identified in Australia that have inherited improved NaCl tolerance from ILL6788 (Materne *et al.* 2006) and investigations are currently being conducted to determine the inheritance and importance of the tolerance. Lentils that are growing slowly are more sensitive to salt than those grown rapidly according to Lachaal *et al.* (2002).

Breeding for increased vigor and the elimination of other abiotic and biotic stresses may thus offer potential to indirectly reduce the effects of salinity.

5.3. Sodicity

Increasing soil sodicity (10–25 exchangeable sodium percentage, ESP) reduced plant height, leaf area, leaf dry weight, total biomass production and seed yield in lentil, and reduced both the nitrate reductase activity in the leaf and the total concentration of nitrogen in India (Singh *et al.* 1993). PL-406 was most tolerant to sodicity in that study. Gupta and Sharma (1990) found lentil had a sodicity threshold of 14.0 %. When compared to the ESP threshold for wheat (40.2 %), it is clear that pulses are more sensitive to high ESP than cereals. A possible reason for the difference has been suggested to be the cumulative effect of ionic imbalance and water uptake (Gupta and Sharma 1990).

6. CONCLUSIONS

Abiotic stresses have a major impact on lentils worldwide, but when compared to biotic stresses, research and progress have generally been more limited. However, there is now a greater understanding of the general adaptation of lentil and the impacts of drought and waterlogging, high and low temperatures and the major soil toxicities in lentil. In many cases agronomic practices can also add value in addressing abiotic constraints of lentil. In some cases such as with winter hardiness, tolerance to boron and salinity and potentially waterlogging, genetic variability has been identified and selected genotypes will enable the expansion of suitable planting areas and increase the reliability in yield of lentil production. However, genetic improvement for drought, frost and heat has proven more difficult, except where escape is the primary tolerance mechanism. For these, it is likely that further, in depth, studies regarding environment and genetic (multiple gene) components will be required. The identification of the major genetic components governing these important traits, that function in specific environments, will lead to their selection through advanced breeding technologies and the production of superior cultivars.

REFERENCES

Agrawal, S.C., Khare, M.N. and Kushwaha, L.S. (1976) Effect of sowing dates on the collar rot of lentil caused by Sclerotium Rolf's Sac. JNKVV Research Journal 10: 172–173 CAB abstracts

Ali, A., Johnson, D.L., Stushnoff, C. (1999) Screening lentil (*Lens culinaris*) for cold hardiness under controlled conditions. Journal of Agricultural Science 133: 3, 313–319

Ali, M., Saraf, C.S., Singh, P.P., Rewari, R.B. and Ahlawat, I. P. S., S. (1993) Agronomy of lentil in India. In: Lentils in South East Asia, Proceedings of the seminar on lentil in South Asia, 11–15 March 1991, New Delhi, India, pp 103–127 (Eds W. Erskine and M. C. Saxena)

Ashraf, M. and Waheed, A. (1990) Screening of local/exotic accessions of lentil (*Lens culinaris* Medic.) for salt tolerance at two growth stages. Plant and Soil 128: 167–176

Ashraf, M. and Waheed, A. (1993) Responses of some local/exotic accessions of lentil (*Lens culinaris* Medic.) to salt stress. Journal of Agronomy and Crop Science 170: 103–112

Ashraf, M. and Waheed, A. (1998) Genetic basis of salt (NaCl) tolerance in lentil. LENS Newsletter 25: 15–22

Bagheri, A., Paull, J.G. and Rathjen A.J. (1994) The response of Pisum sativum L. germplasm to high concentrations of soil boron. Euphytica 75: 9–17

Bejiga, G. (1991) Effect of sowing date on the yield of lentil (*Lens culinaris* Medik.) Journal of Agronomy and Crop Science 167: 135–140

Bejiga, G. and Anbessa, Y. (1995) Waterlogging tolerance in lentil. LENS Newsletter 22: 8–10

Bhattarai, A.N., Bharati, M.P. and Gyawali, B.K. (1988) Factors which limit the productivity of cool season food legumes in Bangladesh. In: World Crops: Cool season food legumes, pp 230–234 (Ed R. J. Summerfield). Kluwer Academic Publishers

Buddenhagen, I.W. and Richards, R. A. (1988) Breeding cool season food legumes for improved performance in stress environments. In: World Crops: Cool season food legumes pp 81–95. (Ed R. J. Summerfield) Kluwer Academic Publishers

Cartwright B., B.A. Zarcinas & A.H. Mayfield (1984) Toxic concentrations of boron in a red-brown earth at Gladstone, South Australia. Australian Journal of Soil Research 22: 261–272

Chauhan R.P.S. and Asthana A.K. (1981) Tolerance of lentil, barley and oats to boron in irrigation water. Journal of Agricultural Science, Cambridge 97, 75–78.

Chen CC. Miller P. Muehlbauer F. Neill K. Wichman D. McPhee K. 2006) Winter pea and lentil response to seeding date and micro- and macro-environments. Agronomy Journal 98(6):1655–1663

Dannel, F., Pfeffer, H. and Romheld, V. (2002) Uptake on boron in higher plants – uptake, primary translocation and compartmentation. Plant Biology 4: 193–204

Erskine, W. (1985) Perspectives in lentil breeding. In: Faba Beans, Kabuli Chickpeas and Lentils in the 1980's, pp 91–100 (Eds M. C. Saxena and S. Varma). Aleppo, Syria: ICARDA

Erskine, W. (1996) Seed-size effects on lentil (*Lens culinaris*) yield potential and adaptation to temperature and rainfall in West Asia. Journal of Agricultural Science, Cambridge 126: 335–341

Erskine, W. and Saxena, M. C. (1993) Problems and prospects of stress resistance breeding in lentil. In: Breeding for stress tolerance in cool-season food legumes. pp 51–62. (Eds K. B. Singh and M. C. Saxena) John Wiley and Sons

Erskine, W., Adham, Y. and Holly, L. (1989) Geographic distribution of variation in quantitative traits in a world lentil collection. Euphytica 43: 97–103

Erskine, W., Myveci, K., and Izgin, N. (1981) Screening a world lentil collection for cold tolerance. LENS Newsletter 13: 19–27

Eujayl, I.; Erskine, W.; Baum, M.; Pehu, E. (1999) Inheritance and linkage analysis of frost injury in lentil. Crop Science 39: 3, 639–642

Erskine, W., Tufail, M., Russell, A., Tyagi, M. C., Rahman, M. M. and Saxena, M. C. (1994a) Current and future strategies in breeding lentil for resistance to biotic and abiotic stresses. Euphytica 73: 127–135

Erskine, W., Hussain, A., Tahir, M., Bahksh, A., Ellis, R. H., Summerfield, R. J., and Roberts, E. H. (1994b) Field evaluation of a model of photothermal flowering responses in a world lentil collection. Theoretical and Applied Genetics 88: 423–428

Gahoonia, T. S., Omar Ali, Sarker, A., Rahman, M.M. and Erskine, W. (2005) Root traits, nutrient uptake, multi-location grain yield and benefit-cost ratio of two lentil (*Lens culinaris*, Medikus.) varieties. Plant and Soil. 272: 153–161

Gupta S. K. and Sharma S. K. (1990) Response of crops to high exchangeable sodium percentage. Irrigation Science 11: 173–179.

Hamdi, A. and Erskine, W. (1996) Reaction of wild species of the genus Lens to drought. Euphytica. 1996. 91: 2, 173–179

Hamdi, A., Kusmenoglu, I. and Erskine, W. (1996) Sources of winter hardiness in wild lentil. Genetic Resources and Crop Evolution. 1996. 43: 1, 63–67

Hezewijk, M. J. van, Pieterse, A. H., Saxena, M. C. and Borg, S. J. ter (1987) Relationship between sowing date and *Orobanche* (broomrape) development on faba bean (*Vicia faba* L.) and lentil (*Lens culinaris* Medikus) in Syria. Proceedings of the 4th international symposium on parasitic flowering plants. 1987, pp 377–390. CAB abstracts

Hobson, K., Armstrong, R., Connor, D., Nicolas, M. and Materne, M. (2003) Genetic variation in tolerance to high concentrations of soil boron exists in lentil germplasm. In: "Solutions for a better environment". Proceedings of the 11th Australian Agronomy Conference. 2–6 February 2003, Geelong, Victoria. Australian Society of Agronomy

Hobson, K.B., Armstrong, R.D, Nicolas, M.E, Connor, D.J. and Materne, M.A. (2004) Boron tolerance of lentil – highlights of a research program. In: "New directions for a diverse planet". Proceedings of the 4th International Crop Science Congress 26 Sep–1 Oct 2004, Brisbane, Australia.

Hobson, K., Armstrong, R., Nicolas, M.,Connor, D. and Materne, M. (2006) Response of lentil (*Lens culinaris*) germplasm to high concentrations of soil boron. Euphytica 151, 371–382.

Jana, M. K. and Slinkard, A. E. (1979) Screening for salt tolerance in lentils. LENS Newsletter 6: 25–27

Jayasundara, H. P. S., Thomson, B. D. and Tang, C. (1998) Responses of cool season grain legumes to soil abiotic stresses. Advances in Agronomy 63: 77–153

Kannaiyan, J. and Nene, Y. L. (1975) Note on the effect of sowing dates on the reaction of twelve lentil cultivars to wilt disease. Madras Agricultural Journal 62: 240–242. CAB abstracts

Karaman A, Kusmenoglu I, Aydin N, Aydogan A, Erskine W, Muehlbauer FJ (2004) Genetics of winter hardiness in 10 lentil recombinant inbred line populations. Crop Science 44:5–12

Knights, E. J. (1987) Lentil: A potential winter grain legume crop for temperate Australia. Journal of the Australian Institute of Agricultural Science 53: 271–280

Knight, T. L., Martin, R. J. and Harvey, I. C. (1989) Management factors affecting lentil production in mid Canterbury. Proceedings Annual Conference – Agronomy Society of New Zealand. 1989, 19, pp 17–24. CAB abstracts

Kusmenoglu, I. and Aydin, N. (1995) The current status of lentil germplasm exploitation for adaptation to winter sowing in the Anatolian highlands. In Autumn-sowing of lentil in the highlands of West Asia and North Africa, pp 63–71 (Eds J. D. H. Keatinge and I. Kusmenoglu) Ankara: CRIFC

Lachaal, M., Grignon, C. and Hajji, M. (2002) Growth rate affects salt sensitivity in two lentil populations. Journal of Plant Nutrition 25: 2613–2625

Lawn, R. J., Summerfield, R. J., Ellis, R. H., Qi, A., Roberts, E. H., Chay, P. M., Brouwer, J. B., Rose, J. L. and Yeates, S. J. (1995) Towards the reliable prediction of time to flowering in six annual crops. IV. Applications in crop improvement. Experimental Agriculture 31: 89–108.

Leport, L., Turner, N. C., French, R. J., Tennant, D., Thomson, B. D. and Siddique, K. H. M. (1998) Water relations, gas exchange and growth of cool-season grain legumes in a Mediterranean-type environment. European Journal of Agronomy 9: 295–303

Maher, L., Armstrong, R. and Connor, D. (2003) Salt tolerant lentils – a possibility for the future? In: "Solutions for a better environment". Proceedings of the 11th Australian Agronomy Conference. 2–6 February 2003, Geelong, Victoria. Australian Society of Agronomy

Malhotra, R. S. and Saxena, M. C. (1993) Screening for cold and heat tolerance in cool-season food legumes. In Breeding for stress tolerance in cool-season food legumes. Ed Singh, K. B.; Saxena, M. C. John Wiley & Sons Ltd, Chichester, UK: 1993. 227–244.

Materne, M. A. (2003) Importance of phenology and other key factors in improving the adaptation of lentil (Lens culinaris Medikus) in Australia. Thesis presented for the degree of Doctor of Philosophy at The University of Western Australia, School of Plant Biology and Centre for Legumes in Mediterranean Agriculture (CLIMA), Faculty of Natural and Agricultural Sciences

Materne, M., McMurray, L., Nitschke, S., Regan, K., Heuke, L., Dean, G. and Carpenter, D. (2002) The future of Australian lentil production. In: Proceedings of Lentil Focus 2002, 14–18 (Ed JB Brouwer) Horsham, Victoria, Australia

Materne, M., Regan, K., McMurray, L., Nitschke, S., Dean, G., Heuke, L., and Matthews, P. (2006) Breeding for NaCl tolerance and improved adaptation in lentil. In: 'Breeding for success: Diversity in action' (Ed C.F. Mercer) Proceedings of 13th Australasian Plant Breeding Conference, Christchurch, new Zealand 18–21 April 2006. pp 1198–1203

McWilliam, J. R. (1986) The national and international importance of drought and salinity effects on agricultural production. Australian Journal of Plant Physiology 13: 1–13

Mishra, J. S., Singh, V. P. and Bhan, V. M. (1996) Response of lentil to date of sowing and weed control in Jabalpur, India. LENS Newsletter 23: 18–23

Mittal, R. K. (1997) Effect of sowing dates and disease development in lentil as sole and mixed crop with wheat. Journal of Mycology and Plant Pathology 27: 203–209. CAB abstracts.

Moody, D. B., Rathjen, A. J., Cartwright, B., Paull, J. G. and Lewis, J. (1988) Genetic diversity and geographic distribution of tolerance to high levels of soil boron. In: Proceedings of the 7th International Wheat Genetics Symposium, Cambridge, UK, p 859–865

Muehlbauer, F. J. and McPhee, K. E. (2002) Future of North American lentil production. In: Proceedings of Lentil Focus 2002, pp 8–13 (Ed JB Brouwer) Horsham, Victoria, Australia

Muehlbauer, F. J. and Slinkard, A. E. (1983) Lentil improvement in the Americas. In: Proceedings of the international workshop on Faba beans, Kabuli chickpeas and lentils in the 1980's. pp 351–366. (Eds M. C. Saxena and S. Varma) ICARDA, 16–20 May 1983, Aleppo, Syria.

Muehlbauer, F. J., Cubero, J. I. and Summerfield, R. J. (1985) Lentil (*Lens culinaris* Medic.) In: Grain Legume Crops. pp 266–311. (Ed R. J. Summerfield and E. H. Roberts) Collins, London

Muehlbauer, F. J., Kaiser, W. J., Clement, S. L. and Summerfield, R. J. (1995) Production and breeding of lentil. Advances in Agronomy 54: 283–332

Murinda, M. V. and Saxena, M. C. (1983) Agronomy of faba beans, lentils, and chickpeas. In: Proceedings of the international workshop on Faba beans, Kabuli chickpeas and lentils in the 1980's. pp 229–244. (Eds M. C. Saxena and S. Varma) ICARDA, 16–20 May 1983, Aleppo, Syria.

Nuttall, J. G., Armstrong, R. D., and Connor, D. J. (2001) Understanding subsoil water-use on southern Mallee soils: I. Spatial characteristics of subsoil constraints. In: Proceedings of the 10th Australian Agronomy Conference pp 51. Hobart. Tasmania (28 January–1 February 2001)

Nuttall, J., Armstrong, R. and Connor, D. (2003a) The effects of salinity, sodicity and soluble boron on wheat yields in the Victorian southern Mallee. In: "Solutions for a better environment". Proceedings of the 11th Australian Agronomy Conference. 2–6 February 2003, Geelong, Victoria. Australian Society of Agronomy

Nuttall J.G., R.D. Armstrong, D.J. Connor & V.J. Matassa (2003b) Interrelationships between edaphic factors potentially limiting cereal growth on alkaline soils in north-western Victoria. Australian Journal of Soil Research 41: 277–292.

Rahman, M. M. and Mallick, R. N. (1988) Factors which limit cool season food legume productivity in Bangladesh. In: World Crops: Cool season food legumes, pp 230–234 (Ed R. J. Summerfield). Kluwer Academic Publishers

Rai, R., Nasar, S. K. T., Singh, S. J. and Prasad, V. (1985) Interactions between Rhizobium strains and lentil (*Lens culinaris* Linn.) genotypes under salt stress. Journal of Agricultural Science 104: 199–205

Ralph W. (1991) Boron problems in the southern wheat belt. Rural Research 153: 4–8.

Rathjen A.J., J.D. Brand, C.-Y. Liu, J.G. Paull & D. Cooper (1999) Breeding for tolerance to soil toxicities. *In* 11th Australian Plant Breeders Conference, Adelaide, 19–23 April 1999, 1999. Eds P. Longridge, A. Barr, A. Auricht, G. Collins, A. Granger, D. Handford & J. Paull. pp 34–40. CRC for Molecular Plant Breeding.

Ryan J, Singh M, Yau SK (1998) Spatial variability of soluble boron in Syrian soils. Soil & Tillage Research 45, 407–417.

Sakal, R., Singh, A. P. and Sinha, R. B. (1988) Differential reaction of lentil cultivars to boron application in calcareous soil. LENS Newsletter 15: 27–29

Sakar, D., Durutan, N. and Meyveci, K. (1988) Factors which limit the productivity of cool season food legumes in Turkey. In: World Crops: Cool season food legumes, pp 137–145 (Ed R. J. Summerfield). Kluwer Academic Publishers

Saxena, M. C. (1983) Food legume improvement program at ICARDA – An overview. In: Proceedings of the international workshop on Faba beans, Kabuli chickpeas and lentils in the 1980's. pp 1–13. (Eds M. C. Saxena and S. Varma) ICARDA, 16–20 May 1983, Aleppo, Syria

Saxena, M. C. and Hawtin, G. C. (1981) Morphology and growth patterns. In: Lentils. pp 39–52 (Eds C. Webb and G. Hawtin) Commonwealth Agricultural Bureaux, ICARDA

Saxena, N. P., Johansen, C., Saxena, M. C. and Silim, S. N. (1993) The challenge of developing biotic and abiotic stress resistance in cool-season food legumes. In: Breeding for stress tolerance in cool-season food legumes. pp 245–270. (Eds K. B. Singh and M. C. Saxena) John Wiley and Sons

Shah, M. and Muhammad, Y. N. (1976) Effect of salinity on the chemical composition of three cultivars of lentil (*Lens esculenta*). Journal of Agricultural Research, Pakistan 14: 65–74. CAB abstracts

Siddique, K. H. M., Loss, S. P., Regan, K. L. and Jettner, R. L (1999) Adaptation and seed yield of cool season grain legumes in Mediterranean environments of southwestern Australia. Australian Journal of Agricultural Research 50: 375–387

Silim, S. N., Saxena, M. C. and Erskine, W. (1991) Effect of sowing date on the growth and yield of lentil in a rainfed Mediterranean environment. Experimental Agriculture 27: 145–154

Silim, S.N., Saxena, M.C. and Erskine, W. (1993) Adaptation of lentil to the Mediterranean environment. I. Factors affecting yield under drought conditions. Experimental Agriculture 29: 9–19

Singh, B. B., Tewari, T. N. and Singh, A. K. (1993) Stress studies in lentil (*Lens esculenta* Moench). III. Leaf growth, nitrate reductase activity, nitrogenase activity and nodulation of two lentil genotypes exposed to sodicity. Journal of Agronomy and Crop Science 171: 196–205 CAB abstracts

Singh, G. and Dhingra, K. K. (1980) Effect of sowing dates and varietal reaction on the incidence of lentil rust. Journal of Research, Punjab Agricultural University 17: 233–235

Smartt, J. (1984) Evolution of grain legumes. I. Mediterranean pulses. Experimental Agriculture 20: 275–296

Srivastava, S. P., Joshi, M., Johansen, C. and Rego, T. J. (1999) Boron deficiency of lentil in Nepal. LENS Newsletter 26: 22–24

Srivastava, S. P., Bhandari, T. M. S., Yadav, C. R., Joshi, M. and Erskine, W (2000) Boron deficiency in lentil: yield loss and geographic distribution in a germplasm collection. Plant and Soil 219: 147–151

Stoilova, T. (2000) Evaluation of lentil germplasm accessions for winter hardness in Bulgaria. Bulgarian Journal of Agricultural Science. 6: 61–164

Summerfield, R. J., Muehlbauer, F. J. and Short, R. W. (1989) Controlled environments as an adjunct to field research on lentils (*Lens culinaris*). V. Cultivar responses to above- and below-average temperatures during the reproductive period. Experimental Agriculture 25: 327–341

Turner, N. C., Wright, G. C. and Siddique, K. H. M. (2001) Adaptation of grain legumes (Pulses) to water-limited environments. Advances in Agronomy 71: 193–231

Weigand, S., Pala, M. and Saxena, M. C. (1992) Effect of sowing date, fertilizer and insecticide on nodule damage by *Sitona crinitus* Herbst (Coleoptera: *Curculionidae*) and yield of lentil (*Lens culinaris* Medik.) in northern Syria. Zeitschrift fur Pflanzenkrankheiten und Pflanzenschutz. 99: 174–181. CAB abstracts.

Yau, S. K. (1999) Boron toxicity in lentil: yield loss and variation between contrasting lines. LENS Newsletter 26: 14–17

Yau, S. K. (2002) Comparison of European with West Asian and North African winter barleys in tolerance to boron toxicity. Euphytica 123: 307–314

Yau, S. K. and Erskine, W. (2000) Diversity of boron-toxicity tolerance in lentil growth and yield. Genetic Resources and Crop Evolution 47: 55–61

CHAPTER 20

INSECT PESTS OF LENTIL AND THEIR MANAGEMENT

PHILIP C. STEVENSON[1], M. K. DHILLON[2], H. C. SHARMA[2], AND M. EL BOUHSSINI[3]

[1]*Natural Resources Institute, University of Greenwich, Chatham, ME4 4TB, UK and Royal Botanic Gardens, Kew, Surrey, TW9 3AB, UK*
[2]*Genetics Resources Divisions International Crops Research Institute for the Semi-Arid Tropics (ICRISAT), Hyderabad 502324, Andhra Pradesh, India*
[3]*International Center for Agricultural Research in the Dry Areas (ICARDA), Aleppo, Syria*
Email: p.c.stevenson@gre.ac.uk

Abstract: Lentil is one of the world's most important food plants and is particularly so in North Africa and South Asia and parts of North America, Europe and Australia. Consequently the crop is exposed to a broad spectrum of insect species in a wide variety of locations. The management of insect pests of the crop is crucial to optimizing production. The major insect pests of lentil in the field are aphids (*Aphis craccivora* & *Acyrthosiphon pisum*), leaf weevils (*Sitona* spp.), Lygus bugs, (*Lygus* spp.), and the Cutworm, (*Agrotis ipsilon*). Several other insect species are considered as minor field pests which are also noteworthy and include Thrips (*Thrips, Kakothrips,* & *Frankiniella*), Bud weevils (*Apion arrogans*), the pea moth, (*Cydia nigricana*), pod borers, (*Helicoverpa armigera* & *Heliothis* spp.), Lima-bean pod borer, (*Etiella zinckenella*), root aphids (*Smynthurodes betae*) and leaf miners (*Liriomyza* spp. and *Phytomyza* spp.). The most serious and frequently encountered insect pests of the stored grain are *Bruchus ervi* and *B. lentis* with *Callosobruchus chinensis* and *C. maculatus* also widespread. This chapter describes the morphology, lifecycle and crop damage caused by each of the insects pest species on lentil and provides detailed descriptions of management options for each species with references for each recommended action. For most insect species the use of pesticides is the primary management option. However, for some species, there are known sources of host plant resistance, as well as other integrated pest management options including biological control (e.g., beneficial insect predators and biological pesticides) and cultural practices, that can be used to help manage the pests and where known these are also described

S.S. Yadav et al. (eds.), Lentil: An Ancient Crop for Modern Times, 331–348.

1. INTRODUCTION

Lentil, *Lens culinaris* (Medikus) is an important pulse crop which is grown in North America, southern Europe, North Africa, West Asia (including Turkey), the former USSR countries of central Asia, and northern and central parts of India. In India, it is grown on about 1.29 million hectares with an average production of 0.5 million tones, and productivity level of 624 kg/ha (Yadava and Ahmad, 2000). Ethiopia contributes more than 66% of the area and about 60% of the production of lentil in sub-Saharan Africa (Bejiga et al., 2000). In North America where production is concentrated in Canada and Washington State (43% of US production) and yields of around 1300 kg Ha^{-1} $annum^{-1}$ are achieved, measures to control insect pests are effective and well established. Poor crop management and abiotic and biotic stresses reduce yields and are discussed in detail elsewhere in this book. Among the biotic constraints, insect pests play a major role in yield reduction. About three dozen insect pests have been reported to infest lentil under field and storage conditions (Hariri, 1981), out of which 21 species have been reported from India alone (Lal, 1992). However, only some of these are economically important, and require control measures. The field insect pests of note include aphids, *Sitona* weevils, *Lygus* and stinkbugs, cutworms, thrips, bud weevils, and pod borer. During storage, several species of seed beetles inlcuding *Bruchus* spp. and *Callosobruchus* spp. can cause severe damage. The pest status of each species largely depends on the location with different countries and regions reporting different insects. For example, *Aphis craccivora* Koch (Thakur et al., 1984), phycitid, *Etiella zinckenella* Treit. (Singh and Dhooria, 1971; Hammad, 1978) and bruchid, *Callosobruchus chinensis* Linn. (Staneva, 1982) have been reported as the major pests of lentil in India and *Sitona crinitus* Herbst, *Bruchus lentis* Froel., and *E. zinckenella* have been identified as the most important harmful insects of lentil in Turkey (Tamer et al., 1998). Whereas, aphids (*Acyrthosiphon pisum* Harris and *A. craccivora*), unvoluntine bruchid (*Bruchus lentis*), thrips (*Thrips tabaci* and *T. angusticeps*), and leaf weevils (*S. lineatus* L.) are the key pests of lentil in Castilla La Mancha (central Spain) (Perez Andueza et al., 2004) and Lygus bugs are major pests of lentils in the North West of North America. Each species or group of insects (usually related species or genera) are described along with information about biology, damage and control. Photographs are not presented of each pest insect species described since there are more than adequate insect image pages available on the world wide web.

2. MAJOR INSECT PESTS OF LENTIL

2.1. Aphids, *Aphis Craccivora & Acyrthosiphon Pisum*

Cowpea or groundnut aphid, *Aphis craccivora* Koch, and the pea aphid, *Acyrthosiphon pisum* Harris (Homoptera: Aphididae) are the major Hemipteran insect species of lentil in America, Europe, Africa, Australia, and Asia (Manero and L'Argentier 1987; Bejiga and Anbessa, 1993; Weigand et al., 1994; Muehlbauer, 1996; Yadava and Ahmad, 2000; Perez Andueza et al., 2004).

Aphis craccivora is small (2 mm long), soft-bodied, shiny and black, with dull greyish lightly powdered nymphs, and *A. pisum* is slightly larger at 3 to 4 mm long and green in color with characteristically long legs. It lives throughout the year without producing sexual forms. The alates (winged females) that are dispersed largely by wind, reproduce parthenogenetically producing viviparous colonies of apterae (wingless females), which later revert to winged forms for dispersion depending upon overcrowding and deterioration of the host plant. *A. pisum* generally take 9 to 11 days to reach the adult stage and then begins producing live young. An apterous female aphid contains the developing embryos of its grandchildren! This combination of the telescoping of generations and a very short life cycle leads to very rapid population increases under favorable conditions. *A. craccivora* and *A. pisum* are both polyphagous species, with a preference for herbaceous Leguminosae. They feed on the young shoots, leaves, inflorescences and fruits and, in herbaceous plants, also on the stems. Lentil is perhaps not the most suitable host for *A. craccivora* however since the net reproductive rate and post-reproductive period, adult longevity, survival, fecundity, and the lifespan were significantly longer on lentils than on other crops studied by Wale et al. (2000). Daily nymph production was significantly correlated with the minimum temperature on lentil. In autumn, changing photoperiod for *A. pisum* triggers the production of sexual males and females. After mating, the sexual female lays a few large, over-wintering eggs. Each egg hatches into a female nymph who is the first individual of a new clone.

Aphids damage lentil plants directly by feeding on them and more seriously transmit plant viruses. Both nymphs and adults suck the plant sap from the tender leaves, stems and pods, and mostly colonize on the young leaves and growing points, which become characteristically deformed. Host reaction to insect feeding are not characteristic but large populations on young plants can prevent their normal growth, affecting yield. The infestations of crops are always initiated by alate forms produced on preinfested plants. Yield can be drastically reduced, and if infestations are early and severe, plants can be killed, and the diseases transmitted by these aphids are covered elsewhere in this book. Of note however are Alfalfa and Cucumber mosaic viruses (CMV); diseases with broad host spectra that are serious problems in lentils (Latham et al., 2004) in Australia and CMV occurs frequently along with Pea seed-borne mosaic virus on lentils in Pakistan (Mammouk et al., 2001). While these viruses are transmitted via the seed (Latham and Jones, 2001; Jones and Coutts, 1996), the insects are by far the primary route for disease spread in the crop and thus a primary target for control. *Aphis craccivora* also transmits lentil tobacco streak virus (Lal, 1994). Aphids, leaf miners, and semi-loopers were the first dominant group of insect pests during the vegetative and flowering stages (Bhatnagar and Seghal, 1990).

Host plant resistance to bean aphid, *A. craccivora*, and biological control with predatory coccinellids has a potential scope for its management (Sharma and Yadav, 1993) but currently there are no alternative control measures or cultural practices for the management of aphids in lentils. While aphids are attacked by a number of natural enemies, coccinellids especially prevent their rapid reproduction rate and may reduce infestation levels sufficiently. In the event of severe infestations before or at flowering, aphids require control using chemical sprays the most widespread of which is dimethoate (various trade names at 0.5 kg. ha^{-1}) with 14 day PHI. Application in the US and Canada are usually made by air to 100% of the crop (Bragg and Burns, 2006) and subsequent losses to aphids are near 0%. Dimethoate is usually applied for aphids and *Lygus* bugs (discussed below) at the same time. A single spray should be enough for 3 weeks effective control but if reinfestation occurs before pod maturity a second spray may be given. Aphids may infest lentil fields only at a later stage of crop development (podding), where no spray is necessary. Other insecticides (e.g., fenvalerate, malathion, disulfoton, carbaryl, methomyl, methyl parathion, and endosulfan) are registered for control of pea aphid and have been tried by growers. However, none provide cost-effective control comparable to dimethoate. Efforts are, however, being invested in developing alternatives to dimethoate, which, like other organophosphates, is considered environmentally too hazardous (Bragg and Burns, 2001). Seed treatment with imidacloprid was shown to reduce Bean leaf roll virus (BLRV, family Luteoviridae), Faba bean necrotic yellows virus (FBNYV, genus Nanovirus) and Soybean dwarf virus (SbDV, family Luteoviridae) in faba bean and lentil, using artificial inoculation with the aphid vector, *Acyrthosiphon pisum* (Harris). The treatment increased yields of susceptible lentil varieties but not that of lentil varieties reported to be resistant (Makoouk and Kumari, 2001). The U.S. Dry Pea and Lentil Commission and the Washington State Commission on Pesticide Registration have funded research into environmentally more benign chemicals for the control of pea aphid in dry peas, which could translate to lentil. The products with greatest potential include bifenthrin, cyfluthrin, imidacloprid and lambda-cyhalothrin as floral applications made at 50 percent bloom. In addition there is research ongoing to develop thiomethoxam as a seed treatment for the United States and Canada (Bragg and Burns, 2001).

2.2. Leaf Weevils, *Sitona* spp.

Sitona crinitus Herbst (*Sitona macularius* Marsh.) (Coleoptera: Curculionidae) is one of the main insect species attacking lentil in southern Europe, North Africa, West Asia (Turkey, Syria, Lebanon, Jordan), and the former USSR (Solh et al., 1986; El Damir et al., 1999; Perez Andueza et al., 2004). Other related species that attack lentil include *S. lineatus* L. and *S. limosus* Rossi. The adult leaf weevils are distinguished by their grey-brown body (3–4 mm), a pronotum that has three longitudinal light lines and elytra that have three rows of dark and white spots. Females oviposit spherical yellow eggs, which become black with age. The larvae are white, with brown head capsules, and pupae are also white. Over-wintered adults of *S. crinitus* appear in late March and feed on young shoots and leaves (Kaya and Hincal, 1987), while the larvae appear when the climatic conditions are suitable, and have root nodules as food. Females of *S. lineatus* reportedly lived significantly longer than males when maintained as single reproductive pairs, but there were no significant differences between female and male lifespans when adults were kept in groups of 13 reproductive pairs (Schotzko and O'Keeffe, 1988). The spring migration state of the adults and the number of months spent in hibernal quiescence (dormancy) also had a significant effect on adult lifespan. In the Mediterranean region, where hot and dry summers prevail, the adults aestivate in the soil and start emerging in December/January (El Damir et al., 2001). Sometimes, the neonate adults have also been observed to emerge in May, where the lentil matures during this period, and might also feed little on other crops, but then enter the soil to aestivate until early winter. The adults of the previous generation will die in April/May. Thus, there is only one generation per year, and the adults live for almost one year. After mating the females oviposit eggs on the soil around the lentil plants or loosely on the leaves, which later fall to the ground and this can continue for several months with each female laying several hundreds of eggs. Temperature appears to be a critical factor in the emergence of the larvae from eggs and a study by El Damir et al. (2004) revealed that temperatures $< 10\,°C$ induce quiescence in eggs of *S. crinitus*. When the larvae do hatch they move into the soil and infest the nodules of the plants and each larva may consume many nodules during development until pupation, which also occurs in the soil. The larval and pupal periods last for between 5–6 and 3–4 weeks, respectively, dependent on local climatic conditions, notably temperature.

While both the adults and larvae of *Sitona* sp. damage the crop, larvae are the main problematic stage. The adult weevils feed on the foliage in a characteristic manner, making semicircular notches from the leaf edges early in the season but this does not usually affect yield. If the insect populations are very high and the growth of the lentil seedlings is retarded by unfavorable environmental conditions then the plant can not compensate the damage to foliage quickly and the crop can consequently suffer economically important damage. The larvae of *Sitona* spp., however, feed on the root nodules and consequently are by far the most serious problem since this affects the capacity of the plant to fix N_2 (Solh et al., 1986). Nodule damage has also been reported to be significantly higher in early-sown than in late-sown lentils (Weigand et al., 1992) thus a late sown crop could lead to reduced damage through escape. Mineral nitrogen does not compensate for the damaged nodules, and fail to supplement fixed nitrogen for yield increase. In severe attacks the foliage can assume the yellow appearance like nitrogen deficiency characteristics. *Sitona crinitus*, *S. macularius*, and *Apion arrogans* are efficient vectors of broad bean stain comovirus (BBSV), while *S. limosus* transmit broad bean mottle bromovirus (BBMV) in lentil (Makkouk and Kumari, 1995) thus potentially pose an additional threat to crop security for farmers and emphasize the importance of managing this pest.

There is a significant and positive correlation between visual damage score and nodule damage by *Sitona* sp. ($r = 0.69$ and 0.75), where visual damage can help evaluate large numbers of genotypes for *Sitona* resistance at the same time under field conditions (El Damir et al., 1999) or determine levels of infestation. The white *Sitona* larvae inside the nodules or empty nodules can be found in uprooted lentil plants. Carbofuran and aldicarb reduce nodule damage significantly

(Solh et al., 1986), and yield increases due to carbofuran application are generally higher in early than in a late sown crop. Seeds of lentil can be treated with Promet^R (Furathiocarb) @ 12 ml/kg seed or granular insecticides may also be applied (e.g., Carbofuran) at planting. The use of carbofuran @ 1.5 kg a.i. ha^{-1} improved both nodulation and grain production (Islam and Afandi, 1982). Seed treatment with Promet (furathiocarb) effectively controls *Sitona*, increases grain and straw yields, and is less disruptive to the environment than insecticide sprays (Weigand et al., 1992, 1994). Rhizobial inoculation and phosphorus application increases lentil productivity in arid locations under Mediterranean environments (Al Karaki, 1996). If no preventive control is taken, and is observed with high infestation, Imidan^R (Phosmet) @ 1 kg a.i. ha^{-1} can be sprayed. This is, however, less effective than granular application and seed treatment. Carbofuran increases nodule mass by significantly reducing *Sitona* nodule damage, and seed protein content slightly, which increases overall protein yield (Islam et al., 1985). Alternatively, chlorpyriphos @ 720 g a.i. ha^{-1} or malathion @ 1 300 g a.i. ha^{-1}, or a systemic insecticide, oxydemeton methyl @ 265 g a.i. ha^{-1} can also be applied in cases of severe infestation (Erman et al., 2005). Application of oxydemeton methyl reduced 41.2 to 52.6 % damage on nodules, and increased seed yield (729 and 1461.6 kg/ha), biological yield (1825.5 to 3521.6 kg ha^{-1}), pods/plant (23.8 to 25.7), 1000-seed weight (47.7 to 48.4 g), plant height (26.1 to 30.0 cm), branches/plant (7.3 to 8.9), root dry weight/plant (0.106 to 0.136 g), and shoot dry weight/plant (0.859 to 1.056 g) over the unsprayed control. Cultivar, 'Yerli Kirmizi' showed low nodule feeding (0.8 to 1.9) by the Sitonid pest, and increased seed yield (712.6 to 1393.3 kg ha^{-1}) over 'Sazak 91 (537.4 to 1301.3 kg ha^{-1}). The water extract of *Melia azedarach* dry fruits extracts at 50 g l^{-1} significantly reduced *S. crinitus* adult damage on lentil leaves for one week (El Damir et al., 2000).

The weevils emerge from the soil after aestivation so crop rotation can reduce the likelihood of successful recolonisation and subsequent infestation to some extent. However, while crop rotation should be encouraged it should also be noted that *Sitona* weevils are strong fliers and can migrate considerable distances thus reinfestation from distance will also occur (Beniwal et al., 1993). Furthermore, weedy fields are more prone to *Sitona* damage so the concurrent management of weeds can help to reduce the problem of Sitonid weevils. Thus early sowing combined with the control of *Sitona* and weeds (cyanazine and pronamide), and P application gives higher net return with virtually no risk of economic loss to the farmer (Pala and Mazid, 1992: and elsewhere in this volume). Formononetin and associated metabolites in red clover act as chemical defenses against adult *S. lepidus* and the distribution of this pest in forage legumes could be manipulated through improvement in root health (Gerard et al., 2005). Isoflavonoid enzymes such as isoflavone reductase have the potential to be selected for high levels of these compounds for less susceptibility of lentils to *Sitona* (Robeson, 1978; Jung et al., 2000). However, no sources of resistance to *Sitona* have yet been found in the lentil germplasm (Erskine et al., 1994). The CryIII toxin expression in nodules resulted in significant reductions in nodule-feeding damage by *S. lineatus* on *Pisum sativum* and *S. hispidulus* on *Medicago sativa* (Bezdicek et al., 1994), and could be the new strategic component to produce *Sitona*-resistant lentils. However, since little progress has been made in this respect since then it is likely the technology is either unsuitable or not needed. Bt-toxins from the Centre for Legumes in Mediterranean Agriculture (CLIMA) collection have been used to screen adult insects and larvae of *Sitona* at ICARDA but the strains were only found effective against adult insects, not against larvae (Baum, 2000). There is presently no transgenic material available to or being used by farmers to manage *Sitona* weevils.

2.3. Lygus Bugs, *Lygus* spp.

Lygus bugs are a major pest in lentil production particularly in the North West of North America. In northern Idaho, more than 20 species of lygus bug have been identified on 70 species of plants with the most abundant being *L. hesperus* and *L. elisus*. The adult bugs are about 6 mm long, flattened, oval, and of various colors from pale green to yellowish brown. Lygus bugs survive the winter protected in ground litter, crop residues and buildings. They emerge soon after the snow melts in spring, and feed on winter annuals and the buds of flowering shrubs. Adults lay eggs in the spring and feed on various plants (Fuchs and Hirnyck, 2000). In Canada, over-wintering adults can be abundant in canola especially if the crops are in bud or flower and other hosts are not yet available. Eggs hatch into nymphs in about 10 days and reach maturity within one month and only a single generation develops on lentils. Immature *Lygus* bugs (nymphs) are light green and wingless. Several black spots, usually five, appear on their backs as they moult or mature through five instars before becoming adults. Wing buds are evident in fourth and fifth instars. In late summer, the new generation adults disperse from mature canola fields into later maturing hosts, such as alfalfa, and continue feeding until they migrate to over-wintering sites. In the South, the new generation adults first appear by about the end of June (Fuchs and Hirnyck, 2000, Bragg and Burns, 2000).

Finding *Lygus* successfully in the field is critical to prevent damage to the crop. Because lentils lie close to the ground sweeping with a net is an ineffective scouting practice and will often miss the presence of *Lygus*. Economic thresholds have been established for lygus bug control. When lentils are in bloom, and podding has begun, sweep nets can be used to determine quantity of adult bugs with 1 lygus bug for every 3 sweeps considered worthy of insecticide treatment. Close examination of the plants, however, is the only way to find *Lygus* bugs, which are usually found under the curly leaves of the lentil plants in the daytime, and rarely seen on the visible portions of the crop. Any presence of the bugs just before or during bloom justifies treatment according to the lentil industry (Fuchs and Hirnyck, 2000, Bragg and Burns, 2000). *Lygus* bugs pierce tender leaves, stems, buds, petioles, and developing seeds but are considered to be a serious pest of lentils primarily owing to the seed damage known as chalky spot syndrome which is characterized by pitted, crater-like depressions in the seed coat with or without a discolored chalky appearance. Chalky spot results in significant economic losses to the producer by reducing market price/value. Lentils with more than 3.5% chalky spot damage are graded and their value is discounted according to the level of the symptoms. Yield reduction due to either direct feeding or chalky spot can range up to 50%, and without control measures reductions in yield are about 30% (Fuchs and Hirnyck, 2000).

Lygus bugs have several natural control agents including a fairy wasp, in the family *Mymaridae*, that appears to remain unidentified according to the literature. It parasitizes the eggs of the bug (Jones, 1999). In addition a parasitic wasp, *Peristenus pallipes* attacks lygus nymphs but its effectiveness is not well documented (Baird, 2000). A European species, *P. digoneutis* has been introduced into alfalfa fields in eastern North America where it parasitizes about 40% of the tarnished plant bugs (Day and Mahr, 1999, Jones, 1999). One of the few parasitoids of lygus adults is a tachinid fly, *Alophorella sp*., and nabid plant bugs, big-eyed bugs and spiders occasionally prey on young lygus bug nymphs. Cultural control programs for lygus bug are only partially effective because the target insect is supported by a variety of hosts. The continuity of plant hosts support lygus bugs throughout their life cycle. Disturbing habitat by disking near fencerows and mowing roadsides can potentially lower lygus bug numbers, but also will injure over-wintering populations of beneficial insects (Fuchs and Hirnyck, 2000). Treatment for *Lygus* bugs almost invariably takes place when treatment for pea aphid is carried out in North America to where its pest status is restricted. This usually occurs at 50 percent bloom, and the rate of dimethoate used for aphid control (0.5 lb a.i./ha) is adequate for *Lygus* control (Bragg and Burns, 2001).

2.4. The Cutworm, *Agrotis Ipsilon*

Agrotis ipsilon (Hufnagel) (Lepidoptera: Noctuidae) is a worldwide polyphagous pest, the larvae of which attack leaves, stems, and roots of many agricultural crops, including lentil although among leguminosae it is a more serious pest on soybean. It is also a pest on cauliflower, cotton, maize, strawberry, tomato and grapevine, which gives an indication of the breadth of its food source. The scientific name comes from the marking on the forewing of the adult, which resembles the Greek character ipsilon. The adult moths are grey-brown with a 40 mm wingspan, forewings are light brown, patterned with an ipsilon shape, and the hind wings are creamy white with brown edges. The females' lay over 1000 eggs in clutches of a dozen or so on leaves and even on soil. The eggs hatch after about 5 days (Beniwal et al., 1993). The young larvae, which like the adults are nocturnal, are green or greyish, becoming dark green or grey with age, and marked with two bright lines. Larvae remain below the surface of the ground, under clods of soil, or other shelters during the day. The first two larval stages feed on the foliage of the plant. The third and later stages often become cannibalistic and thus adopt solitary habits (Hill, 1983). They grow to as much as 50 mm in length when they can be found coiled round a damaged plant. Mature larvae bury deeply into the ground and pupate within cells, from which the adults emerge. Depending on the climate there might be one or several generations per year.

The older larvae cut the plant above the root crown. Most of the plant is not consumed after cutting, and larvae move to another plant leaving the earlier one to wither and dry. Some species feed on the upper leaves before moving to the soil surface. The abundance of *A. ipsilon* in some areas is partly affected by rainfall. In the drier areas (e.g., Syria) infestations are lower in years with high rainfall. In fact, flooding fields is recommended as a control measure in some cases. Deep ploughing of fields between crops turn up larvae and pupae to the soil surface making them susceptible to predators and sun. Weed hosts on outlying areas are often preferred sites of oviposition and serve as food for the cutworms in off-season. Cutworms are often difficult to control, especially when populations are epidemic in proportion. Large populations may cause severe crop damage with indications that the pest is the black cutworm. Unfortunately, by the time the pest is identified, the cutworms would have already developed into a life stage, which is not as susceptible to insecticides as the early larval stages. The sporadic nature of cutworm populations can make preventive treatments futile. And of course their soil-dwelling habits often beneath heavy foliage make control difficult with insecticides since they do not to reach the target (Hill, 1983). One way to control cutworms is to broadcast a poison bait prepared with wheat bran, cotton, or groundnut cake, moistened with water and trichlofon (Dipterex), carbaryl (Sevin) and Parathion @ 10 kg/ha in the evening (Bakr, 1994).

3. MINOR PESTS

3.1. Thrips, *Thrips, Kakothrips*, and *Frankiniella*

Thrips (*Thrips tabaci* Lindeman, and *T. angusticeps* Uzel) are the key pests of lentil in Castilla La Mancha (central Spain) (Perez Andueza et al., 2004), and Turkey (Tunc et al., 1999). However, several other species of thrips such as *Kakothrips robustus* (Uzel), *T. angusticeps*, and *Frankiniella* spp.. (Thysanoptera: Thripidae) have also been reported to infest lentil in most lentil-growing areas, but rarely cause serious damage (Beniwal et al., 1993). Thrips are minute (1 to 2 mm), elongate insects with four extremely slender wings in which the developmental stages resemble the adults, but are lighter coloured and wingless. Thrips attack leaves, flowers and pods by puncturing the plant organs and sucking up the sap, causing silvery blotches and dashes. As the attack increases the leaves and flowers become distorted and under heavy infestation can occasionally cause serious economic damage.

Removing volunteer soybean plants from the lentil crop is critical, since volunteer soybean plants have been reported to be the source for the thrips outbreak in lentil (Singh and Singh, 1994). Early maturing and small seeded genotypes of lentil have been reported to be more susceptible to thrips in Bangladesh (Sardar and Ahmad, 1991) but generally thrips are not a major field pest of lentil. Under greenhouse conditions thrips can appear in high numbers and cause severe damage on lentil plants grown for experimental purposes. Insecticides, Zolon, Nogos, and Sevin @ 0.025% effectively controls thrips in lentil (Sardar, 1990). Application of an insecticide for sucking insects (e.g., deltamethrin, malathion, dimethoate or endosulfan) will otherwise also provide good control (Beniwal et al., 1993).

3.2. The Bud Weevil, *Apion Arrogans*

Bud weevils infest several leguminous crops including lentil. Of particular relevance to the Mediterranean region is *Apion arrogans* Wenck. and *A. trifolii* (L.) from Europe, former USSR, and Southwest Asia. Adult *A. arrogans* weevils are about 3 mm long and the snout is characteristically longer than the body. The species also characteristically has long legs. The adults have dark blue elytra with black head, thorax, legs and abdomen. Larvae are yellow and legless. There is no published information on the life cycle of *A. arrogans*. Adult weevils feed on lentil leaves making small holes, but the main source of damage is caused by the larvae feeding on the buds and flowers whereby the ovules are destroyed (Beniwal et al., 1993). The bud weevil, *A. arrogans* has also been reported to transmit broad bean mottle virus (BBMV) in lentil (Makkouk and Kumari, 1989, 1995). Buds and flowers dry up and drop off. Inside the buds the larvae of *A. arrogans* can be found, which feeds on the developing seeds. Infestation levels are usually not high enough to warrant control measures, but in the occasional years when localized population densities are high methidathion at 0.5 kg a.i. ha^{-1} and monocrotophos at 2 ml L^{-1} can provide control.

3.3. The Pea Moth, *Cydia nigricana* (Fab.)

The pea moth is primarily a pest of field pea in North America, Europe and the Mediterranean, but it may also attack lentil. The adult is small compared to many Lepidoteran pests of Legumes (15 mm wingspan) and is brown with short, black and white lines along the front edge of the of the forewings. The larvae are pale off white with dark spots and short hairs and up to 12 mm long. The larvae over winter in a silk cocoon below the soil surface and pupate in spring. Adults appear at flowering and lay eggs on plants, after which larvae bore into the pods. Fully grown larvae drop to the soil to aestivate/hibernate (Beniwal et al., 1993). The larvae eat the seeds, and the damage often remains undetected until the pods are opened. Infestations are not severe enough to require control measures. However, since this insect directly affects the seeds and thus yield, insecticide application might be necessary sometimes. Sprays of methidathion (0.5 kg a.i./ha), deltamethrin (38 g a.i./ha), and endosulfan (6 ml/L) at the time of flowering/early pod setting can provide adequate control (Darty and Wimmer, 1983).

3.4. Pod Borers, *Helicoverpa armigera* (Hb.) & *Heliothis* spp. (Lepidoptera: Noctuidae)

Pod borers are arguably the most economically significant insect pest in the world owing to their widespread occurrence and broad spectrum of hosts. They are however only a minor pest of lentil in West Asia and the Indian subcontinent and do not present a serious threat to yield as they do in other crops such as chickpea (Stevenson et al., 2005). The adults are large and brown and up to 20 mm long, are active at night and lay hundreds of eggs singly on the underside of leaflets. Larvae can reach 40 mm, and have different and quite attractive coloration, but mostly green. Fully grown larvae, usually 6th instars drop to the soil to pupate. The larvae cause damage to the leaves with young instars scraping the surface of leaflets and feeding on flowers, while older larvae feed on foliage and more damagingly on pods. Control measures are rarely needed, however, as with *Cydia* spp. since this insect directly affects the seeds and consequently yield, insecticide application may occasionally be necessary using methidathion (Supracide^{R} @ 0.5 kg a.i./ha), deltamethrin (Decis @ 38 g a.i./ha) (Beniwal et al., 1993) and endosulfan (Thiodan 35 @ 3 ml/L) at the time of flowering/early pod-setting (Stevenson et al., 2005). Technologies for managing *H. armigera* with biopesticides such as *Helicoverpa armigera* Nucleopolyhedrosis Virus (HaNPV) are also proven and provide a viable alternative to chemical control strategies for this pest (Jayaraj et al., 1987, Cherry et al., 2000).

3.5. Lima-Bean Pod Borer, *Etiella Zinckenella* (Treit.) (Lepidoptera: Pyralidae)

The Lima-bean pod borer is an important insect pest of leguminous crops including lentil in USA, Europe, North and East Africa, Southwest Asia, India, and Pakistan. The mean egg, larval, pre-pupal and pupal life stage periods of *E. zinckenella* on lentil have been reported to be 5.4, 17.2, 2.3 and 13.8 days, respectively (Jaglan et al., 1995, 1996) and the fecundity per female is reportedly in the region of 60 eggs. The adult moths are 10 to 12 mm long with a wingspan of 22 to 28 mm. The forewings are brown-gray with a white anterior margin, while the hind wings are lighter. The larvae attain 10 to 12 mm length, and are greenish, with a brown line and head capsule. Mating takes place at night or in dark places, and females survive longer than males (Jaglan et al., 1995, 1996) albeit only for one week. Eggs are laid near the calyx of the flowers or on pods. The larval period lasts for 2 to 3 weeks after which the larvae pupate in the soil. Larvae are also known to diapause in winter. It completes 3 to 5 generations per year. The larvae feed on the soft green seeds in the pods even up to whole pod destruction but usually infestations rarely require control measures. Host plant resistance can contribute towards the pest management of *E. zinckenella* in lentil since some variation in infestations has been reported. For example, short duration genotypes have been reported to have higher larval population of *E. zinckenella* (Dashad et al., 2005). Minimum pod infestation was observed in LH 90–39, and was categorized as least susceptible (Jaglan et al., 1993), while LL 147 was considered tolerant to *E. zinckenella* damage (Brar et al., 1989). Short duration genotypes have been reported to have higher larval population of *E. zinckenella* (Dashad et al., 2005). Cultivars P 927 and P 202 proved to be resistant to the borer with a substantial increase of

52.9 and 43.5% yield over L 9–12, respectively (Chhabra and Kooner, 1970). Since this insect directly affects the seeds and thus yield, insecticide application might be necessary sometimes. Sprays of methidathion (Supracide^{R} @ 0.5 kg a.i./ha), deltamethrin (Decis @ 38 g a.i./ha), endosulfan (Thiodan 35 @ 6 ml/L) at the time of flowering/early pod-setting can provide adequate control.

3.6. The Root Aphid, *Smynthurodes betae* Westwood (Homoptera: Aphididae)

Root aphids have been reported infesting lentil from Iran (Rezwani, 1995), Turkey and Syria (Bayaa et al., 1998). In some years yellow patches can be found in lentil fields that are caused by this and are visible as white woolly aphids on the roots. It is a heteroecious species with a 2-year cycle and forms yellowish to red galls on the primary host pistachio by rolling the leaf edge into a spindle near the leaf base. The root aphid is rarely regarded as economically important.

3.7. Leaf Miners, *Liriomyza* and *Phytomyza*

Several species of leaf miner such as, *Liriomyza* spp., *Phytomyza* spp. etc. (Diptera: Agromyzidae) have been reported occasionally as pests of lentil in West Asia, North Africa, and South America (Tamer et al., 1998, Weigand et al., 1994). *Liriomyza* spp. notably has attained a status as an important pest of lentil in Egypt (Ismail et al., 1995, Attia, 1997). The larvae feed inside the leaves and produce discoloured mines in which larvae can be seen during heavy infestations, the leaves are curved upward forming a cup shape, and sometimes lead to defoliation. The fully-grown larvae pupate in the soil and sometimes in the cup shaped leaves. The adult females lay eggs in the leaf tissues. Infestations are rarely heavy enough to warrant control measures.

4. STORAGE PEST INSECTS

The most serious and frequently encountered insect pests of stored lentils are *Bruchus ervi* Froel. (Coleoptera: Bruchidae), occurring in Europe, North Africa, and Southwest Asia; and *B. lentis* Fab., in the USA, Europe, North Africa, Southwest Asia, and India although other species do occur including *Callosobruchus chinensis* (L.), and cowpea seed beetle, *C. maculatus* (F.) (Coleoptera: Bruchidae). A 10 year study (1991–2000) of around 2517 samples of *Lens* spp. from over 40 countries that were processed for quarantine clearance at the National Bureau of Plant Genetic Resources, New Delhi, revealed through X-ray screening for infestations that one of these 4 species of bruchids occurred in about 30% of the samples. The studies also revealed the presence of *B. tristiculurs* Fahraeus, *Callosobruchus analis* (Fabr.) (Bhalla et al., 2004).

4.1. *Bruchus* spp.

B. ervi and *B. lentis* are both about 3 to 3.5 mm long and differ in that the elytra of the former are black with light brown hairs and whitish spots whereas *B. lentis* adults have dense and reddish grey hairs on the back, marked with several whitish spots. The larvae of both are light yellow with dark brown heads. The adults infest fields at flowering, where they feed on nectar and pollen. Yellow transparent eggs are oviposited on the young pods and upon hatching, the larvae penetrate the pod and feed on developing seeds. The larval period may last 6 weeks where upon the larvae eat an exit hole leaving only a thin circular window of epidermal membrane intact. After pupation the emerging adult opens this membrane and leaves the pod. Adults re-enter the seed or remain in other protected places for hibernation until the flowering the following season and some may remain in the dry seed until the seeds are planted. Thus there is only one generation per year and no eggs are laid on dry seeds, and there is no reproduction during storage. Fumigation of infested seeds with phosphine (Phostoxin) before storage can control infestations in the following crop. Every effort should be made to avoid planting infested seeds. Applications of two sprays (one before 50% flowering, and second 15 days later) of either of the insecticides, endosulfan (Thiodan 35) @ 4 ml/L, alpha capamethrin (Fasctac EC 10) @ 0.25 ml/L, or Methyl parathion (Metyphon EC 50) @ I ml/L, provide adequate control from lentil seed beetle (Beniwal et al., 1993).

4.2. *Callosobruchus* spp.

Both species of *Callosobruchus* are widespread, and have been reported from all continents with sub-tropical or tropical conditions (USA, Mediterranean, Asia, and Australia). The adults are similar in length to *Bruchus* but *C. chinensis* has characteristic triangular white spot at the base of the thorax, and elytra are rust coloured with two brown spots, while those of *C. maculatus* are black tipped with a large round spot. These species mainly occur in the stored seed but occasionally infest fields that are close to harvest. Although a pest of stored lentil *Callosobruchus* spp. prefer other legumes such as chickpeas and garden pea and this is attributed to the levels of moisture, proteins and phenols in the seed-coat (Bhattacharya and Banerjee, 2001). The eggs are laid on the seed coat with several eggs per seed and the emerging larvae hatch through the base of the egg and bore straight through the seed coat. The white larvae develop and pupate inside the seed with one generation complete in 3 to 4 weeks. Thus, these insects reproduce very rapidly and can consequently cause considerable damage to stored products including lentil.

Stores should be clean from all residues of earlier stored products and may be de-infested with malathion. The seed should also be free from straw, stones, pebbles, and flour and other implements such as threshers and vehicles should be properly cleaned. Stored seeds can be enclosed in polythene and fumigated with phospine (Phostoxine) which controls all insect stages, leaves no residue so is safe for seeds stored for food, and does not adversely affect taste or germination. Seeds stored for planting can also be treated with Actellic @ 4 to 10 ppm a.i. (0.5 g/kg seed) or malathion @ 10 ppm a.i., which will protect the seeds for several months. Mixing seeds with olive oil and salt (5 ml and 20 g/kg seed) or Neem seed oil (3 ml/kg seed) can provide adequate control for a period of 3 to 4 months (Beniwal et al., 1993). In addition, extracts of *Clerodendron siphonanthus* reportedly reduces the number of eggs laid on the seed surface of stored lentil and may provide an alternative pesticidal plant option for farmers (Pandey and Khan, 1998). *Callosobruchus chinensis* is also susceptible to biological control measures and specifically the parasitic wasp, *Dinarmus basalis* (Rond.), which can completely control the pest among red lentil when introduced at between 30 to 50 pairs in a 50 m^2 room depending on the time of year (Islam and Kabir, 1995). A dose of 1 kGy of gamma radiation completely killed *C. chinensis* within a week, and was indicated to be a suitable alternative control measure for *C. chinensis*. The dose was also sufficient for the management of other pests including *Tribolium castaneum* Herbst) and *Rhizopertha dominica* Fabricius (Roy and Prasad, 1993). There are countless other non chemical approaches for controlling species of *Callosobruchus* in the literature but most reports are on the control of *C. maculatus* infesting cowpea. Of all those reported, Neem is probably the most accepted alternative to fumigation and chemical control (Lale and Mustapha, 2000).

REFERENCES

Al Karaki, G. 1996. N Effects of phosphorus, *Rhizobium* and *Sitona* weevil control on seed growth, nodulation and yield of lentil under Mediterranean rainfed conditions. *Legume Research* 19: 201–210.

Attia, M.B. 1997. Ecological and control studies on the serpentine leaf miner, *Liriomyza trifolii*, Burgess, Diptera: Agromyzidae on lentil in Sharkia governorate, Egypt. *Minufiya Journal of Agricultural Research* 15: 1047–1057.

Baird, C.R. 2000. Using Parasites to Manage *Lygus* in the Pacific Northwest. Proceedings, Lygus Summit, 28 November 2000. Visalia, CA, University of CA Division of Agriculture and Natural Resources. http://www.uckac.edu/cottonipm/PDF_files/summit_papers/baird.pdf (accessed 17 August 2006)

Bakr, M.A. 1994. Plant protection of lentil in Bangladesh [*Lens culinaris*]. In: W. Erskine and M.C. Saxena (eds.), Lentil in South Asia, Proceedings of the Seminar on Lentil in South Asia, Indian Council of Agricultural Research, New Delhi, India: International Center for Agricultural Research in the Dry Areas (ICARDA), Aleppo (Syria), p. 177–186.

Baum, M. 2000. Pulse Transformation technology transfer. Unpublished report. http://www.aciar.gov.au/web.nsf/projectprint/ACIA-6SF23N?opendocument. Last accessed 17 August 2006.

Bayaa, B., Kumari, S.G., Akkaya, A., Erskine, W., Makkouk, K.M., Turk Z.O. and Zberk, I. 1998. Survey of major biotic stresses of lentil in South-East Anatolia, Turkey. *Phytopathologia Mediterranea* 37: 88–95.

Bejiga, G. and Anbessa, Y. 1993. Genetics and breeding research in lentil [*Lens culinaris*]. In: First National Cool-season Food Legumes Review Conference, 16–20 December, 1993, Addis Abeba, Ethiopia, South Africa.

Bejiga, G., Degago, Y. and Knight, R. 2000. Region 4: Sub-Sahara Africa. In: Linking research and marketing opportunities for pulses in the 21st Century, Proceedings of the Third International Food Legumes Research Conference, Adelaide, Australia, 22–26 September 1997. *Current Plant Science and Biotechnology in Agriculture* 34: 99–105. Kluwer Academic Publishers, Dordrecht, Netherlands.

Beniwal, S.P.S., Bayaa, B., Weigand, S., Makkouk K.H. and Saxena M.C. (1993) Field guide to lentil diseases and insect pests. http://www.icarda.org/Publications/Field_Guides/Lentil/Lentil.htm.

Bezdicek, D.F., Quinn, M.A., Forse, L., Beck, D.P., Weigand, S., Muehlbauer, F.J. (ed.) and Kaiser, W.J. 1994. Cloning *Bacillus thuringiensis* toxin genes for control of nodule-feeding insects. In: Expanding the production and use of cool season food legumes, Proceedings of the Second International Food Legume Research Conference on pea, lentil, faba bean, chickpea, and grasspea, Cairo, Egypt, 12–16 April 1992, pp. 738–752.

Bhalla, S., Kapur, M.L., Singh, C., Gupta, K., Kumar, N. and Lal, B. 2004. Interception of bruchids in imported lentil (Lens spp) germplasm. *Indian Journal of Agricultural Sciences* 74 (6): 332–333.

Bhatnagar, A. and Seghal, V.K. 1990. Incidence and seasonal occurrence of insect fauna associated with lentil crop in northern India. *Lens Newsletter* 17: 21–26.

Bhattacharya, B., Barik, A. and Banerjee, T.C. 2001. Bioenergetics and water balance in *Callosobruchus maculatus* (F.) (Coleoptera : Bruchidae) larval populations. *Oriental Insects* 37: 423–437.

Bragg, D. and Burns, J.W. 2000. Crop Profile for Lentil in Washington State. http://www.ipmcenters.org/cropprofiles/docs/WALentils.html (accessed 16 August 2006).

Bragg, D. and Burns, J.W. 2001. Pea Aphid and Pea Seed Weevil Control in Dry Field Peas: Effects on Yield and Yield Components from Insecticide Application Timings http://variety.wsu.edu/Updates/2001_Pea_Report.pdf (accessed 16 August 2006).

Brar, J.S., Verma, M.M., Sandhu, T.S., Singh, B. and Gill, A.S. 1989. LL 147 variety of lentil (*Lens culinaris* L.). *Journal of Research, Punjab Agricultural University* 26: 170.

Cherry, A.C., Rabindra, R.J., Grzywacz, D., Kennedy and Sathiah, R. 2000. Field evaluation of *Helicoverpa armigera* – NPV formulations for control of chickpea pod borer, *H. armigera* (Hubn.) on chickpea (*Cicer arietinum* var Shoba) in Southern India. *Crop Protection* 19: 51–60.

Chhabra, K.S. and Kooner, B.S. 1970. Field resistance in some lentil cultivars against pod borer *Etiella zinckenella* Treit. *LENS Lentil Experimental News Service* (Canada) 7: 46–49.

Darty, J.M. and Wimmer, F. 1983. Lentil: Control of the pea midge and the peamoth (*Contarinia lentis*). *Phytoma* 347: 29–31.

Dashad, S.S., Kumar, Y. and Dahiya, B. 2005. Evaluation of small seeded lentil genotypes of different maturity groups against *Etiella zinckenella* Tr. *Research on Crops* 6: 332–336.

Day, W. and Mahr, S. 1998. *Peristenus digoneutis*, parasite of tarnished plant bug. *Midwest Biological Control News Online*. http://www.entomology.wisc.edu/mbcn/kyf510.html (17 August 2006).

El Damir, M., El Bouhssini, M. and Al Salti, M.N. 1999. A simple screening technique of lentil germplasm for resistance to *Sitona crinitus* H. (Coleoptera: Curculionidae) under artificial infestation. *Arab Journal of Plant Protection* 17: 33–35.

El Damir, M., El Bouhssini, M. and Al Salti, M.N. 2000. Effect of *Melia azedarach* (L.) (Meliaceae) fruit extracts on *Sitona crinitus* (H.) (Coleoptera: Curculionidae) adults feeding. *Arab Journal of Plant Protection* 18: 64–67.

El Damir, M., Al-Salti, M.N. and El Bouhssini, M. 2001. Ecology and biology of *Sitona crinitus* H. in northern Syria. *Bassel Al-Assad Journal for Engineering Sciences* 14:109–119.

El Damir, M., El Bouhssini, M. and Al Salti, M.N. 2004. Embryo development and egg hatching of *Sitona crinitus* Herbst (Coleoptera: Curculionidae) under constant temperature regimes. *Pakistan Journal of Biological Sciences* 7(7): 1191–1193.

Erman, M., Yardm, E.N. and Kulaz, H. 2005. Effect of cultivars and insecticides on sitonid weevil, *Sitona crinitus* (Coleoptera: Curculionidae), and on yield, yield components and nodulation of lentil (*Lens culinaris*). *Indian Journal of Agricultural Sciences* 75: 204–206.

Erskine W, Tufail M, Russell A, Et Al. 1994. Current and Future Strategies in Breeding Lentil for Resistance to Biotic and Abiotic Stresses. *Euphytica* 73 (1–2): 127–135.

Fuchs, S.J. and Hirnyck, R.E. 2000. Crop Profile for Lentil in Idaho State. http://www.ipmcenters.org/cropprofiles/docs/IDlentils.html (accessed 17 August 2006).

Gerard, P.J., Crush, J.R. and Hackell, D.L. 2005. Interaction between *Sitona lepidus* and red clover lines selected for formononetin content. *Annals of Applied Biology* 147 (2): 173–181.

Grzywacz, D., Rabindra, R.J., Brown, M., Jones, K.A. and Parnell, M. 2002. The *Helicoverpa armigera* NPV production manual. NRI Report 2706, University of Grenwich. 130pp.

Hammad, S.M. 1978. Pests of grain legume and their control in Egypt. In: Pest of grain legumes: Ecology and control. Academic Press Inc., New York, pp. 135–137.

Hariri, G. 1981. Insect and other pests. In: C. Webs and G. Hawtin (eds.), Lentils. Commonwealth Agricultural Bureau, England, pp. 173–189.

Hill, D.S. 1983. *Agrotis ipsilon* (Hfn.). In: Agricultural Insect Pests of the Tropics and Their Control, 2nd Edition. Cambridge University Press, Cambridge, London, New York, New Rochelle, Melbourne, Sydney, pp. 357–358.

Islam, W. and Kabir, S.M.H. 1995. Biological-Control Potential of *Dinarmus-Basalis* (Rond) (Hymenoptera, Pteromalidae), A Larval-Pupal Ectoparasitoid of the Pulse Beetle, *Callosobruchus chinensis* (L). *Crop Protection* 14 (6): 439–443.

Islam, M.S., Solh, M.B. and Saxena, M.C. 1985. Effect of fertilization, inoculation, and carbofuran on nodulation, yield, and protein content of lentil [*Lens culinaris*]. *Lens Newsletter* 12: 32–36.

Islam, R. and Afandi, F. 1982. Effect of some insecticides on nodulation and yield of two lentil cultivars. *Lentil Experimental News Service ICARDA (Syria)* 9: 24–25.

Ismail, I.I., Sharaf, E.I. and Din, A.A.A. 1995. Susceptibility of lentil varieties and strains to infestation with *Bemisia tabaci* Genn *Aphis craccivora* (Koch.) and *Liriomyza congesta* Becker in Egypt. *Zagazig Journal of Agricultural Research* 21: 269–277.

Jaglan, M.S., Sucheta, Khokhar, K.S. and Solanki, I.S. 1993. Screening lentil for susceptibility to *Etiella zinckenella* Treitschke infestation [*Lens culinaris*]. *Lens Newsletter* 20: 13–14.

Jaglan, M.S., Sucheta, Khokhar, K.S. and Kumar, S. 1995. Biology of lentil pod borer, *Etiella zinckenella* Treitschke on lentil and pea. *Annals of Biology*, Ludhiana 11: 224–228.

Jaglan, M.S., Sucheta and Khokhar, K.S. 1996. Lentil pod borer (*Etiella zinckenella* Treitschke), biology on lentil and pea. *Lens Newsletter* 23: 48–51.

Jayaraj, S., Rabindra, R.J. and Santharam, G. 1987. Control of *Heliothis armigera* (Hubner) on chickpea and lablab bean by nuclear polyhedrosis virus. *Indian Journal of Agricultural Science* 57: 738–741.

Jones, J. 1999. Lygus Bugs in Canola. Agdex 622–20. May 1999, http://www1.agric.gov.ab.ca/$department/deptdocs.nsf/all/agdex741?opendocument (accessed 17 August 2006).

Jones, R.A.C. and Coutts, B.A. 1996. Alfalfa mosaic and cucumber mosaic virus infection in chickpea and lentil: incidence and seed transmission.*Annals* of Applied Biology 129: 491–506.

Jung, W., Yu, O., Lau, S.M.C., O'Keefe, D.P., Odell, J., Fader, G. and McGonigle, B. 2000. Identification and expression of isoflavone synthase, the key enzyme for biosynthesis of isoflavones in legumes. *Nature Biotechnology* 18 (2): 208–212.

Kaya, N. and Hincal, P. 1987. Population fluctuations and damage by the lentil leaf weevil (*Sitona crinitus* Herbst.) (Coleoptera: Curculionidae) in Denizli. Turkiye I. Entomoloji Kongresi Bildirileri, 13–16 Ekim, 1987, Ege Universitesi, Nornova, Izmir, pp. 259–266.

Lal, S.S. 1992. Insect pest of lentil and their management review. *Agricultural Review* 13: 225–232.

Lale, N.E.S. and Mustapha, A. 2000. Efficacy and acceptability of neem (*Azadirachta indica* A. Juss) seed oil and pirimiphos-methyl applied in three storage devices for the control of *Callosobruchus maculatus* (F.) (Coleoptera : Bruchidae). *Journal of Plant Diseases and Protection* 107 (4): 399–405.

Latham, L.J. and Jones, R.A.C. 2001. Alfalfa mosaic and pea seed-borne mosaic viruses in cool season crop, annual pasture, and forage legumes: susceptibility, sensitivity, and seed transmission. *Australian Journal of Agricultural Research* 52 (7): 771–790.

Latham, J.J., Jones, R.A.C. and Coutts, B.A. 2004. Yield losses caused by virus infection in four combinations of non-persistently aphid-transmitted virus and cool-season crop legume. *Australian Journal of Experimental Agriculture* 44 (1): 57–63.

Makkouk, K.M. and Kumari, S.G. 1989. *Apion arrogans*, a weevil vector of broad bean mottle virus [*Vicia faba*]. *Fabis Newsletter* 24: 26–27.

Makkouk, K.M. and Kumari, S.G. 1995. Transmission of broad bean stain comovirus and broad bean mottle bromovirus by weevils in Syria. *Journal of Plant Diseases and Protection* 102: 136–139.

Makkouk, K.M. and Kumari, S.G. 2001 Reduction of incidence of three persistently transmitted aphid-borne viruses affecting legume crops by seed-treatment with the insecticide imidacloprid (Gaucho (R)). *Crop Protection* 20: 433–437

Makkouk, K.M., Bashir, M., Jones, R.A.C. and Kumari, S.G. 2001. Survey for viruses in lentil and chickpea crops in Pakistan. *Journal of Plant Diseases and Protection* 108: 258–268.

Manero, E.A. de and L'Argentier, S.M. de. 1987. Aphididae and Thripidae injurious to lentil crop in the Jujuy Province. *Revista de Investigacion Centro de Investigaciones para la Regulacion de Poblaciones de Organismos Nocivos* 5: 17–26.

Muehlbauer, F.J. 1996. Advances in the production of cool season food legumes. *American Journal of Alternative Agriculture* (USA) 11: 71–76.

Pala, M. and Mazid, A. 1992. On-farm assessment of improved crop production practices in northwest Syria. 2. Lentil. *Experimental Agriculture* 28: 185–193.

Pandey, S.K. and Khan, M.B. 1998. Inhibitory effect of biopesticide on development of pulse weevil, *Callosobruchus chinensis* (L.) on lentil, *Lens esculentis* through injection method. *Journal of Advanced Zoology* 19: 94–98.

Perez Andueza, G., Mozos Pascual, M. de los and Portillo Rubio, M. 2004. Main pests of lentil (*Lens culinaris* Medikus) in Castilla La Mancha (central Spain): crop losses and influence on yield parameters. *Boletin de Sanidad Vegetal, Plagas* 30: 763–772.

Rezwani, A 1995. The sugar beet root aphids of Iran. *Journal of Entomological Society of Iran* 15: 45–51.

Robeson, D.J. 1978. Furanoacetylene And Isoflavonoid Phytoalexins in *Lens culinaris*. *Phytochemistry* 17 (4): 807–808.

Roy, M.K. and Prasad, H.H. 1993 Gamma-Radiation in the Control of Important Storage Pests of 3 Grain Legumes. *Journal of Food Science and Technology (Mysore)* 30 (4): 275–278.

Sardar, M.M.A. 1990. Insect pests reaction to various genotypes of lentil and bean and intervention of control tactics in Bangladesh. *Bangladesh Agricultural University Research Progress* (Bangladesh) 4: 8–14.

Sardar, M.M.A. and Ahmad, M. 1991. Genotypical evaluation of major pests on lentil and lablab bean in Bangladesh. *Bangladesh Agricultural University Research Progress* (Bangladesh) 5: 225–231.

Schotzko, D.J. and O'Keeffe, L.E. 1988. Effects of food type, duration of hibernal quiescence, and weevil density on longevity of *Sitona lineatus* (Coleoptera: Curculionidae). *Journal of Economic Entomology* 81: 1631–1636.

Sharma, R.P. and Yadav, R.P. 1993. Response of lentil varieties to the incidence of bean aphid (*Aphis craccivora* Koch.) and its predatory coccinellids. *Lens Newsletter* 20: 60–62.

Singh, H. and Dhooria, M.S. 1971. Bionomics of pea pod borer, *Etiella zinckenella* (Treit). *Indian Journal of Entomology* 33: 123–130.

Singh, K.J. and Singh, O.P. 1994. Role of volunteer soybean plants in the outbreak of thrips, *Megalurothrips distalis* (Karny), on lentil in India [*Lens culinaris*]. *Lens Newsletter* 21: 41–42.

Solh, M.B., Itani, H.M. and Kawar, N.S. 1986. The effect of *Sitona* weevil on nodulation and yield of lentils and the implication of certain control measures. *Lebanese Science Bulletin* 2: 17–27.

Staneva, E. 1982. Studies on the food plants of the cowpea weevil, *Callosobruchus maculates* F. (Coleoptera: Bruchidae). *Revista Zashchita* 19: 111–119.

Stevenson, P.C., Pande, S., Neupane, R.K., Chaudary, C.N., Bourai, V.A., Narayana Rao, J. and Grzywacz, D. 2005. The adoption of Integrated Crop Management Technologies for Chickpea by Poor Farmers in Nepal. pp. 135–142. In Pande, S., Stevenson P.C., Neupane, R.K. and Grzywacz, D. (Eds). Policy and strategy for increasing income and food security through improved crop management of chickpea in rice fallows in Asia. Summary of a NARC-ICRISAT-NRI Workshop, 17–18 November 2004, Kathmandu, Nepal. Patancheru 502 324, Andhra Pradesh, India: International Crops Research Institute for the Semi-Arid Tropics. 252 pp.

Tamer, A., Has, A., Aydemir, M. and Caliskaner, S. 1998. Faunistic survey studies on harmful and beneficial insects on food legumes (lentil, chickpea, bean) in Central Anatolia Region. *Bitki Koruma Bulteni* 38: 65–90.

Thakur, B.S., Verma, R., Patitunda, A. and Rawat, R.R. 1984. Chemical control of aphid, *Aphis craccivora* Koch on lentil. *Indian Journal of Entomology* 46: 103–105.

Tunc, I., Vierbergen, G. (ed.) and Tunccedilla, I. 1999. Thrips infestations on field crops in Turkey. In: Proceedings, Sixth International Symposium on Thysanoptera, 27 April-1 May, 1998, Akdeniz University, Antalya, Turkey. Akdeniz University, Faculty of Agriculture, Department of Plant Protection, Antalya, Turkey, pp. 145–150.

Wale, M., Jembere B. and Seyoum, E. 2000. Biology of the pea aphid, *Acyrthosiphon pisum* (Harris) (Homoptera: Aphididae) on cool-season legumes. *Insect Science and its Application* 20: 171–180.

Weigand, S., Lateef, S.S., El Din, N.E.S., Mahmoud, S.F., Ahmed, K., Ali, K., Muehlbauer, F.J. (ed.) and Kaiser, W.J. 1994. Integrated control of insect pests of cool season food legumes. In: Expanding the production and use of cool season food legumes, Proceedings of the Second International Food Legume Research Conference on pea, lentil, faba bean, chickpea, and grasspea, Cairo, Egypt, 12–16 April 1992, pp. 679–694. Kluwer Academic Publishers Group, Dordrecht, Netherlands.

Yadava, C.P. and Ahmad, R. 2000. Insect pests of lentil and their management. *Applied Entomology* 2: 82–95.

CHAPTER 21

QUALITY SEED PRODUCTION

ZEWDIE BISHAW[1], ABDUL A. NIANE[1] AND YANTAI GAN[2]

[1]*International Center for Agricultural Research in the Dry Areas (ICARDA), P.O. Box 5466, Aleppo, Syria*

[2]*Agriculture and Agri-Food Canada, Box 1030, Swift Current, SK, S9H 3X2, Canada*

E-mail: z.bishaw@cgiar.org

Abstract: Modern varieties are backbone of formal seed industry. The availability, access, and use of quality seed of adaptable crop varieties, are critical in increasing agricultural productivity, ensuring food security, and improving farmers livelihoods. However, research in legumes particularly lentil is relatively new compared to cereals both at IARCs and NARS (Aw-Hassan *et al.*, 2003). The impact of investments in agricultural research can be realized only if farmers have better access to high quality seed of the new crop varieties. Since seed quality is one of the main factors that affects crop production potential it should reach farmers in a good quality state. Seed quality comprises many aspects where four key attributes are explicitly identified: genetic, physical, physiological and health quality. However, seed quality can be affected by environmental conditions under which the crop is grown and the cultural practices used for production. Maintaining seed quality is essential if the variety is to meet the expectation of farmers and consumers. Seed producers should be aware of the technical and regulatory requirements for growing a crop for seed, and ensure that all operations are carried out strictly under specific guidelines in a timely fashion. The quality of the seed can be ensured by following a combination of key technical procedures and regulatory measures. The quality control assurance system by establishing administrative guidelines and technical procedures plays a supervisory role for smooth operation and implementation of the program and for enforcing the regulatory measures to maintain the quality of seed produced. Limited choice of improved varieties, lack of sufficient quantity of seed, mechanization problems in developing countries, and high seed production costs, are some of the major constraints hindering the development of an effective and efficient lentil seed industry. This chapter covers key components such as variety maintenance and technical aspects of seed production, seed processing, seed storage, and seed quality assurance

S.S. Yadav et al. (eds.), Lentil: An Ancient Crop for Modern Times, 349–383.

1. INTRODUCTION

Lentil has been cultivated for millennia in the dry areas of south west Asia dating as far back as 7000 BC. It has spread from the Near East to the Mediterranean, Africa, Asia, Europe, Eurasia, and the Americas. Lentil crop is best adapted to the cooler temperate zones of the world, or the winter season in the Mediterranean climates. It has been introduced in the farming systems of developed countries as part of agricultural diversification program where cereals have traditionally been the main crops for production and export. Although Canada and USA are becoming major global players in production and marketing of lentil, Asia remains the major producer and importer of lentil, with India being the single most important country in terms of production and consumption.

Modern varieties are a backbone of the formal seed industry. The availability, access, and use of quality seed of adaptable crop varieties, are critically important in increasing agricultural productivity, ensuring food security, and improving farmers livelihoods. However, research in legumes particularly lentil is relatively new compared to cereals both at IARCs and NARS (Aw-Hassan *et al.*, 2003). International agricultural research centers in partnership with national research institutes have developed several lentil varieties which are adapted to local conditions with high and stable grain yield, better grain and nutritional quality, and tolerance to biotic and abiotic stresses (Aw-Hassan *et al.*, 2003). The impact of investments in agricultural research can be realized only if farmers have better access to high-quality seed of these new crop varieties.

With current systems, however, the availability of quality seed of cool season food legumes in general and of lentil in particular does not meet the annual national requirements. Limited choice of improved varieties, lack of sufficient quantity of seed, mechanization problems in developing countries, and high seed production costs, are some of the major constraints hindering the development of an effective and efficient lentil seed industry. Lentil is considered as 'low value' seed crop, because it is self-pollinated, making it easy for farmers to save their own seed and the profit margin for attracting private sector investment is low. The net result is that the formal seed industry pays less attention to supplying lentil seed.

2. DEFINING QUALITY SEED

Seed is a primary input in crop production, whether agriculture is practiced at commercial or subsistence levels, by large or small-scale producers, or in favorable or less favorable environments. It is also a means for delivering new agriculture-based technologies to farmers. Since seed quality is one of the main factors that affects crop production potential it should reach farmers in a good quality state.

'High quality seed' can be broadly defined as *'seed of an adapted variety with high genetic varietal, species, and physical purity; high germination and vigor; free from seed-borne pests (fungi, bacteria, viruses, insects, nematodes, parasitic weeds); and properly cleaned, treated, tested and labeled'*. Seed quality is a multiple

concept comprising of several components each with a relative importance under a sets of circumstances.

In general, seed quality comprises many aspects where four key attributes may be explicitly identified:

1. *Genetic quality*—the inherent genetic make up of the variety contained in the seed which provides the potential for higher yield, better grain quality, and greater tolerance to biotic or abiotic stresses;
2. *Physiological quality*—the viability, germination and vigor of seed which determines the potential germination and subsequent seedling emergence and crop establishment in the field;
3. *Physical quality*—freedom from contamination with other crops, common and particularly noxious and parasitic weed seeds, seed size, seed weight and seed lot uniformity;
4. *Health quality*—absence of infection/infestation with seed-borne pests (fungi, bacteria, viruses, nematodes, insects, etc).

However, seed quality can be affected by environmental conditions under which the crop is grown, the cultural practices used for production such as soil conditions, nutrient deficiency, water stresses, extreme temperatures and pests as well as the handling operations. Lentils produce good quality seed in semi-arid or drier areas. High humidity and excessive rainfall during the season encourages vegetative growth and reduce yield and seed quality. Excessive drought and/or high temperatures during the flowering and pod-fill period reduce yields.

Testa quality is one of the principal components of legume (lentil, chickpea and faba bean) seed quality and is influenced by harvest. Since legume embryos are surrounded by a testa, mechanical damage influences the conditions of the testa quality. The amount of cracking which results from a given handling treatment is dependent upon seed moisture content. Therefore, appropriate measures should be used to produce quality seed.

3. REQUIREMENTS FOR SEED PRODUCTION

Seed is a living biological product that requires special attention and care to ensure the varietal, physical, physiological and health quality. Maintaining seed quality is essential if the variety is to meet the expectation of farmers and consumers. Seed producers should be quality oriented and aware of the legal and administrative requirements for growing a crop for seed, and ensure that all operations are carried out strictly under specific guidelines in a timely fashion. The quality of the seed can be ensured only by following the combination of key regulatory control, administrative measures, and technical procedures.

The regulatory measures include a framework of: (1) establishing variety release systems to allow only superior varieties enter commercial seed production, (2) defining the category for seed production by limiting the number of generation to minimize contamination, (3) setting appropriate field and seed standards that need to be met during seed multiplication, and (4) establishing a mechanism (or agency) to oversee the enforcement of regulatory framework. The quality control agency by establishing administrative guidelines and technical procedures plays a supervisory role for smooth operation and implementation of the program and for enforcing the regulatory measures to maintain the quality of seed produced. The administrative measures may include the registration of seed growers, producers, processors, and suppliers as well as the implementation of the certification scheme through field inspection and seed testing through a robust quality assurance system.

Seed production should be strictly monitored throughout the entire crop growth period, from planting through harvesting, cleaning, storage and marketing. Key technical components for producing quality seed may include: (i) selection of production sites to avoid high risk environments; (ii) selection of clean field to eliminate volunteer plants from preceding crops (previous cropping) and to avoid build-up of noxious weeds and soil-borne pests; (iii) isolation from sources of contamination (genetic or physical); (iv) roguing to remove off-type contaminants; (v) maintaining the cleanliness of farm machinery during planting, harvesting and transportation; (vi) maintaining the cleanliness of processing machinery to avoid admixtures during cleaning, treatment, (vii) provision of adequate storage facilities to protect against damage from insect pests and maintain quality; and (viii) production arrangements by selecting specialized contract seed growers.

The following sections deal with variety release and variety maintenance as well as techniques in seed multiplication, seed cleaning, seed treatment, seed storage, and seed quality assurance.

4. VARIETY TESTING AND RELEASE

A new and potential promising variety, once identified by agricultural research institutes, it is customary to release and make it available to farming communities. The term 'variety release' encompasses several broadly interrelated activities from identifying promising lines and submission for further testing by a competent authority to releasing a new variety to end users by making available the breeder seed for further multiplication. It is a mechanism to safe guard against release of varieties with uncertain performance to protect the farmers and the industry. Consequently, new lentil varieties must pass through a series of simultaneous evaluation i.e. registration and performance testing before they are officially released to enter large-scale commercial seed production and supply.

4.1. Variety Testing

Performance testing are often referred to as 'variety trials' which focus on the value for cultivation and use (VCU) of the variety, *i.e.*, the benefit of the new variety to crop producers (farmers) and end users (industry, consumers). Performance trials are conducted at multi-locations in different agro-ecological zones with at least a minimum of over two years to assess the adaptation and performance of the variety. The effects of different crop management practices are also assessed. In VCU trials, new varieties are compared with existing standard commercial varieties to measure their merits. Performance trials are usually run for three consecutive years. In some countries, however, the variety is further tested in on-farm verification trials under farmer management conditions before final release (e.g. Ethiopia).

A registration testing is a descriptive assessment, establishing the distinctness, uniformity and stability (DUS) of the variety. Morphological, physiological, cytological, chemical characters or presently molecular techniques are used to establish the varietal identity by assessing the distinctness, uniformity and stability. In these tests, new varieties are compared with a wide range of existing varieties. At the end, the variety description is prepared and differences with other varieties established. The registration testing is usually conducted for a minimum of two consecutive seasons in at least one location where the variety will be released and commercialized. The International Union for the Protection of Plant Varieties (UPOV) has published test guidelines (http://www.upov.int) for morphological description of lentil (UPOV, 2003).

4.2. Variety Release

Variety release is a culmination of several interrelated activities where a decision could be taken to approve a new variety for commercial use based on the results of registration and performance tests. Almost all countries have a variety release procedure in place whether that is done by an ad hoc committee or legally sanctioned independent authority. The varieties that meet the requirements for registration and performance are officially released and the owner of the variety makes breeder seed available for commercial seed multiplication and marketing.

The way variety release procedures and protocols are organized and conducted and extent the breeders are involved is described as compulsory or voluntary system. In a compulsory system (e.g. EU), the governments strictly regulate the introduction of new varieties, prohibiting seed production and marketing until the variety is tested and meet all the requirements by a competent government agency and approved by the release committee (Gisselquist and Srivastava, 1997). In a voluntary system (e.g USA), the responsibility of releasing a variety is at the discretion of the breeders or breeding institutions which introduce and commercialize the variety. The breeder will be responsible to provide supporting data to prove the merit of the variety on voluntary basis.

4.3. Variety Registers

The new variety, upon approval, will be listed in a variety register to inform the stakeholders i.e. seed producers, farmers and the industry. The list could be informative or recommendatory. The register is periodically updated removing obsolete varieties and defining varieties that are currently eligible for commercial seed production at national or regional levels. Many countries have a national variety register (e.g. Crop Variety Register in Ethiopia) whereas the OECD has a common variety catalogue (http://www.oecd.org) which enables the variety to be produced and marketed in all member countries participating in the OECD Seed Scheme.

4.4. Harmonizing Variety Release System

Variety release procedure is a collective term that refers to the release type, the attached terms and conditions, the protocols and administrative procedures used in releasing a new variety for seed production and distribution to farmers (Delouche and Gomaa, 1999). In many developing countries the NARS receive almost similar breeding lines supplied through a network of IARCs. Despite certain similarities in the agroecology, farming system and germplasm there is no mechanism for sharing of data in making decisions for variety release even among neighboring countries.

In principle the procedures for variety testing, registering, and releasing are essentially similar in most countries. This provides an opportunity for developing testing protocols, sharing data and developing flexible and harmonized variety release schemes within regional or international contexts to make available a wider choice of varieties to farmers. In Turkey, foreign registered varieties from member countries of EU, OECD and UPOV are exempted from DUS testing and accepted as part of the variety release procedure. It is highly desirable for policy and regulatory frameworks to encourage countries to move from mandatory to voluntary and from single-to multi-country lists in variety release system (Gisselquest, 1997); this will increase the choice and movement of varieties to harness the impact of plant breeding research at national and/or international levels.

5. SEED CATEGORY FOR PRODUCTION

In a formal sector a limited generation system, though of different nomenclature, is used for seed production to minimize the risk of contamination (Table 1). In this

Table 1. Comparative nomenclature in selected countries of West Asia and North Africa

Generation	OECD	AOSCA	Egypt	Ethiopia	Morocco	Syria
First generation	Breeder	Breeder	Breeder	Breeder	Epis-lignes (G_0)	Nucleus
Second generation	Pre-basic	Foundation	Foundation	Pre-basic	Prébase ($G_{1,}$ G2, G_3)	Foundation
Third generation	Basic	Registered	Registered	Basic	Base (G_4)	Registered
Fourth generation	Certified 1	Certified	Certified	Certified 1	Reproduction 1 (R_1)	Certified
Fifth Generation	Certified 2	–	–	Certified 2	Reproduction 2 (R2)	–

Source: ICARDA, 2002; AOSCA (http://www.aosca.org); OECD (http://www.oecd.org)

chapter, the OECD nomenclature will be used. *Breeder seed* is the initial source of seed, produced by the breeder (his agent) or plant breeding institution. *Pre-basic seed* is the progeny of breeder seed and is usually produced under the supervision of a breeder or his designated agency. *Basic seed* is the progeny of breeder or pre-basic seed and is produced under the supervision of a breeder or his designated agency. *Certified seed* is the progeny of basic seed and produced on contract by selected seed growers under the supervision of the public or private seed enterprise. Certified seed can be used to produce further generations of certified seed, or can be planted by farmers for grain production. In a formal sector, both basic and certified seed are produced under the certification scheme controlled by a quality control agency.

6. CONTAMINATION OF VARIETY

Seed production follows a generation system where a small quantity of 'parental material' or 'nucleus seed' received from breeders is systematically multiplied into larger quantity of certified seed for distribution to farmers (Table 1). In the process the seed is prone to several deteriorative factors reducing the genetic, physical and health quality due to progressive increase in the degree of contaminants over the years. Generally three types of contamination can be distinguished, *viz*: (i) genetic contamination, (ii) mechanical contamination, and (iii) pathological contamination.

Genetic contamination arises from residual segregation, spontaneous mutations or undesirable natural out-crossing with other varieties, species or wild relatives as the case may be. Self-fertilizing crops like lentil often have small percentages of cross-pollination resulting in genetic contamination of the seed crop, whereas undesirable cross-pollination is a common feature in all cross-fertilizing crops, unless fields are adequately isolated. The spontaneous mutation rate is generally low, usually recessive and often difficult to detect.

Mechanical contamination results from a very wide range of sources but is mainly due to insufficient cleanliness of fields (cropping history, crop rotation) and equipment (planters, combines, cleaners, vehicles, bags, stores) and inadequate measures to avoid such contamination (field selection, physical isolation, cleanliness).

Pathological contamination occurs through infection particularly with seed-borne diseases which are exclusively transmitted by seed. Pathological contamination is usually caused by pathogens from the same variety, other varieties, other crops or weeds which spread the disease to normal plants or contaminate the seed lot. It could also happen due to gradual loss of tolerance of the variety to plant diseases.

7. VARIETY MAINTENANCE

Seed production follows a generation system (Table 1) where a small quantity parental seed, received from breeders, is multiplied into larger quantities of certified seed for marketing to farmers. In the process, seed is constantly prone to deteriorative factors during each cycle of multiplication irrespective of the quality of

Year 0	Select 500 plants of typical of each variety
	Harvest and thresh each selected plant separately
	Sow seed from each plant in a single progeny row
	Inspect progeny-rows and discard all rows with off-types
	Select again 500 plants for next cycle of maintenance
	Bulk harvest remaining seed to constitute Breeder Seed
Year 1	Sow 500 progeny rows and repeat the above cycle = Maintenance
Year 2	Sow breeder seed bulk to produce Pre-basic Seed = Multiplication

Figure 1. Variety maintenance and breeder seed production of lentil

the starting material which could reduce its genetic, physical and health quality. Reducing potential contamination and ensuring quality particularly varietal purity and identity is of paramount importance. Therefore, new lots of breeder seed must be regularly produced in a process called variety maintenance and used to initiate production of later generations (e.g., pre-basic, basic, certified).

Laverack (1994) defined variety maintenance as *'the perpetuation of a small stock of parental material through repeated multiplication following a precise procedure'*. Production of new breeder seed stock as source for starting new cycle of seed multiplication is necessary as long as the variety is under commercial production. Legume variety maintenance procedures have been described by Bouwman (1992), Drijfhout (1981), Julen (1983), Sharma (1987), and Singh and Saxena (1999). For lentil as in chickpea there are several options as follows:

Purification of varieties: The best available seed field is selected, and obvious contaminants are removed from the field (negative mass selection). Plants not conforming to the variety are rogued out, the crop is bulk harvested and the seed is then used for further multiplication. This purification aims at producing a very clean starting stock for further multiplication, and is used in situations where there is no organized maintenance procedure in place.

Mass selection: In this method the individual best plants from a field are selected. These selected individual plants are bulk harvested and the rest of the field is discarded. The bulked seed is used for further multiplication.

Plant-to-row: It is considered to be the best method for variety maintenance of lentil (Figure 1). Single plants typical of the variety are selected and harvested and kept separately. These seed are planted in rows (plant rows); during production, off-type rows and rows with off-type plants are discarded. Those rows which only conform to the varietal description are maintained and bulk-harvested as breeder seed. When there is doubt as to varietal purity, plant rows may be individually harvested and planted as small plots for further observation. Within plots, negative mass selection is carried out before plots are bulk harvested as breeder seed.

7.1. Production Arrangements

The best method of variety maintenance is to produce enough breeder seed to satisfy estimated requirements for the lifetime of a variety (Bouwman 1992), because this method involves less work and minimizes the risk of contamination. However, this breeder seed must be stored under optimum conditions for medium to long-term storage (4 °C and 40% RH). This method is not appropriate for lentil, because tits multiplication rate is low and the seed material is bulky. A suggested alternative is to multiply breeder seed at regular intervals, after initially producing a sufficiently large quantity to meet seed needs for 4 to 5 years (Julen, 1983). This breeder seed must be stored under relatively low temperature and relative humidity (20 °C and 50% RH or 10 °C and 60% RH). The best approach is to produce new lots of breeder seed annually.

It is important that breeder seed must be planted in at least two different locations to avoid risking loss of the complete generation. Crop management practices and procedures are similar to those used for other seed classes (see 'Seed Production'), but for breeder seed, the best possible practices must be applied. Breeder seed is the earliest and lowest-volume, and any mistake made in this stage will be difficult to correct at a later stage.

7.2. Responsibility for Variety Maintenance

In the private sector, the variety development-seed production-use continuum is closely linked as success is measured on large-scale commercialization of the variety for return on investment. In contrast, plant breeders particularly from the public sector, often lose interest once a variety is officially released because of lack of associated benefits. Consequently, in many developing countries where the public plant breeding is predominant neither the research institutions nor the public seed producing agencies give the required attention to variety maintenance and breeder seed production. Hence, the availability and access to breeder seed becomes a bottleneck in adoption of new varieties by farmers. In Morocco, for example, a separate Unit within the Institut National de la Recherche Agronomique is responsible for producing breeder seed of public varieties. In Ethiopia, the Ethiopian Seed Enterprise is responsible for production of pre-basic and basic seed on its own basic seed farms. These approaches overcome the problem of seed availability for the major crops, but seldom include variety maintenance and breeder seed production of legumes in general and lentil in particular. Laverack (1994) described different arrangements and management options for breeder seed production that are useful and adopted in developing countries.

8. SEED PRODUCTION

Seed production practices for legume crops have been described by Agrawal (1985), Doerfler (1976), Erskine *et al.* (1988), Saxena and Singh (1987), Singh (1986), and Wellving (1984). Bishaw *et al.* (2006) provided detailed basic techniques for seed production, which can be used as the basis for lentil seed production. The most important aspects, as related to lentil, are discussed below.

8.1. Previous Cropping

In seed production, requirements for previous cropping on the seed field specify the crops that should not be grown for a specified time preceding the production of the seed crop to avoid varietal admixture or contamination with volunteer plants. Appropriate previous cropping will avoid volunteer plants which may reduce the varietal purity, and consequently the buildup of diseases and noxious weeds. For lentil, the land selected to produce pre-basic and basic seed should be free of any other lentil variety for at least two years. The field in which the previous crop was of the same variety and of the same or higher seed category can be used for pre-basic and basic seed production, but diseases and weed problems may be expected to occur. For certified seed, only one year between two crops of different varieties is required.

Fields planted with vetches (Vicia, Lathyrus) in the previous year should be avoided, because the seed of such species is of similar size and difficult to remove during seed cleaning. In areas where ascochyta blight and anthracnose are prevalent, a three to four year rotation is recommended. Parasitic weeds such as broom rape (Orobanche spp.) remain major problems for lentil seed production. For example, broom rape seeds remain viable in soil for over 10 years and may have several hosts. Since an effective crop rotation may be difficult to implement, fields with a history of orobanche infestation should be avoided.

8.2. Seedbed Preparation

Cool season food legumes are very sensitive to water logging, particularly during the seedling stages. Therefore, adequate provision of surface drainage is very important in land preparation. Lentil can grow in all types of soil, but dry sandy soils are most suitable for seed production compared to fertile soils which promote vegetative growth at the expense of seed yield (Doerfler, 1976). A well-tilled field may promote better root and nodulation of seedlings. Good seedbed preparation is also a prerequisite for combine harvesting.

Firm, weed-free seed beds on well-drained soils are essential for lentil seed production. Fields with stones should be avoided and rolled down after planting but before emergence to reduce risk of erosion in case the soil is dry and to ensure smooth and level surface for easy harvesting. Rolling up to five to seven node stage after crop emergence appeared to have no effect on the crop whereas rolling thereafter may damage plants, increase spread of foliar diseases and reduce yield. A seeding depth of three to eight cm is recommended because of shallow root depth. It may fail to emerge if planted deep due to small seed size or if there is an extensive soil crustation.

8.3. Planting Methods

Lentil seed is susceptible to mechanical damage during planting operations. A dry seed with less than 14% moisture is brittle and can easily crack or split leading to reduced germination. It is recommended to moisturize the seed before planting to reduce mechanical injury. Lentil can be planted by a cereal drill (Erskine *et al.*, 1987). Generally for seed production, row planting is preferable to broadcasting, as it requires less seed, and facilitates mechanical weed control, roguing and field inspection (Galanopoulou *et al.*, 1996). In lentil, drilling gave better seed yield compared to conventional methods (Saxena, 1981). Planters should be properly cleaned between sowing of different varieties to avoid admixture or mechanical contamination.

8.4. Seed Rates

Seed rates differ from variety to variety, depending on seed size and pre-seed germination. Optimum plant density should be used based on local recommendation for better weed competition and yield. Low seed rates or wider row spacing may result in a more open crop canopy which reduces the risk of poor pod set, incidence of foliar disease and lodging particularly in wetter climates. However, for a new variety the multiplication rate (yield per unit of seed planted) is more important than maximum yield. The higher the multiplication rate, the faster the new variety is multiplied and the sooner it is made available to farmers. Using lower seed rates (and having larger initial amounts of breeder seed) is the best management strategy for achieving faster seed multiplication. At ICARDA, unpublished data on the relationship between seed rate and yield showed that decreasing the seed rate increases the multiplication rate in lentil.

8.5. Planting Dates

The optimum planning dates should be followed according to local recommendations. Early planting may increase plant height allowing pods to be positioned higher on the stem which will facilitate harvest. In contrast, late planting dates produce shorter plants and increasing the proportion of late maturing pods which increases harvest looses.

8.6. Rhizobium Inoculation

In legumes, root nodules are highly specialized structures formed as a result of interactions between the host plant and the invading *Rhizobium*. Lentil seed must be inoculated at planting time with the specific *Rhizobium* (Gan et al., 2005), particularly if they are planted in new lentil growing areas. Under favorable growing conditions, lentils can fix a significant portion of their nitrogen requirements ($> 70\%$). Low available soil N and good soil moisture and temperature at planting encourages N fixation.

If the seed is dressed with chemicals, inoculation should be made after drying the treated seed. Granular inoculants are preferred because the granules are separated from the seed and are less affected by chemical seed treatment. The inoculants are also sensitive to direct contact with fertilizers. If fertilizers and inoculants are both applied, fertilizers should be banded adjacent to but not in direct contact with the seed. After inoculation the seed must be planted immediately as delays or exposures to high temperature, drying winds, direct sunlight or planting in dry seedbed kills the bacteria and reduce the efficacy of the inoculants.

8.7. Fertilization

Since legumes are known to fix atmospheric nitrogen, their N fertilizer requirements are less compared to other field crops. Generally nitrogen fertilization is not required except where small dosage is recommended as a starter fertilizer (Gan et al., 2005). Also, high nitrogen levels, in the soil or applied, reported to reduce nodulation and nitrogen fixation and may delay maturity in some cases.

Adequate levels of phosphorus, potassium, calcium and other nutrients are required to ensure proper plant growth and development. Fertilizer application should be based on soil analysis to determine the nutrient requirements of the crop under particular growing conditions. It is indicated that lentil has a relatively high requirement for phosphorus for it promotes the development of extensive root systems, vigorous seedlings and N fixation.

8.8. Weed Control

The use of clean seed is not only an effective means of producing weed free quality seed, but also a means to avoid the introduction of weed species into production fields. It also involves selecting clean fields, cultivating to remove weed seedlings, and roguing to remove troublesome weeds (Table 2). Lentil is a very poor competitor to weeds so selection of a mostly weed-free field is essential. A pre-seeding "burn-off" treatment to the field using non-selective herbicide (such as glyphosate) has been shown to be the most efficient weed control tactics in lentil seed production in Canada and the USA (Table 2). Moreover, good weed control requires a long-term strategy involving the entire cropping systems. Early seeding allows better competition with weeds. Post emergence harrowing can be used to control weeds when the crop is very short (< 10 cm). Basler (1981) reported that annual weed species such as vetches (*Vicia* and *Lathyrus* spp) and rough bedstraw (*Galium tricorne*) are similar in appearance, size and time of maturity to lentils and consequently they can not be weeded efficiently by manual labor. Moreover, volunteers of rape seed, mustard and cereals (wheat, barley) are difficult to remove and should be avoided or controlled in the field.

Broom rape (*Orobanche* spp.) is one of extremely destructive parasitic weeds with no single effective method for control. Lentil is susceptible to several species such as *Orobanche. crenata*, *O. aegyptica* and *O. ramosa* (Basler, 1981). Khalil *et al.* (2004) suggested an integrated approach for orobanche control which include the use of resistant/tolerant varieties, crop rotation (including trap or catch crops), land preparation (deep plowing, zero tillage), delayed sowing, fertilization (ammonium sulphate), frequent irrigation or flooding, biological control (*Phytomyza*

Table 2. Chemical herbicides for weed control in lentil seed production

Weed growth stages	Chemicals			Target weeds
	Trade name	Active ingredient	Rate (kg a.i. ha^{-1})	
Pre-emergence	Gesagard	prometryne	0.75–1.0	Dicotyledons
	Kerb	pronamide	0.5	Monocotyledons
Post-emergence (early growth)	Broadstrike	flumetsulam	0.02	Monocotyledons
	Challenge	aclonifen	0.6	Monocotyledons
Post-emergence (mid growth)	Fusilade	fluazifop-P-butyl	0.25–0.5	Monocotyledons
	Focus-Ultra	cycoxydin	0.25–0.50	Monocotyledons
	Agil	Propaquizafop	0.1–0.20	Monocotyledons
Parasitic weeds	Cadre-Oroban	Imazapic	3–5 g $ha^{-1} \times 2$	*Orobanche*
	Stomp	Pendimethalin	1.0	*Cuscuta*
	Persuit	Imazathapyr	0.2	*Cuscuta*

Source: Pala, M (Personal Communication)

orobanchia), solarization and chemical control (glyphosate, imazapic, imazethapyr). Baya and Yahyaoui (2005) suggested an integrated orobanche management for lentil which includes the use of delayed sowing, early maturing variety adapted to late sowing in combination with either two post emergence applications of imazapic (5 g a.i. ha^{-1}) or two post emergence applications of imazethapyr (15×2 g a.i. ha^{-1}) at two week interval. The knowledge of orobanche phenology, however, is essential for effective chemical control because of limited margin of application.

Fields infested with orobanche should not be used for seed production; if orobanche is found in seed fields; the affected plants should be uprooted, removed and burned. If a strip of *Orobanche* is found, the area should be chemically treated to sterilize the soil temporarily. There is urgent need for an integrated control strategy combining cultural practices, chemical herbicides and legislative procedures to limit the distribution of parasitic weeds with certified seed.

8.9. Disease and Other Pest Control

Legumes are affected by several fungal and viral diseases and many of them are seed-borne which contributed to their local and international spread and survival (Kaiser *et al.*, 2000). They reviewed fungal and viral diseases affecting legumes i.e. faba bean, chickpea, lentil, pea and, lupine. Some of the following sub-sections pertinent to lentils heavily draw from this article. Although efforts have been made to develop resistant varieties to economically devastating fungal diseases there is little progress in finding varieties completely resistant or immune to the fungal pathogens. An integrated pest management approach is encouraged where tolerant varieties must be combined with production technologies which are difficult to implement at the farm level.

8.9.1. Ascochyta

Ascochyta blight (*A. lentis*) has been reported as the most widespread and potentially devastating disease of lentil worldwide (Kaiser *et al.*, 2000). It is reported that cool wet weather condition as conducive to infection, disease development and spread. Use of resistant varieties and chemical seed treatment appeared to effective measures for disease control.

8.9.2. Lentil wilt

Wilt caused by *Fusarium oxysporum* is a seed- and soil-borne disease in both lentil and chickpea. Wilt is common in vertisols when temperatures are relatively high, often above 30 °C. Fields or areas infested with *Fusarium* should be avoided for seed multiplication. Crop rotations with cereals should be followed to prevent the buildup of the pathogen. Infected plants should be destroyed and seeds should be dressed with an appropriate fungicide, particularly in areas where wilt is a problem.

De and Chaudhary (1999) found that the use of biological control combined with seed treatment provides better wilt control and increase in yield compared to biocontrol or chemical seed treatment alone. For example, the use of *Bacillus subtilis* with carboxin (Vitavax) resulted in 66.8% wilt control and 145.4% increase in seed yield of lentil in variety PPDL 2, whereas in separate combination with *Gliocladium virens*, *Trichoderma harzianum* or *T. virde* they reduced wilt by 79% and increased yield by 140, 224 and 241%, respectively.

8.9.3. Viruses

Accordingly about 45 viruses have been reported infecting these five crops globally; and about 50% of viruses affecting legumes are seed-borne. From 12 viruses reported in lentil six (Alfalfa mosaic alfamovirus, Bean leaf roll luteovirus, broad bean stain comovirus, Cucumber mosaic cucumovirus,) Pea enation mosaic enamovirus and peas seed-borne mosaic ptyvirus) are prevalent worldwide whereas others are restricted to some regions (Kaiser *et al.*, 2000). It was indicated that at present, cultural practices such as varying sowing dates, rouging infected plants early in the season and using border of plants which are not hosts to virus as the only effective measures.

8.9.4. Field pests

Specific and broad-spectrum pesticides and insecticides are available for effective control of field and storage pests of lentil (Table 3). In lentil the source of *Bruchus* infestation comes from infested seeds scattered in the field during harvesting or from infested seed used for planting the crop (Pajni *et al.*, 1996). Therefore, field spraying may be necessary to control infestations of*Bruchus* spp. Spraying the seed production fields with insecticides during the time of oviposition can prevent infestation, but is expensive unless other insects have to be controlled at the same time.

Table 3. Chemicals for control of insects in lentil seed production

Active Ingredient	Type	Chemical Group	Target pest
Deltamethrin	insecticide	Pyrethroid	Leaf minor, bruchids, pod borers
Methidathion	insecticide	Organophosphate	Leaf minor, bruchids, pod borers
Imidacloprid	insecticide	Chloronicotinyl	Aphids
Pirimiphos-methyl	insecticide	organophosphate	Bruchids in seed and grain store

8.10. Isolation

Isolation involves growing a seed crop away from any source of contamination whether it is genetic, mechanical or pathological. These can be achieved through spatial isolation—separating the seed field by space or distance, or through temporal isolation—by planting the seed crop which flowers or matures at a different time schedule or during off-season. Minimum isolation distances are usually prescribed based on field size, crop pollination habit, direction and speed of wind, and presence of natural barriers. Lentil is selffertilizing, with a very low percentage (0.5%) of out-crossing (Wilson and Law, 1972). Therefore, a physical distance of 1–2 m between two fields is sufficient. However, slightly longer isolation distances are recommended; *i.e.*, 5 m for pre-basic seed and 3 m for basic and certified seed. Contamination can also be reduced by growing different generations of the same variety side by side. For example, in some North African countries, G_0 is planted in a field surrounded by G_1 and then by G_2, G_3 and G_4 (Table 1) to avoid contamination. Since micro-organisms and insects (virus vectors) can move over longer distances, isolation will not prevent the problem of viral diseases.

8.11. Roguing

Roguing is defined as the systematic examination of seed fields and removal of undesirable plants which may contaminate the seed crop (Gregg *et al.*, 1990a; Laverack and Turner, 1995). Roguing maintains varietal purity and, to some extent, assures freedom from seed-borne diseases. The following contaminants are removed: (1) offtypes and other variety crop plants, (2) other crop species which have similar seed size (vetches), (3) weed plants whose seed are not easily separated in cleaning, (4) parasitic weeds such as *orobanche* spp., and (5) plants infected with seed-borne fungal diseases and viruses.

Roguing should be carried out at appropriate growth stages of the crop. For lentil, the morphological plant characteristics used for roguing such as plant height, vegetation color and fruit characteristics are not clearly and easily distinguishable at late flowering, pod setting and maturity. At these growth stages the crop stand is too dense for individual plants to be distinguished by a roguing crew. Thus rouging should be carried out at flowering stage. It should be stressed that roguing of breeder and pre-basic seed is more practical than large areas of later generations of basic and certified seed.

8.12. Harvesting

The main risks for seed quality during harvesting are mechanical damage of seed if the machinery is not adjusted properly and physical admixture with seed of other varieties, if the machinery is not cleaned properly. In general legume seeds are more prone to mechanical damage than most field crops, and should be harvested with care (proper combine speeds, cylinders, sieves and air blast). Combining should be done in the early morning when seed moisture content is higher, so as to minimize seed cracking and loss.

Lentil plants are more difficult to harvest because of shorter plant height. However, fields that were rolled after planting are easier for mechanical harvesting using a combine equipped with a flex header or with a pick-up reel and vine lifters when seed and pods are fully mature or after desiccation. Excessively dry seed will chip and peel during threshing. Drum speed and concave width have to be adjusted to avoid splitting and decorticating of seed. The cylinder speed should be adjusted between 250 to 500 rpm and the distance between concave and drum put to near its maximum. Standard cereal sieves and wind settings are acceptable for lentils, but also upper sieve of 9 mm and lower sieve of 3 mm is recommended (Erskine *et al.*, 1988). A lower ground speed may also be required when harvesting lentils compared to cereals.

Lentil seeds, due to their lens shape, are more susceptible to mechanical damage than seeds with more rounded shape such as pea, faba bean or chickpea (Muehlbauer *et al.*, 1985). Delaying harvest subjects seed to more deteriorative field stress and causes greater loss of quality (increase physiological age). Lentil should be harvested when the crop reaches 100% pod maturity or otherwise the pods dehisce and drop if harvest is delayed (Erskine *et al.* 1988). Moreover, delaying harvest leaves the mature seeds under weathering conditions which may reduce seed germination significantly (Ellis *et al.*, 1987).

9. SEED PROCESSING

Seed processing includes all steps involved in preparing harvested seed for marketing; *i.e.*, drying, pre-cleaning, cleaning, upgrading, treating and packaging. If necessary immediately after harvest, the seed is cleaned and dried to remove excess moisture. For example, lentil seed is harvested at relatively higher moisture content (18–20%) and dried down to 14% to avoid seed damage and harvest losses. The dried seed is cleaned to remove undesirable contaminants such as: (1) inert matter i.e. plant parts, soil particles and stones; (2) weed seed; (3) other crop seed; (4) other variety seed; and (5) seed of the variety which are immature, shriveled, broken, damaged or deteriorated (Thomson, 1979). Upgrading includes removing poor seed and applying chemical treatment. Cleaning and upgrading is based on physical differences between good seed and undesirable contaminants. Separation is based on differences in length, width, thickness, weight, shape, surface texture, and color (Boyd *et al.*, 1975). Detailed information on processing are available in Brandenburg (1977a, b), ISTA, 1977, FAO (1981) Gregg *et al.* (1983) and van der Burg (1986).

9.1. Machines for Cleaning

Farmers traditionally use air and/or sieves to clean their harvest for seed. Modern seed plants apply some of the same traditional principles, but in much large-scale mechanical operation. Seed is cleaned using screens, cylinders and air blast. Generally, different machines are combined in a specific sequence to make a proper separation. The choice of machines depends on type of contaminants and the quality standard that must be achieved (Boyd *et al.*, 1975). For lentil, the screens, cylinders, gravity table and air blast are most important. Lentil seed is susceptible to mechanical damage during cleaning and handling, therefore, it requires using the smallest possible number of machines.

Pre-cleaner: These are high-capacity machines of one or two sieves with large round holes and a powerful air blast to remove larger contaminants. A pre-cleaner may be required for lentil.

Fine cleaner: The fine cleaner, also known as air-screen cleaner, is the most important and basic machine in seed processing plant. It separates seed according to width, thickness, shape and terminal velocity by using a combination of several screens and air blast.

Indented cylinder: Commercial processing plants have indented cylinders in the standard processing sequence, separating on the basis of differences in seed length. Indented cylinders are essential for lentil. It is used to remove short or long impurities.

Gravity separator: The gravity, density separator can be used to separate stones and soil particles which have the same size as the seed. The gravity is also very useful to remove bruchid-infested seed from lentil seed lots.

Scarifier: Hardseededness, impermeability of the seed coat to water, is common in almost all legumes including lentil. A scarifier scratches the seed coat to allow water uptake and improves germination.

Elevators and conveyors: During processing, seed must flow efficiently from one machine to another and should not be damaged. Elevators move seed vertically; conveyor belts move seed horizontally or at an angle. Choosing the right elevators and conveyors for lentil seed is important, because they are sensitive to mechanical damage. Pipes feeding seed into bins or machines should be sloped at 45° to ensure that seed do not jam up in the pipes. For legumes in general there is a need to fit 'impact-absorbing' materials (rubber) or equipment such as 'bean ladders' at places where seed fall for more than 1 m, and/or strike hard surfaces.

Cleanliness of equipment: Cleanliness in seed plants is essential to avoid mechanical admixture. Seed cleaning machines must be thoroughly cleaned between seed lots, using an air compressor and industrial vacuum cleaner. Screens, decks and cylinders should be removed and individually cleaned, and the machines should run empty to remove hidden seed. Mechanical admixture can also be reduced by careful planning, such as by cleaning early-generation seed immediately after cleaning a later generation of the same variety to reduce the chances of contamination by seed of another variety.

For cleaning lentil, the screens, cylinders and gravity table are very essential. Lentil has 'flat round' seeds and – in the airscreen cleaner – a round sieve separates on diameter and an oblong sieve separate on the thickness. For example, from experiences at ICARDA the following are recommended for large seeded (top screens with round holes of 8 mm and bottom screens with 5.5. to 6.5 mm) and small seeded (top screens with round holes of 5 mm and bottom screens with 3.0 to 3.5 mm) lentils. *Vicia, Lathyrus* and *Galium* seeds have rather similar seed shape as lentil and are therefore difficult to separate. A large proportion of these impurities can be removed by carefully choosing the right sieves.

Mechanically harvested lentils may have many impurities which justify the use of indented cylinder. An indented cylinder will remove smaller and shorter impurities, which escaped separation in the air screen cleaner. In short grain application (5.5 mm), shorter impurities than the lentil seed are removed whereas in long grain application (6.5 to 7.5 mm), some of the larger vetches, cereals (mainly barley) and un-threshed lentil pods may be removed.

The gravity table can also be used to separate stones and soil particles which have the same size. The machine is very useful to remove bruchid infested seeds; seeds without seed coat; and *Galium* seeds, which have the same size as the smaller lentil seeds.

9.2. Seed Treatment

Legume seeds may carry fungi, bacteria, viruses, nematodes, and insects, either on the seed surface or internally in the seed coat. Seed infection may lead to low germination, reduced stand establishment, severe yield loss, or total crop failure in some circumstances. For example, *A. lentis* have been associated with reduced seedling emergence in lentil (Morrall and Sheppard, 1981). Seed, moreover, provides an important means of carry over and dispersal of plant pathogens. Seed health is one of the most important quality attributes. Chemical seed treatment is a standard procedure for disease control in many crops. Chemicals may be used in different formulations, *e.g.*, dust, wettable powder for slurry treatment, or liquid concentrates. Liquids or slurries are preferred over dust formulations, because of ease of exact measurement, better coating of the seeds, and less potential hazard for operators.

9.2.1. Ascochyta blight

Thiabendazole (Tecto WP) at 3g a.i. per kg was found to control *Ascochyta lentis* on lentil (Kaiser and Hannan, 1987). Vitavax-200 (Thiram and Carboxin) is used at the rate of 2–3 g per kg as a wide spectrum seed treatment chemical at ICARDA. Beniwal *et al.* (1989) reported effective control of ascochyta blight by 57 and 96%, respectively by sun drying of lentil seed after 10 and 30 days of exposure under polyethylene sheet cover, although reduction in germination was observed in all treatments except 10 days exposure without polyethylene sheet. They recommended as an alternative seed treatment in situations where chemical seed treatment is not readily available, but may not applicable to commercial agriculture.

9.2.2. *Wilt, root rot and damping-off*

Fungi causing root-rot and damping-off are *Pythium* spp., *Rhizoctonia solani*, *Fusarium oxysporum*, *F. solani*, *Aphanomyces euteiches*, *Thielaviopsis basicola*, and others. The following fungicides (or their combinations) have been found effective: Captafol, Chloroneb, Thiophanate-methyl, PCNB, ETMT, Triforine, Captan, Dowco 444, Prothiocarb, Metalaxyl, Thiabendazole, Benalaxyl, Thiram (Papavizas and Lewis, 1975; Kraft, 1982: Trapero-Casas *et al.* 1990).

Thiram at 2g a.i. kg^{-1} seed seems to be cost-effective seed treatments if problems with seed rot and/or damping-off diseases are encountered in faba bean, chickpea and lentil. *F. oxysporum*, the pathogen causing wilt in legumes, is difficult to control by seed treatment. More important is good rotation. In practice non-systemic fungicides can control other fungi, because fungal pathogens in legumes are mostly located in the seed coat (Ellis and Paschal, 1979; Tu, 1988). In legumes the side effects of seed treatment on nodulation has to be considered. Ram *et al.* (1984) found that rhizobium inoculation followed by treatment with Dithane M-45 gave the best results in chickpea. Chemicals reported to have a negative effect on nodulation are oxycarboxin, PCNB, and copper fungicides, while the reports on Captan, Carboxin, Chloranil, Dichlone, Mercurials and Thiram are inconsistent (Agrawal and Sinclair, 1987).

9.2.3. *Field Insect pests*

Some field pests, e.g. wireworms or nematodes may be controlled by pesticide applied as seed treatment (Table 4). This is generally more cost-effective than application of the chemical in the field. In lentils, *Sitona* spp. can be controlled by seed treatment with Promet (furathiocarb) at 12 ml kg^{-1} seed (ICARDA, 1990).

9.2.4. *Equipment for seed treatment*

In general chemical seed treatment is part of cleaning operation. However, even with excellent chemicals and equipment, it is difficult if not impossible, to achieve 100% uniform coverage of seed with chemicals. During the treatment process, some seed always escape proper treatment, while others receive more than the recommended dosage. Choosing the right equipment and calibrating it properly can minimize these problems.

Table 4. Fungicides and insecticides for lentil seed treatment

Fungicides	Active ingredients	Rate kg^{-1} seed	Target disease
Vitavax–200	Thiram + Carboxin	2–3 cc	Wide spectrum
Tecto WP	Thiabendazole	2 g	Wide spectrum
Insecticides			
Lindane			Wireworms
Carbofuran			*Sitona* spp., nematodes
Furathiocarb			*Sitona* spp.

For large-scale treatment, machines have been developed that ensure automatic measuring of chemicals. In Gustafson treaters, for instance, the weight of seed, measured in a weigh pan, is used to operate the chemical measuring system. By adjusting a counterweight, a fixed quantity of seed is treated with a fixed quantity of chemical, measured in standard cups and operated by tripping (dumping) the weigh pan. Most commercially available treaters are designed for slurry or liquid applications. The solid particles in slurry may settle out as sediment; so a stirring device is indispensable in the treater tank. These particles also may clog nozzles, so the mixture is applied from small measuring buckets on an endless chain. Liquid formulations are usually sprayed on seed, as in Mist-O-Matic treaters, to assure even coverage of the seed.

9.3. Monitoring Seed Quality

Seed plants should have adequately-equipped and well-staffed internal seed quality control facilities. The laboratory should carry out simple purity, germination and moisture tests, which are necessary to monitor quality of incoming seed material, the cleaning and treating process, and during storage. The laboratory can ensure that contractual agreements have been observed by the growers to effect payment, the need for drying incoming seed lots, selection of appropriate machines for cleaning, and to achieve the desired quality standards.

10. SEED STORAGE

Seeds attain its maximum potential germination and vigor at physiological maturity, dependent on production environment. Legume seed reach physiological maturity at seed moisture contents ranging from 45–50% (Elllis *et al.*, 1988). Loss of vigor and germination is a natural phenomenon; and is manifested in various physiological and biochemical events. This appears to be associated with loss of membrane integrity, changes in molecular structure of nucleic acid, and reduction in enzyme activity which results in reduced rate of germination, increase in number of abnormal seedlings, lower vigor and field emergence (Roberts and Dsei-Bonsu, 1988). McDonald *et al.* (1986) also described the physiology of seed deterioration. Deterioration moves inexorably toward death, never stopping; it neither reversed nor eliminated with loss of vigor preceding germination.

Long dry conditions during seed maturation in the field are very important for seed quality, while unfavorable weather conditions (rain, high humidity, high temperature) or delaying harvest will have a negative influence on seed quality. Once the seed is harvested, it is important to provide proper storage conditions to retard the rate of seed deterioration and minimize losses of physiological quality. The principles and practices of storage were described by Justice and Bass (1978) and Delouche *et al.* (1973), whereas requirements for storage of cool season food legumes are summarized by Delouche (1988) and Ellis *et al.* (1988). Some mathematical models have been developed to relate the viability of seed to their storage environment, and they are also used to facilitate prediction of seed longevity (Ellis and Roberts, 1980).

Roberts (1972) classified seed into two major groups based on their physiological behavior, as either orthodox or recalcitrant seed. Orthodox seed desiccate on the mother plant and can be dried to low moisture contents without damage, and (when dry) are tolerant to temperatures far below zero. Decrease in seed moisture and temperature increases the longevity of orthodox seed. Recalcitrant seeds usually lose viability upon desiccation, and die if their moisture content is reduced below a certain level (e.g. 15%). Lentil belongs to the orthodox group, and its seed can be kept viable for many years, depending upon storage conditions. Lentil seed is considered moderate to relatively good storer (Delouche, 1988). In India, studies on lentil seed storage showed that under insect free ambient conditions germination started to decline after 17 months of storage and can be lowered to 25% after 37 months (Agrawal, 1980).

10.1. Temperature and Moisture Content

Seed ageing is a function of time as well as temperature and moisture content; increasing either of these factors reduces the survival period (Ellis *et al.*, 1988). A combination of moisture and temperature results in a synergistic effect on seed deterioration, but can compensate for each other. Delouche (1988) stated that seeds with moisture contents of 14–16% can be stored well for a year or so at 10 °C or below, whereas seeds with less moisture (10% or less) survive well even at high temperatures of 30 °C or above. Harrington's classical rules-of-thumb describe the effects of temperature and moisture on seed deterioration, which can be combined and have geometric effects, as follows (Harrington and Douglas, 1970):

- *Seed life is doubled for every decrease of 5 °C in storage temperature when temperatures are between 0 °C and 50 °C.*
- *Seed life is doubled with every decrease of 1% in seed moisture content when seed moisture content is between 5 and 14%.*

A computer program based on the seed survival equation developed by Ellis and Roberts (1980) is used to predict the seed storage period of several crops, including cool season food legumes (Kraak, 1992). This program can be used to calculate (1) initial viability, (2) viability after storage, (3) storage period, (4) seed moisture content, or (5) temperature during storage, if three of the five parameters are known and one can be made variable.

10.2. Fungi, Mites and Insects

Storage fungi can severely reduce seed quality by (1) decreasing germination, (2) heating, (3) developing mustiness and caking, and (4) total decay. Such fungi do not cause damage to seed during storage if seed moisture content is in equilibrium with a relative humidity of 65–70% or less. As a general guideline, most storage fungi, mites and insects do not develop below 0 °C, 5 °C, and 15 °C respectively.

Infestations of storage pests are one of the serious problems in lentil seed. Most serious are bruchids; when the seed is infested there is complete loss of viability within two to four months of storage.

10.3. Mechanical Damage

The amount of cracking (of testa) which results from a given handling treatment is dependent upon seed moisture content. Mechanically-injured seed are less storable; they deteriorate faster, and are more susceptible to damage by storage fungi and seed treatment. Thus, seed have to be very carefully harvested, threshed, processed and handled. Lentil seed are most susceptible to physical injury, especially when seed moisture content is less than 14%.

10.4. Seed Storage Period

Lentil seeds are considered moderate to relatively good storers within orthodox seed group (Delouche, 1988). Lentil seed can be stored from planting to harvesting for a period of few months or as carry-over seed for more than one season. The latter is true particularly in situations where there are recurrent natural disasters to overcome seed shortages. Moreover, early generation seed can be kept even for longer period for various reasons. The length of time seed is stored can be classified into three categories: short-term, medium-term and long-term storage.

Short-term Storage: For short-term storage (up to a maximum of nine months), seed storage conditions or facilities that fall between an average of about 30 °C with 50% relative humidity and 20 °C with 60% RH are satisfactory for most kind of seed, including lentil (Delouche, 1988). The equilibrium seed moisture content for lentil is about 14% (Ellis *et al.*, 1988).

Medium-term Storage: In many formal seed production programs, nearly 100% of early generation seed and 20–25% of certified seed is carried-over through one growing season to the second planting season (up to 18 months) as a guarantee against crop failure or other disasters. According to Ellis *et al.* (1988), for such storage periods of lentil seed, the recommendations are 30 °C with 40% relative humidity (9% to 10% seed moisture content), 20 °C with 50% RH (10% to 12% m.c.), or 10 °C with 60% RH (12% to 14% m.c.).

Long-term Storage: Sometimes breeder seed may be produced at several-year intervals or only once for the lifetime of the variety, and the seed must be stored for longer periods. For a storage period of 4–6 years, a temperature of 10 °C and 45% RH (9.5% to 10.5% m.c.) is recommended. In general, authentic samples of new varieties, breeding materials and seed for genetic conservation are also stored for longer periods. FAO recommends a storage temperature of −18 to −20 °C, with 5% seed moisture content (in equilibrium with 10–15% RH for starchy seed, and 20–25% RH for oily seed).

10.5. Safe Seed Storage Conditions

Storage facilities must protect seed from damage and deterioration, and maintain seed quality as expressed by vigor, germination, physical purity and identity. If seed moisture does not exceed 10% during storage, seed may be stored under rather temperate ambient conditions for at least 18 months and still meet the minimum germination requirements for seed certification. Since relative humidity and storage temperature are the two most important factors influencing seed viability during storage, seed stored in places with low relative humidity and low temperature, will maintain viability longer. Temperature of 20 °C is considered safe for seed storage, with an upper limit of 30 °C as a reasonable compromise (Agrawal, 1976). Relative humidity of 70% is usually the maximum level for safe storage. To avoid losses and to keep seed free from insect pests during storage, careful planning and management is essential to maintain clean, cool and dry conditions.

10.6. Safe Seed Storage Facilities

Seed storage facilities must be specially designed, equipped and managed to provide clean, cool, dry and safe conditions. Well-constructed, ventilated stores are adequate for short-term storage of lentil seed in most production areas (Delouche, 1988). Good seed storage should have no windows, and must be in a dry, well-drained area. The floor should be one meter above ground level, or at truckbed height, with a built-in vapor barrier equivalent to 0.25 mm polyethylene sheeting installed with hot-brushed bitumen. There should be only one or two doors, in the middle of the building's short sides (ends), to minimize floor space lost to aisles for handling seed. The storages must be rain-, moisture-vapor and insect-proof. There should be no cracks in walls or floors, to facilitate cleaning and to eliminate cover for insect pests. Roof and walls should join without cracks, to prevent the entry of birds and pests. There should be a rat-proof lip extending out about 35 cm around the building at about one meter height.

Seed storages must be designed and constructed so as to minimize the entry of solar radiation and outside heat. Walls and ceiling should provide enough thermal insulation to minimize solar heat gain. Non-insulated wall construction may be adequate, but the ceiling or roof usually requires some thermal insulation. Roof, walls and doors should be painted with a light-colored reflective paint, to reduce solar heat intake. An extensive roof overhang will shade and cool the walls, and protect the ventilation openings from rain. Orienting the building from east to west will also minimize solar heat on the longer side walls. An exhaust fan may be used for ventilation when outside temperature is lower than that inside the seed storage, but relative humidity of the outside air should also be kept in mind when planning to ventilate the seed storage. All ventilation openings should be screened with 6 mm wire mesh, and located so as to remove hot air from the upper part of the building and moist air from near the floor level.

11. SEED STORAGE PESTS

There is a wide range of storage pests, including rodents, birds, insects, mites, fungi and bacteria. Losses due to storage pests vary according to climatic conditions, crop, and storage facilities. Quantitative storage losses are estimated to reach up to 30% worldwide. Qualitative losses (losses in viability of seed) are more difficult to estimate, and the consequences of these losses may be more drastic. Preventing losses of stored seed deserves special attention. Information on control of storage pests can be found in Bond (1984), Diekmann (1988), Gwinner *et al.* (1990), and Sauer (1992).

11.1. Fungi and Bacteria

The fungi that infect seeds in storage are not seed-transmitted fungi, which cause plant diseases in the field and are carried on or in seed. Bacteria normally do not affect stored seed, unless seed moisture content is very high and temperature is raised by fungal infection. Storage fungi require a high moisture content to grow, (minimum about 12–14% seed moisture) so they do not play an important role in dry climates. The most important genera are *Aspergillus* and *Penicillium* spp. These are mostly saprophytes; *i.e.*, they are unable to attack living tissue. They grow on dead cells of the seed surface, where they produce toxins, cause seed decay and may kill the embryo. The best way to control storage fungi is to maintain low moisture/humidity in seed and storage facilities. Low temperature slows growth of fungi. Fungicidal seed treatment often does not give the expected results, because of lack of free water, which is required for many fungicides to become effective.

11.2. Rats and Mice

Rodents damage seed, soil seed with urine and excrement, damage bags, fumigation sheets, electrical wires, buildings and carry diseases which are dangerous for human beings. Moreover, rodents have a high reproductive capacity, are extremely adaptable and very clever. Effective control depends on the rodent species, but unfortunately rodents are not seen easily and can be indirectly identified through: (1) shape, size and place of droppings, (2) type of damage and (3) footprints. The brown rat (*Rattus norvegicus*) has round droppings; the roof rat (*Rattus rattus*) has more oblong, banana-shaped droppings; mice (*Mus musculus*) droppings are much smaller. Rodents damage bags; the presence of spilled seed under a stack (pallet) indicates their presence. The behavior and areas where they live can also be used to identify the species: rats are suspicious and mistrust all new things, while mice are curious, investigating everything. *Rattus rattus* lives in roofs of the storage and runs around in storage without using the same runs. *Rattus norvegicus* lives outside the store, always using the same runs to go into the store as well as around within the store. Rats generally run along walls, pallets, bags etc. Mice run around in all directions, but live in a very restricted area (1 m^3) and may never be seen by the storekeeper.

It is important to make daily inspections for the presence of rodents and to identify the species based on droppings, damage to bags, foot prints, etc. Preventive measures are most important, particularly the cleanliness of the seed storage and its surroundings, which should be kept spotlessly clean and dry, without vegetation which provides hiding places for brown rat. To avoid entry of rats and mice into the storage, all openings should be screened with wire mesh and storages should, as much as possible, be rat-proof. Biological control measures such as keeping cats in seed storages, and traps in and outside the store should be used.

11.2.1. Rats

If a high population of rats is present, use acute poisons which have a very high toxicity and kill almost on the spot. Rats do, however, relate sickness or death with the poison and will not touch these baits anymore. To avoid bait shyness, pre-baiting is used, whereby rats are first attracted by high-quality baits without poison. After a few days, the poison is added to the baits. Because not all individuals will be killed and survivors will inform the rest of the rat family, the baits should be changed after a few days (4–5 days). The most-used poison is zinc phosphide

Chronic poisons are used for long-term control. The 'old' chemicals need several intakes by the rat (up to seven times) to kill, whereas the 'new' generation of chemicals requires only one intake. The effect of the chemical is not immediate, but takes place hours later (the rat becomes sleepy and dies) and no bait shyness is developed. These chemicals are anticoagulants and the rat is killed through internal bleeding. A well-known 'several-intake' chemicals is Warfarin; and Brodifacoun is a well known 'one-intake' chemical.

11.2.2. Mice

Mice are more difficult to control because they live in very small areas ($1\,m^3$). A dense distribution of bait is required. Widely used is Calciferol.

11.3. Insects and Mites

Insects can be very destructive to the seed. Different species feed on food legumes, and their different life cycles determine the appropriate control measures. There are many sources of infestation; the most important is contamination from infested seed already in the storage. Frequently, the storages have hidden places where insects survive, such as crevices, corners, spilled seed outside the storage, empty sacks with some leftover seed, etc. It is extremely important to control these areas. It is desirable to detect infestation at early stages, before the insect population builds up. There are different methods to detect infestations before they become clearly visible; some of them are simple (flotation of grains), others are more sophisticated (x-ray). However, insect infestation can be detected mainly by visual inspection.

Callosobruchus and *Bruchidius* spp. are typical storage pests of lentils. (Bushara, 1988) which may cause complete loss of viability within 2 to 4 months of storage if appropriate control action is not taken. Among the most important lentil pests the *Bruchus* spp. (*B. lentis* and *B. ervi* are prevalent in Europe, North Africa, Near East, Central Asia) while the *Callosobruchus* spp. (*C. Maculatus* and *C. chinensis*) has world wide distribution. In many legumes infestation with storage pests like *Callosobruchus* takes place before harvest except in chickpea (Reed *et al.*, 1987). Resistance to *Callosobruchus* has been reported by many workers (Bushara, 1988).

Field infestation of lentil by *B. dentipes* is common whereas storage pests do not attack chickpea seeds in the field. Visual inspection of the seeds after harvest can give an indication on the level of infestation. The seeds show characteristic dark spots (holes) through which the newly hatched larvae penetrate into the seeds. If high infestation levels are detected, fumigation might be applied. This helps to stop larval feeding and maintain germination. Moreover, the source of infestation for the next season, the adults that are carried with the seeds to the fields, will be reduced. *Callosobruchus* spp. lay eggs on the dry seeds and several generations develop in the store through re-infestation. A complete seed lot may be damaged if no appropriate action is taken. For the 'true storage pests', there are many sources of infestation, the most important is contamination from stored infested seeds. The stores also have hidden places such as crevices, corners, spilled seeds outside the store, empty sacks with some leftover seeds, etc where insects survive.

The optimum condition for maximum activity of insects is in the range of 30–35°C and 60–80% relative humidity. Insect infestation can be reduced or eliminated by proper drying and storing seed at low temperatures, or can be controlled by proper sanitation and pesticides. In general, development of insects is faster with higher temperatures, but there are differences among species. About 35°C is considered the maximum temperature for development, and about 38°C the maximum for survival. Below 0 °C, most insects cannot survive for more than 2–3 weeks. Also, a certain moisture level is required for insect development. Seed moisture of 10–11% is, in general, considered to be the minimum for insect development. The number of insects increases with increasing moisture, up to a level where too many microorganisms develop (about 15–16% seed moisture). Mites mainly play a damaging role by transmitting spores of storage fungi and causing skin irritation and allergies in persons handling infested seed.

11.4. Spraying Protective Insecticides

Distinction should be made between insecticides that (1) kill the insect population and have a longer-lasting effect (pyrethroids, organophosphorous insecticides) and (2) those which kill insects but have no residual effect (fumigants). These pesticides should be used as complementary to each other, and in combination with storage sanitation. Storage insects are mostly controlled chemically where a wide range of products are available (Actellic with Malathion, K-othrine, etc.). Some alternative methods have also been tried, e.g., the use of olive oil and salt mixtures, or neem extracts (ICARDA, 1990) or traditionally used by farmers (Reed *et al.*, 1987).

11.4.1. Spraying Insecticides

Insecticides may be applied in different ways, i.e., dusting, spraying, fogging, or evaporation. Dusting does not require much equipment; a powder formulation is either (1) mixed with the seed, (2) applied in layers ('sandwich method'), or (3) dusted over stacks. The latter can only prevent re-infestation (e.g., after fumigation), since most powder insecticides do no penetrate into the stacks to control internal infestation. For spraying, either a suspension (solid insecticide, usually a wettable powder formulation, suspended in water) or a solution (liquid insecticide diluted in water) can be used.

Sprays can be applied with knapsack sprayers. For suspensions, care should be taken that the particles remain suspended and do not sink to the bottom of the sprayer. This requires special stirring devices or frequently shaking the spray container. Fogging is a technique used especially in storages. Fog droplets are much finer than those in spraying. Special fogging equipment is required. Evaporation can be used, in special cases, to control flying insects (moths). This technique requires volatile insecticides, and well-closed stores.

Based on efficacy, low toxicity, long-lasting effect and minimum side-effects on seed viability, the following spraying scheme is recommended for stored lentil seed:

- Use, as a preventive measure, spray insecticide once every three to four months
- Alternate applications of Actellic with Malathion and K-othrine
- Use fumigation with Phostoxin when insects are detected

11.4.2. Fumigation

When insects are found in a seed lot, the entire lot should immediately be fumigated. The main advantage of fumigation is that all stages of the insect including eggs, larvae and pupae, are controlled, along with other storage pests including rodents. Mainly two products are used in fumigation: methyl bromide and aluminum phosphide. These two fumigants are active in the gaseous phase, have good penetration into piles of sacks or seed stored in bins, but are hazardous to human beings.

Methyl bromide: Its only advantage as compared to Phostoxin is its quick action. Fumigated seed stacks can be aerated after 12–24 hours of methyl bromide fumigation. Methyl bromide is often used in seaports because storage in ports is expensive so fumigation must be done as quickly as possible. Methyl bromide is: (1) is extremely toxic and accumulates in the body of human beings; (2) is odorless and colorless and difficult to detect without proper devices; (3) residues will remain in the seed; (4) is heavier than air, so fans are needed to re-circulate the gas; and (5) has an adverse influence on seed germination. Because of its toxicity to humans and other harmful characteristics, methyl bromide is banned in many countries.

Phostoxin: It releases a gas called phosphine which has excellent penetration capacity because of the small size of its gas molecules. Phosphine penetrates bags, carton boxes, and other containers. It has no influence on germination, so seed can be treated repeatedly. Phostoxin is inflammable at normal temperatures and care should be taken while fumigating.
Conditions for fumigation: Smaller quantities of seed can be fumigated in airtight fumigation chambers. Complete fumigation of seed storages is not practical because it is impossible to make it sufficiently airtight. Normally, large quantities of seed are fumigated in individual stacks under airtight fumigation sheets. It is most important to hermetically seal the area under fumigation. It is also a good practice to treat seed stacks that have been fumigated with an insecticide of some persistence, because fumigation does not have a lasting effect and re-infestation may take place immediately after fumigation. Based on the characteristics of the fumigants, aluminum phosphide is recommended for use in seed; methyl bromide is not recommended.

12. SEED QUALITY ASSURANCE

A quality assurance system ensures that the seed quality meets required seed production standards. The national quality assurance system is a combination of technical and administrative procedures and guidelines supported by legislation intended to maintain and ensure that seed offered for sale meet established standards of genetic, physical, physiological and health quality. It involves varietal certification through field inspection of growing crops, laboratory testing of seed quality attributes to ensure meeting standards prescribed by legislation including labeling and sealing of seeds offered for sale. It ensures that seed sold to farmers is of the designated variety and of the desired specific quality attributes. All seed quality control and certification programs have the main features of: setting field and seed standards, monitoring, supervising and enforcing seed quality standards during production and marketing. Seed quality control in developing countries has been described by Van Gastel *et al.* (2002).

12.1. Field Standards

The standards suggested here are based on those used in several countries of West Asia and North Africa (ICARDA, 2002) and those given by Doerfler (1976). Field standards for lentil seed production are usually set for: (1) off-types and other varieties, (2) other crops, (3) parasitic plants (*Orobanche*), virus-infected plants, and *Ascochyta* (Table 5).

12.2. Inspecting Seed Fields

Field inspection should be made at the time when potential contamination is likely to happen and contaminants are easiest to identify. During the inspection, the

Table 5. Suggested field standards for lentil seed production

Factor	Pre-basic seed	Basic seed	Certified seed I	Certified seed II
Offtypes & other varieties (max., %)	0.1% (1:1000)†	0.3% (1:333)	0.5% (1:200)	1.0% (1:100)
Other crops (max., %)	0.1% (1:1000)	0.3% (1:333)	0.5% (1:200)	1.0% (1:100)
Orobanche (max., %)	0.05% (1:2000)	0.1% (1:1000)	0.2% (1:500)	0.3% (1:333)
Ascochyta on pods (max., %)	0.3% (1:333)	0.4% (1:250)	0.5% (1:200)	1.0% (1:100)
Total diseases (max., %)	1% (1:100)	1.0% (1:100)	1.0% (1:100)	2% (1:50)

† Numbers in parenthesis represent the ratio of one contaminant plant to the specified number of lentil crop plants in which one contaminant is allowed.

inspector verifies whether or not standards are met as prescribed in the regulations. At least two inspections should be made; one during flowering and another toward crop maturity. An additional inspection for Ascochyta blight should be made in the vegetative stage.

The Association of Official seed Certifying Agencies (AOSCA), as described by Revier and Young (1970) and modified by Gregg *et al.* (1990b), is suggested to be made. It consists of two steps: '*Field Overview*' and '*Field Inspection Sample*'. In the field overview, the field is generally observed to assess the uniformity of crop stand and quality. The field inspection sample involves inspecting randomly-selected representative areas of the field and counting the actual number of each contaminant in a statistically determined number of plants. The numbers of contaminants found are compared with the standards; if above the tolerance level, the field is rejected.

12.3. Seed Standards

To evaluate seed quality, standards for quality attributes should be established (Hampton, 1998). Laboratory tests are conducted to ensure that standards have been met. Suggested standards for lentil seed are given in Table 6, based on Moroccan national seed standards (ICARDA, 2002) and those suggested by Doerfler (1976).

12.4. Seed Quality Tests

Laboratory seed testing are intended to assess the quality of seed before planting to minimize the risk of crop failures. Seed quality is evaluated by standardized tests performed on representative samples taken from the seed lot (ISTA 2003), Although seed quality is composed of different attributes which are important, only

Table 6. Laboratory quality standards for lentil seed production

	Pre-basic seed	Basic seed	Certified seed 1	Certified seed 2
Physical purity				
Purity (min., %)	98	98.0	98.0	97.0
Other crop seed (max., %)	0.2	0.2	0.2	0.5
Weed seeds (max., %)				
Orobanche No of seeds 100 g^{-1}	1	2	2	2
Germination				
Germination (min., %)	85	85.0	85.0	85.0
Hard seed expected to germinate (max., %)				
Insects				
Life insects	0	0	0	0
Bruchid-damaged seed (max., %)	3	3	3	5
Diseases				
Ascochyta (max., %)	1.0	1.0	1.0	2.0
Seed moisture (max., %)	12.0	12.0	12.0	12.0

a selected few, *viz*: physical purity, germination and moisture content, are routinely evaluated in the laboratory (Hampton, 2002). For lentil, at least the following tests should be conducted: (a) physical purity, (b) germination, (c) moisture content, and (d) seed health.

Seed of the parasitic weed *Orobanche* may be mixed with crop seed or adhere to the surface of crop seed, and be planted with the crop and thus infest clean fields. Since seed of parasitic weeds are very small, they may not be easily detected in the purity analysis. Therefore, a washing test may be carried out, in which the seed sample is submerged in water (100 g seeds in 100 ml of water with a few drops of household detergent to eliminate surface tension), shaken thoroughly and poured over a sieve covered with a white filter paper. The number of *orobanche* seed can then be determined under a stereo microscope.

Hardseededness, impermeability of the seed coat to water, is a common phenomenon in all food legumes. The development of seed coat impermeability is associated with the dehydration of seed during later stages of maturation, particularly low relative humidity of the air during ripening (Agrawal, 1985; Ellis *et al.*, 1988). Hardseededness is a substantial problem in drier regions; for example, in India 9–43% hard seed have been reported for lentil (Anonymous, 1984). However, in lentils the percentage of hard seeds falls from a maximum immediately after harvest to practically zero after five to six months (Agrawal, 1985).The number of hard seed is determined during the germination test.

Ascochyta transmission by seed is of major importance, as compared to other sources of inoculum, and special seed health tests are required. Two test methods, agar plate test and blotter test, can be used. The latter is recommended by ISTA and is easy and inexpensive, whereas the former is more sensitive.

12.5. Control Plot Tests

A well developed certification schemes usually include field plots to provide additional checks on varietal identity and purity and occurrence of weeds and seed-borne diseases. Post-control plots are planted with samples taken from seed lots that were approved (certified) in the previous season. The main benefit of control plots is to ascertain whether or not the certification system works satisfactorily. A sample from each approved basic seed lot and 10–20% of certified seed lots is taken and planted in post-control test plots to confirm that varietal characters have remained unchanged during the seed multiplication process. Procedures for post-control plots are available in OECD seed scheme (OECD, 2000).

12.6. Lot Numbering

An effective seed quality assurance program must use a lot numbering system which can track each lot back to producers, processors or distributors, and to the seed used to plant the crop. Each seed lot must be given a unique number which provides information on production year, field or grower, crop, processor, and seed class.

12.7. Managing Seed Quality Assurance

The seed certification authority should remain independent of seed production so it can impartially serve both seed producers and seed users. However, with the emergence of a diverse seed industry, certification should be flexible and decentralized with more responsibility given to seed producers than is provided in comprehensive and compulsory certification schemes. Tripp *et al.* (1997) suggested that the certification agency performs its task with greater efficiency by using appropriate criteria in a transparent manner through the participation of stakeholders.

12.8. Harmonizing Seed Certification Scheme

Procedures for seed certification are essentially similar in most countries, with the possible exception of seed classes where there is variation in nomenclature. Moreover, most national seed regulations and standards are similar in many ways. Countries should work together to develop a flexible regionally-harmonized or internationally-acceptable seed certification scheme, for the benefit of the national seed industry development and to ensure adequate supply of high-quality seed to farmers.

REFERENCES

Agarwal, V.K. and J.B. Sinclair. 1987. Principles of Seed Pathology, Vol. II. CRC Press, Boca Raton, FL. 168 pp

Agrawal, P.K. 1985. Seed production technology for chickpea (*Cicer arietinum*) and lentil (*Lens culinaris*). *In* Faba beans, Kabuli Chickpeas, and Lentils in the 1980s (Saxena, M.C. and Varma, S., eds.). ICARDA, Aleppo, Syria. 271–279 pp.

Agrawal, P.K. 1980. Relative storability of seeds of ten species under ambient conditions. Seed Research 9: 94–99.

Agrawal, P.K. 1976. Identification of suitable seed storage places in India on the basis of temperature and relative humidity conditions. Seed Research. 4: 6–11.

Anonymous. 1984. Annual progress report. Division of Seed Science and Technology. New Delhi, India.

Aw-Hassan, A., K. Shideed, A. Sarker, R.Tutwiler and W. Erskine. 2003. Economic impact of international and national lentil improvement research in developing countries. *In* Evenson, R.E. and Gollin, D. (eds.) Crop Variety Improvement and Its Effect on Productivity: The Impact of International Agricultural Research. CABI, Wallingford, UK. 275–291 pp.

Basler, F. 1981. Weeds and their control. *In* Lentils (Webb, C and Hawtin, G. eds.). CAB, Farnham, UK. 143–154 pp

Baya, B. and A. Yahyaoui. 2005. Keys to Orobanche Management. ICARDA, Aleppo, Syria. 8 pp.

Beniwal, S.P.S. Seid Ahmed and Negussie Tadesse. 1989. Effect of sun drying of lentil seeds on the control of *Ascochyta lentis*. *Lens* 16(2) 27–28.

Bishaw, Z., Abdoul Aziz Niane and A.J.G. van Gastel. 2006. Technical Guidelines for Quality Seed Production. ICARDA, Aleppo, Syria. 23pp

Bond, E.J. 1984. Manual of fumigation for insect control. Rome, Italy.

Bouwman, A.J. 1992. Maintenance breeding and multiplication of pea and faba bean cultivars. Maintenance of protein peas (*Pisum sativum*) and field beans (*Vicia faba*). *Euphytica* 61: 213–215.

Boyd, A.H., G.M. Dougherty, R.K. Matthews, K.W.and Rushing. 1975. Seed drying and processing. *In* Cereal Seed Technology (Feistritzer, W.P., ed.). FAO, Rome, Italy. 60–86 pp.

Brandenburg, N.P. 1977a. The principles and practice of seed cleaning: Separation with equipment that senses dimensions, shape, density and terminal velocity of seeds. *Seed Science and Technology* 5(2): 173–186.

Brandenburg, N.P. and J.K. Park. 1977b. The principles and practice of seed cleaning: Separation with equipment that senses surface texture color, resilience and electrical properties of seeds. *Seed Science and Technology* 5(2): 187–197.

Bushara, A.G. 1988. Insect depredation during storage. *In* World Crops: Cool Season Food Legumes (Summerfield, R.J., ed.). Kluwer Academic, Dordrecht, The Netherlands.367–378 pp.

De, Rajib K. and R.G. Chaudhary. 1999. Biological and chemical seed treatment against lentil wilt. *Lens Newsletter*. 26. (1/2): 28–31

Delouche, J. and A.A.Gomaa. 1999. Policy procedures for release of new publicly developed crop vareites in Egypt. Agricultural Policy Reform Project, Report No. 62. MALR, Cairo, Egypt. 66 pp

Delouche J.C. 1988. Seed storage practices and problems for cool season legumes. *In* World Crops: Cool Season Food Legumes (Summerfield, R.J., ed.). Kluwer Academic Publishers, Dordrecht, The Netherlands. 331–339 pp.

Delouche, J.C., R.K. Matthes, G.M. Dougherty and A.H. Boyd. 1973. Storage of seed in sub-torpical and tropical regions. *Seed Science and Technology* 1: 671–700.

Diekmann, M. 1986. Seed treatment. *In* Seed Production Technology (Srivastava, J.P. and Simarski, L.T., eds.). ICARDA, Aleppo, Syria. 219–225 pp.

Diekmann, M. 1988. Control of storage pests. *In* Quality Seed Production. (van Gastel, A.J.G. and Kerley,J., eds.). ICARDA, Aleppo, Syria. 73–81 pp.

Doerfler, Th. 1976. Seed production guide for the tropics. GTZ, Eschborn, Germany.

Drijfhout, E. 1981. Maintenance breeding of beans. *In* Seeds, FAO Plant Production and Protection Paper 39. Proceedings of FAO/SIDA Technical Conference on Improved Seed Production 2–6 January 1981, Nairobi, Kenya.

Ellis, R.H., P.K. Agrawal and E.E. Roos. 1988. Harvesting and storage factors that affect seed quality in pea, lentil, faba bean and chickpea. *In* World Crops: Cool Season Food Legumes (Summerfield, R.J., ed.). Kluwer Academic Publishers, Dordrecht, The Netherlands. 303–329 pp.

Ellis, R.H., T.D. Hong and Roberts E.H. 1987. The development of desiccation-tolerance and maximum seed quality during seed maturation in six grain legumes. *Annals of Botany* 59:23–29.

Ellis, R.H. and E.H. Roberts. 1980. Improved equations for prediction of seed longevity. *Annals of Botany* 45, 13–30.

Ellis, M.A. and E.H. Paschal. 1979. Effect of fungicide seed treatment on internally seed borne fungi, germination and field emergence of pigeon pea. *Seed Science and Technology* 7: 75–81

Erskine, W., K.B. Singh. and L.D. Robertson. 1988. Seed production of food legumes in the Mediterranean area. *In* Seed Production in and for Mediterranean Countries (van Gastel, A.J.G. and Hopkins, J.D., eds.). ICARDA, Aleppo, Syria. 120–130 pp.

Erskine, W., K.B. Singh, L.D. Robertson, M.C. Saxena and J. Diekmann. 1987. Mechanization of field experimentation in faba bean, kabuli chickpea and lentil. *In* Mechanization of Field Experiments in Semi-arid Areas (Karisson, D., ed.), Proceedings of the IAMFE/ICARDA Conference, 23–27 May 1987, Aleppo, Syria. ICARDA, Aleppo, Syria. 169–176 pp.

FAO (Food and Agriculture Organization). 1981. Cereal and grainlegume seed processing. Technical Guidelines. FAO, Rome, Italy.

Galanopoulou, S., M. Facinelli and F. Lorenzetti. 1996. General agronomic aspects of seed production. *In* Seed Science and Technology, Proceedings of a Train-the-trainers Workshop Sponsored by Med-campus Programme (EEC) (van Gastel, A.J.G., Pagnotta, M.A. and Porceddu, E, eds), 24 April to 9 May 1993. Amman, Jordan. ICARDA, Aleppo, Syria. 311 pp.

Gan, Y., K.G. Hanson, R.P. Zentner, F. Selles, and C.L. McDonald. 2005. Response of lentil to microbial inoculation and low rates of fertilization in semiarid Canadian prairies. Can. J. Plant Sci. 85: 847–855.

Gisselquist, D. 1997. Private commercial varieties and seeds: opportunities and obstacles.*In* Tripp, R (ed.) New Seed and Old Laws: Regulatory Reform and Diversification of National Seed Systems. ODI, London, UK. 174–184

Gisselquist, D. and J. Srivastava (ed.). 1997. Easing barriers to movement of plant varieties for agricultural development. World Bank Discussion Paper No. 367. The World Bank, Washington D.C., USA. 148 pp.

Gregg, B.R. 1983. Seed processing in the tropics. *Seed Science and Technology* 11: 19–39.

Gregg, B., A.J.G. van Gastel, B. Homeyer, K. Holm, A.S.A. Gomaa and M., Salah Wanis. 1990a. Roguing seed production fields. NARP Publication No. 40.

Gregg, B., A.J.G. van Gastel, B. Homeyer, K. Holm, A.S.A. Gomaa, M. Salah Wanis, A.E. Ghanem, A. Abdel Monem, A. Gouda, and O. Shehata. 1990b. Procedures for Inspecting Wheat Seed Fields. NARP Publication No. 39, Cairo, Egypt

Gwinner, J., R. Harnisch and O. Mück. 1990. Manual on the prevention of post-harvest grain losses. GTZ, Postfach 5180, D-6236 Eschborn, Germany.

Hampton, J.G. 2002. What is seed quality? *Seed Science and Technology* 30: 1–10.

Harrington, J.F., and J.E. Douglas. 1970. Seed storage and packaging. National Seeds Corporation and Rockefeller Foundation. New Delhi, India. 222 pp.

ICARDA. 1990. Annual report legume improvement program. Aleppo, Syria.

ICARDA. 2002. WANA Catalogue of field and seed standards. Network Publication No. 25/02. ICARDA, Aleppo, Syria.

ISTA (International Seed Testing Association) 2003. International Rules for seed testing and Supplement. ISTA, Bassersdorf, Switzerland.

ISTA (International Seed Testing Association). 1977. Seed cleaning and processing. *Seed Science and Technology* 5(2).

Julen, G. 1983. Maintenance of varietal purity for seed production. More Food from Better Technology, FAO, Rome, Italy.

Justice, O.L. and L.N. Bass. 1978. Principles and practices of seed storage. USDA Agriculture Handbook 506. U.S. Government Printing Office, Washington, DC.

Kaiser, W.J., M.D. Ramsey, K.M. Makkouk, T.W. Bretag, N.A. Açikgöz, J. Kumar and F.W. Nutter, Jr. 200. Foliar diseases of cool season food legumes and their control, 437–455. *In* Linking Research and Marketing Opportunities for Pulses in the 21st Century. Proceedings of the Third International Food Legumes Research Conference, 22–26 September 1997, Adeladie, South Australia (Knight, R. ed.). Kluwer Academic Publishers, Dordrecht, The Netherlands.

Kaiser, W.J. and R.M. Hannan. 1987. Seed treatment fungicides for control of seed borne *Ascochyta lentis* on lentil. *Plant Disease* 71: 58–62

Khalil S., M. Kharrat, R. Malhotra, M. Saxena and W. Erskine. 2004. Breeding faba bean for *Orobanche* resistance. *In* Integrated Management of *Orobanche* in Food Legumes in the Near East and North Africa. Proceedings of the Expert Consultation on IPM for *Orobanche* in Food Legume Systems in the Near East and North Africa, 7–9 April 2003, Rabat, Morocco (eds. Dahan, Rachid and El-Mourid, M.). ICARDA, Aleppo, Syria. 1–18 pp

Kraak, H.L. 1992. A computer program to predict seed storage behavior. *Seed Science and Technology* 20: 337–338.

Kraft, J.M. 1982. Field and greenhouse studies on pea seed treatments. *Plant Disease* 66: 798–800

Laverack, G. K. 1994. Management of breeders seed production. *Seed Science and Technology* 22. 551–563.

Laverack, G.K. and M.R. Turner. 1995. Roguing seed crops for genetic purity: a review. *Plant Varieties and Seeds* 8: 29–45.

Wilson, V. E. and Law, A. G. 1972. Natural crossing in *Lens esculenta* Moench. *Journal of the American Society for Horticultural Science*: 79:142–3.

McDonald (Jr), M.B. and C.J. Nelson, C.J. (eds.). 1986. Physiology of deterioration, CSSA Special Publication Number 11. CSSA, Inc. Madison, Wisconsin, USA.

Morrall, R.A.A. and J.W. Sheppard. 1981. Ascochyta blight of lentils in western Canada. *Canadian Plant Disease Survey* 6:7–13

OECD. 2000. OECD schemes for varietal certification of seed moving in international trade: OECD seed schemes 2000. OECD, Paris, France. 215 pp

Papavizas, G.C. and J.A. Lewis. 1975. Effect of seed treatment with fungicides on bean root rots (*Fusarium solani*, *Rhizoctonia solani*, *Thielaviopsis basicola*). *Plant Disease Reporter* 59: 24–28

Ram, G., B.S. Chandrakar, M.K. Misra, and R.K. Katre. 1984. Effect of seed treatment with *Rhizobium* and Dithane M-45 on plant height, nodulation and yield of chickpea. *Indian Journal of Agricultural Sciences* 54: 214–216

Reed, W. C. Cardona, S. Sithanantham and S.S. Lateef. 1987. The chickpea insect pests and their control. *In* The Chcikpea (Saxena, M.C. and K.B. Singh, eds.). CAB International,Wallingford, UK. 283–318 pp

Revier, P.R. and A.W. Young. 1970. Field inspection techniques in seed production. Department of Agronomy, Texas Technological College, Lubbock, Texas.

Roberts, E.H. 1972. Storage environments and the control of viability. *In* Viability of Seeds (Roberts, E.H., ed.). Chapman and Hall Ltd., London.

Roberts, E.H. and K. Osei-bonsu. 1988. Seed and seedling vigour. *In* World Crops: Cool Season Food Legumes (Summerfield, R.J., ed.). Kluwer Academic Publishers, Dordrecht, The Netherlands. 897–910 pp.

Sauer, D.B. 1992. Storage of cereal grains and their products. American Association of Cereal Chemists, St. Paul, Minnesota, USA.

Saxena, M.C., J. Diekmann, W. Erskine and K.B. Singh. 1987. Mechanization of harvest in lentil and chickpea in semi-arid areas. In: Mechanization of Field Experiments in Semi-arid Areas, Proceedings of the IAMFE/ICARDA Conference, 23–27 May, 1987, Aleppo, Syria. ICARDA, Aleppo, Syria. 211–228 pp.

Sharma, S.P. 1987. Breeder seed production in self-pollinated crops. *In* Techniques in Seed Science and Technology (Agrawal, P.K. and Dadlani, M. eds.). South Asian Publishers

Singh, K.B. 1986. Principles and techniques for seed production in chickpea. *In* Seed Production Technology (Srivastava, J.P. and Simarski, L.T., eds.). ICARDA, Aleppo, Syria. 247–256 pp.

Singh, K.B. and M.C. Saxena. 1999. Chickpeas. *In* The Tropical Agriculturist (Coste, R. ed.). Macmillan Education Ltd, London. 134 pp.

Thomson, J.R. 1979. An introduction to seed technology. Leonard Hill, Glasgow, UK. 252 pp.

Trapero-Casas, A., W.J. Kaiser, and D.M. Ingram 1990. Control of *Pythium* seed rot and pre-emergence damping-off of chickpea in the U.S. Pacific Northwest and Spain. *Plant Disease* 74: 563–569

Tu, J.C. 1988. Control of bean anthracnose caused by the delta and lambda races of *Colletotrichum lindemuthianum* in Canada. Plant Disease 72: 5–8

Van der Burg, W.J. 1986. An introduction to seed cleaning. *In* Seed Production Technology (Srivastava, J.P. and Simarski, L.T., eds.). ICARDA, Aleppo, Syria. 107–126 pp.

Van Gastel, A.J.G., B.R. Gregg and E.A. Asiedu (2002). Seed quality control in developing countries. *Journal of New Seeds 4(1/2)*: 117–130.

UPOV (International Union for the Protection of new Varieties of Plants). 2003. Lentil: Guidelines for the Conduct of Tests for Distinctness, Homogeneity and Stability. TG/210/1, UPOV, Geneva, Switzerland. pp 21–22.

CHAPTER 22

DRYING AND STORING LENTILS: ENGINEERING AND ENTOMOLOGICAL ASPECTS

P.K. GHOSH[1], D.S. JAYAS[1], C. SRIVASTAVA[2], A.N. JHA[2]
[1]*Department of Biosystems Engineering, University of Manitoba, Winnipeg, Manitoba, Canada, R3T 5V6;*
[2]*Division of Entomology, Indian Agricultural Research Institute, New Delhi 110 012, India*
Email: Digvir_Jayas@umanitoba.ca

Abstract: For effective storage and end-uses, different drying and storage techniques and principles need to be followed based on the requirements and problems associated with the end use characteristics of lentils. This chapter deals with established drying and storage methods of lentils with emphasis on the mathematical models and associated drying and storage related thermo-physical properties of lentils. Potential of other methods of drying lentils has been mentioned. Effects of drying and storage on lentil quality have been assessed and recent techniques to measure these quality parameters have been discussed. This chapter also deals with the insect free storage of lentil for seed as well as for food purpose. Various insect management strategies for the storage of lentils have been discussed. Apart from conventional storage methods, recent control measures which can be used for lentil storage have also been discussed

1. INTRODUCTION

Lentil (*Lens culinaris* Medik.) is an important cool season legume crop grown extensively in Canada, India, Turkey, US and Australia. World lentil production was 4.15 million tonnes (Mt) in 2005–06 (AAFC 2006) consisting of 70% red, 25% green and 5% brown or other colors. Canada is the largest lentil producer (1.3 Mt) followed by India (1.0 Mt). Canada is also the largest lentil exporting country (70% of its total lentil production) after Turkey to almost all over the world (Europe, Middle East, Africa, South America, North America and Asia) (AAFC 2006). Canada and US are the main producers for green lentils whereas other countries mainly produce red lentils. Several varieties of lentils are grown extensively in three Prairie provinces (Saskatchewan, Manitoba and Alberta) of western Canada

S.S. Yadav et al. (eds.), Lentil: An Ancient Crop for Modern Times, 385–414.

Table 1. Lentils varieties grown in western Canada (compiled from CGC 2005 and AAFC 2006)

Color	Size	Variety
Green	Small	Eston, Milestone, Viceroy
	Medium	Richlea, Vantage, Meteor
	Large	Laird, Glamis, Grandora, Plato, Sedley, Soereign
Red		Blaze, Crimson, Robin, Redcap, Redberry, Rouleau, Rosetown

Table 2. Major constituents of Canadian lentils (Compiled from CGC 2005)

Constituents (% dry basis)	Green			Red
	Small Lentils	Medium lentils	Large lentils	
Protein	25.9	25.7	26.5	28.7
Starch	48.2	48.7	47.9	46.1

(Table 1). Since lentils have a higher protein content than cereal grains and have lower fat content, human consumption of lentils is increasing rapidly. The relative amounts of protein and fats in lentils grown in Canada are given in Table 2. This chapter summarizes the requirements for proper drying and storing of lentils.

2. HARVESTING

Lentil plants are usually grown in dark brown soil zone. The plants have an irregular growth habit with ripe pods and flower buds on the plant at the same time. To prevent shattering losses of the ripe pods, lentils are recommended to be harvested at 18–20% w.b. (wet mass basis) field average moisture content followed by drying to 13–14% w.b. moisture content to maximize yield of high quality (uniform color) seed and for safe storage without subsequent mechanical damage or breakage (Tang et al. 1990a,b, 1992). Sometimes, swathing or chemical desiccation is applied to dry the seeds uniformly in the field before harvesting which has no negative impact on seed germination or cooking time (Tang et al. 1992). The major constituents of lentils are protein (28.6% on dry basis), ash (3.1%), crude fibre (4.4%), ether extract (0.7%), total carbohydrates (63.2%) (Bhatty 1988, Tang et al. 1990a). Pirman and Stibilj (2003) indicated that overall fat content in lentils is approximately 1–3%, in which proportion of two major essential fatty acids (linoleic acid and α-linolenic acid) is about 1:0.33. Cooking reduces the lentil fats by about 50%. Therefore, cooking time reduction is an important issue to avoid the physical, chemical and physico-chemical changes such as partial starch gelatinization, protein denaturation and Maillard reaction in lentils. Iliadis (2001) reported that early harvesting of lentils (plants mostly yellow and partly green) is desirable to reduce the cooking time. Dry climatic weather during crop growth also shortens the cooking time.

3. STORAGE OF LENTILS

Lentils need to be stored after harvesting until they are consumed. Storage time can vary from few months to more than a year. Long term storage can produce 'darkening' of lentil seeds due to possible oxidation of tannins in the seed coat and therefore it reduces the quality and market value (Anonymous 1992). The Prairie Agricultural Machinery Institute (Saskatchewan, Canada) reported that market value drop was relatively higher in small-size seeds (for example, 'Eston' cultivar) than the large-size seeds (for example, 'Laird' cultivar), an estimated annual loss of $1.6 million. Lentils are, therefore, suggested to be dried to a safe moisture content of about 13% w.b. and to be cooled to within 5 °C of average ambient temperature while storing them in dark conditions to maintain color and seed viability (Anonymous 1992). Lentils producers are, therefore, recommended to seal all bin holes to prevent light entry and to run the drying fan occasionally to cool the grains (Anonymous 1992). Peace et al. (1988) reported that if lentils are stored under moderate conditions (12% relative humidity and 20 °C) no significant changes in the protein quality can be detected even after three years of storage.

4. MATHEMATICAL MODELS OF HOT AIR DRYING

Drying is commonly practiced to reduce the moisture content of lentils to a safe storage level. However, drying induces hydrothermal stresses and shrinkage and therefore seeds become susceptible to mechanical damage. Mathematical modeling of the drying process is therefore essential to predict moisture and temperature distribution within the seeds and to locate the points of high temperature and moisture gradients in the seeds. An essential transport property related to moisture distribution during drying of seeds is moisture diffusivity. Tang (1993a, 1994) presented the following moisture diffusivity equations for three major structural components of lentils (cotyledons, hilum and seedcoat) (cv. Laird) dried from 19.7% w.b. moisture content at 30–40 °C, 20–30% relative humidity, and $0.3\,m\,s^{-1}$ air velocity:

$$D_{cot\ yledons} = 0.0192\ e^{-(3887/T)} \tag{1}$$

$$D_{hilum} = 6.44\ e^{-(3901/T)} \tag{2}$$

$$D_{seedcoat} = 63.2\ e^{-(9741/T)}\ e^{[-0.549+(284.5/T)]M} \tag{3}$$

where D is the diffusivity, m^2h^{-1}; T is the absolute temperature, K; and M is the moisture content, % d.b.

Tang and Sokhansanj (1994) developed a three-component drying model for lentils where the cotyledons were considered as a homogeneous slab in which moisture transfer was modeled by a one-dimensional diffusion equation and the hilum and seedcoat were modeled considering two parallel routes of moisture transport. The equations were in the following form:

Moisture transfer through cotyledons:

$$(4)\qquad M_i = C_{b,i} + (M_{i-1} - C_{b,i})\frac{8}{\pi^2}\sum_{n=0}^{\infty}\left[\frac{1}{(2n+1)^2}e^{\frac{-(2n+1)^2\pi^2 D_{cotyledons}\Delta t_i}{4a^2}}\right]$$

where M_i is the transient moisture content at each time interval Δt_i, % d.b.; $C_{b,i}$ is the boundary condition at time t_i; $D_{cotyledons}$ is the diffusivity obtained from eq (1); and a is the equivalent half-thickness of the cotyledon slab.

Moisture transfer through hilum and seedcoat:

$$(5)\qquad \left(\frac{J_{hilum}}{m_d} + \frac{J_{seedcoat}}{m_d}\right)_i = \left[6568.2\, e^{-(3901.1/T)} + 3.24\times 10^9 e^{-(9741.4/T)} e^{\{-0.549+(284.5/T)\}M_i}\right](C_{b,i} - M_e)$$

where

$$(6)\qquad M_e = \left[\frac{-\ln(RH/100)}{\exp(10.5 - 0.018T)}\right]^{-0.47}$$

where J_{hilum} and $J_{seedcoat}$ are the rate of moisture transfer through the hilum and seedcoat, respectively; m_d is the lentil dry mass; and T is the temperature, K.

Karatas (1997) estimated the effective moisture diffusivity of green, red and a Turkish variety of lentils during drying from 45 to 60 °C by Fick's law. Three falling rate periods were found in which moisture diffusivity showed a stepwise increasing trend from the first to third falling rate period. Moisture diffusivity in the third falling rate period was comparatively higher than the first two falling rate periods for three different lentils. It was reported that drying rate was sensitive to temperature above 45 °C. Further the activation energy for moisture removal process ranged between 3.38 to 13.21 kcal g^{-1} Mol^{-1} for the different lentil varieties.

Carmo and Lima (2005) have modeled transient moisture transfer during drying of lentils considering shrinkage. Model simulation was performed for unsteady-state diffusion phenomena in an axisymmetric oblate spheroid with convective boundary condition at the surface. Their model was of the following form:

Moisture transfer equation:

$$(7)\qquad \frac{\partial M}{\partial t} = \nabla.(D\nabla M)$$

with boundary condition:

$$(8)\qquad D\nabla M + h_m(M - M_e) = 0 \qquad \text{at surface}$$

Table 3. Coefficients of Eqs. (10) and (11) (compiled from Carmo and Lima 2005)

Temperature (°C)	Relative humidity (%)	M_e (%), d.b.	A	B (m^2 s^{-1})	C (m^2 s^{-1})
23	15	8.5	0.0625	7.27	−1.47
23	30	10.7	–	3.95	1.13
23	50	13.9	–	3.41	3.67
40	5	6.1	0.9239	10.15	−1.19
40	30	9.3	1.1184	10.47	−0.02
40	50	12.1	–	13.80	1.88
60	5	5.2	0.8808	16.77	0.86
60	30	7.9	1.0241	18.52	1.74

and initial condition:

$$M(t=0)=M_o \quad (9)$$

where M is the moisture content, d.b., decimal; M_o is the initial moisture content; M_e is the equilibrium moisture content (Table 3); D is the diffusivity, m^2 s^{-1}; and h_m is the convective mass transfer coefficient, m s^{-1}.

Shrinkage was modeled using the following equation:

$$\frac{V_t}{V_o}=1-A(M_o-MR) \quad (10)$$

where V_t is the volume at time t, m^3; V_o is the initial volume, m^3; A is the shrinkage coefficient (Table 3); and MR is the moisture ratio $(M_t–M_e)/(M_o-M_e)$.

Diffusivity was modeled as:

$$D=B\,MR+C \quad (11)$$

where B and C are constants, m^2 s^{-1} (Table 3).

5. OTHER METHODS OF DRYING LENTILS

Cenkowski and Sosulski (1997) investigated the applicability of infrared drying of lentils of 19–39% w.b. moisture content as an alternative to conventional drying process. The infrared treatment generated heat (about 130–150 °C) using near-infrared rays with wavelengths of about 1000–3500 nm and the process could reduce about 8–20 percentage points of moisture content very quickly (within 55–85 s). This process was efficient and effective for reduction in cooking time and for starch gelatinization and solubilization. Cooking time was reduced primarily due to strong dependence of the porosity and water diffusivity to the moisture content (Scanlon et al. 2005).

González and Pérez (2002) found that microwave irradiation and extrusion cooking of lentils could reduce the moisture content, crude protein, crude fibre, water absorption, solubility, swelling, retrogradation of starch with an increase in ash content, reducing sugar and absolute density of the lentil seeds.

6. DRYING AND STORAGE RELATED PROPERTIES

6.1. Initial Moisture Content Determination

Tang and Sokhansanj (1991) recommended an oven drying procedure for determining moisture content of unground lentils. According to them, a 16 g sample of whole kernel lentil should be dried at 130 °C in an oven for 20 h for moisture content determination.

6.2. Equilibrium Moisture Content

Knowledge of equilibrium moisture content of lentils is essential for efficient operations of the drying systems and proper storage. Cenkowski et al. (1989) proposed the following equations obtained from former ASAE Standard D245.4 (presently D245.5) (ASAE 1995) for equilibrium moisture content (M_e) determination of lentils from data over a relative humidity (RH) range of 0.20–0.80 and temperature (T_C) range of 5–50 °C:

The modified Henderson equation:

$$\text{(12)} \qquad RH = 1 - \exp\left[(0.000005T_C + 0.00043)\{T_C + (-0.75T_C + 17.4)\}(100\,M_e)^{1.57\exp(0.0061T_C)}\right]$$

and the Chung equation:

$$\text{(13)} \qquad RH = \exp\left[\frac{-(2.4T_C + 97.1)}{T_C + 10}\exp(-15.1\,M_e)\right]$$

Tang et al. (1992) indicated the following equation for determining the equilibrium moisture content of lentils (cv. Laird) corresponding to the environment:

$$\text{(14)} \qquad M_e = [-\exp(5.72 - 0.0176\,T_C)/\ln(RH)]^{-0.465}$$

Menkov (2000) proposed the following equations for determining the equilibrium moisture content of lentils (cv. Larisa) over a temperature range of 5–60 °C and a relative humidity range of 0.11–0.88:

The modified Oswin equation:

$$\text{(15)} \qquad M_e = (A + BT_C)\left(\frac{RH}{1 - RH}\right)^C$$

Table 4. Parameter values of Eqs. (15) and (16) (compiled from Menkov 2000)

	Modified Oswin			GAB				
	A	B	C	A	B	C	h_1	h_2
Adsorption	11.78	−0.06	0.39	7.21	0.41	0.0009	1563.04	23606.26
Desorption	13.90	−0.08	0.35	9.12	0.32	0.0007	1937.14	24419.88

The GAB (Guggenheim-Anderson-de Boer) equation:

$$(16) \qquad M_e = \frac{A\left[B\exp\left(\frac{h_1}{RT}\right)\right]\left[C\exp\left(\frac{h_2}{RT}\right)\right](RH)}{\left[1-\left[B\exp\left(\frac{h_1}{RT}\right)\right](RH)\right]\left[1-\left[B\exp\left(\frac{h_1}{RT}\right)\right](RH)+\left[B\exp\left(\frac{h_1}{RT}\right)\right]\left[C\exp\left(\frac{h_2}{RT}\right)\right](RH)\right]}$$

where T is the absolute temperature, K; A,B,C, h_1, h_2 are coefficients (Table 4); and R is the universal gas constant (J mol^{-1} K^{-1}).

Barrozo et al. (2000) proposed the following equation for determining equilibrium moisture conditions between lentils and air:

$$(17) \qquad M_e = \left[\frac{-\exp(-5.08\times10^{-3}\,T_C+5.42)}{\ln(RH)}\right]^{1/2.17}$$

7. PHYSICAL PROPERTIES

Knowledge of physical properties is important for designing equipment for drying, storage and other processing operations of lentils. Measurement of seed dimensions, size and shape is essential for designing the cleaners and graders. Bulk density, kernel density and porosity are important considerations for designing the dryers and storage structures since they directly influence the airflow distribution in grain mass. Angle of repose and coefficient of friction are important for the calculation of bin wall pressure and proper design of storage structures (Amin et al. 2004). Several researchers have reported the dimensions (diameter, thickness, 1000 kernel weight) and bulk properties (porosity, bulk and particle density, angle of repose, frictional coefficient) of lentils (Sokhansanj et al. 1990, Irvine et al. 1992a,b, Tang and Sokhansanj 1993a, Amin et al. 2004). Table 5 provides the bulk properties data of two major Canadian lentil varieties.

Also, measurement of frictional characteristics is important for correct design of storage and handling facilities under varying levels of lateral pressure that is common during unloading. Tables 6 and 7 provide the frictional coefficient of lentils against different surfaces as a function of practical storage moisture content and lateral pressure. Frictional coefficients increased with increase in both the moisture content and the lateral pressure.

Table 5. Bulk properties of two major Canadian lentil varieties at various moisture contents (compiled from Sokhansanj et al. 1990 and Irvine et al. 1992a)

Moisture Content (% w.b.)	Bulk density (kg m^{-3})	Particle density (kg m^{-3})	Angle of repose (°)			Coefficient of friction					
			Emptying angle	Filling angle (left)	Filling angle (right)	Galvanized steel	Plywood-grain parallel to flow	Plywood-grain perpendicular to flow	Steel-trowelled concrete	Wood-floated concrete	Corrugated steel
'Laird'											
11.4	–	1430	–	–	–	–	–	–	–	–	–
11.5	759	1426	–	–	–	–	–	–	–	–	–
11.7	804	–	21.6	23.1	24.5	0.20	0.15	0.21	0.27	0.26	0.37
13.1	–	1433	–	–	–	–	–	–	–	–	–
13.8	789	–	23.9	23.3	24.1	0.25	0.24	0.29	0.34	0.31	0.41
15.2	–	1409	–	–	–	–	–	–	–	–	–
15.9	782	–	23.7	22.4	24.9	0.25	0.19	0.26	0.32	0.28	0.38
17.7	767	–	23.8	22.8	24.2	0.30	0.22	0.27	0.34	0.33	0.39
18.0	–	1393	–	–	–	–	–	–	–	–	–
'Eston'											
11.4	825	–	25.7	27.3	28.5	0.24	0.17	0.25	0.32	0.31	0.45
11.5	762	1395	–	–	–	–	–	–	–	–	–
11.6	825	1410	25.9	26.8	27.2	0.25	0.19	0.25	0.29	0.29	0.46
12.3	814	–	27.8	27.8	28.6	0.26	0.23	0.27	0.36	0.34	0.47
13.3	–	1406	–	–	–	–	–	–	–	–	–
14.5	800	–	28.6	27.9	28.3	0.29	0.26	0.31	0.37	0.36	0.49
15.2	–	1395	–	–	–	–	–	–	–	–	–
15.8	791	–	27.8	28.0	28.7	0.31	0.24	0.29	0.37	0.34	0.47
18.0	783	–	28.5	26.9	27.9	0.33	0.24	0.31	0.41	0.38	0.47
18.2	–	1392	–	–	–	–	–	–	–	–	–

– indicates data are not available.

Table 6. Frictional coefficients of lentils (compiled from Irvine et al. 1992b)

Variety	Surface	Moisture content (% w.b.)	Pressure, kPa		
			10	30	50
'Eston'	**Galvanized steel**	8.0	0.14	0.15	0.15
		12.5	0.17	0.15	0.16
		15.0	0.14	0.14	0.16
		18.9	0.16	0.16	0.18
	Plywood-grain perpendicular to flow	8.0	0.23	0.20	0.21
		12.5	0.19	0.18	0.19
		15.0	0.21	0.20	0.21
		18.9	0.21	0.21	0.21
	Plywood-grain parallel to flow	8.0	0.14	0.16	0.18
		12.5	0.16	0.17	0.19
		15.0	0.17	0.17	0.17
		18.9	0.17	0.17	0.18
'Laird'	**Galvanized steel**	7.8	0.11	0.13	0.15
		12.5	0.19	0.16	0.16
		13.7	0.20	0.15	0.15
		17.4	0.13	0.15	0.16
	Plywood-grain perpendicular to flow	7.8	0.16	0.18	0.22
		12.5	0.18	0.20	0.20
		13.7	0.24	0.23	0.24
		17.4	0.20	0.21	0.22
	Plywood-grain parallel to flow	7.8	0.14	0.16	0.18
		12.5	0.15	0.17	0.18
		13.7	0.22	0.20	0.19
		17.4	0.19	0.19	0.18

Table 7. Static and Dynamic coefficient of friction of BARI lentil-1 on various surfaces at various moisture contents (Compiled from Amin et al. 2004)

Moisture Content (% w.b.)	Coefficient of friction							
	Galvanized steel		Plywood		Concrete		Glass sheet	
	Static	Dynamic	Static	Dynamic	Static	Dynamic	Static	Dynamic
10.3	0.40	0.37	0.39	0.36	0.46	0.43	0.37	0.35
14.2	0.41	0.37	0.39	0.37	0.47	0.44	0.38	0.36
18.1	0.42	0.38	0.40	0.38	0.48	0.45	0.39	0.37
21.0	0.43	0.40	0.42	0.40	0.49	0.47	0.40	0.38

Amin et al. (2004) explained the relationship between seed dimensions and bulk properties and seed moisture content for Bangladesh-grown lentils (cv. BARI lentil-1) over a moisture content (m) range of 10.3–21% w.b.:

$$\text{Bulk density (kg.m}^{-3}) = 889.90 - 5.961\,\text{m} \tag{18}$$

$$\text{Particle density (kg.m}^{-3}) = 1326.20 - 5.286\,\text{m} \tag{19}$$

$$\text{Porosity}(\%) = 32.03 + 0.23\,\text{m} \tag{20}$$

$$\text{Angle of repose}(^\circ) = 21.64 + 0.28\,\text{m} \tag{21}$$

$$\text{Diameter (mm)} = 3.63 + 0.02\,\text{m} \tag{22}$$

$$\text{Thickness (mm)} = 1.89 + 0.03\,\text{m} \tag{23}$$

$$\text{1000 ker nel weight (g)} = 14.4 + 0.53\,\text{m} \tag{24}$$

Sokhansanj et al. (1990) reported the dimensions and porosity of lentils at a moisture content of 11.5% w.b. (Table 8).

7.1. Damage during Handling

Grains are usually dropped from a considerable height in grain handling facilities and typical drop heights vary from 3 m in grain bins to more than 50 m in large concrete grain terminals. Therefore, physical damage can be significant which reduces the grain quality and results in economic loss. Bergen et al. (1993) reported that significant damage could occur in dry 'Laird' lentils (10.4% m.c.) at a drop height of 18.3 m on concrete floors compared to steel and plywood floors.

7.2. Airflow Resistance

Lentils are generally dried and cooled using ambient air to maintain safe storage moisture and temperature. Knowledge of airflow resistance through bulk lentils is thus important. Following equation was developed for airflow resistance through

Table 8. Dimensions and porosity of two major Canadian lentil varieties at 11.5% w.b. moisture content (compiled from Sokhansanj et al. 1990)

Parameter	Laird	Eston
Major diameter (mm)	6.99 ± 0.32	5.0 ± 0.24
Minor Diameter (mm)	6.66 ± 0.31	4.66 ± 0.21
Thickness (mm)	2.77 ± 0.15	2.48 ± 0.16
Porosity (%)	46.5	45.4

lentils (cv. Laird) at 11.4% moisture content for airflow range between 0.0028–0.5926 $m^3\ s^{-1}\ m^{-2}$ (Sokhansanj et al. 1990):

$$\Delta P = \frac{AQ^2}{\ln(1+BQ)} \tag{25}$$

where P is the pressure drop per unit depth of grain, Pa m^{-1}; A and B are constants (Table 9); and Q is the airflow rate ($m^3\ s^{-1}\ m^{-2}$).

Sokhansanj et al. (1990) further reported that small seeds, dense filling and presence of fines increased the airflow resistance in bulk lentils whereas wet lentils reduced airflow resistance. The resistance to airflow in the horizontal direction was 0.3–0.66 times that in the vertical direction for lentils for an airflow range of 0.0019–0.28 $m^3\ s^{-1}\ m^{-2}$ (Sokhansanj et al. 1990, Alagusundaram et al. 1992).

Jayas and Mann (1994) presented the airflow resistance data of 'Laird' and 'Eston' lentils over an airflow range of 0.004–0.35 $m^3\ s^{-1}\ m^{-2}$ with the following equation:

$$\frac{\Delta P}{L} = (M \times \tilde{n})Q^N \tag{26}$$

where ΔP is the pressure drop, Pa; L is the bed depth, m; ñ is a modifier scaled to data for wheat for easiness in fan selection and assessment of filling depths in grain bins (Table 10); Q is the airflow, $m^3\ s^{-1}\ m^{-2}$; and M and N are constants (Table 10).

Li and Sokhansanj (1994) proposed the following generalized equation for airflow resistance through bulk lentils (cv. Laird) considering bulk porosity, particle dimension and air properties:

$$\Delta P = f \frac{1-\epsilon}{\epsilon^3} \frac{2\rho_a Q^2}{d_p} \tag{27}$$

Table 9. Values of constants for Eq. (25)

Airflow range ($m^3 .s^{-1} .m^{-2}$)	A	B
0.0028–0.1605	45502	27.32
0.1605–0.5926	53544	35.20
0.0028–0.5926	54312	36.79

Table 10. Values of M, N, and ñ used in the Eq. (26)

Airflow rate ($m^3\ s^{-1}\ m^{-2}$)	M	N	ñ	
			Eston	Laird
0.004–0.05	5178	1.11	0.75	0.68
0.05–0.35	6368	1.67	2.14	2.30

where

(28) $$\in = 1 - \frac{\rho_b}{\rho_k}$$

(29) $$f = \frac{533.2}{(Re)} + 6.678$$

(30) $$Re = \frac{\rho_a Q d_p}{(1 - \in)\mu_a}$$

f is the coefficient of friction; $\in$ is the porosity; ρ_a is the air density, kg m^{-3}; ρ_b is the bulk density, kg m^{-3}; ρ_k is the kernel density, kg m^{-3}; Q is the airflow, m^3 s^{-1} m^{-2}; d_p is the volume equivalent diameter, m; Re is the Reynolds number (varied from 0.03 to 500 corresponding to airflow range of 0.754×10^{-4} m^3 s^{-1} m^{-2} to about 0.9 m^3 s^{-1} m^{-2}); and μ_a is the air viscosity, Pa s. The similar equation was recommended to be used for pressure drop calculation for lentils mixed with fines with a modified equivalent particle diameter of the following form:

(31) $$d_p^* = d_p(1 - f_m)$$

where

(32) $$f_m = \frac{m_f}{m_t}$$

d_p^* is the modified equivalent particle diameter, m; f_m is the fine content, decimal; m_f is the mass of fines; and m_t is the total mass of fines and lentils.

In an attempt to study the nature of the movement of CO_2 in bulk lentils (porosity 0.40) resulting from natural convection during controlled atmosphere (CA) storage with temperature gradient range of 20–56 °C, Bundus et al. (1996) found that temperature gradient and location of CO_2 introduction have no influence on the direction of the convective flow of CO_2 through the lentil bulk.

8. THERMAL PROPERTIES

Since lentils are heat sensitive, knowledge of heat transfer during production, handling and processing is important. Mass transfer is also associated with drying. Therefore, thermal properties should be measured to predict the moisture and temperature changes in lentils during the process operations. Thermal properties play an important role in designing the drying or cooling systems and in predicting temperature distribution when using these systems. Alagusundaram et al. (1991) proposed the following equation for measurement of thermal conductivity as a function of temperature and moisture content of lentils (cv. Laird) from data over a moisture content range of 9 to 23% w.b. and temperature range of −28 to 29 °C:

(33) $$K = 0.193 + 10 \times 10^{-4} T_C + 1.52 \times 10^{-3} m$$

where K is the thermal conductivity, W m^{-1} K^{-1}; T_C is the temperature, °C; and m is the moisture content, % w.b. The values of K ranged from 0.187 to 0.249 W m^{-1} K^{-1} for the range of temperature and moisture content studied. Linearity of the above equation (Eq. 33) indicates that thermal conductivity increases with an increase in both the temperature and the moisture content. Tang et al. (1991a) proposed the following empirical model to determine the specific heat (C_p) of lentils ('Laird') over a temperature range of 10–80 °C and moisture content range of 2–35% d.b.:

$$(34) \qquad C_p = 0.5773 + 0.00709\ T_C + (6.22 - 9.14\ M)\ M$$

They calculated that the specific heat of lentils ranged from 0.8–2.2 kJ kg^{-1} K^{-1} from the above relation (Eq. 34).

9. QUALITY

The major quality parameters associated with the production, processing and consumption of lentils are color, cooking time, dehulling efficiency, firmness of cooked seeds, 1000 kernel weight, protein content, seed size distribution, starch content, and water absorption (Bhatty 1988, CGC 2005). Studies dealing with quality and composition of lentils were reviewed exhaustively by Bhatty et al. (1988). Nutritional quality of lentils is strongly associated with its principal constituents, i.e. protein and starch and seed grading is determined largely by color, presence of foreign materials and damaged seeds (Table 11). All these parameters have an overall effect on the cooking quality.

9.1. Effects of Drying and Storage on Quality

Tang et al. (1990a, 1993b) studied the effects of artificial drying of lentils on breakage susceptibility, cooking quality and seed viability. Lentils with initial moisture content less than 15% w.b. are highly susceptible to breakage; however

Table 11. Canadian grading of lentils (compiled from Bhatty 1988)

Grade	Color	Physical conditions	Damaged (%)	Foreign materials (%)
No. 1	Good natural color	Sound, well-matured	2.0	0.2
No. 2	Reasonable good natural color	Moderately immature, lightly discolored	3.5	0.5
Extra No. 3	Fair color	Immature, moderately discolored	5.0	1.0
No. 3	Fair color		10.0	1.0

moisture content and drying temperature have no effect on the cooking quality. They further recommended that drying should not be performed at temperature higher than 70 °C to avoid the germination loss. Breakage susceptibility is negatively correlated to drying temperature and positively correlated to drying time. Whereas germination loss is positively correlated to drying temperature, seed moisture, and drying time and negatively correlated to relative humidity. Effects of artificial drying and six-month storage on the seed breakage, germination and cooking quality of lentils were further evaluated by Tang et al. (1991b). Storage of lentils increased breakage susceptibility and reduced overall cooking quality whereas germination was not significantly affected by the storage time. To avoid possible germination loss, two processing conditions were recommended: drying of seeds with 20% moisture content at less than 60 °C and drying of seeds with 16% moisture content up to 70 °C. Qualitative changes of stored lentils are periodically checked for seed conductivity, pH, free fatty acids, viscosity, appearance, flavor, taste, mold, bacterial count, and germination potential (Bhatty 1988, Mills et al. 1999).

9.2. Machine Vision Systems

There is an increasing demand that the types (variety) of various lentils should be properly identified from the bulk to meet specific needs of the market. Since color of lentil is critical for its marketability and different types of lentils have different color and size, machine vision systems have potential for lentil type identification and color-based gradation relying on two major seed feature types: morphology and color (Shahin and Symons 2003a). A minimum acceptable color for each of the lentil types can thus be set as an industry standard for lentil class identification from the bulk samples. Official lentil gradation system is based mainly on visual inspection of color.

Shatadal et al. (1995) introduced a color camera-based machine vision technology to classify non-touching small and large-sized lentils based on their morphological features (length, width, area, perimeter, maximum and minimum radii, rectangular aspect ratio, thinness ratio, radius ratio, area ratio and a ratio of mean to standard deviation of all the radii). They indicated a need for using color dependent features along with the morphological features for increasing classification accuracy. An automated seed presentation device was designed and fabricated by Jayas et al. (1999) to automatically pick up bulk grains and separate seeds based on their types for use in machine vision identification of grain.

Mills et al. (1999) recommended the use of chromometer to denote brightness, greenness-redness and yellowness-blueness to classify nine categories of lentil samples based on their appearance: green with occasional brown spots, green with some brown discoloration of individual seeds, green with more brown discoloration of individual seeds, green with considerably more brown discolored seeds, green/brown with many brown seeds, green/brown with considerable number of brown seeds, brown/green with brown seeds, brown with many split seeds, and pale fawn/green in a 50:50 mixture. They measured pH, conductivity of the concentration of electrolytes in solution as a result of ion-leakage through the cell membranes and germination percentage of theses lentils which varied from 5.92–6.41, 55–361 μS m^{-1}, and 4–89%, respectively. Brown lentils with many splits showed low pH and germination with high conductivity whereas green lentils with occasional brown splits showed high pH and germination, and low conductivity.

Since machine vision technology has strong potential for its adoption by the industries there is a need to develop easy and accurate machine vision system for lentil class identification. Shahin and Symons (2001) developed a less expensive commercial flatbed scanner-based machine vision system for predicting lentil color and grade. This scanner-based technology is portable, less expensive and can be attached to a notebook computer for image analysis. But drawback of this technology is variability in image information acquired by using different commercial scanners. However, this problem was solved by calibrating images obtained from different commercially available flat-bed scanners against the images of a Q60 (Kodak, Canada) color chart (Shahin and Symons 2003b). The Canadian Grain Commission has also developed a software package (LentilScan©) for accurate lentil type identification using machine vision technique (Shahin and Symons 2003a).

9.3. Shrinkage Effect

Shrinkage is another effect of drying and makes the seeds vulnerable to breakage. Therefore, accurate prediction of shrinkage is necessary to avoid seed loss due to breakage. The following equations were developed for calculating the shrinkage effect of whole and milled lentils for data over a temperature range of 30–70 °C and relative humidity range of 4–40% (Tang and Sokhansanj 1993b):

(35) $$\text{whole lentils}: V_t = V_o e^{-0.00933(M_o - M)}$$

(36) $$\text{milled lentils}: V_t = V_o e^{-0.0113(M_o - M)}$$

where V_t is the seed volume at time t, m^3; V_o is the initial volume, m^3; M is the moisture content at time t, % d.b. and M_o is the initial seed moisture content, % d.b.

9.4. Dehulling Efficiency

Dehulling efficiency is an important quality parameter which determines the throughput. In industrial application, lentils are generally hydrated for a short time followed by a quick drying and tempering before final dehulling. Erskine et al. (1991) demonstrated that dehulling efficiency decreases with increase in seed hydration time irrespective of seed sizes with the following equations:

(37) $$\text{Dehulling efficiency of large seeds}(\%) = 83.69 - 0.312 \times \text{immersion time}$$

(38) $$\text{Dehulling efficiency of small seeds}(\%) = 83.87 - 0.160 \times \text{immersion time}$$

The dehulling efficiency is not affected by drying temperature but drying time affects the dehulling efficiency by the following equations:

(39) $$\text{Dehulling efficiency of large seeds}(\%) = 83.56 - 0.14\,t + 0.00077\,t^2$$

(40) $$\text{Dehulling efficiency of small seeds}(\%) = 83.64 - 0.07\,t + 0.00039\,t^2$$

where t is the drying time, min. Dehulling efficiency can be increased by increasing the tempering time.

10. INSECT INFESTATION OF LENTILS

Between harvesting and consumption lentils are highly susceptible to insect damage if proper storage conditions are not met and losses may go up to 50% resulting in substantial monetary loss. In most of the tropical countries, one fifth of the harvested pulses are stored in large-scale warehouses in urban areas whereas farmers in rural areas retain the bulk of their produce. The damage results in produce loss and deterioration in quality due to the presence of insects and their excreta inside the seeds (Caswell 1981). The extent of damage to lentils is usually very high both qualitatively and quantitatively.

The most important pests of stored legumes belong to the family Bruchidae of the order Coleoptera. According to their infestation behaviour they can be categorized as (a) those that breed on and infest the grain in store, (b) those that infest the grain in fields and are brought into store with harvested grain but do not breed in store, and (c) those that breed on and infest both in field and stored grain. Most of the problem insects pests of lentil belong to group (c) although the majority of damage occurs in storage. The important pulse beetles of the Bruchidae family for lentils are: *Callosobruchus maculatus*, *C. chinensis*, *C. analis*, *Zabrotes subfasciatus*, *Bruchus lentis and B. ervi.* These pests are reported to occur in Lentils (Bhalla et al. 2006).

Pulses are usually consumed in split form after removal of their testa. Generally the pulse beetles as such attack only the whole grain but not their processed form. But besides the Bruchidae the following cereal grain insects can also damage the processed stored pulses/ split pulses or as secondary infestations: *Trogoderma granarium*, *Rhizopertha dominica*, *Tribolium castaneum*, *Latheticus oryzae*, *Oryzaephilus surinamensis*, *Cadra cautella*, *Corcyra cephalonica*, *Lasioderma serricornae*, *Stegobium paniceum*, *Cryptolestes ferrugenius* (Khanna and Singh 2002).

Bruchids are small insects with small head and short antennae. Adult bruchids fly from the infested stores to nearby fields where pulses grow. As the pods develop, the female lays eggs either on the outside or within the pods or directly upon the exposed seeds. The eggs hatch into the grubs that enter into the soft developing seeds through the pod wall. Since the grubs are tiny, the holes through which they enter into the seed could not be seen with naked eyes. Generally pulses mature faster than the grubs, thus in mature grains larvae continue to feed. Most species develop entirely within a single seed. Larvae feed inside the seed and pupate inside and develop into adults. The beetles have a sharp pair of mandibles through which they cut circular flaps in the seed coating and make a small hole which is the first visual evidence of the presence of beetle in the seeds. The adults emerge through these holes and come up by removing the lid. If they are undisturbed, they remain in the seeds without flying out. The rate of breeding depends upon several ecological factors, i.e. temperature and relative humidity of the air and storage methods. Most of the species are multi-voltine. After completing few generations in the storage, beetles rush to the field by instinct and start life cycle afresh in the field and infest crop.

Stored lentils are infested by two principal species of *Callosobruchus* beetles: the Adzuki bean seed beetle [*C. chinensis* (L.)] and cowpea seed beetle [*C. maculatus* (F.)] (Coleoptera: Bruchidae). Both species are widespread and found in all continents with subtropical or tropical conditions (USA, Mediterranean, India and Australia). *C. chinensis* is very common in stored lentils. The adults are about 2.5 mm to 3.5 mm long. *C. chinensis* has a characteristic triangular white spot at the base of the thorax and the elytra have two distinct dark brown spots on a rust background. In *C. maculatus* the elytra have a large rounded spot at midlength and the tips are black. There are similar black areas on the abdomen exposed beyond the wing covers. These species are rarely found in the field laying eggs on the developing pods. They mainly occur during storage in the stored seeds. The females lay eggs on the seed coat (several eggs per seed). The young larvae hatch through the base of the egg and bore straight through the seed coat. The white larvae develop and pupate inside the seed. One generation is completed in 3 to 4 weeks. Thus, these insects reproduce rapidly and can damage large quantities of stored lentil seed. *Callosobruchus* spp. is mainly found in stored seeds. Infested seeds have white eggs attached and later show cylindrical windows or adult emergence holes (Beniwal et al. 1993).

The other most common and serious univoltine species of bruchids infesting lentil are *Bruchus ervi* Froel. (Coleoptera: Bruchidae), occurring in Europe, North Africa and Southwest Asia, and *Bruchus lentis* Froel. in the USA, Europe, North Africa, Southwest Asia and India. *B. ervi* is a slightly elongated beetle with a body length of 3 mm to 3.8 mm. The elytra are black with light brown hairs and whitish spots. The larvae are white-yellowish with a dark brown head. Adults of the other major *Bruchus* pest species for lentil *B. lentis* are 3 mm to 3.5 mm long and have dense, grey reddish hairs on the back, marked with several whitish spots. In both the species the adults move into lentil fields at the time of flowering where they feed on nectar and pollen. Ovary development and copulation occur after feeding. The females glue their elongated yellow transparent eggs to the outside of the young pods. Upon hatching, the tiny larva penetrates the pod wall, burrows through the pod until it reaches the developing seed, enters the seed and feeds on its contents. The larva grows slowly and requires about 6 weeks until it is fully grown. Before pupation the larva eats an exit passage to the surface of the seed, leaving only a thin circular window (the epidermal membrane) intact. After pupation the emerging adult opens this membrane and leaves the seed. Adults may re-enter the seed for hibernation or hibernate in other protected places, until lentil flowering next season. Some adults may remain in the dry seeds until the seeds are planted. Thus there is only one generation per year, no eggs are laid on dry seeds and there is no reproduction during storage. The first symptom is a small dark pinhole or penetration dot on the seed at the time of harvest. If the seeds are opened the larva can be found inside. Later a window appears on the seed and most of the seed is eaten out by the *Bruchus* larva (Beniwal et al. 1993)

11. INSECT CONTROL STRATEGIES

All the control strategies mentioned here are generally used to manage the insect pests of stored grain. Most of the mentioned management methods have been used to check the insects of pulses in storage. Thus although all methods are not having examples to control storage insects of lentil but they can be used for lentil as they are being used for insects of pulses in storage.

11.1. Harvesting

Time and frequency of crop harvesting is important in controlling storage insect pests. It is reported that timely harvesting can reduce infestation of *C. maculatus* by 50 to 90% (Paddock and Reinhard 1919). This prevents egg laying on matured crops. Field infestation occurs due to migrating bruchids from infested grains in storage.

11.2. Hygiene in Store

Processing plants and store premises should be thoroughly cleaned on a regular basis. Web, dirt, debris, spillage should be collected and burnt or buried to destroy surviving insect-pests. The storage facility should be painted with white color at least once in a year. Cleanliness or sanitation is the most important step towards prevention of insect infestation. Dusts, grain, and chaffs should be removed from the storage area as well as threshing yard and transport system before bringing or storing the new crops. Spillage of seeds during inspection or drawing samples should be avoided. Spilled seed should immediately be removed as it attracts wandering insects that may infest and lay their eggs. Hygiene of store is one of the most important aspects for ensuring control of insect infestation.

11.3. Bag Storage

The filled seed/storage bags should always be stacked on wooden pallet (platform). This prevents moisture pick-up by the seeds from the floor and facilitates proper cleaning and circulation of fumigant or fresh air. Use of steel hooks should be avoided for seed bags as these damage bags and cause spilling of seeds and facilitate entry by insects. Aisle (moving) space should be maintained between bag-stacks and walls for inspection or sampling. In fact only 50 to 60% space is utilized for proper storage of bags. It has been observed that if the old and new seeds are stored separately, management of insect control can be simplified because different crop seeds are normally attacked by different group of pests.

11.4. Physical Control

The use of inert dust is a traditional method to control storage insect pests from post harvest materials. Occasional use of abrasive mineral dusts, natural dessicants like wood ash and various plant materials with repellent or insecticidal properties is well known and documented (Golob and Webley 1990). Non toxic grain protectants, inert dusts have been employed to control different type of insect pests. Its application removes the epicuticular lipid layer that causes desiccation. Rock phosphate of 1% level is also used to control insects in horse bean (*V. faba*), cowpea (*V. unguiculata*) and lentil (Anonymous 1979). Magnesium carbonate, crystalline silica, talc, amorphous silica, walnut ash, cow dung ash, rice husk ash, Diatomaceous Earth (DE) at 1 to 2% (by mass) and sand treatment have been found to be effective against bruchids. Attapulgite based Activated Clay (ABCD) at 0.5 to 1% gave very good results to protect pulse grain from *Callosobruchus sp* (Varma and Siddiqui 1977). One of the important alternatives to control storage insect pest is use of diatomaceous earth (DE) which are formed from fossils of diatoms (diatoms are unicellular algae which occurred during Eocene and Miocene periods) (Vayias and Athanassiou 2004). They are mainly composed by amorphous hydrated silica. The bodies of insects trap DE particles inflict damage to cuticle and insects die due to water loss and desiccation. There are several formulations with DE dusts approved for treating stored wheat, but these dusts are not commonly used because they adversely affect the physical properties of the grain such as test weight and flowability (Subramaniam and Roseli 2000). They are chiefly of natural origin with very low mammalian toxicity.

11.5. Chemical Control

The disinfestation of insect pests in post harvest materials through prophyllactic treatments still relies strongly on the use of synthetic insecticides, which leads to series of problems. The synthetic chemical insecticides are effective for a longer time and degrade slowly, thus giving longer lasting protection. However, this persistence means that they pose a threat to the health of consumers. as well as the environmental consequences of persistent chemicals (Fishwick 1988, Wright et al. 1993) and partly because many storage insect species have developed resistance against most of the insecticide.

Beetles under ambient storage conditions infest legumes heavily. In the humid tropical areas, insect infestation in stored material is more or less inevitable for multi-voltine pest species due to favourable and stable environments of food storage system. Technical advances in synthetic chemistry led to the confidence that synthetic contact pesticides and fumigants would provide permanent solution to the storage-related problems. The general confidence in the efficacy of synthetic pesticides extended a common assumption that the technology is safe. On the basis of research various sources recommended Malathion for control of storage insects. Champ and Dyte (1976) took the global survey of monitoring the pesticide resistance and reported the occurrence of Malathion resistance against many storage insects. Later on, Malathion was replaced by the synthetic Pyrethroid Deltamethrin which gave good results but insects also developed resistance to this pesticide (Lorini and Galley 1999). Organophosphates viz., Chlorpyrprifos-methyl, Pirimiphos-methyl, Fenitrothion and Dichlorvos etc. are also used in many countries (Tyler and Binns 1977, Yadav 1980, Yadav et al. 1980, Yadav and Singh 1994). As the mixing of insecticides with edible commodity is always discouraged in India, these insecticides are used for prophylactic treatment and their residual toxicity is of great importance (Pradhan et al. 1971). Among the other synthetic pyrethroids, Cyfluthrin and Bifenthrin have emerged to control the storage insects infesting grains/seeds in storage (Zettler and Arthur 2000, Daglish and Wallbank 2003). The residual toxicity of insecticides depends on physical, biological and environmental factors such as application rate, specific insecticide and its formulation, time interval in which insects are exposed to the insecticides, surface substrates, target insect species, temperature and relative humidity of the air where insects are exposed. Although very few reports are available for insecticide resistance against pulse beetles continuous use may lead to its occurrence.

11.6. Plant Products

Plant products that are traditionally used and produced by the farmers in developing countries and plant based pesticides have gained considerable importance as they are quite safe and promising (Jilani et al. 1988). It is an old age practice of traditional farmers in the tropics to mix a local plant with the grain legume. The ideal alternative agent would be a natural compound that has repellency or toxic effects on the insect pests preventing their development and that does not affect the consumers or non-target organisms and the environment. Botanical insecticides have a broad spectrum of pest control. They tend to be breakdown readily, especially in UV light so may be less hazardous to the environment, although this can mean their persistence is poor. Some are less toxic to the mammals (Adler 2000). As a caution, a complete toxicological assessment should be carried out on plant products to ensure that they are safer. Many of the plant products are unique in their action and can be easily processed and used. The main advantage of botanical insecticides is that the crude formulation can be easily produced by farmers and small-scale industries and are potentially less expensive (Talukedar and Howse 1995). There are about 189 plant families and over 2400 plants with several active ingredients/extracts that are known to have insecticidal properties. Mixing of local plants with grains has been in use with traditional farmers. The chemical composition of a plant product and its efficiency depends on several factors: plant species, plant part, season (temperature, photoperiod, hygrometry), method of harvesting, geographical location, pedological conditions, and the method for isolating the plant products. Therefore, extracts of the same species from different geographical locations and from various plant parts can be different in chemical composition (Belmain and Stevenson 2001). Boeke et al. (2001) reviewed the possibility of applying various plant products in the form of powder, ash, volatile oil, non-volatile oil and other extracts against beetles. Plant powder can affect all the stages of the developing beetle and in most cases do not have direct effect on the stored products. The ash hinders adult movement thus hampering oviposition. Isman (2006) provides an excellent overview of the current status of botanical insecticides.

11.7. Application of Oils

11.7.1. Edible and Non-Edible Oils

Practice of adding a small quantity of plant oil (1 to 15 ml kg^{-1} of grain) or mineral oil to grain is a common practice for insect control in storage in India and Africa (Periera 1983). This method is convenient and inexpensive for the protection of stored seeds in households and on small farms. Vegetable oils obtained from groundnut, mustard, rapeseed, soybean, cotton seed, neem, palm, sesame, safflower, castor, karanj, or coconut are reported to protect pulses from damages caused by beetle attack. Several authors have reported the use of plant oils against *Callosobruchus* adults and oviposition by females. Crude oil gave better results in comparison to purified oils (Shaaya and Ikan 1979). Schoonhoven (1978) found that crude oils provide significantly better protection to stored beans from bruchid attack than the purified oils and they increased adult mortality and reduced oviposition, egg hatching, and the number of adult progeny. Studies of the effect of mineral oils showed that the viscosity of oil reduces the insect activities (Calderon 1979). Biological activities of the edible oils are attributable to both their physical and chemical properties such as viscosity, volatility, specific gravity and hydrophobicity. Fatty acid chain length may also attribute to the biological activity (Shaaya et al. 1976).

The oil smear on the grain surface results in preventing egg laying and reducing oviposition and adult mortality. It also causes reduced egg hatching, interference with larval development and reduced adult progeny. The oil coating is believed to block the oxygen supply to the embryo and interfere with the normal respiration resulting in suffocation and finally causes insect death. In case of eggs, oil caused coagulation of protoplasm by penetrating through micropyle. Oil treatment does not generally have adverse effects on the seed viability, seed palatability, cooking quality and physical appearance. Although different oils showed variable efficacy, neem oil showed very good results up to 3 months with an application rate of 8 ml kg^{-1} while the rate was 5 ml kg for groundnut seeds for protection up to 6 months (Periera 1983). Singh et al. (1978) found that the action of the oil within the eggs may be due to both chemical toxicity and the physical properties of the oil. Groundnut oil enters into the eggs of *C. maculatus* through the micropyle and stops the protoplasmic movement and thereby causes the coagulation of the protoplasm. Thus partially or fully formed larvae die within minutes of the entry of oil. In addition to reduced oviposition, higher adult and larval mortality due to vegetable oils have also been reported (Van Huis 1991). He also reported that vegetable oils did not affect the viability, palatability, cooking quality and physical appearance of seeds.

11.7.2. Essential Oils

Aromatic spices and herbs contain volatile compounds, which are known to possess insecticidal activity (Miresmailli et al. 2006). These allelochemical compounds are mainly essential oils. Plant essential oils and their major components, various monoterpenoids, show contact, antifeedant, repellent and insect growth regulatory activity. Contact as well as fumigants insecticidal activity of plant essential oils are reported against storage pests e.g., *Acorus* oil, clove oil, turmeric oil, sweet flag oil, jatropha oil, *Ocimum* oil and *Eucalyptus* oil. Most of the essential oil constituents are monoterpenoids, which are secondary plant chemicals considered to be of importance for perfume and food industry. Apart from this essential oils showed insecticidal properties for *Callosobruchus* spp. (Deshpande and Tipnis 1977). *Acorus* oil caused reduction in fecundity and regression in the terminal follicle of the vitellarium in treated females of *C. chinensis* (Saxena et al. 1976). Repellency of essential oil was also reported in *Acorus calamus*, *Adhatoda vasica* and *C. chinensis* (Kokate et al. 1985). The mode of action of essential oils as insecticides is probably complex with combined physical and chemical action. Essential oils may block respiration or the water balance of eggs and developing embryos (Messina and Renwick 1983, Tikku et al. 1981). Volatile oils are mainly effective against adult beetles either as repellents or as toxicants and do not need necessarily to be in direct contact with stored products. But the disadvantage is that these oils evaporate quickly. Fumigant activity of essential oil vapours against various storage insects suggest that such substances often behave as conventional fumigants exhibiting strong adverse effects on both immature and adult stages (Stamopoulos 1991, Papachristos and Stamopoulos 2002). These natural substances have a relatively low mammalian toxicity (Isman 2000) and degrade rapidly in the environment.

11.8. Insect Growth Regulators

Insect growth regulators (IGR) are compounds which interfere with insect metabolism in a manner that affects growth. They restrict development of insects by producing supernumerary larvae, abnormal larval or pupal intermediates, abnormal pupae and adults. These compounds have low mammalian toxicity thus can be used in storage. They are lethal to eggs, larvae and pupae of stored product beetles

(Mian and Mulla 1982). They can inhibit the formation of chitin required to make a new cuticle at each moult. They can also replace or disrupt the production of juvenile hormone, which controls insect metamorphosis and development. Generally, insect growth regulators do not kill adult insects but prevent juvenile stages from completing their development. The percentage control of adult emergence reflects the cumulative mortality sustained by all developmental stage, i.e. eggs, larvae, pupae and adults. Different workers have evaluated Methoprene, Fenoxycarb, Diflubenzuron against storage insects (Mian et al. 1990). Although no direct report is available to protect lentil from bruchid infestation through insect growth regulators but Daglish et al. (1993) reported to have good efficacy of juvenile hormone agonists against *C.maculatus and C. phaseoli* in mung bean.

11.9. Biological Control

A large number of natural enemies of bruchids have been identified with the aim to utilize them to check infestation in stored pulses. Parasites and predators have been studied in relation to different species of bruchids (Van Huis 1991). Several groups of parasitoids from Braconidae, Trichogrammatidae, Eulophidae, Eupeloidae, Tonymidae, Peteromalidae, Eurytomitae and mites depredated on bruchids in field and storage system.

11.10. Pheromones

Pheromones and other semiochemicals (behavioral chemicals) play important roles in the lives of stored product insects and hold great potentials as tools for pest management. In storage, pheromones may be used either for monitoring or suppression of insect pests. Suppression can be achieved through mass trapping by luring insects to source with chemosterilants or pathogen or disrupting mating in storage by permeating them with sex pheromone. Pheromone identification and laboratory studies for bruchid seed beetles have been reported by Phillip (1994). Bioassay-directed isolation and fractionation recently resulted in the identification of five female-produced sex pheromone from *Callosobruchus maculatus* (Phillip et al. 1996).

11.11. Hermetic Control

It is an ancient, simple, cheap and effective method of insect management. The natural respiration of insects and grains results in depletion of oxygen level and increases in carbon dioxide so that insects' development is arrested. But the storage facility must be completely air tight..Crude structures, e.g., Pusa bin and Pusa kothar based on airtight principles serve the purpose of insect control. However, hermetic storage takes a longer time to kill all insects.

11.12. Fumigation

As most of the insects in storage are internal feeders, fumigation is the most popular method to control them but most of the storage insects have developed resistance against phosphine, the most widely used fumigant (Taylor 1989, Collins et al. 1998, Rajendran 2000, Nayak et al. 2002). Several earlier fumigants have been withdrawn due to operator hazards, carcinogenicity or lapse of registration and only phosphine and methyl bromide are used. Due to the combined advantages of low cost, ease of use and acceptance as a residue free treatment, phosphine is widely used in most countries. There are several reports that most of the storage insects have developed resistance to phosphine particularly in Indian subcontinent (Tyler et al. 1983, Bhatia 1986). According to the decision of the Montreal Protocol, methyl bromide, a proven ozone depleterin the atmosphere, was recommended to be phased out by 2005 in advanced countries and by 2015 in developing countries (TEAP 2000) Thus there is a necessity for other alternative methods to control storage insects.

11.13. Modified Atmosphere (MA)

Disinfestation of stored grain using modified atmosphere involves the alteration of inter-granular storage gases (O_2, N_2, CO_2). Modified atmosphere may be achieved in several ways such as adding CO_2 or N_2 or reducing O_2 and leaving no air (vacuum). Carbon dioxide is more effective method than use of N_2 because CO_2 stimulates insect respiration by displacing O_2 (Bell et al. 1980, Jayas and Jeyamkondan 2002). Nitrogen kills insects only by displacing oxygen to level of less than 1% in the air. Modified atmosphere is lethal to insects and can be obtained using nitrogen provided a hypoxic atmosphere of 1% or less is maintained whereas, when flushed with CO_2, insects are under stress due to both hyper-carbia and hypoxia. There are reports about the toxic effects of CO_2 rich atmosphere on several insect pests of stored products (Annis 1987). The toxicity of CO_2 to insects is known to vary among species, developmental stages and age groups. Parameters of the physical environment such as temperature, humidity and CO_2 levels in storage also influence the toxicity. Apart from the lethal effects, CO_2 at sub-lethal levels may affect reproduction and developmental processes in insects and the extent of adverse effects are dependent on the species, life stages exposed environmental factors and type of treatment (Bailey and Banks 1980, Nicolas and Sillians 1989).

There are several reports available about the sub-lethal effects of controlled atmospheres influencing the overall fitness of the exposed insect population. A delay in developmental processes corresponding to exposure time in *Callosobruchus chinensis* (L.) under high CO_2 concentrations of 40–53% has been observed (Oosthuizen and Schmidt 1942). Increase in exposure time and increasing CO_2concentration to 100% caused a progressive and significant decrease in fecundity during the first 40 h in *Callosobruchus maculatus* (Dawson 1995). The MAs may provide an alternative to phosphine but not to methyl bromide due to comparatively longer exposure time requirement to kill storage insects. However, quick action can be achieved with the application of CO_2 under reduced pressure or hypoxia. In an endeavor to find an alternative to the conventional CO_2 application, CO_2 under high pressure followed by quick decompression was introduced (Gerald et al. 1988). The greatest advantage of this method as a pest control measure is its short lethal exposure time. Several workers reported that the exposure time of CO_2 at high pressure on different insect pests of cereal grains can be reduced to less than an hour, regardless of the species or developmental stages (Reichmuth 1991, Nakakita and Kawashima 1994, Locatelli et al. 1999).

Stahl et al. (1985) attributed the lethal action to increased solubility of CO_2 in the insect body fluids under high pressure, causing a subsequent decrease in pH. An increase in the uptake of CO_2 under high pressure causes rapid expansion and rapid evaporation from the body liquids when the pressure is reduced, resulting in lesions in the cell membranes of the adults and larva (Reichmuth and Wohlgenmuth 1994). The integument of the insects exposed to the treatment was severely damaged due to the expansion of the internally dissolved CO_2 in the body when the gas pressure was rapidly reduced to atmospheric pressure (Nakakita and Kawashima 1994). The efficacy of CO_2 increases with temperature and decreases with increase in humidity.

REFERENCES

AAFC (2006) Lentils: Situation and outlook. Bi-weekly Bulletin. Agriculture and Agri-Food Canada

Adler C, Ojimelukwe P, Tapondjou AL (2000) Utilisation of phytochemicals against stored product insets. Integrated Protection of Stored Product, IOBC Bulletin 23, pp 169–175

Alagusundaram K, Jayas DS, Chotard F, White NDG (1992) Airflow pressure drop relationships of some speciality seeds. Sciences des Aliments 12: 101–116

Alagusundaram K, Jayas DS, Muir WE, White NDG (1991) Thermal conductivity of bulk barley, lentils, and peas. Trans ASAE 34: 1784–1788

Amin MN, Hossain MA, Roy KC (2004) Effects of moisture content on some physical properties of lentil seeds. J Food Engng 65: 83–87

Annis PC (1987) Towards rational controlled atmosphere dosage schedules: A review of current knowledge. In: Donahaye, E and Navarro, S (eds.) Proceedings of Fourth International Working Conference on Stored-Product Protection. Tel Aviv, Israel, pp 128–148

Anonymous (1979) Grain legume: Processing and storage problems. Food and Nutrition Bulletin I(2). United Nations University, Tokyo, Japan

Anonymous (1992) Lentil storage. Research Update 678. Prairie Agricultural Machinery Institute, Humboldt, Saskatchewan

ASAE (1995) ASAE Standard D245.5. Moisture relationships of plant-based agricultural products. American Society of Agricultural and Biological Engineers. St. Joseph, Michigan

Bailey SW, Banks HJ (1980) A review of recent studies of the effects of controlled atmospheres on stored product pests. In: Shejbal J (ed.) Proceedings of an International Symposium on Controlled Atmosphere Storage of Grains. Vol. 1. Elsevier Scientific Publishing Company, Amsterdam, pp 101–118

Barrozo MAS, Oliveira DT, Sancineti GP, Rodrigues MV (2000) A study of the desorption isotherms of lentils. Brazilian J Chem Engng 17: 105–109

Beniwal SPS, Baya'a B, Weirard S, Makkouk KH, Saxena MC (1993) Field guide to lentil diseases and insect pests. International Center for Agricultural Research in the Dry Areas Aleppo, Syrian Arab Republic

Bell CH, Spratt EC, Mitchell DJ (1980) The effect of nitrogen and carbon dioxide on eggs of *Ephestia cautella* and *E. kuhniella* Zeller (Lepidoptera: Pyralidae). Bulletin of Entomological Res 70: 293–298

Belmain SR, Stevenson PC (2001) Ethnobotanicals in Ghana: reviving and modernizing an age-old practice. Pesticide Outlook 6: 233–238

Bergen GA, Jayas DS, White NDG (1993) Physical damage to peas and lentils due to free fall. Canadian Agric Engng 35: 151–155

Bhalla S, Gupta K, Kapur ML, Lal B, Khetrapal RK (2006) Phytosanitary risk of bruchids in lentil imported into India. EPPO Bulletin 36: 25–29

Bhatia S K (1986) Pesticide resistance in Agricultural Pests in India. Proc. Indian Natn Sci Acad B32: 148–164

Bhatty RS (1988) Composition and quality of lentil (*Lens culinaris* Medik): a review. Canadian Inst Food Sci Tech J 21: 144–160

Boeke SJ, van Loon JJA, van Huis A, Kossou KD, Dicke M (2001) The use of plant material to protect stored leguminous seeds against seed beetles: A review. Backhiys Publishers, The Netherlands, pp 108

Bundus CL, Jayas DS, Muir WE, White NDG, Ruth D (1996) Average convective-pore velocity of carbon dioxide gas through grain bulks. Canadian Agric Engng 38: 91–98

Calderon M (1979) Mixing chickpeas with paraffin oil to prevent *Callosobruchus maculatus* Infestation. Progress Report for the year 1978/79. Stored Products Division, Special Publication No. 140, Ministry of Agriculture, Israel

Carmo JEF-do, Lima AGB-de (2005) Drying of lentil including shrinkage: a numerical simulation. Drying Tech 23: 1977–1992

Caswell GH (1981) Damage to stored cowpeas in the northern part of Nigeria. Samaru J Agric Res 1: 11–19

Cenkowski S, Sosulski FW (1997) Physical and cooking properties of micronized lentils. J Food Process Engng 20: 249–264

Cenkowski S, Sokhansanj S, Sosulski FW (1989) Equilibrium moisture content of lentils. Canadian Agric Engng 31: 159–162

CGC (2005) Quality of western Canadian pulse crops 2005. Canadian Grain Commission, Winnipeg, Manitoba

Champ BR, Dyte CE (1976) Report of the FAO global survey of pesticide susceptibility of stored grain pests. FAO Plant Production Protection Series, No 5, Rome, pp 297

Collins PJ (1998) Resistance to grain protectants and fumigants in insect pests of stored products in Australia. In: Banks HJ, Wright EJ, Damceski KA (eds) Stored grain in Australia. CSIRO Stored Grain Research Laboratory, Canberra, pp 55–57

Daglish GJ, Erbacher JM, Eelkema M (1993) Efficacy of protectants against *Callosobruchus phaseoli* in mungbeans. J Stored Prod Res 29: 345–349

Daglish GJ, Wallbank BE (2003) An update on development of bifenthrin as a grain protectant. In: Wright EJ, Webb MC, Highley E (eds.) Proceedings of the Australian postharvest Technical Conference. CSIRO Stored grain Research Laboratory, Canberra, Australia

Dawson C (1995) The carbon dioxide induced anesthesia on fecundity of *Callosobruchus maculatus* (Coleoptera: Bruchidae). J Stored Prod Res 31: 49–54

Deshpandey RS, Tipnis HP (1977) Insecticidal activity of *Ocimum basilicum* Linn. Pesticides 11: 11–12

Erskine W, Williams PC, Nakkoul H (1991) Splitting and dehulling lentil (*Lens culinaris*): effects of seed size and different pretreatments. J Sci Food Agric 57: 77–84

Fishwick FB (1988) Pesticide residues in grain arising from post harvest treatment. Aspects Appl Biol 17: 37–46

Gerald D, Kraus J, Quirin KW (1988) Residue-free insect control using natural carbon dioxide under high pressure. Gordian 88: 90–94

Golobe P, Webley DJ (1980) The use of plants and minerals as traditional protectants of stored products. Tropical Products Institute, G138: 1–32

González Z, Pérez E (2002) Evaluation of lentil starches modified by microwave irradiation and extrusion cooking. Food Res Intl 35: 415–420

Iliadis C (2001) Effects of harvesting procedure, storage time and climatic conditions on cooking time of lentils (*Lens culinaris* Medikus). J Sci Food Agric 81: 590–593

Isman MB (2000) Plant essential oils for pest and disease management. Crop Protection 19: 603–608

Isman MB (2006) Botanical insecticides, deterrents, and repellents in modern agriculture and an increasingly regulated world. Annual Review of Entomology 51: 45–66

Irvine DA, Jayas DS, White NDG, Britton MG (1992a) Physical properties of flaxseed, lentils, and fababeans. Canadian Agric Engng 34: 75–81

Irvine DA, Jayas DS, Britton MG, White NDG (1992b) Dynamic friction characteristics of bulk seeds against flat vertical surfaces. Trans. ASAE 35: 665–669

Jayas DS, Jeyamkondan S (2002) Modified atmosphere storage of grains, meats, fruits, and vegetables. Biosystems Engng 82: 235–251

Jilani G, Khan MI, Ghiasuddin AL (1988) Studies on insecticidal activity of some indigenous plant materials against the pulse weevil *Callosobruchus analis* (F.) *Pakistan.* J Entomol 3: 21–32

Jayas DS, Mann DD (1994) Presentation of airflow resistance data of seed bulks. Appl Engng Agric 10: 79–83

Jayas DS, Murray CE, Bulley NR (1999) An automated seed presentation device for use in machine vision identification of grain. Canadian Agric Engng 41: 113–118

Karatas S (1997) Determination of moisture diffusivity of lentil seed during drying. Drying Tech 15: 183–199

Khanna SC, Singh S (2002) Stored legume by ecofriendly methods. In: Prasad D, Puri SN (eds.) Crop Pest and Disease Mangement: Challenges for the millennium. Jyoti Publishers, New Delhi, India,,pp 134–137

Kokate CK, D'Cruz JL, Kumar RA, Apte SS (1985) Anti-insect and juvenoidal activity of phytochemicals derived from Adathoda vasica Nees. Indian J Nat Prod 1: 7–9

Li W, Sokhansanj S (1994) Generalized equation for airflow resistance of bulk grains with variable density, moisture content and fines. Drying Tech 12: 649–667

Locatelli DP, Suss L, Frati M (1999) The use of high pressure carbon dioxide (20 bar) to control some insects of food stuffs. In: Jin Z, Liang Q, Liang Y, Tan X, Guan L (eds.) Proceedings of seventh International Working conference on Stored-Product Protection. Sichuan Publishing House of Science and Technology, Beijing, China, pp 671–675

Lorini I, Galley DJ (1999) Deltamethrin resistance in *Rhyzopertha dominica*, a pest of stored grain in Brazil. J Stored Prod Res 35: 37–45

Menkov ND (2000) Moisture sorption isotherms of lentil seeds at several temperatures. J Food Engng 44: 205–211

Messina FJ, Renwick JAA (1983) Effectiveness of oils in protecting stored cowpea from the cowpea weevil (Coleoptera: Bruchidae). J Econ Entomol 76: 634–636

Mian LS, Mulla MS (1982) Residual activity of insect growth regulators against stored-product beetles in grain commodities. J Econ Entomol 75: 599–603

Mian LS, Mulla MS, Hussain N (1990) Insect growth regulators as control agents against stored product insects. Sarhad J Agric 6: 287–298

Mills JT, Woods SM, Watts BM, Lamari L, White NDG (1999) Comparison of techniques to measure seed color and its relationship to other quality parameters in stored lentil (*Lens culinaris* Medik.). Seed Sci Tech 27: 1015–1028

Miresmailli S, Bradbury R, Isman MB (2006) Comparative toxicity of *Rosmarinus officinalis* L. essential oil and blends of its major constituents against *Tetranychus urticae* Koch (Acari: Tetranychidae) on two different host plants. Pest Management Sci 62: 366–371

Nakakita H, Kawashima K (1994) A new method to control store product insects using carbon dioxide with high pressure followed by sudden pressure loss. In: Highley E, Wright HJ, Champ BR (eds) Proceedings of sixth International Working conference on Stored-Product Protection. CAB International, Canberra, Australia, pp 126–129

Nayak MK, Collins PJ, Pavic H (2002) Phosphine resistance in psocids:challenges ahead! In: Wright EJ, Banks HJ, Highley E (eds) Stored grain in Australia 2000. CSIRO Stored Grain Research Laboratory, Canberra, pp 113–118

Nicolas G, Sillians D (1989) Immediate and latent effects of carbon dioxide on insects. Annual Rev Entomol 34: 97–116

Oosthuizen MJ, Schmidt UW (1942) The toxicity of carbon dioxide to the cowpea weevil. J Entomol Soc S Africa 5: 99–110

Paddock FB, Reinhard HJ (1919) The cowpea weevil. Bull Texas Agric Exp Station. No. 656

Papachristos DP, Stamopoulos DC (2002) Repllent, toxic and reproduction inhibitory effects of essential oil vapours on *Acanthoscelides obectus* (Say) (Coleoptera: Bruchidae). J Stored Grain Prod 38: 117–128

Peace RW, Keith MO, Sarwar G, Botting HG (1988) Effects of storage on protein nutritional quality of grain legumes. J Food Sci 53: 439–441

Periera J (1983) Effectiveness of six vegetable oils as protectants of cowpeas and bambara groundnuts against infestation by *Callosobruchus maculatus* (F.) (Coleoptera: Bruchidae). J Stored Prod Res 19: 57–62

Phillip TW (1994) Pheromones of stored-product insects: current status and future prospectives. In: Highley E, Wright EJ, Banks HJ, Champ BR (eds.) Proceedings of the 6th International Working Conference on stored-product protection, Canberra, Australia, pp 479–486

Phillip TW, Phillip JK, Webster FX, Tang RT, Burkholder WE (1996) Identification of sex pheromones from cowpea weevil, *Callosobruchus maculatus* and related studies with *C.analis* (Coleoptera: Bruchidae). J Chemical Ecology 22: 2233–2249

Pirman T, Stibilj V (2003) An influence of cooking on fatty acid composition in three varieties of common beans and in lentil. European Food Res Tech 217: 498–503

Pradhan S, Agarwal NS, Thomas PM (1971) Policy regarding mixing of pesticides in food and feed grains. Entomologists Newsletter 1: 25–28

Rajendran S (2000) Phosphine resistance in stored grain insect pests in India. In: Jin Z, Liang Q, Liang Y, Tan X, Guan L (eds) Proceedings of seventh International Working conference on Stored-Product Protection. Sichuan Publishing House of Science and Technology, Beijing, China, pp 635–641

Reichmuth C, Wohlgenmuth R (1994) Carbon dioxide under high pressure of 15 bar and 20 bar to control the eggs of the Indian meal moth *Plodia punctella* (Hubner) (Lepidoptera: Pyralidae) as the most tolerant stage at 25 °C. In: Highley E, Wright HJ, Champ BR (eds) Proceedings of Sixth International Working Conference on Stored-Product Protection. CAB International, Canberra, Australia, pp 126–129

Reichmuth C (1991) New techniques in fumigation research today. In: Fleurat-lessard F, Ducom P (eds) Proceedings of fifth International Working conference on Stored-Product Protection. Bordeaux, France, pp 701–725

Saxena BP, Koul O, Tikku K (1976) Non-toxic protectant against the stored grain insect pests. Bull Grain Tech 14: 190–193

Scanlon MG, Cenkowski S, Segall KI, Arntfield SD (2005) The physical properties of micronised lentils as a function of tempering moisture. Biosys Engng 92: 247–254

Schoonhoven AV (1978) Use of vegetable oils to protect stored beans from bruchid attack. J Econ Entomol 71: 254–256

Shaaya E, Ikan R (1979) Insect control using natural product. In: Gessbuhler H (ed) Advances in Pesticide Science, Part 2. Pergamon Press, Oxford, pp 303–306

Shaaya E, Grossman G, Ikan R (1976) The effect of straight chain fatty acids on growth of *Sitophilus oryzae*. J Entomol 11: 81–91

Shahin MA, Symons SJ (2001) A machine vision system for grading lentils. Canadian Biosys Engng 43: 7.07–7.14

Shahin MA, Symons SJ (2003a) Lentil type identification using machine vision. Canadian Biosys Engng 45: 3.05–3.11

Shahin MA, Symons SJ (2003b) Color calibration of scanners for scanner-independent grain grading. Cereal Chem 80: 285–289

Shatadal P, Jayas DS, Hehn JL, Bulley NR (1995) Seed classification using machine vision. Canadian Agric Engng 37: 163–167

Singh SR, Luse RA, Leuschner K, Nangju D (1978) Groundnut oil treatment for the control of *Callosobruchus maculatus* (F.) during cowpea storage. J Stored Prod Res 14: 77–80

Sokhansanj S, Falacinski AA, Sosulski FW, Jayas DS, Tang J (1990) Resistance of bulk lentils to airflow. Trans ASAE 33: 1281–1285

Stahl E, Rau G, Adophi H (1985) Entwesung von Drogen durch kohlendioxide Ddrukbehand lum (PEX_verfahren). Pharmazeutisch Industrie 47: 528–530

Stamopoulos DC (1991) Effects of four essential oil vapours on the oviposition and fecundity of *Acanthoscelides obectus* (Say) (Coleoptera: Bruchidae): Laboratory evaluation. J Stored Prod Res 27: 199–203

Subramaniam B, Roseli R (2000) Inert dusts. In: Subramaniam B, Hanngstrum DW (eds) Alternatives to Pesticides in Stored-product IPM. Kluwer Academic Publishers, Boston, Massachusetts, pp 321–397

Talukedar FA, Howse PE (1995) Evaluation of *Aphanomixis polystachya* as repellent, antifeedant, toxicant and protectant in storage against *Tribolium castaneum* (Herbst.). J Stored Prod Res 31: 55–61

Tang J, Sokhansanj S, Slinkard AE, Sosulski FW (1990a) Quality of artificially dried lentil. J Food Process Engng 13: 229–238

Tang J, Sokhansanj S, Sosulski FW, Slinkard AE (1990b) Effect of swathing and moisture content on seed properties of laird lentil. Canadian J Plant Sci 70: 1173–1178

Tang J, Sokhansanj S (1991) Determination of moisture content of whole kernel lentil by oven method. Trans ASAE 34: 255–256

Tang J, Sokhansanj S, Yannacopoulos S, Kasap SO (1991a) Specific heat capacity of lentil seeds by differential scanning calorimetry. Trans ASAE 34: 517–522

Tang J, Sokhansanj S, Sosulski FW, Slinkard AE (1991b) Lentils quality-effects of artificial drying and six-month storage. Canadian Inst Food Sci Tech 24: 8283–8286

Tang J, Sokhansanj S, Sosulski FW, Slinkard AE (1992) Effect of harvest method on moisture content and quality of lentil seeds. Canadian J Plant Sci 72: 451–456

Tang J, Sokhansanj S (1993a) Moisture diffusivity in Laird lentil seed components. Trans ASAE 36: 1791–1798

Tang J, Sokhansanj S (1993b) Geometric changes in lentil seeds caused by drying. J Agric Engng Res 56: 313–326

Tang J, Sokhansanj S (1993c) Drying parameter effects on lentil seed viability. Trans ASAE 36: 855–861

Tang J, Sokhansanj S (1994) A model for thin-layer drying of lentils. Drying Tech 12: 849–867

Tang J, Sokhansanj S, Sosulski FW (1994) Moisture-absorption characteristics of laird lentils and hardshell seeds. Cereal Chem 71: 423–428

Taylor RWD (1989) Phosphine-a major fumigant at risk. Intl Pest Control 31: 10–14

Tyler PS, Taylor RW, Reeds DP (1983) Insect resistance to phosphine fumigation in food warehouses in Bangladesh. International Pest Control 25: 10–13

TEAP (2000) Montreal protocol on substances that deplete the ozone layer: UNEP Technology and economic Assessment Panel. April 2000 Rep. United Nations

Tikku K, Koul O, Saxena BP (1981) Possible mode of action of vegetable oils to protect *Phaseolus aureus* Roxb. from bruchid attack. Science and Culture 41: 103–105

Tyler PS, Binns TJ (1977) The toxicity of seven organophosphorous insecticides and lindane to eighteen species of stored product beetles. J Stored Prod. Res 8:13–19

Van Huis A (1991) Biological methods of bruchid control in the tropics: A review. Insect Sci Appl 12: 87–102

Varma BK, Siddiqui MKH (1977) Control of storage pests through inert dusts. Indian Farming 5, 21, 25

Vayias BJ, Athanassiou CG (2004) Factors affecting the insecticidal efficacy of the diatomaceous earth formulation SilicoSec against adults and larvae of the confused flour beetle, *Tribolium confusum* DuVal (Coleoptera: Tenebrionidae). Crop Protection 23:565–673

Wright CG, Leidy RB, Dupree J (1993) Cypermethrin in the ambient air and on the surface of rooms treated for cockroaches. Bull Env Cont Toxic 51: 356–360

Yadav TD (1980) Efficacy of insecticidal dusts against developing stages of *Callosobruchus maculatus* and *Callosobruchus chinensis* (Linn.) Indian J Ent 42: 798–802

Yadav TD, Singh S (1994) Persistence toxicity and efficacy of four insecticides as jute fabric treatment to protect cereal and legume seeds. Indian J Ent 56:146–155

Yadav TD, Pawar CS, Khanna SC, Singh S (1980) Toxicity of organophosphorous insecticides against stored product beetles. Indian J Ent 42: 28–33

Zettler JL, Arthur FH (2000) Chemical control of stored product insects with fumigants and residual treatments. Crop Protection 19: 577–582

CHAPTER 23

LENTIL GROWERS AND PRODUCTION SYSTEMS AROUND THE WORLD

SHYAM S. YADAV[1], A. H. RIZVI[1], M. MANOHAR[1], A. K. VERMA[1], R. SHRESTHA[2], CHENGCI. CHEN[3], G. BEJIGA[4], W. CHEN[5] M. YADAV[6], AND P. N. BAHL[7]

[1]*Pulse Laboratory, Division of Genetics, Indian Agricultural Research Institute, New Delhi 110012, India*
[2]*Division of Agronomy, Nepal Agricultural council, GPO box 404, Kathmandu, Nepal*
[3]*Department of Agronomy, Montana State University, Bozeman, Montana, USA*
[4]*Green Focus Ethiopia, PO Box 802, Code No. 1110, Addis Ababa, Ethiopia*
[5]*USDA-ARS, Grain Legume Genetics and Physiology Research Unit, 303 Johnson Hall, Washington State University, Pullman, WA 99164-6402, USA*
[6] *Weatherford College, Weatherford, TX 76086, USA*
[7]*A– 9, Nirwan Vihar, New Delhi 110092, India*
Email: shyamsinghyadav@yahoo.com

Abstract: Taking into consideration different ecologies, regions, countries and continents lentils are adapted throughout world. Its cultivation has been taken up by large, medium and small farmers mainly under rainfed but also in irrigated ecosystems. The lentil growers in different countries face the challenges of biotic and abiotic stresses more or less of the same magnitude which are responsible for the low productivity and stagnation in the production. Marketing and trade arrangements and distortions can also produce enormous impacts in some regions. Examples of the main production systems from around the world are given. Within these productivity varies greatly from country to country and there are wide gaps between developed and under developed nations. Such gaps in the productivity can be minimize with the introduction of modern techniques. National and international research organizations are working on various aspects of lentil improvement and these programs have come out with excellent technologies for lentil growers. Their applicability and adoption has varied around the world. The availability of high yielding, resistant, quality and widely adapted cultivars with an appropriate agronomic package is not generally a problem in any part of the world. The chain of quality seed production is also being maintained and improving day by day. The various organizations are involved in technology transfer to farmers

S.S. Yadav et al. (eds.), Lentil: An Ancient Crop for Modern Times, 415–442.

1. INTRODUCTION

Lentil, the seed of *Lens culinaris* Medik *(previously Lens esculenta* Moench. *Ervum lens*), is a small annual of the vetch tribe. It was probably one of the first plants brought under cultivation by mankind and consequently the country of origin of the lentil is not known. Lentils have been found in the Bronze Age lake dwellings are referred to as red pottage in the Old Testament (www.1911encyclopedia.org, 2006). It is now grown in India, Persia, Syria, Egypt, Nubia and North Africa, and in Europe, along the coast of the Mediterranean, and as far north as Germany, Holland and France. Production systems are highly variable depending on local factors. Lentils are used in a diverse range of traditional foods such as parched seed for long journeys in Egypt, Syria and other Eastern countries and as a chief ingredient in the Spanish puchero. For this purpose they are usually sold in the shelled state Figure 1 (www.1911encyclopedia.org, 2006).

Lentils are propagated by seed and grown for human consumption. They are frequently consumed in parts of Europe and a staple food throughout much of the Middle East and India. Lentils are dried in the field and rewetted rather than consumed fresh. Lentil is often used as a meat substitute and has a range of types including green lentils (used with the seed coat on), red lentils (generally smaller and decorticated) and a range of specialty types (brown, black speckled etc). Diversity of type and use thus also contributes to a diversity of production and marketing systems worldwide. In this chapter the major factors influencing that diversity will be discussed along with major illustrative case studies of different regions.

2. GLOBAL SITUATION

2.1. Production

Lentils are grown in the cooler temperate zones of the world or in the winter season in countries, such as India and Australia, which have a warm winter and a hot summer. World lentil production has ranged from 2.5 million tonnes (Mt) in 1989–1990 to reach a peak of 4.17 Mt in 2004–2005 (FAOSTAT, 2007). Agriculture and Agrifood Canada estimates 75% of world lentil production as the red type, 20%

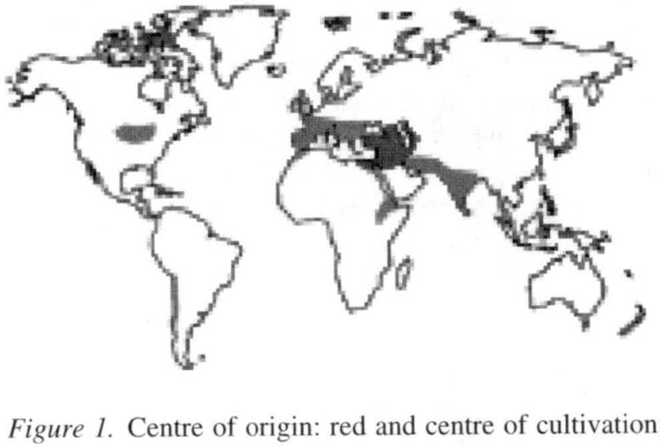

Figure 1. Centre of origin: red and centre of cultivation: green of lentil

green type and 5% brown and other types. Canada and the US produce mainly the green type whereas the rest of the world produces mainly the red type (AAFC, 2006).

A variety of lentils exist with colours that range from yellow to red-orange to green, brown and black which may be solid or speckled in colour. The colours of the seeds when removed from the pods also vary, and there are large and small varieties. They are sold in many forms, with or without the pods, whole or split. About half of the worldwide production of lentils is from India and Canada. Most Indian lentils are consumed domestically, while Canada is the largest export producer of lentils in the world (exporting 500,000 Mt 2004/5; FAOSTAT 2007) and Saskatchewan is the most important producing region in Canada. The next largest producer is Turkey which consumes, imports and exports lentils depending on type and year. Eastern Washington (especially the Palouse Region) is the most important producing region in the United States (AAFC, 2006).

2.2. Potential Lentil Producing Countries

Around the world lentil cultivation differs drastically due to differences in consumption pattern, agro-climatic conditions, growing environments, local demands, export situations and farmers interests, resources and knowledge. Consequently production and productivity differ greatly from country to country.

The major lentil producing countries are India, Canada, Turkey, USA, Nepal, Australia, Syria, China, Bangladesh, Iran, etc. The countries like Ethiopia, Morocco, Pakistan, Spain, France, Russia, Mexico, Peru, Yemen and Argentina also contributing as a important producer in the world. In these countries, yields have ranged from 637 to 1263 kg/ha while the highest yield, 5000 kg/ha was recorded in Germany. World production of lentil increased by about 240% over the past 25 years (FAOSTAT, 2007). All these countries contributing 3.385 million tonnes (99.2%) while the total world production was approximately 3.414 million tones on an average during 2001–2005 (Table 1).

3. FACTORS INFLUENCING THE CULTIVATION PATTERN

The major factors which influencing its cultivation at global level are farmers land holding, cropping pattern, agro-ecosystem, crop losses due to biotic and abiotic stresses, international marketing, improved cultivars, quality seeds availability, integrated management, polices etc. Many of these aspects have been discussed in previous chapters. This chapter will try to integrate to cover the total picture for lentil growers in different regions looking first at the overall regional differences and then looking at the major regions in more detail.

3.1. Cropping Seasons

Agro-climatic conditions and cropping systems in which lentil is cultivated vary across growing regions. In the sub-tropics like the Indian sub-continent the crop

Table 1. Lentil production in major growing countries during 2001–05

Country	2001	2002	2003	2004	2005	Average
India	915,200	974,400	880,000	1,100,000	1,000,000	973,920
Canada	566,300	353,800	519,900	962,000	1,277,900	735,980
Turkey	520,000	565,000	540,000.	540,000	555,000	544,000
Australia	266,000	67,000	103,811	52,300	210,000	139,822
Nepal	143,084	148,384	149,963	158,671	160,716	152,164
USA	131,450	113,760	110,770	189,690	234,190	155,972
China	125,000	125,000	132,000	145,000	135,000	132,400
Syria	177,467	132,805	168,437.	125,300	153,665	151,535
Bangladesh	126,000	115,000	116,000	122,000	121,000	120,000
Iran	104,399	117,000	120,000	125,000	125,000	118,279
Ethiopia	65,250	38,430	35,275	35,275	63,357	47,517
Morocco	12,890	41,670	33,620	35,520	35,000	31,740
Pakistan	26,900	26,200	29,200	31,100	25,900	27,860
Spain	19,088	22,467	20,696	27,378	6,000	19,126
Argentina	1,800	3,000.	2,000	2,000	2,000	2,160
Russia	4,900	8,110	11,970	7,400	5,530	7,582
Mexico	2,873	8,518	4,008	6,632.	7,711	5,948
France	6,279	8,018	9,026.	8,054	15,542	9,384
Peru	5,559	4,933	4,590	3,538	3,011	4,326
Yemen	4,929	4,465	4,270	5,450.	5,450.	4,913
% of World Total	**99.2**	**99.0**	**99.0**	**99.2**	**99.3**	**99.1**
World	**3,252,056**	**2,907,133**	**3,025,569**	**3,711,114**	**4,172,408**	**3,413,656**

Source: FAOSTAT 2007

is planted at the end of the rainy season under relatively high temperatures. The vegetative growth is on the residual soil moisture in the winter season. Flowering occurs in the cooler temperatures. Rainfall during this period is limited/scanty or no rain. Podding begins as the temperatures start rising and the crop faces increasing drought stress at this critical stage. In some parts of India winter rains may alleviate the moisture stress. Occasional excessive precipitation may induce excessive vegetative growth and lodging. The crop is harvested in the dry season at the beginning of summer.

In the temperate climates in West Asia, North Africa, Southern Europe, USA and Canada chickpea is planted in the winter and spring under low temperatures and vegetative growth occurs in later spring and the summer. There is some rainfall during flowering and podding. The crop matures in autumn when temperatures begin to fall (Chen 2006, Montana State University, Montana, USA, personal communication). Harvest time temperatures are down also in some regions in West Asia and North Africa. Early frost may discolour and damage the maturing seed leading to lower quality produce.

In West Asia and North Africa some lentil is seeded as a winter crop before snowfall. As soon as the snow melts in the spring, the seeds emerge, plants grow, flower and pod in a much better moisture regime than the summer crop. Seed yields are often much higher under such cultivation. However, higher moisture conditions and better crop canopy favour development of fungal diseases and the crop is often damaged. Thus, variations in cropping system and crop season differ region to region and influence the crop productivity drastically. Under such seasonal variations in cropping system the lentil growers are compelled to harvest this crop with great variations in productivity.

3.2. Farmers' Holdings

Ninety percent of lentil growers in the developing world have small holdings, generally one to 10 ha. The land is often double cropped. Most operations are manual and seed is mostly consumed on farm or sold to local markets. The farmers in the developed world (USA, Canada, Australia, Spain etc.) grow lentil as a commercial crop on large farms often larger than 200 ha. The harvest is mainly exported. All the field, seed handling and cleaning operations are mechanical. Thus, the varietal requirements are different not only for machine operations but also for marketing. Red types are preferred over green types in the Indian sub-continent and in other Asian countries like Iran, Iraq, Yemen, Jordan, Syria etc. They thus attract a higher market price. Much of the crop in the Indian subcontinent is for consumption as dhal and red types are preferred. Therefore, larger farmers in India also prefer red types for marketing. The land holdings of farmers in different continents differ greatly which affect both the crop cultivation system and crop production. With these variations in land holdings the situation of lentil growers differ widely which then influences global cultivation systems.

3.3. Crop Losses

The lentil crop faces biotic and abiotic stresses during growth. The major constraints to increased productivity include: drought, *Fusarium* wilt, *Ascochyta* blight and rust. As a second tier crop research that addresses these problems is limited Rough estimates suggest that in India alone damage due to biotic and abiotic stresses causes losses that exceed 45% annually. In other countries damage due to these stresses is also high. In Australia more than 35% loss occurs annually due to drought. In the USA and Canada losses due to *Ascochyta* blight are very high ranging between 35 to 45%. Like wise losses due to *Fusarium* wilt and *Ascochyta* blight are very high ranging between 40–60% in Turkey. Stable, multiple resistant, high yielding cultivars are needed by farmers to increase lentil production. Early maturity, increased seed yield, improved harvestability, and disease resistance are required to stabilize and expand the area under this crop.

3.4. Biotic Factors

Several fungal diseases are recorded on lentil, but lentils are generally not considered to be disease prone. They are seriously affected by fewer diseases than *Phaseolus* and *Pisum*. One of the worst diseases in Asia is *Fusarium* wilt, caused by *Fusarium oxysporum* f. sp. *lentis*, which is favoured by light and dry soils. Rust, caused by *Uromyces fabae*, is favored by high humidity and moderate temperatures (17–25 °C) and is important in Morocco, Ethiopia, Pakistan and South America. Infection of susceptible cultivars could lead to total failure as observed in Morocco and Ethiopia. Other fungal diseases recorded on lentil include, *Ascochyta* blight, caused by *Ascochyta lentis*, which is an important foliar disease in many parts of the world including Canada. Resistance sources have been identified in North America and elsewhere, for *Fusarium* wilt, rust and viruses, and are being used extensively in crossing programs. Among the viruses, pea seed borne mosaic virus (PSbMV) is potentially dangerous because it is seed borne and is transmissible by aphids (Muehlbauer et al., 1995). Aphids are serious insect pests on lentil and sometimes cause total failure of lentil production as has been observed in Ethiopia. Lentils are also attacked by a bacterium, *Mycobacterium insidiosum*; by parasitic flowering plants, *Cuscuta* sp. and *Orobanche* sp.; by viruses, pea enation mosaic, bean yellow mosaic; and by nematodes (Kaiser et al., 1994; van Emden et al., 1988). Insects of economic importance in the United States Lygus bugs (*Lygus* sp.) and aphids. In Southeast Asia, the main insect pests are the gram caterpillar, *Heliothis obsoleta*; white ants. *Clotermes* sp.; the gram cutworm, *Ochropleura flammatra*; and the weevil *Callosobruchus analis* (Duke, 1981). Other important insect pests throughout the world include *Frankliniella* spp. (Bud thrips), *Bruchus ervi* and *B. lentis* (Bean seed beetle), *Callosobruchus chinensis* (Adzuki bean seed beetle), *Sitonia lineatus* (bean weevil), *Cydia nigricana* (Pea moth), *Agriotis* spp. (Cutworms), and *Heliothis armigera* (African bollworm) (van Emden et al., 1988).

3.5. Market Influencing Factors

- **Weather fluctuations:** The farming communities in Asian and African countries are generally less educated and do not have sufficient information on the weather forecasting and fluctuation in climatic conditions of cropping seasons. The unfavorable conditions during cropping seasons affect the crop badly and yield losses are very high which differ region to region and year to year. Due to this situation the G × E interactions are very high. It is essential to provide detailed weather related information to farmers so that they can adjust themselves for good harvest. However, it is worth noting that weather induced fluctuations in lentil yield in Australia are also very high in spite of high education levels and good weather information.
- **Information flow regarding the supply levels:** In under–developed countries, the international marketing information regarding demand, supply, pricing, traders position etc. are not properly assessed and also not supplied by the appropriate organizations to farmers in advance. This situation is responsible for very high price fluctuations in different countries. Thus it is important to maintain proper information flow regarding these factors. It is also imperative to watch price movements of the substitute pulses which influence both supply and price of the commodity in the market.

- **Production level in the main exporting countries:** The total production of the crop in exporting countries will greatly affect international marketing. If the environments and cropping seasons are not favorable to the crop and total production is below the expected or targeted level than export will decline and the prices in the international market will be greatly affected.

4. REGIONAL CASE STUDIES OF LENTIL GROWERS

Below are described the production and marketing systems for several countries exhibiting the diversity of production systems worldwide. These include; small holders in India (world's largest area); large mechanised growers in the USA (growers backed up by the US Farm Security and Rural Investment Act of 2002), Canada (world's largest exporter) and Australia; medium sized farms in Turkey,

4.1. Indian Growers

In India, lentils are cultivated on about 1.5 million ha (up 40% in the last 25years), producing approximately 0.99 million tonnes (~ 0.7 t/ha). Lentils are mostly grown in dryland areas but some short duration lentil varieties are also grown in flood plains as an inter-crop and under intensive farming systems. The lentil crop suffers from both biotic and abiotic stresses during the cropping season which affects its production adversely. Due to the impact of these stresses the productivity of lentil has stagnated over the last 20 years and approximately 670 kg/ha is being harvested at the national level. In India lentil is cultivated in all the states, however, the important states are U.P., M.P., Bihar, West Bengal, Rajasthan, Assam, Haryana, Maharashtra, Punjab, etc. Together these states have about 1.468 million hectare area under lentil cultivation (99% of Indian total). More or less the same pattern is also true for production of lentil (Table 2).

Indian lentil growers are generally small farmers, with 1–5 ha of land holdings. The majority of growers in central India grow lentils after the rainy season on conserved rainwater as a sole crop. However, Indian farmers particularly in the northern and central India grow lentil in different cropping system. Being a winter it fits well in a rice base cropping system. The grower plants lentil after rice, maize, millet, sorghum etc. harvestings it as a sole crop under both rainfed and irrigated agro-eco systems. Some growers fit lentils into an inter cropping system with sugarcane, chickpea, mustard etc.

Most Indian lentils are whole or split red or yellow types from Turkey, Canada or domestic production. In general, lentils sold at a lower price than pigeonpea. Red lentils are harvested in March/April, so September–October is seen to be a good time for exports of Canadian red lentils. The Muslim market (Gulf area and eastern India) is the largest consumer of red lentils in the South Asia region. Split lentils could hold a large potential in India and other countries as a replacement for pigeonpea *dhal*, (split pigeonpea), particularly in years when summer pulse production is poor. Seed yields range from 450–675 kg/ha in dry areas but may increase to 2000 kg/ha

Table 2. Lentil area, production and productivity in important Indian states during 1995–2005

STATES		95–96	96–97	97–98	98–99	99–2000	2000–01	2001–02	2002–03	2003–04	2004–05
ASSAM	**A**	0.162	0.193	0.193	0.253	0.214	0.203	0.213	0.200	0.220	0.205
	P	0.087	0.102	0.102	0.133	0.111	0.107	0.112	0.110	0.120	0.115
	Y	537	528	528	526	519	527	526	550	545	561
BIHAR	**A**	1.672	1.733	1.739	1.835	1.754	1.722	1.726	1.796	1.710	1.787
	P	0.971	1.637	1.285	1.843	1.738	1.690	1.378	1.566	1.598	1.270
	Y	581	945	739	1004	991	981	798	872	935	711
HARYANA	**A**	0.111	0.111	0.138	0.073	0.079	0.074	0.089	0.054	0.063	0.086
	P	0.077	0.090	0.119	0.054	0.051	0.049	0.078	0.051	0.051	0.082
	Y	694	811	862	740	646	662	876	944	810	953
M. P.	**A**	4.997	5.121	4.794	5.130	5.348	4.892	5.002	4.669	4.788	4.977
	P	2.29	2.459	2.061	2.450	2.793	2.065	2.404	1.81	2.404	2.477
	Y	458	480	430	478	522	422	481	388	502	498
MAHARASHTRA	**A**	0.104	0.101	0.070	0.074	0.085	0.060	0.080	0.060	0.080	0.055
	P	0.049	0.049	0.030	0.031	0.04	0.02	0.04	0.02	0.03	0.014
	Y	471	435	429	419	471	333	500	333	375	255
PUNJAB	**A**	0.059	0.057	0.052	0.043	0.039	0.047	0.043	0.040	0.036	0.033
	P	0.038	0.033	0.034	0.028	0.026	0.031	0.030	0.027	0.023	0.016
	Y	644	579	654	651	667	660	698	675	639	485
RAJASTHAN	**A**	0.178	0.318	0.332	0.496	0.411	0.217	0.170	0.059	0.261	0.288
	P	0.148	0.323	0.400	0.526	0.506	0.258	0.154	0.057	0.288	0.286
	Y	831	1016	1205	1060	1231	1189	906	966	1103	993
U.P.	**A**	4.874	5.540	5.057	5.455	5.855	6.517	6.250	5.839	5.567	6.130
	P	3.170	4.492	3.654	3.954	4.846	4.081	5.000	4.499	5.048	5.014
	Y	650	811	723	725	828	626	800	771	907	818
WEST BENGAL	**A**	0.441	0.513	0.512	0.519	0.820	0.760	0.714	0.687	0.694	0.627
	P	0.304	0.432	0.350	0.356	0.675	0.685	0.375	0.432	0.532	0.379
	Y	689	842	684	686	823	901	525	629	767	604
ALL INDIA	**A**	12.604	13.694	12.896	13.885	14.616	14.778	14.664	13.770	13.964	14.730
	P	7.137	9.620	8.039	9.378	10.789	9.152	9.744	8.732	10.378	9.942
	Y	566	702	623	675	738	619	664	634	743	676

A = Area in lakh hectare; P = Production in lakh tones; Y = Yield in Kg/ha

Source: Directorate of Economics & Statistics, Ministry of Agriculture, Government of India

with irrigation, and yields over 3,000 kg/ha have been recorded. The straw-to-seed ratio in one cultivar was about 1.2:1 and in studies conducted on 28 cultivars in New Delhi, India, pulse yields ranged from 558 to 1,750 kg/ha, while dry matter yields ranged from 2,667 to 3,550 kg/ha (Duke, 1981).

4.1.1. Plantings

Generally, lentil plantings go from mid October in central India to the end of November in northern India. Most lentil growers in central India complete their plantings on conserved moisture in the month of October when day temperature is generally high, which causes development of soil born diseases. However, lentil farmers in northern India plant in mid November with pre sowing irrigations. Thus soil borne diseases have a low incidence at the early growth stage. Sowing is carried out with different methods like seed drills, desi plough, broadcasting etc. throughout India. Being a crop of marginal lands and small size of land holding generally farmers avoid the application of fertilizers. However, 100 kg diammonium phosphate (DAP) per hectare is a national recommendation for both irrigated and rainfed plantings (AICRP report 1997).

4.1.2. Weed and irrigation management

Weed infestation is generally low in rainfed crops, however, under irrigation various weeds compete with the lentil crop. One or two hand weedings are carried out by progressive farmers/growers, but many farmers/growers do not bother with manual weeding. Due to lack of knowledge and resources farmers do not utilize weedicides or herbicides for this crop. In central India farmers are unable to provide irrigation to this crop, however, in northern India generally farmers protect the crop from moisture stress with the help of one or two irrigations.

4.1.3. Disease management

Under Indian conditions both biotic and abiotic stresses like *Fusarium wilt*, root rots, *Aschochyta* blight, drought, high and low temperatures, water logging, salinity etc. are major problems for the cultivation of this crop (AICRP report 2004). Lentil growers in India generally do not understand and implement advanced management strategies for these stresses. Therefore, area, production and productivity are stagnating. However, recently national research centers like the Indian Agricultural Research Institute and State Agricultural Universities have collected desirable resistant germplasm and initiated aggressive breeding programs for the development of resistant cultivars. Fortunately, recently more then 25 resistant cultivars have been released for commercial cultivation under different agro-climatic conditions. Resistant cultivars are being promoted on farmers field for commercial cultivation. The production of quality seed for these cultivars has been assigned to various seed producing agencies like the State Farm Corporation of India for production, marketing and popularization.

4.1.4. *Harvesting threshing and yields*

Harvesting is generally carried out in the months of March and April throughout India. Most of the lentil growers use manual harvesting due to non-availability of harvesters and small size of holdings. Lentil threshing is generally completed by threshers and bullocks in the month of April–May. The mechanization for these operations is still minimal in India.

Lentil yields on farmers fields under rainfed plantings in central and south India generally range between 600–650 kg/ha. Under irrigated conditions the yield generally obtained is about 1200 kg/ha. In northern India the yield levels are high under both rainfed and irrigated environments. The average yield under rainfed environments ranges between 1400–1800 kg/ha and under irrigated environments ranging between 2000–2200 kg/ha in northern India. The lentil yields on research farms with the adoption of integrated management technologies are approximately 2500 kg/ha (AICRP report 2006).

4.1.5. *Marketing*

Lentil growers generally sell their surplus produce in the local *mandies*/markets to the small traders and these small traders sell to the big traders. The big traders act on behalf of the wholesale traders. In this system the maximum profit during trading goes to middlemen and poor small farmers get a low price for their produce. This unorganized marketing is not providing enough support and encouragement to small farmers. It is not possible for small farmers to store their produce properly as they do not have proper storage facilities. Therefore, they sell their produce immediately after harvesting and threshing to the traders. Thus, in this system the lentil growers fetch up minimum/lowest price of their produce. Considering all these situations the profitability of lentil growers is negligible under Indian conditions. However, grower cooperative marketing groups do exist in some regions that may produce better return to their growers.

4.1.6. *Gaps in productivity*

The lentil crop occupies an important position among Indian pulses because of its regular consumption and utilization in daily life. The national average yield is low because of influences of various biotic and abiotic stresses during the cropping season. However, a comparison of lentil productivity at the national level, on farmers fields and at research organizations shows opportunities for improvement (Figure 2).

The national average yield is stagnating at 700 kg/ha, however, large scale farmer demonstrations are able to harvest about 1400 kg/ha (up from 1100 kg/ha in 1995/6). The average yield under multilocation yield trials on research farms was between 1900 and 2500 kg/ha. At some locations reaching more than 3000 kg/ha. Thus there exist wide gaps between national average and large-scale demonstrations on farmers fields and large scale demonstrations and research farms. Therefore it is important to develop and set in place methods to minimize these large yield gaps.

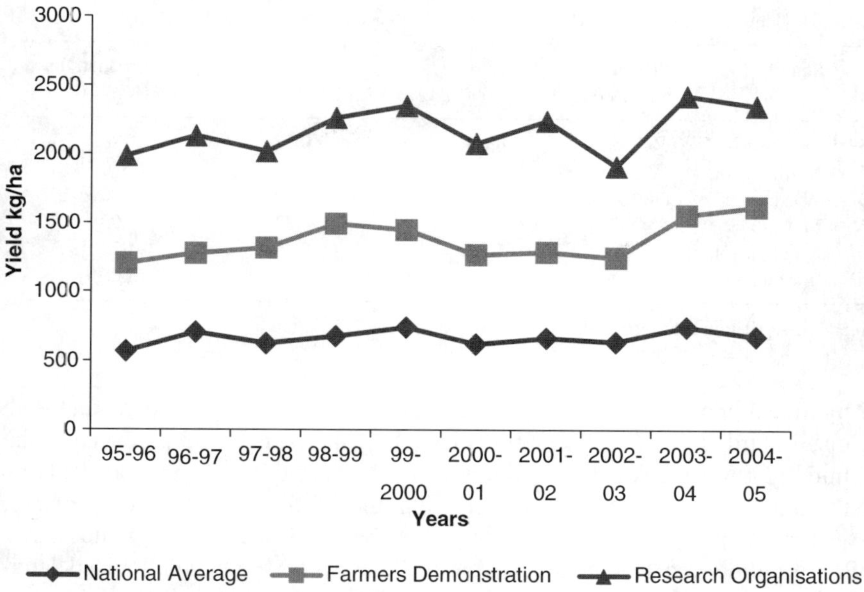

Figure 2. Gaps in lentil productivity between national average, farmers demonstration and research organizations in India during 1995–2005

4.2. USA Growers

US lentil growers differ greatly from their Indian counterparts. They are generally large scale farmers planting lentil on an average of about 1000–2000 acres. These growers are generally knowledgeable and apply substantial amounts of inputs like quality seeds, weedicides, herbicides and insecticides during the crop season. These farmers are well equipped with modern machinery and possess or have access to good storage.

Lentil is an important rotational crop in the cereal based production systems in the US Pacific Northwest. Initially production of lentil in the USA concentrated in the Palouse region of eastern Washington and northern Idaho. The very first seeding of lentil was in 1916 and the 1917 lentil crop was sold for about 20 cents per kg. The Palouse region has been the major lentil production region of USA producing more than 95% of US lentils for the ensuing decades. The city of Pullman in eastern Washington proudly claims itself the lentil capital of USA and holds its annual National Lentil Festive in late August. However, production of lentil has expanded to northern tier states including Montana and North Dakota (Table 3). Present production area is estimated to be about 200,000 hectares yielding roughly about 15 thousand metric tons in the US (Janzen et al., 2006).

Lentil production in the US has progressed significantly despite a short production history. First, in the production process the stationary thresher was phased out and replaced with combines in the early 1940's. Second, the yield has increased steadily

Table 3. The lentil area production and productivity in US States during 2005–06

Year	State	Planted Area('000 ha)	Harvested Area ('000 ha)	Production (MT)	Productivity (Kg/ha)
2005	Montana	60.0	58.4	0.75	1,280
2005	North Dakota	60.0	58.4	0.79	1,350
2005	Washington	34.0	33.6	0.30	900
2006	Montana	56.0	53.6	0.32	600
2006	North Dakota	64.0	59.2	0.49	820
2006	Washington	30.8	30.4	0.30	1,000

Source:U.S. Agricultural Statistics (http://www.nass.usda.gov/QuickStats)

and significantly over the decades thanks to the research efforts by federal and state scientists. Third, several types of lentils have been introduced and produced in the US. Initially it was the small-to-medium sized "Persian" type. Then it was the large sized "Chilean" type. Now lentils are grown in the US for more than seven market classes for exporting to different countries. Last but not least, recently winter-hardy lentil was introduced to commercial production, which has revolutionized lentil production in the Palouse region. The winter-hardy cultivar 'Morton', a product of the USDA research program located in Pullman, WA., can be direct seeded into cereal stubbles in the autumn. It provides advantages of preventing soil erosion and has much higher yield potential than spring-sworn lentils due to its vegetative growth during early rainy season. However production is also strongly affected by US farm policy.

4.2.1. US Farm Security and Rural Investment Act of 2002 (FSRIA)

Lentils, dry peas and small chickpeas were included under the loan program for the first time in 2002. The loan provides a floor price to the producer for lentils because if the price is lower than the loan rate, the producer is eligible for a loan deficiency payment. This made it easier for producers to obtain operating loans. The loan rate for lentils was US$11.94/cwt (100 pounds) for crop years 2002 and 2003, and will be US$11.72/cwt for 2004 to 2007. The base quality levels for the 2002 crop year was No. 1 grade, but was lowered starting with the 2003 crop year to No. 3 grade, which makes it easier for lentils to qualify for loan deficiency payments (LDP).

Loans made under the program for 2002–2003 were US$0.36 M and US$2.06 M for the first 11 months of 2003–2004. LDPs were US$2.38 M for 2002–2003, but there were no LDPs for the first 11 months of 2003–2004 as the posted county prices were above the loan rate. For 2002–2003, the average LDP was US$1.25/cwt and was paid for about 75% of US lentil production. US lentil production in 2003–2004 occurred in the states of Washington (37%), Idaho (26%), Montana (11%) and North Dakota (26%). The medium green and brown types accounted for most of the production, but the US also produced large green, small green and red types. The largest buyer of US produced lentils is the United States Department of Agriculture (USDA), which uses them for food aid programs.

US seeded area increased by 6% from 2001 to 2002, when the loan program was introduced for lentils. The seeded area increased by 10% from 2002 to 2003 and is forecast to increase by 12% from 2003 to 2004. The US Congressional Budget Office forecast a doubling of seeded area from 2001 to 2007. Although the rate of increase in seeded area for 2002 and 2003 was lower than the Budget Office forecast, including lentils under the loan program has supported increased seeding. Increased lentil production in the US (responding to the FSIRA) is expected to pressure world prices. If US production doubled, that is a 4% increase in world production and a 13% increase in lentils available for export with more sold commercially rather than to the USDA. World prices, would come under pressure but producers in the US will be protected from lower prices by the loan rate. Most of the increase in US production is expected to be in Montana and North Dakota, as there is more competition for land from other crops in Washington and Idaho. Production of lentils is not expected to spread to other states as they are either too hot or too wet for lentil production

4.2.2. Cropping pattern

Fields may be plowed and harrowed to a fine texture or direct seeded. Well drained soils on south-and east-facing slopes are preferred because these slopes get warmer during the early part of the growing season and as a result crop emergence will be faster (Muehlbauer et al., 1981; Muehlbauer et al., 1985). Other constraints in the US include preference for slightly acidic soils, moderate rainfall and avoidance of even short term waterlogging. Seed may be broadcast but is mostly sown in drills at the rate of 25–90 kg/ha, in rows 20–30 cm distance apart. Muehlbauer et al. (1981) reported that in the United States lentils are often planted with small grain equipment in rows spaced 15–18 cm distance apart. Lentils are a cool season legume species and as such are grown as a summer annual in temperate US climates. As young lentil plants are tolerant of light frost, early spring planting is possible in North Dakota.

In the United States, lentils are usually harvested with swathers that cut and windrow the crop for drying. After a 5–10 day drying period, the lentil crop is harvested by combine harvesters used for cereals but with slower cylinder speeds are used for lentils. Also, in USA, the lentil crop is cleaned by air and sieves and other seed processing equipment to remove inert materials and produce graded products. For red lentils, 50% of the product should remain on a 4.35 mm round hole perforated screen and 100% to remain on a 3 mm screen, while for the yellow lentils they should remain on 7 mm and 5 mm screens (Muehlbauer et al., 1985). In Russia, problems with mechanical harvesting were overcome by intercropping with a fiber plant Camelina that served as a support. Lentils are usually stored in bulk bins or elevators.

4.2.3. Planting and weed control

Lentil may be inoculated with rhizobium but other seed treatments with insecticides and fungicides are not recommended because these compounds can interfere with the nodulation process. Lentils are seeded in late April or early May into a firm, level seedbed to avoid machine harvesting difficulties associated with their low height. Weed control is common in the USA as lentil seedlings are not competitive with many broadleaf and grass weeds. A diverse range of integrated control methods are used. Sowing populations and spacings are aimed at crowding out weeds. Other methods include; choice of land with low weed populations, tillage and chemical control before seeding (eg Glyphosate) and chemical control after sowing (eg Sethoxydim with crop oil for grass weeds). Harrowing or rotary hoeing lentil fields after the lentil seedlings have emerged is not advised because seedlings are fragile and more easily injured than weeds.

4.2.4. Diseases and insects

Some diseases which may affect lentils include *Sclerotinia*, *Ascochyta*, *Fusarium* root rot and *Rhizoctonia* root rot. The only effective method of controlling these pests is crop rotation. However, breeding programs are developing resistant and tolerant cultivars. It is recommended that fababean, field bean, field pea, mustard, canola, rapeseed, soybean, sunflower, sugar beet and potato not be produced in consecutive growing seasons, because they are susceptible to the same diseases. Corn and small grains work well in rotations with lentils. Lentils generally do not suffer enough damage by insects to warrant pesticide application. Some insects which may damage lentils include aphids, thrips, Lygus bugs, seedcorn maggots and wireworms.

4.2.5. Fertilizer

Lentil are generally sown without nitrogen relying on residual soil N and *Rhizobium* bacteria for fixation. However, if the soil contains less than 2 percent organic matter 20 to 30 pounds of nitrogen per acre may be used to get the plants off to a healthy start. Lentils will not fix atmospheric nitrogen if excess nitrogen is available in the soil. Also, excess nitrogen will cause excessive vegetative growth, reducing seed yield. To attain optimum yields, 30 to 60 pounds per acre of phosphorus and 180 to 240 pounds of potassium may be applied if needed.

4.2.6. Harvest

When planted early, most varieties of lentils should reach maturity within 85 days of emergence. All US harvesting is mechanical with a range of set ups used. Lentils may be combined directly with a flexible cutterbar if the fields are uniformly ripe. Most lentils are swathed when pods on lower branches are brown to yellow-brown in color. Generally they are swathed when there is a dew or at least during the cooler portions of the day to avoid shattering losses. However, large areas to harvest may make this difficult.

Because the lentil has a weak stalk, it tends to lodge thus in a mechanised system machinery adapted to the plant is needed. Whether the crop is combined directly or swathed, pickup guards on the cutterbar and a pickup reel should be used. Depending on the weather, lentil windrows may take about a week to dry down.

Lentil seeds may be stored in a bin at 14 percent moisture; however, they should be combined at 18 to 22 percent moisture to decrease the number of cracked seeds. Natural air drying works best. If artificial heating is used, lentil seeds should not be heated above 110 F, to reduce seed coat cracking. Compared to wheat, lentils are easy to thresh. Cylinder speeds are slower than that used for wheat, and concaves wider. Approximately the same fan and sieve settings are used for lentils as for wheat.

4.2.7. Markets

During the 1980s, the North America's lentil industry grew substantially, but world market share shifted between Canada and the United States. Canadian share of the world lentil market grew from 1.2 percent in 1977 to 3.9 percent in 1985 to 30.6% in 2005. During the same time frame, the U.S.'s world share of lentil production fell from 5.2 percent to 2.8 percent rebounding (post 2002, FSRIA) to 5.6% in 2005 (FAOSTAT 2007). Nearly all lentils are produced in a six-county area of Washington and Idaho; counties include Whitman and Spokane in Washington and Benewah, Kootenai, Latah and Nez Perce in Idaho.

Several important issues concern U.S. lentil producers. Prices for lentils are among the most variable of any field crop grown in Washington and Idaho. Also, most U.S. consumers are unaware of lentils and, as a result, consume very little. Thus, a high percentage of U.S. production must be exported, forcing producers to rely on the volatile export market moderated, however, since 2002 by the FSRIA. The price for lentil seed ranges from $14 to $32 per hundredweight.

4.3. Canadian Lentil Growers

The average size of land holdings are extremely large in Canada and consequently the lentil growers grow lentils on large areas. Mostly approximately 1000–2000 hectares annually. All the operations are fully mechanized and growers use high levels of inputs. Production systems are thus very similar to those already described for the USA.

4.3.1. Production pattern

Canada is the largest exporter and second largest producer of lentils in the world (Table 1) The value of Canadian exports has averaged $230 million(M) during 2000–04. Canadian lentil production has increased in response to market signals and contributed to the diversification of crop production in the Prairie Provinces, especially in Saskatchewan. The increase in lentil production has proven to be valuable in crop rotations, which help to control weeds, diseases and insects and improve soil texture and fertility. The increased production also contributed to the expansion of the pulse crops handling, marketing and processing industry, which increased employment opportunities in rural areas. During the past 10 years, lentil production has been concentrated in Saskatchewan, which accounted for more than

95% of Canadian production. The balance was produced in Alberta and Manitoba. (www.pulsecanada.com)

In the Prairie Provinces of Canada, lentils are best suited to the Brown and Dark Brown soil zones, but can be grown successfully in the Black soil zone in years without excessive moisture. Lentils work well in a rotation with cereals, such as spring or durum wheat. Growers rely on inoculation with rhizobium and N fixation which may also benefit other crops in the rotation. Lentils require 90–100 days to mature and should be seeded as soon as the soil temperature is greater than 5° Celsius.

Canadian production reached a record of 914,000 tonnes (t) in 2000–2001, but fell sharply in the following three years, peaked again at 1,277,000 t in 2005 and fell again in 2006. Falls were due to one or more of the following factors: lower seeded area, drought, excessive rainfall during the harvest and excess carryover stocks (FAOSTAT 2007). Canada is the main producer of the green type of lentils in the world, accounting for about 70% of world production. However, production of the red type has been increasing and Canada has become a significant producer (Table 4). Canadian production of dark green speckled and brown types is small, accounting for only about 2% of total Canadian lentil production. The Canadian lentil harvest generally occurs during the period from mid-August to early October.

Most of the lentils produced in Canada have a green seed coat and yellow cotyledon. They are normally referred to as large green, medium green and small green, based on the seed size. The large green type includes the Laird, Glamis, Sovereign, Grandora, Plato and Sedley varieties. Their seed size is 60–70 grams/1000 seeds. The medium green type includes the Richlea and Vantage varieties, with seed size of 50–55 grams/1000 seeds. The small green type includes the Eston, Viceroy and Milestone varieties, with seed size of about 35 grams/1000 seeds. Canadian red types of lentils have a brown or pale green seed coat with red

Table 4. Canadian Lentil production, productivity and supply during 2000–05

Crop year	2000–01	2001–02	2002–03	2003–04	2004–05
Planted Area (kha)	699	708	601	554	696
Harvested Area (kha)	688	664	387	536	680
Yield (t/ha)	1.33	0.85	0.91	0.97	1.00
Production (thousand tonnes)					
Large Green	440	235	185	270	350
Medium Green	120	55	40	70	85
Small Green	180	110	38	60	80
Red	155	155	85	110	150
Others	19	11	6	10	15
Total Production	**914**	**566**	**354**	**520**	**680**
Imports	5	6	9	5	5
Total Supply	999	828	494	580	700

Source: Statistics Canada and Agriculture and Agri-Food Canada Forecast, May 2004

cotyledons. The red type varieties include Crimson, Redwing, Redcap, Redberry, Robin and Blaze, with seed size of 30–40 grams/1000 seeds.

4.3.2. Marketing

All of the lentils produced in Canada are sold on the open market to dealers. With the increase in production, the number of dealers across the Prairie Provinces who buy, clean and ship lentils to domestic and export customers has increased to about 50. The dealers range from large corporations to small family-owned businesses. In recent years, producers have invested in several new plants, which handle pulse crops, including lentils. There are several processing plants in Saskatchewan capable of de-hulling and splitting red and green types of lentils for the world market.

Lentils are shipped to ports mainly bagged in containers, although bulk shipments have been increasing with the building of suitable handling facilities. From the ports to overseas customers, they are shipped mainly bagged in containers, although some are also shipped bulk in containers or bulk inside the hold of ships. Most of the Canadian lentils are exported through the ports of Vancouver and Montreal. In addition to whole lentils, Canada also exports split lentils. The export of split lentils has been increasing, as Canadian splitting capacity expanded through the construction of new plants.

4.3.3. Exports

Canada exports about 75% of its production, while most other major producers export a relatively small portion of their production (except, Australia, Syria, Turkey, USA). Canadian lentil exports are dispersed throughout the world. Although the large green type of lentils is exported all over the world, the main destinations are north-western and southern Europe, northern Africa, South America, and Central America. The medium green type is exported mainly to the US, north-western Europe, Spain and northern Africa. The small green type is exported mainly to Morocco, Greece, Italy, Egypt, and Mexico. The red type is exported mainly to southern Asia, the Middle East and northern Africa. The dark green speckled type is exported mainly to France and the brown type mainly to Spain.

4.3.4. Prices

Canadian prices are largely determined in the international markets because Canada exports about 75% of its production. Since Canada produces most of the green type of lentils in the world, while it is a relatively small producer of the red type (changing in 2007), the level of production in Canada has much more influenced green type prices than red type prices. The substitution of one type of lentil with another is very limited. Therefore, it is common for wide price spreads to exist between different types of lentils. Since there is no future market for lentils, prices are negotiated directly between dealers and customers, based on supply and demand factors for each type of lentil, for immediate delivery or for delivery at some future date. Some lentils are grown under production contracts, which guarantee a price for part of the production, but most are sold on the spot market.

4.4. Turkey Lentil Growers

4.4.1. *Production and trade*

Domestic consumption of lentil is very high and approximately 50% to 60% of Turkish production is consumed locally. However, Turkey remains a net exporter of lentils. Domestic production is largely focused on red lentils, of which a significant quantity is produced solely for the export market. The bulk of Turkish red lentils are grown in South-Eastern Anatolia. Domestic production of green lentils, on the other hand, is significantly less (35,000 t pa). However, the desire for green lentils is increasing therefore, imports (35,000 t pa from Canada) are required to meet local green lentil demands. Green lentils are grown primarily in north-central Anatolia province. The climatic conditions, domestic requirement and the interest of farming community shows that lentil is stable crop for cultivation. The production figure during 2001 to 2005 in table 1 shows that lentil has a very stable production of about half million tonnes each year.

The lentil growers condition in Turkey differ from other regions being neither big land holders (like Canada, USA) nor small farmers (like India) with 80% of farm area in farms exceeding 5 hectares. The growing environments are also harsh due to moisture stress and low temperature and mechanization of lentil cultivation has not been a great success. The implementation of integrated management for lentil cultivation also has a low profile. Therefore their national average yield is stagnating at around 1000 kg/ha. Attempts to incorporate lentils into modern farming developments such as The Southeastern Anatolia Project have not succeeded.

Considering the growing conditions farmers are also not happy with lentil cultivation due to poor profitability. . Turkey has not modernised its lentil production thus growers are often low in knowledge and have low quality seeds of improved resistant cultivars, poor management of stresses, and weeds, poor irrigation management and infrastructure, inadequate machinery for planting, harvesting and threshing. Storage and marketing infrastructure are also not of a modern type. Traders take a substantial proportion of the crop value in buying and selling of lentil in domestic and international marketing and poor farmers are thus getting low prices.

4.4.2. *Imports by Turkey*

Turkish imports fluctuate widely 0 t in 1993, 140,000 t in 2000, 5,600 t in 2004. Despite the challenges currently faced within the Turkish lentil market, it is premature to disregard Turkey as a market for the future. Several issues, though, must be resolved before Turkey becomes a viable market once again. Turkey must create greater economic and political stability to install greater confidence among international exporters. Implementation of sudden volatile tariff and non-tariff barriers to foreign imports negatively affects confidence of international exporters and investors. However, given Turkey's geographical location and its trade relations with the Middle East, Europe, and the Commonwealth of Independent States, it will remain a key trade factor for the future. Furthermore, should Turkey become a member of the European Union, it would provide a lucrative market for agricultural products via a re-export program.

4.5. Lentil Revolution in Nepal

Lentil is the major grain legume crop of Nepal occupying 60% of area and production under grain legumes (Table 5). It is the cheapest source of protein for poor and middle class family in Nepal, and lentil dhal (decorticated or split seed) requires less time for cooking as compared to the other pulses and thus economize in fuel consumption. Apart from its importance in the cropping system, lentil dhal is in high demand in domestic and international markets.

4.5.1. Production

Lentil is traditionally grown in the lowland plains (Terai and inner Terai, 95% of area and production) during the winter months (November to March), and more recently production has expanded to the mid-hills (800–2,000 m, 4%) from October to March/April (Neupane and Shrestha 1991) and the high hills (above 2,000 m, 1%) from March to September (Dhital *et al.* 1994). There was a sharp increase in lentil area from 68,000 ha in 1961 to 188,895 ha in 2005 (Figure 3). Similarly, total annual production has increased from 30,500 to 160,716 tonnes during the same period (Figure 3). An increase in price, cultivation in the rice fallow, and a decline in the area under grasspea contributed to area expansion in lentil (ABPSD 2005; NARC 2000; Neupane and Shrestha 1991). A ban on the marketing of grasspea by the Nepalese government in 1991/92 due to its neurotoxin, ODAP (β-N-oxalyl-L-2-3-diamino-propionic acid) content in the seed (NARC 2000) caused a drastic reduction in area (253,660 ha in 1985/86 to 5,863 ha in 2005) sown to grasspea (ABPSD 2005; Yadav 1996). In addition, the high cost of cultivation for wheat, and severe pod borer (*Heliothis armigera*) and *Botrytis* gray mold (*Botrytis cineria*)

Table 5. Area and production of grain legumes in Nepal, 2004–06 [Source: ABPSD (2005)]

Crops	Area ('000 ha)	Production ('000 t)	Yield (kg/ha)
Lentil (*Lens culinaris* Medikus)	188.9	160.7	851
Chickpea (*Cicer arietinium* L)	11.8	10.4	887
Grasspea (*Lathyrus sativus* L.)	5.9	4.5	761
Pigeonpea (*Cajanus cajan* L.)	19.3	17.8	923
Blackgram (*Vigna mungo* L.)	32.4	28.0	864
Soybean (*Gycine max* L.)	22.6	19.8	879
Horsegram (*Macrotyloma uniflorum* L.)	7.8	5.7	725
Others+	28.1	24.3	865
Total	316.8	271.3	856

+includes field pea (*Pisum* spp.), broad bean (*Vicia faba* L.), dry bean (*Phaseolus* spp. L.), cowpea (*Vigna unguiculata* L.), rice bean (*Vigna umbellata*) and mungbean (*Vigna radiata*) etc.

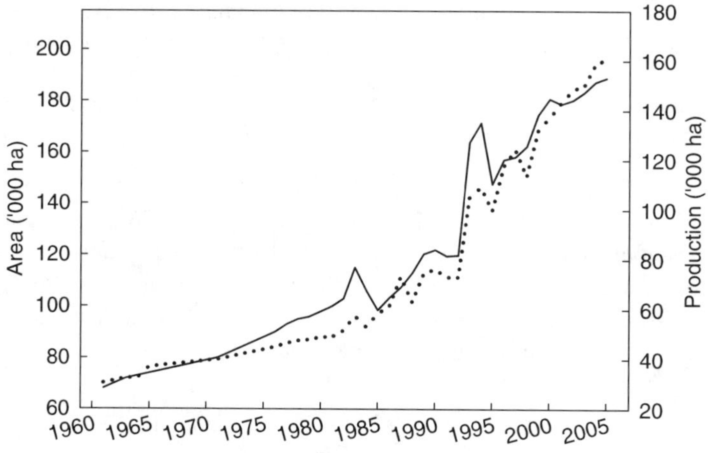

Figure 3. Area (——) and production (·····) of lentil in Nepal from 1961 to 2005 [Source: ABPSD (2005) FAOSTAT data (2004a)]

problems in chickpea production are the other contributing factors leading to farmers to growing more lentils.

4.5.2. Lentils in the cropping system

Lentil is usually grown in rotation with rice either as relay crop (*utera* or relay system, zero tilled) or immediately after the rice has been harvested and the soil cultivated (paira or post-rice system) /or as a mixed crop with wheat (*Triticum aestivum* L.), barley (*Hordeum vulgare* L.), mustard (*Brassica juncea* L.), linseed (*Linum usitatissimum* L.), grasspea (*Lathyrus sativus* L.) or field pea (*Pisum sativum* L.) (Shrestha *et al.* 2005). In the inner Terai, relay sowing of lentils in autumn sown maize is an emerging pattern, while intercropping of lentils in autumn planted sugarcane is gaining popularity in sugarcane belts (RK Neupane, Kathmandu, personal communication, 2006). There is scope for inclusion of lentil in between rice and spring maize/cowpea in river basin areas of mid-hills. Pande and Joshi (1995) indicated 10–40% more rice yield when grown after lentils or chickpea compared with when grown after wheat or land was left fallow.

4.5.3. Market

Lentil exports from Nepal account for about 2% of world exports (FAOSTAT data 2004b). There is a high demand on the international market with major export markets in Bangladesh (TPC 2005). In 2004/05, Nepal exported 2,708 MT of lentil (valued NRs 103,939,000) to Bangladesh. The national average retail price of lentil dhal has increased from NRs 16.71/kg in 1990/91 to NRs 44.86/kg in 2004/05, which is about 24% cheaper than blackgram and pigeonpea (ABPSD 2005).

4.5.4. *Production constraints*

Lentil growers are challenged with both biotic and abiotic stresses in Nepal. The average yield of lentil is 0.85 t/ha (Table 4), while seed yields of 1.5 t/ha and 2.5–3.0 t/ha are recorded from well-managed farmers' fields and experimental plots, respectively. This yield gap between the potential and national average yield is due to a number of biotic, abiotic and socioeconomic constraints. Excess water during germination and early establishment (particularly in low lying paddy field with occurrence of late rain), water deficit during flowering and pod filling (terminal drought), cultivation in marginal lands (nutrient poor and acidic soils), and use of no or minimal inputs (weeds infestation, no or very little fertilizer application and lack of proper irrigation system) are an important factors contributing poor yield and poor quality produce.

The rainfall pattern in Nepal is bimodal, where about 80% of annual rainfall occurs during summer months (May to September), and the remaining 20% occurs during winter (October to April) (Manandhar and Shakya 1996). Winter rainfall (October to January) can vary from 11 mm in 1994/95 to 204 mm in 1988/89 (Shrestha 1997; 1998). Therefore, soil water deficits during sowing affect crop establishment, and result in unstable and variable yields, while terminal drought results in reduction in biomass accumulation, partitioning, pod fill duration and causes forced maturity and therefore reduce crop yields (Shrestha *et al.* 2005). Lentil plants may be exposed to waterlogged condition in the early stage particularly in heavy soils of low infiltration rate and in uneven fields resulting in complete crop failure (Schulz 1997; Shrestha 1996).

The major biotic constraints in lentil production are wilt root rot complex, *Stemphylium* blight, *Botrytis* gray mold, storage pest bruchid and weed infestation (in particular with relay lentil). The pathogens associated with root rot/wilt complex or seedling mortality are *Sclerotium rolfsii*, *Fusarium oxysporum*, *Pythium* sp and *Rhizoctonia solani*. *Fusarium* wilt *(Fusarium oxysporum* f. sp. *Lentis*) is an important disease of Nepal and can cause complete crop failure The pathogen attacks the crop from seedling to maturity stage (Joshi 1997), with more severe attacks in seedling stage (Oct.–Nov.) and maturity stage (Feb.–Mar.) (Karki 1991). *Stemphylium* blight has been a serious disease in the main lentil growing pockets of the terai and inner terai in recent years.

In socioeconomic and institutional constraints non-availability of quality seeds of improved varieties, inefficient seed production and distribution systems, and inadequate extension services are major problems at the farm level.

4.5.5. *Scope for future expansion*

About 0.39 million ha of cultivated land is left fallow in winter after rice harvest in the terai, inner terai and hills, mainly due to inadequate soil water status or lack of irrigation facilities (Subbarao *et al.* 2000). Dissemination of improved production technologies (seed priming, plant protection measures etc), recommended (Shikhar, Simal, Shital, Khajura 1, Khajura 2) and promising varieties (ILL 7982, ILL 6829,

ILL 7979, ILL 7723), and effective extension would be helpful in increasing production in newer areas and for existing relay and sequential systems.

4.6. Lentil Production in Ethiopia

Lentil is one of the major highland pulses of Ethiopia that grows in rotation with tef, wheat and barley particularly on the heavy black clay soils (vertisols). It is an important part of the farming system and essential to nutrition in the subsistence farming community in Ethiopia. Currently, lentil is considered as a cash crop that fetches higher price than most of the cereals and pulses (Bejiga, 2006). Generally, lentil production is not mechanized and thus grown only by small farmers with the land holding of 2–10 hectares.

Ethiopia was among the five top lentil-producing countries of the world in the early 1970s (FAOSTAT 2007). However, its production was reduced to average about 26,000 t in the Mid/Late-1980s. This happened due to controlled market and fixed price by the so-called communist military government. However, this decline was arrested and reversed in the early 1990s due to liberalization of market with production over the last 6 years averaging about 50,000 t. This was accompanied by the release and popularization of the new varieties developed through research. Lentil research was formally started in 1972 and Debre Zeit Agricultural Experiment Station was assigned to coordinate its National Research Program which has released; El – 142, R186. Chalew (NEL– 358), Chekol (NEL–2704), Gudo (FLIP 84 –78L), Adaa (FLIP 86 – 41L), Alemaya (FLIP 88 – 63L), Alem Tena and Teshale. Among these EL–142, Chekol and Alem Tena were released for the lowland dry areas. Varieties R186, Chalew, Gudo, Adaa and Alemaya were for the central, northern and south eastern highlands of Ethiopia (Bejiga nad Anbessa, 1998). All these varieties are rust resistant but Adaa is resistant to both rust and *Fusarium* wilt. Variety R186 which, was performing well in the highlands where crop growth period was long was hit by frost in two seasons. Hence it was removed from production and followed by Chalew. However, it was difficulty for the researchers to deliver these varieties into the hands of farmers since Ethiopian farmers keep their own seeds from the previous season's crop.

The Major production constraints of lentil in Ethiopia are rust (Uromyces fabae), Fusarium wilt and root rots while pea aphid is a widespread insect-pest (Bejiga et al., 1996a). Usually rust causes about 25% yield loss in the normal year while 100% crop loss seldom occurs. In 1996/97 cropping season rust devastated the lentil crop in the central highlands of Ethiopia and no one could harvest local varieties that could be used as seed for the 1997/98 crop season. Luckily, The Pulse Section of Debre Zeit Agricultural Research Centre planted Adaa and Gudo on some farmers' plots in this area. These varieties remained green and free of rust in the middle of burned lentil fields. Field days were organized and farmers were convinced that these varieties were resistant to rust. Variety Adaa got a new name 'DIMA' after the name of a farmer that planted it in Gimbchu district of East Shoa in 1996/97. The seed size is larger than the local land races with popular grey colour in the local markets. The demand for the seed of this variety went up beyond the capacity of the Research Center. Off-season irrigation was used to multiply the seeds at the research center and farmers fields where water was available. At the meeting held with farmers in Gimbchu district, it was agreed not to give more than 2 kg seeds of these varieties to an individual farmer to enable all farmers to multiply their own seeds. Hence, Dima and other farmers who would plant these varieties were instructed and distributed the seeds under the supervision of the socio-economics department of the Debre Zeit Agricultural Research Center. In the following three years, Farmer- to –Farmer Seed Multiplication system was effectively used by the Debre Zeit Agricultural Research Center. Dima, Borena and Lemma were the farmers that played a key roll in the promotion of lentil production technology in the central highlands. Variety Alemaya, which was released after Adaa and Gudo, became the most poplar of these varieties because of its unbeatable high yield and good recovery (over 80%) during dehulling. This variety is estimated to cover more than 85% of the lentil farms in the central highlands Gimbchu (Chefe Donsa) and Barehe (Sendafa) districts of East and North Shoa (Bejiga et al., 1995).

Ethiopia has the Ethiopian Seed Enterprise as the major seed producer in the country plus several private companies. However, seeds of crops such as lentil and other legumes were not be produced by any one except breeders' and basic seeds by research centres. However, recently the Farmer Based Seed Multiplication scheme has been used by the Ethiopian Seed Enterprise under the supervision of its technical staff. Currently this approach has become a model and includes several progressive farmers who are involved in the seed production of different crops. Varieties and agronomic packages are available for the major lentil growing areas of Ethiopia.

A very high demand for lentil in the local markets accompanied by high prices encouraged farmers to grow lentils in Ethiopia. Split lentil costs about 1USD/kg. Its domestic price is higher than the export price. The crop has a bright future and hence researchers are working closely with both farmers and local processors in selecting acceptable varieties. Varieties suitable for different agro-ecological zones are now available with a well-established seed multiplication and distribution system (Bejiga et al. 1996b).

4.7. Lentil in Spain

The lentil crop peaked at 94,000 ha in 1987, then fell to 21,000 ha in 1999 gradually rising to 33,000 ha in 2005. This has been matched with a stable import quantity of about 40–50,000 t (FAOSTAT 2007). The reason for the expansion was the good quality and considerable acceptance by the consumer of Spanish lentils. Cubero (1994) has reviewed the situation with respect to Spanish lentils with much of the following information coming from his treatise. Area reductions arose because "agronomic techniques had improved but not the varieties, of which only the local races were known" (Cubero, 1994). Around 250 races were collected approximately the same as in Puerta Romero's collections from the 1950s. Low prices, good promotion of imported products and lack of an adequate response have a contributed to marketing difficulties for Spanish lentils (Cubero, 1994).

While still consumed as a traditional food lentil is no longer an essential food in Spain. Quality lentils (eg. Verdinas varieties) are still sought after and agronomically it is desired because of its toleration of drought and relatively fewer pests than other legumes (Cubero, 1994).

4.7.1. Prospects for improvement

Genetic improvement work on the Spanish lentil has been limited and lacked institutional support. However, Cubero (1994) states "it is essential to continue, nevertheless, with the aim of obtaining more productive high-yield cultivars, particularly from the local race Verdinas if Spanish producers wish to remain competitive with imports". Growers have solved many problems (eg mechanization) associated with growing lentils and pests and diseases are few and thus the crop is desired from an agronomic perspective. However, as elsewhere no resistance has been found against the broomrape *(Orobanche sp.)* (Cubero, 1994).

4.8. Syrian Lentil Growers

Lentil is best adapted to production in the cooler temperate zones of the world, or the winter season in Mediterranean climates such as that prevailing in the Northern farmlands in Syria. Lentil is thus a very important pulse crop in Syria, as it is considered as an important source of dietary protein for its low-to-middle-income population. Lentil is grown to improve economic returns to producers, diversify and lengthen crop rotations, and reduce the requirement for nitrogen fertilizer.

4.8.1. Production

According to the Syrian National Statistics (FAOSTAT 2007), the total area of lentil harvested for the year 2005, was 143,000 ha, yielding a total Lentil Production of 154,000 Tons. A quick look on the National Statistics reveals that between 1980 and 2005, the total area of lentil harvested varied between 60,000 ha (1984) and 123,000 ha with a peak of 188,000 ha reached in 1989.

The total area planted with Lentil varies up and down, in line with the fluctuations of the Lentil Crop Market Price, demand for other crops (eg cumin which has high export prices recently). Lentil is grown in various regions in Syria, but most areas under cultivation are concentrated in:

1. The provinces of Aleppo, Idlib,Al Hassakah and Hama which together account for 80% of the yearly crop.
2. The provinces of Damascus and Horan (As Suwayda) which account for the remaining 20% of the yearly crop.

4.8.2. *Plantings*

In Syria, Lentil is best planted during the period of the year falling between mid November and mid December. This planting period for Lentil can be extended up to mid January, depending on local farming know how and weather conditions permitting. Harvesting usually starts in the month of May and is conditioned by the biological maturity of the Lentils stalks, i.e. harvesting starts before these stalks are too mature and "pod drop" begins. The following Lentil varieties are very popular for cultivation in Syria:

1– "Kurdi" variety : usually planted in the north western and western areas of Syria, around Aleppo and Idlib, where average annual rainfall is about 350 mm.
2– "Idlib" variety # 1 - ICARDA : Green Lentil.

In general, the average Syrian Lentil crop yield varies between 0.7 and 1.2 Ton/ha. However, when the climate conditions are favorable and the seasonal rainfall adequate, the crop yield may rise up to 2 Tons/ha.

4.9. Australian Growers

Australian lentil growers are also large scale producers like the USA and Canada and possess modern equipments and knowledge of modern technology. These farmers cultivate this crop primarily for export sale. The growing region is, however, much more subject to drought than N America resulting in very low yields in some years.

Until 1998 Australia was an importer of lentils ($\sim 2,000$ t pa) and there were opportunities for growers to supply lentils under contract. prices paid for lentils in Australia. However, since then Australia has risen to become a major international exporter (peaking at 242,000 t in 2002 of red lentils. A detailed analysis of the Australian situation is given in chapter 12.

5. INTEGRATED MANAGEMENT

Clearly lentil growers throughout world have different types of cropping systems, influences of stresses and agro-climatic conditions. Under such varied conditions and cropping environments it is important for the lentil growers to introduce integrated management technologies suited to their region for higher production. However. yield gaps in productivity at different levels like national average, farmers demonstration and research farms exist throughout world. Thus it is important to minimize these gaps in productivity by introducing integrated management approaches on farmers field. Some of the major important components of these integrated packages which span many regions are listed below:

5.1. Moisture Conservation

This is an important component of the planting technology because in most of the countries lentil growers cultivate this crop in rainfed environments, therefore moisture conservation before planting is essential.

5.2. Planting of High Yielding Resistant Cultivars

Generally high yielding, resistant cultivars are available in both developed and developing countries through ICARDA and the national agricultural research system. It is important that the lentil growers should plant quality seeds of such cultivars whether in Ethiopia or Canada.

5.3. Nutrient, Weed and Irrigation Management

The essential and appropriate doses of basic nutrients (PK), micronutrients (eg B in Nepal) before plantings are required for good harvest. Similarly good N fixation is needed. Weed control (herbicides as well as other management options) is essential due to weak competitive ability.

5.4. Disease Control

To avoid losses from diseases and insect-pest, farmers should adopt IPM techniques. These include crop rotations, resistant and tolerant varieties, quarantine, applications of fungicides, insecticides etc.

REFERENCES

AAFC (2006) Lentils: Situation and Outlook: Agriculture and Agrifood Canada, Bi-Weekly Bulletin, 19(7) pp 6.

ABPSD (2005) 'Statistical information on Nepalese agriculture (2004–05).' Ministry of Agriculture and Co-operatives, Agri-Business Promotion and Statistics Division, Singha Durbar, Kathmandu, Nepal.

AICRP (All India Coordinated Research Project). 1997. All India Coordinated Research Project on Improvement of MULLaRP. Indian Institute of Pulses Research, Kanpur.

AICRP (All India Coordinated Research Project). 2004. All India Coordinated Research Project on Improvement of MULLaRP. Indian Institute of Pulses Research, Kanpur.

AICRP (All India Coordinated Research Project). 2006. All India Coordinated Research Project on Improvement of MULLaRP. Indian Institute of Pulses Research, Kanpur.

Cubero JI (1994) Traditional varieties of grain legumes for human consumption.]n Neglected Crops: 1492 from a Different Perspective. 1994. J.E. Hernándo Bermejo and J. León (eds.). Plant Production and Protection Series No. 26. FAO, Rome, Italy. p. 289–301.

Dhital BK, Budhathoki CB and Devkota HP (1994) 'Time to planting trails of cowpea and lentil.' Lumle Agricultural Research Centre, Pokhara, Nepal.

Website Directorate of Economics & Statistics, Ministry of Agriculture, Government of India.

Duke J A (1981) Handbook of legumes of world economic importance. Plenum Press, NewYork. P. 52–57.

FAOSTAT data (1996) Agricultural Data: Agricultural production, crops/primary. In Food and Agriculture Organization of the United Nations, Statistics Division (http://faostat.fao.org/)

FAOSTAT data (2004a) Agricultural Data: Agricultural production, crops/primary. In Food and Agriculture Organization of the United Nations, Statistics Division (http://faostat.fao.org/)

FAOSTAT data (2004b) Agricultural Data: agriculture & food trade. In Food and Agriculture Organization of the United Nations, Statistics Division (http://faostat.fao.org/).

FAOSTAT 2007 Agricultural Data: agriculture & food trade. In Food and Agriculture Organization of the United Nations, Statistics Division (http://faostat.fao.org/).

Geletu Bejiga 2006: LENS CULINARIS Medik. In: Plant Resources of Tropical Africa1, Cereals and pulses (Brink, M& G. Belay eds.) Prota Foundation, Wageningen, Netherlands / Backhuys Publishers, Leiden, CTA, Wageningen, Netherlands pp 91–96.

Geletu Bejiga and Yadeta Anbessa.1998: Alemaya Self-Adjusting Lentil variety to different environments in Ethiopia. NVRSRP Newsletter No. 1.,1–4.

Geletu Bejiga, Million Eshete and Yadeta Anbessa 1996a: Improved cultivars and production technology of lentil in Ethiopia. Research Bulletin No.3. Debre Zeit Agricultural Research Center, Alemaya University, Debre Zeit , Ethiopia.

Geletu Bejiga, Seifu Tsegaye and Abebe Tullu, 1995. Stability of seed yield for some varieties of lentil grown in the Ethiopian highlands. Crop Research, 9:337–343.

Geletu Bejiga, Seifu Tsegaye, Abebe Tullu, and W. Erskine. 1996b Evaluation of lentil landraces from different regions of Ethiopia. Genetic Resources Crop Evolution. 43: 293–301.

Janzen E, Flaskerud G, Fisher J and Bartsch E (2006) Pulse Crop Marketing Guide EC 1277. NDSU Extension Service. North Dakota State University, Fargo USA. Pp 36.

Joshi S (1997) Lentil Fusarium wilt in Nepal. In 'NARC/CLIMA lentil workshop'. Rampur, Nepal

Kaiser, W. J., R. M. Hannan, and J. D. Rogers, 1994: Factors affecting growth and sporulations of Ascochyta fabae f. sp. lentis.Plant Dis. 76, 605—610.

Karki PB (1991) Plant Protection of Lentil in Nepal. In 'Lentil in South Asia. Proceedings of the Seminar on Lentil in South Asia, 11–15 March'. (Eds W Erskine and MC Saxena) pp. 187–193. (International Center for Agricultural Research in the Dry Areas, Aleppo, Syria: New Delhi, India)

Manandhar DN and Shakya DM (1996) 'Climate and crops of Nepal.' Nepal Agricultural Research Council/Swiss Agency for Development and Cooperation, Kathmandu, Nepal.

Muehlbauer, F.J., W.J. Kaiser, S.L. Clement, and R.J. Summerfield. 1995. Production and breeding of lentil. Advances in Agronomy 54:283–332.

Muehlbauer, F.J., R.W. Short, R.J. Summerfield, K.J. Morrison and D.G. Swan. 1981. Description and culture of lentils. Cooperative Extension, College of Agriculture, Washington State University and USDA-ARS. EB 0957

Muehlbauer F J, J I Cubero and R J Summerfield (1985) Lentil (*Lens culinaris Medik.*).p. 266–311. In: R J Summerfield and E I I Roberts (eds) Grain Legume Crops. Collins, 8 Grafton Street, London, UK.

NARC (2000) Grasspea. In Nepal Agricultural Research Council (http://www.narc-nepal.org)

Neupane RK and Shrestha R (1991) 'Report on lentil varietal and agronomical research.' Nepal Agricultural Research Council, Lalitpur, Nepal.

Pande S and Joshi PK (1995) 'Constraints and prospects of legumes in northern India and the terai region of Nepal. Field observation and group interviews with farmers.' International Crops Research Institute for the Semi Arid Tropics, Patancheru, India.

Schulz S (1997) Performance and residual effects of leguminous crops in rice based cropping systems of the middle mountains of Nepal. Degree of Doctor of Philosophy thesis, The University of Reading.

Shrestha R (1996) Report on varietal evaluation of lentil in Khumaltar (1995–96). In 'National Winter Crops Workshop'. Regional Agricultural Research Station, Bhairahawa, Nepal (Nepal Agricultural Research Council, Nepal)

Shrestha R (1997) Varietal improvement on lentil in Khumaltar, 1996–97. In 'Winter Crops Workshop'. Nepal Agricultural Research Council, Khumaltar, Lalitpur, Nepal. (Nepal Agricultural Research Council, Nepal)

Shrestha R (1998) Lentil research in Khumaltar. In 'Lentil Review Workshop'. Regional Agricultural Research Station, Nepalgunj, Nepal (CLIMA/NARC/GLRP)

Shrestha R, Siddique KHM, Turner NC, Turner DW and Berger J (2005) Growth and seed yield of lentil (*Lens culinaris* Medikus) genotypes of West Asian and South Asian origin and crossbreds between the two under rainfed conditions in Nepal. *Australian Journal of Agricultural Research* 56, 971–981.

Statistics Canada and Agriculture and Agri-Food Canada Forecast, May 2004

Subbarao GV, Kumar Rao JVDK, *et al.* (2000) The spatial distribution and quantification of rice-fallows in South Asia - potential for legumes. In. (DFID Plant Sciences Research Programme)

TPC (2005) 'A Glimpse of Nepal's Foreign Trade (Statistical Presentation).' Trade Promotion Centre, Pulchowk, Lalitpur, Nepal.

U.S. Agricultural Statistics http://www.nass.usda.gov/QuickStats

Van Emden, H.F., S.L. Ball and M.R. Rao. 1988. Pest disease and weed problems in pea lentil and faba bean and chickpea. p. 519–534. In: R.J. Summerfield (ed.), World Crops: Cool Season Food Legumes. ISBN 90-247-3641-2. Kluwer Academic Publishers, Dordrecht, The Netherlands.

Yadav CR (1996) Genetic evaluation and varietal improvement of grasspea in Nepal. In 'Lathyrus Genetic Resources in Asia: Proceedings of a Regional Workshop'. Indira Gandhi Agricultural University, Raipur, India, 27–29 December 1995. (Eds RK Arora, PN Mathur, KW Riley and Y Adham) pp. 21–27. (IPGRI office for South Asia, New Delhi, India)

www.1911encyclopedia.org 2006

www.pulsecanada.com

Index